Wolfgang Gaul · Dietmar Pfeifer
(Editors)

From Data to Knowledge

Theoretical and Practical Aspects of Classification, Data Analysis, and Knowledge Organization

With 123 Figures
and 57 Tables

Springer

Professor Dr. Wolfgang Gaul
Universität Karlsruhe (TH)
Institut für Entscheidungstheorie
und Unternehmensforschung
Postfach 6980
76128 Karlsruhe, Germany

Professor Dr. Dietmar Pfeifer
Universität Oldenburg
FB6 (Mathematik)
Ammerländer Heerstraße 114-118
26129 Oldenburg, Germany

Cataloging-in-Publication Data applied for

Die Deutsche Bibliothek - CIP-Einheitsaufnahme

From data to knowledge : theoretical and practical aspects of classification, data analysis, and knowledge organization ; with 57 tables / Wolfgang Gaul ; Dietmar Pfeifer (ed.). - Berlin ; Heidelberg ; New York ; Barcelona ; Budapest ; Hong Kong ; London ; Milan ; Paris ; Santa Clara ; Singapore ; Tokyo : Springer, 1995
(Studies in classification, data analysis, and knowledge organization)
ISBN 3-540-60354-9

ISBN 3-540-60354-9 Springer-Verlag Berlin Heidelberg New York

43/2202-5 4 3 2 1 0 - Printed on acid-free paper

Studies in Classification, Data Analysis, and Knowledge Organization

Springer
Berlin
Heidelberg
New York
Barcelona
Budapest
Hong Kong
London
Milan
Paris
Santa Clara
Singapore
Tokyo

Titles in the Series

H.-H. Bock and P. Ihm (Eds.)
Classification, Data Analysis, and Knowledge Organization

M. Schader (Ed.)
Analyzing and Modeling Data and Knowledge

O. Opitz, B. Lausen, and R. Klar (Eds.)
Information and Classification

H.-H. Bock, W. Lenski, and M. M. Richter (Eds.)
Information Systems and Data Analysis

E. Diday, Y. Lechevallier, M. Schader, P. Bertrand, and B. Burtschy (Eds.)
New Approaches in Classification and Data Analysis

PREFACE

Selected papers presented at the 18th Annual Conference of the German Classification Society GfKl (Gesellschaft für Klassifikation) are contained in this volume of 'Studies in Classification, Data Analysis, and Knowledge Organization'. The conference took place at the University of Oldenburg in 1994 under the general subject "From Data to Knowledge" and provided an international forum for participants from theory and practice.

The interdisciplinary character of GfKl – although aspects concerning classification and related areas of data analysis are in the center of interest of the activities of the society – is one of the reasons that within a process entitled "From Data to Knowledge" emphasis was laid as well on theoretical contributions as on applications in various fields including information systems and knowledge organization.

More than 330 participants could choose their individual program from more than 130 presentations included in the scientific program of the conference.

The selected papers of this volume are divided into the following major sections:

- Plenary and Semi Plenary Presentations
- Classification Related Results and Other Aspects of Data Analysis
- Spatial Data Analysis
- Applications in Economics
- Applications in Linguistics
- Applications in Medicine and Biology
- Information Systems and Knowledge Organization

Within the sections the contributions are listed in alphabetical order with respect to the authors' names. Besides the plenary and semi plenary presentations which demonstrate the magnitude of the different scientific directions originating in classification and related areas the arrangement of the papers in the sections mentioned gives an overview of the main topics tackled at the conference. As most contributions contain aspects from different areas an unambiguous assignment of papers to sections is not possible in all cases. However, we appreciate that many contributions show in their application parts that the activities which GfKl supports make an important impact on various practical fields.

This time, spatial data analysis was intentionally separated from other aspects of data analysis as the conference attracted a major group of researchers from this area.

In this context, we gratefully take the opportunity to acknowledge support by

- Deutsche Forschungsgemeinschaft (DFG)
- Land Niedersachsen
- Carl von Ossietzky Universität Oldenburg
- Universitätsgesellschaft Oldenburg e.V.

which made it possible to hold the 18th Annual Conference of GfKl in the way described.

The final version of this volume was put together at the University of Karlsruhe by Frank Wartenberg who did an extremely good job in organizing and supervising typesetting and reproduction of figures. From the students who helped, at least, Frau Marzena Gajowa and Lars Björner should be mentioned.

Last but not least we thank Dr. Schuster from Springer-Verlag for excellent cooperation.

Karlsruhe and Oldenburg, July 1995

W. Gaul and D. Pfeifer

Contents

Plenary and Semi Plenary Presentations

Classification Related Results and Other Aspects of Data Analysis

Spatial Data Analysis

Applications in Economics

Applications in Linguistics

Applications in Medicine and Biology

Information Systems and Knowledge Organization

Plenary and Semi Plenary Presentations

Advances in Cluster Analysis Relevant to Marketing Research

P. Arabie[1] and L. Hubert[2‡]

[1]Faculty of Management, Rutgers University, 180 University Avenue, Newark NJ 07102-1895, USA

[2]Department of Psychology, University of Illinois, 603 E. Daniel St., Champaign IL 61820, USA

Summary: We review the current methodological and practical state of cluster analysis in marketing. Topics covered include segmentation, market structure analysis, a taxonomy based on overlap, connections to conjoint analysis, and validation.

1. Introduction

In considering substantive applications of clustering in marketing research, we update and extend the comprehensive review provided by Punj, Stewart (1983) and even follow their outline at times. Limitations of time preclude our discussing or even citing all the relevant and meritorious literature. Whenever possible, we simply refer to other, current review articles and chapters.

2. Substantive Uses of Cluster Analysis in Marketing

Punj, Stewart (1983) listed the most common applications of cluster analysis in marketing research as: market segmentation, identifying homogeneous groups of buyers, development of potential new product opportunities, test market selection, and as a general data reduction technique. We consider these topics in turn before getting to some more recent patterns of usage.

Wind (1978a, 1978b) presciently laid out the problems of market segmentation (also see Gaul, Baier (1993, Ch. 4)) for decades to follow and noted that as an alternative to *a priori* segmentation (in which management or some other expert decides on a basis for segmentation, generally unassisted by data analysis), clustering is the principal recourse for identifying homogeneous subgroups of markets, consumers, or organizations. The variety of applicable methods of clustering and sophistication of their underlying assumptions has increased enormously (e.g., Mahajan, Jain (1978)) in response to the problems chronicled by Wind (1978a). Among the more noteworthy advances

‡For comments on an early draft of this paper, we are indebted to Rick Bagozzi, Doug Carroll, Geert De Soete, Wayne DeSarbo, Akinori Okada, and Dave Stewart. Much of this work appeared in Arabie, Hubert (1994).

are segmentation based on price sensitivity (Elrod, Winer (1982); Blozan, Prabhaker (1984)), a simultaneous approach to segmentation and market structuring by Grover, Srinivasan (1987), and a simultaneous approach to segmentation and estimation in the framework of conjoint analysis (Ogawa (1987)). (Many more references are given in useful reviews by Beane, Ennis (1987) and Wedel (1993).) Researchers in industrial marketing have also shown increased interest in segmentation (Doyle, Saunders (1985); Robles, Sarathy (1986); de Kluyver, Whitlark (1986)). But to date, there has been insufficient progress on Nicosia, Wind's (1977, p. 102) call for industrial firms to pursue segmentation of buying centers (cf. Arabie, Wind (1994)), despite innovative work by Choffray, Lilien (1978, 1980a, 1980b).

A rather closely related area of applications of cluster analysis in marketing research concerns the identification of homogeneous groups of buyers, particularly according to benefit segmentation (Punj, Stewart (1983)). Noteworthy substantive bases toward which such strategies have been applied include: external information search patterns among purchasers of new automobiles (Furse et al. (1984)), reactions to advertising copy schemes, flavor assortments, and other responses to heterogeneous benefit/feature preferences (Green et al. (1985)), inferred choice routes and strategies (Currim, Schneider (1991)), store "image" (Steenkamp, Wedel (1991)), human values systems (Kamakura, Mazzon (1991)), and brand loyal versus brand switching consumers (Grover, Srinivasan (1992)).

A third area of applications in Punj, Stewart's (1983) taxonomy, the development of potential new product opportunities, has seen Choffray, Lilien's (1980a, 1980b) efforts to bring industrial marketing up to par methodologically with its consumer counterpart. But the latter domain has in recent years seen such a shift of emphasis that its use of clustering in new product development will be discussed below under the heading of "market structure analysis".

Punj, Stewart's (1983) fourth area of applications seeks identification of relatively homogeneous and comparable test markets. It is assumed that management requires such test markets to be matched, so that inferences about the population can be attempted. In spite of a highly promising start (e.g., Green et al. (1967); Day, Heeler (1971); Murphy, Tatham (1979)), this area has seen very little development in recent years. Also, few papers now fall in the miscellaneous, fifth category of using cluster analysis as a "general data reduction technique" (Punj, Stewart (1983, p. 136)). This decline probably attests to increasing sophistication in the marketing research community, the realization of what clustering can and cannot do well, and journal editors' refusal to publish straightforward applications.

Explicitly excluded from Punj, Stewart's (1983, p. 135) review were such applications of clustering and closely related techniques as "an alternative to multidimensional scaling (MDS) and factor analytic approaches to representing similarity data". In the last decade, this area of applications has competed with market segmentation as the focus of heaviest usage and development of techniques tailored for marketing applications. For many of the stimulus domains whose mental representations have been intensively studied by experimental psychologists (e.g., perception of rectangles, similarities of animals, etc.), there is no counterpart to managerial significance or immediate practical implications. The choice of such stimulus domains typically reflects *a priori* theorizing about category structure, underlying dimensions, and embedding metrics, as well as the practical and easy presentation to the subjects making judgments (e.g., see Pruzansky et al. (1982)). In contrast, the majority of product domains to which marketing researchers apply clustering have significance, at least in principle, to management.

To the extent that such academically oriented explorations of stimulus domains have supplied answers, hypotheses, or more refined questions to experimental psychologists, the same relevant techniques might offer marketing researchers a better understanding of how consumers perceive, remember, judge, compare, and select among members of a product domain of managerial significance. Noteworthy papers in the marketing literature resulting from this strategy include Johnson, Fornell (1987), Glazer, Nakamoto (1991), and Hutchinson, Mungal (1992). Also, a considerable variety of applications can be found in proceedings volumes edited by Gaul, Schader (1988b), Schader, Gaul (1990), and Schader (1992).

In addition to such papers, a related tradition, but even more distinctly marketing-oriented, falls under the rubric "competitive market structure analysis". The structure being studied is that of competing brands that can be regarded as substitutes for each other by consumers. Aspects of this process for which structure is sought include customer segments, products/brands, and usage/purchase situations. Managerial implications drawn from the portrayal of such structure include sales promotion and sales force allocation, price discounts, advertising, the design and development of new products, and even antitrust considerations. Because an excellent and comprehensive review (covering both substantive and methodological details) is given by DeSarbo et al. (1993) (also see Rao et al. (1982)), we will give only a brief sketch here.

The tradition of research in market structure analysis that has led to so many developments in and applications of clustering effectively begins with Butler (1976) and the Hendry Corporation's scheme for analyzing consumer behavior for frequently purchased products. While seeking to depict patterns of brand switching and market partitioning inherent in aggregate data, this scheme results in a set of alternative *a priori* hierarchical structures for the product domain, such as those found in Fig. 1 (taken from Wind (1982, p. 77)). In this idealized representation of the margarine market, the top panel suggests that marketing strategy should give more emphasis to the form of the product than to competing brands, whereas the lower panel implies the opposite strategy. That is, the top panel implies that the form of the product is of primary concern to the consumer, while the lower panel implies a primary concern instead with the brand. As Wind (1982, p. 77) notes, consumers in a brand-primary market should exhibit higher brand

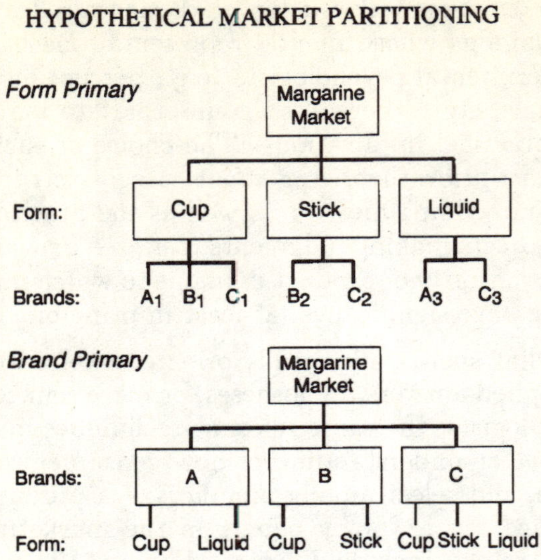

Fig. 1: Form primary versus brand primary market structure (reproduced from Wind (1982, p. 77))

loyalty than those in a form-primary market. Introduction of a new product in a form-primary market should lead to less "cannibalization" than would be found in a brand-primary market.

Rao, Sabavala (1981) made a major contribution by logically embedding the substantive questions underlying market structure analysis in the context of clustering. Those authors began with panel data and extracted from them a matrix, designated here as $\mathbf{S} = \{s_{ij}\}$ $(i, j = 1, \ldots N$, the number of products being considered), where s_{ij} denotes the number of panelists who purchased product i on the first choice occasion but switched to product j on the next occasion. Most forms of clustering ignore the diagonal entries (s_{ii}) in a matrix like $\mathbf{S}$, but departures from symmetry (i.e., the fact that in general $s_{ij} \neq s_{ji}, i \neq j$) are a more persistent problem for most methods of clustering and of multidimensional scaling (see Appendix of Arabie, Soli (1982) for a survey). Another problem of using the raw entries of $\mathbf{S}$ is undue influence of differential market shares among the N products. Rao, Sabavala (1981) therefore devised a transformation of the data that normalizes the entries in $\mathbf{S}$ and is interpreted as a "flow measure":

$$f_{ij} = \frac{s_{ij}s_{..}}{s_{i.}s_{.j}} \tag{1}$$

where $s_{i.}$ is the number of panelists who selected product i on the first choice occasion, $s_{.j}$ is the number of panelists who selected product j on the second choice occasion, and $s_{..}$ is the number of panelists in the homogeneous population.

Rao, Sabavala interpreted their measure as the ratio of the actual number of panelists switching from product i to j to the number expected under the traditional but arguable hypothesis of independence, both between the two consecutive choice occasions and by choice probabilities given by the marginals.

Rao, Sabavala (1981, pp. 85-86) then provided a tightly reasoned argument why aggregate choice processes could be described using hierarchical clustering applied to the **F** matrix[3]. A noteworthy point made by the authors is that brand switching between clusters should be least at the highest levels in a tree like that in Fig. 1 and greatest at the lowest levels.

After the compelling work by Rao, Sabavala, the next logical steps were (a) dealing more adroitly with problems of nonsymmetry and normalization and (b) using types of clustering better suited than traditional hierarchical ones to the data at hand to reveal market structure. DeSarbo, De Soete (1984) used a different normalization[4] (dividing each entry by the product of the row and column arithmetic means, as suggested earlier by Rao et al. (1977)) and, more importantly, devised clustering methods explicitly designed to accommodate nonsymmetric data quite elegantly (for a more recent approach, see DeSarbo, Manrai, Burke (1990)).

In an impressive program of research, Shocker, Srivastava, and their collaborators have made monumental contributions to identifying types of clustering more appropriate than traditional hierarchical ones to the relevant data. Srivastava (1981) emphasized the importance of usage-situational influences on consumers' perceptions of markets. Taking this point further, Srivastava, Alpert (1982) argued that multiple uses of a product, influenced by situational variables, cannot be faithfully represented by hierarchical clustering. The substantive emphasis on "usage situations" that Srivastava, Shocker, and their collaborators have so profitably pursued translates directly into a call for "higher-way" models (i.e., three or more ways, so that usage situations can be accommodated as one or more ways of the design) and corresponding algorithms, as provided, for example, by Carroll et al. (1984) and De Soete, Carroll (1988, 1989).

Srivastava et al. (1984) used the ADCLUS model (Shepard, Arabie (1979); also see Arabie, Carroll (1980) and Arabie et al. (1981)) to portray market structure using *overlapping* clustering. Such clustering allows products to appear in more than one submarket, rather than being confined to exactly one cluster. Srivastava et al. (1984, p. 43) concluded: "While hierarchical market structures may be developed based on measures of substitution-in-use, these structures may be misleading because products can be assigned

---

[3]Actually, only an arbitrarily chosen half of the **F** matrix was used. DeSarbo, De Soete (1984) circumvented this problem with their approach.

[4]A third method of normalization, used by Slater (1984, pp. 121-123) iteratively converts the input matrix to have constant marginals (Deming, Stephan (1940); Fienberg (1970)), possibly after zeroing the diagonal entries. It should be noted that none of the three methods symmetrizes the matrix.

to one and only one branch of a tree (each branch may be viewed as a submarket). This exclusivity requirement is incompatible with the basic reason for examining usage-situational influences. When products (within a set being examined) have multiple uses, each product may be competitive with different subsets of products, depending on the use (usage situation). Accordingly, overlapping clusters become more relevant."

In seeking a tighter logical connection between market structure analysis and the types of clustering employed to depict those structures, Shocker et al. (1990, p. 128) noted: "Several papers reviewing the characteristics of various [clustering] algorithms and the issues related to their use already exist ... Missing from these prior works is explicit discussion of the assumptions that various methods and techniques imply about the phenomena being represented, i.e., market structure." Shocker et al. discussed at the managerial level "options, conceptual issues, and problems that exist in dealing with MSA [market structure analysis]". A more technical and comprehensive discussion of the many techniques that have been developed for such analyses is given by DeSarbo et al. (1993). These two papers are strongly recommended.

Green, Srinivasan (1978) opened a new arena for clustering when they advocated segmentation of consumers by clustering them according to vectors of part-worth utilities derived from conjoint analyses. This recommendation led to a series of papers in which clustering methods (usually either partitioning or hierarchical) were used in further developments of conjoint analysis (Moore (1980); Currim (1981); Ogawa (1987); Kamakura (1988); Green, Krieger (1991)). Some of the methodological issues entailed in this development are considered by Green, Krieger (1985), Green, Helsen (1989), and Green, Srinivasan (1990, p. 8).

As the role of social network analysis becomes more conspicuous in marketing research (especially applied to buying centers; see Wind, Robertson (1982)), it is likely that cluster analysis will be used for many applications in this area (see Iacobucci, Hopkins (1992) and Arabie, Wind (1994) for overviews).

Readers seeking a bibliography of applications of clustering in marketing research should consult Dickinson (1990, pp. 784-796). An annual bibliography, not limited to the marketing literature, is published annually by the Classification Society of North America (Day (1991)) and often contains over 1, 000 references relevant to clustering and related techniques of data analysis (e.g., MDS).

3. Misunderstandings of Clustering and Misuses of (Sometimes) Related Methods

Performing a cluster analysis requires using a method that explicitly and objectively seeks clusters. The data analyst should *not* try visually or manually to "define" clusters on the basis of a spatial representation of the products resulting from an analysis using such continuous models as principal components analysis (PCA), factor analysis, correspondence analysis, or spatial MDS (cf. Stewart (1981, p. 53)). Of course, one may refer to groupings and other spatial patterns in such a configuration of products, but it is misleading to elevate such groupings to the status of "clusters" and suggest that they have any reality other than visual. Such subjective clustering dates back to the early days of factor analysis and has empirically been undermined by Ling (1971) and by Baker, Hubert (1976, p. 877).

Another regrettable legacy from factor analysis, and practiced all too often in marketing research, consists of "tandem" cluster analyses. Concretely, a data analyst wanting to get a clustering of a set of products is faced with a products (rows) by attributes (columns) data matrix (more generally, a two-mode matrix) and becomes concerned with either (a) nonindependence of the columns (i.e., intercorrelations generally not equal to zero) and/or (b) an "excessive" number of columns. An all too common response is to run PCA or a related technique on the matrix, extract a few of the eigenvectors having the largest eigenvalues, and use this "reduced" matrix of components as input for clustering.

Concerning nonindependence, it is difficult to imagine empirical data arising in the behavioral sciences that would have all columns mutually independent, and if one did encounter them, why consider running PCA, cluster analysis, or any other technique that by its very nature assumes that interdependencies are exactly what are to be portrayed? Apropos the concern with too many variables, if the cluster analysis is to be given a chance to reveal structure, why should that structure first be "filtered" through an incompatible, spatial model? Using both actual and simulated data, Chang (1983) showed that using PCA to "reduce" data prior to running a cluster analysis discards relevant "distance" information. Still other statistical arguments are given by Gnanadesikan, Kettenring (1972) and by Dillon et al. (1989). DeSarbo, Jedidi, Cool, Schendel (1990, p. 130) have also noted the number of fairly arbitrary decisions that must be made in the PCA stage of "tandem" cluster analyses. Each of these decisions can have an impact on what ultimately goes into and emerges from the cluster analysis.

The conclusion is straightforward: *"tandem" clustering is an outmoded and statistically insupportable practice.*

When we circulated a preliminary version of Arabie, Hubert (1994) containing the preceding conclusion, several colleagues responded very quickly to the challenge. Green, Krieger (1993) sought to test our "critique of current practice" in a comparative evaluation of alternative approaches to cluster-based market segmentation. The empirical results offer considerable support for our recommendation.

4. Clustering by Partitioning Methods

De Soete, Carroll (1994) provided an explicit methodological alternative to the "tandem" approach with a clustering algorithm that implements MacQueen's (1967) k-means criterion in a low-dimensional Euclidean space. This approach should be of great interest to applied market researchers, especially because the most heavily used (e.g., Grover, Srinivasan (1992)) partitioning approach in marketing applications consists of numerous variants of MacQueen's (1967) k-means method, which is implemented in widely available software packages (e.g., SYSTAT[5]: Wilkinson (1992); ProClus[6]: Bretton-Clark (1993)).

Because the results of partitioning methods tend to be extremely dependent on initial configurations, Punj, Stewart (1983, p. 145) recommended using such hierarchical methods as Ward's (1963) or average-link to get an initial configuration for a partitioning method. Nearly a decade later, that recommendation still seems like a good one, but we would advise additionally using configurations from other clustering methods and/or random configurations.

We note that Ward's method is available in numerous statistical software packages (e.g., SYSTAT, ProClus), but the preprocessing normalization empirically found to be most appropriate (Milligan, Cooper (1988); see discussion below) for the mode not clustered generally requires extra effort from the data analyst. Reassurances about the comparative performance of Ward's (1963) method are given by Klastorin (1985), and Ogawa (1987) has innovatively provided an extension, tailored for use with conjoint analysis (also see Batagelj (1988)).

5. Other Aspects of Clustering Relevant to Marketing Research

Marketing methodologists have devised new statistical procedures that incorporate clustering, to answer traditionally important practical problems that applied statisticians have ostensibly overlooked. An example, much in the spirit of clusterwise aggregation of relations (Gaul, Schader (1988a)) is clusterwise linear regression, which simultaneously clusters observations into a user-specified number of groups and estimates the corresponding regression functions' coefficients, while optimizing a common objective function (DeSarbo, Cron (1988); Kamakura (1988); DeSarbo et al. (1989); Wedel, Steenkamp (1991); Steenkamp, Wedel (1993)). Another such development is Kamakura's clusterwise multinomial logit models (Kamakura, Agrawal (1990); Kamakura (1992)).

[5]Available from SYSTAT, 1800 Sherman Avenue, Evanston, Illinois 60201-3793.

[6]Available from Bretton-Clark, 89 Headquarters Plaza, North Tower, 14th Floor, Morristown New Jersey 07960.

6. Practical Problems and Issues

6.1 Normalization of Variables for Hierarchical Clustering

The option most readily available in most statistical software is standardization of each column to have a mean of zero and variance of unity. But a very important result by Milligan, Cooper (1988) shows empirically that, according to ability to recover known clustering structure, the best of a considerable variety of normalization strategies is instead to divide each variable by its range (possibly after first subtracting the minimum value occurring prior to normalization). This finding held for complete- and single-link, average-link (UPGMA: Sokal, Michener (1958)), and Ward's (1963) methods.

6.2 Cluster Validity

Even if one only considers using hierarchical clustering, different methods will generally produce different solutions based on the same input data. The question naturally arises as to whether the clusters have "reality" or validity vis-à-vis the data (cf. Hartigan (1975, pp. 202 - 203); Dubes, Jain (1979, Ch. 4)). According to Wind (1978b, p. 328), "The validity of segmentation research is by far the most crucial question facing management."

Among the issues most commonly investigated in studies of cluster validation are (a) selection of indices of cluster structure and their distributions and (b) determining the appropriate number of clusters. The first problem appears to have been solved. Concluding a comparative study of the coefficients regarded for either theoretical or empirical reasons to be forerunners for measures of relatedness between partitions in cluster validation, Milligan, Cooper (1986, p. 457) stated: "...it would appear that of the five indices the Hubert, Arabie (1985) adjusted Rand measure seems to be the index of choice for clustering validation research."

Milligan, Cooper ((1985, 1988); also see Milligan (1994)) have also studied the problem of choosing the "correct" number of clusters extensively. When we asked Glenn Milligan for a summary of the current status of research on this problem, we received the following response (Milligan (1992), personal communication): "My current advice is to try two or three of the top-rated methods from Milligan, Cooper ((1985); Cooper, Milligan, (1988)). When consistent results are found, they provide substantial support for the selection of the number of clusters. This situation has occurred for a few of my correspondents. If there is partial agreement, opt for the larger number of clusters. Future research will help confirm whether clusters should be merged. If there is no consistency, then I stress that any selected solution should be clearly interpretable within the context of the area of research. If not, I strongly encourage individuals to consider the hypothesis of no clustering in the data."

An alternative, solution by fiat, is suggested by de Kluyver, Whitlark (1986, p. 280), who note that "To be managerially relevant, the number of clusters must be small enough to allow complete strategy development. At the same time, each cluster or segment should be large enough to warrant such strategic attention and to be reachable, and defensible against competitors."

An exemplary study of cluster validation in marketing research is given by Helsen, Green (1991).

6.3 Variable Selection and Weighting

Because marketing researchers are often faced with data having an excessive number of variables (e.g., in survey research), approaches to selection and weighting of variables are very important to segmentation research. Accordingly, DeSarbo et al. (1984) devised an approach for "synthesized clustering" in which the variables in a two-mode (products $\times$ variables) matrix were iteratively and differentially weighted according to their relative importance to the emergent k-means (MacQueen (1967)) cluster structure. This procedure was extended from partitions to ultrametric trees by De Soete et al. (1985), who also sketched details for further extensions to path-length, multiple, and three-way trees, some of which were implemented later by De Soete, Carroll ((1988); also see De Soete (1994)). De Soete provided both an algorithm (1986) and software (1988) for optimal variable weighting in fitting either an ultrametric or path-length tree to a single two-mode matrix. Fowlkes et al. (1987, 1988) devised a forward selection procedure for variables in a two-mode matrix intended for complete-link hierarchical clustering as well as other methods. For a summary of the (mixed) results of evaluations of these approaches, see Arabie, Hubert (1992, p. 181).

6.4 Constrained Clustering

Mahajan, Jain (1978, p. 341) stressed the importance of being able to impose constraints on partitioning approaches to market segmentation. In the clustering literature, the constraint most commonly considered is contiguity in a plane (see Gordon (1980, 1981, pp. 61 - 69); Finden, Gordon (1985); Murtagh (1985); Legendre (1987); Legendre, Legendre (1993)). DeSarbo, Mahajan (1984) contributed an algorithm allowing constraints for either a partition or overlapping clusters and applied it to the problem of defining sales territories. Similarly, DeSarbo (1982) in generalizing the ADCLUS (Shepard, Arabie (1979)) model also provided an algorithm allowing some very useful constraints in fitting (possibly) overlapping clusters. Other types of constraints for ultrametric and additive tree representations are given by De Soete et al. (1987).

7. Software

Even an integrated statistical software package as good as SYSTAT (Wilkinson (1992)) or the more specialized NT-SYS (Rohlf (1992)) cannot readily support large-scale segmentation analyses involving many thousands of respondents. Fortunately, clustering software designed for marketing applications (e.g., Mark^{2}Man: Gaul, Baier (1993); ProClus: Bretton-Clark (1993)) is available.

References:

ARABIE, P., CARROLL, J. D. (1980): MAPCLUS: A mathematical programming approach to fitting the ADCLUS model. *Psychometrika*, 45, 211-235.

ARABIE, P., CARROLL, J. D., DESARBO, W., WIND, J. (1981): Overlapping clustering: A new method for product positioning. *Journal of Marketing Research*, 18, 310-317.

ARABIE, P., HUBERT, L. (1992): Combinatorial data analysis. *Annual Review of Psychology*, 43, 169-203.

ARABIE, P., HUBERT, L. (1994): Cluster analysis in marketing research. In R. P. Bagozzi (ed.): *Advanced methods in marketing research*, Oxford: Blackwell, 160-189.

ARABIE, P., SOLI, S. D. (1982): The interface between the types of regression and methods of collecting proximity data. In R. G. Golledge and J. N. Rayner (eds.): *Proximity and preference: Problems in the multidimensional analysis of large data sets.* University of Minnesota Press., Minneapolis MN, 90-115.

ARABIE, P., WIND, J. (1994): Marketing and social networks. In J. Galaskiewicz and S. S. Wasserman (eds.): *Advances in social and behavioral sciences from social network analysis.* Sage, Newbury Park, CA.

BAKER, F. B., HUBERT, L. J. (1976): A graph-theoretic approach to goodness-of-fit in complete-link hierarchical clustering. *Journal of the American Statistical Association*, 71, 870-878.

BATAGELJ, V. (1988): Generalized Ward and related clustering problems. In H. H. Bock (ed.): *Classification and related methods of data analysis.* North-Holland, Amsterdam, 67-74.

BEANE, T. P., ENNIS, D. M. (1987): Market segmentation: A review. *European Journal of Marketing*, 21, 20-42.

BLOZAN, W., PRABHAKER, P. (1984): Notes on aggregation criteria in market segmentation. *Journal of Marketing Research*, 21, 332-335.

BRETTON-CLARK, INC. (1993): ProClus. Bretton-Clark, Morristown, NJ.

BUTLER, D. H. (1976): Development of statistical marketing models. In *Speaking of Hendry.* Hendry Corporation, Croton-on-Hudson, NY.

CARROLL, J. D., CLARK, L. A., DESARBO, W. S. (1984): The representation of three-way proximities data by single and multiple tree structure models. *Journal of Classification*, 1, 25-74.

CHANG, W.-C. (1983): On using principal components before separating a mixture of two multivariate normal distributions. *Applied Statistics*, 32, 267-275.

CHOFFRAY, J.-M., LILIEN, G. L. (1978): A new approach to industrial market segmentation. *Sloan Management Review*, Spring, 17-29.

CHOFFRAY, J.-M., LILIEN, G. L. (1980a): Market planning for new industrial products. Wiley, New York.

CHOFFRAY, J.-M., LILIEN, G. L. (1980b): Industrial market segmentation by the structure of the purchasing process. *Industrial Marketing Management*, 9, 331-342.

COOPER, M. C., and MILLIGAN, G. W. (1988): The effect of measurement error on determining the number of clusters in cluster analysis. In W. Gaul and M. Schader (Eds.): *Data, expert knowledge and decisions.* Springer-Verlag, Berlin,

319-328.

CURRIM, I. S. (1981): Using segmentation approaches for better prediction and understanding from consumer mode choice models. *Journal of Marketing Research*, 18, 301-309.

CURRIM, I. S., SCHNEIDER, L. G. (1991): A taxonomy of consumer purchase strategies in a promotion intensive environment. *Marketing Science*, 10, 91-110.

DAY, G. S., HEELER, R. M. (1971): Using cluster analysis to improve marketing experiments. *Journal of Marketing Research*, 8, 340-347.

DAY, W. H. E. (ed.) (1991): *Classification Literature Automated Search Service*, 20.

DE KLUYVER, C. A., WHITLARK, D. B. (1986): Benefit segmentation for industrial products. *Industrial Marketing Management*, 15, 273-286.

DE SOETE, G. (1986): Optimal variable weighting for ultrametric and additive tree clustering. *Quality and Quantity*, 20, 169-180.

DE SOETE, G. (1988): OVWTRE: A program for optimal variable weighting for ultrametric and additive tree fitting. *Journal of Classification*, 5, 101-104.

DE SOETE, G. (1994): Variable selection and weighting in cluster analysis. In P. Arabie, L. Hubert and G. De Soete (eds.): *Clustering and classification*. World Scientific, River Edge, New Jersey.

DE SOETE, G., CARROLL, J. D. (1988): Optimal weighting for one-mode and two-mode ultrametric tree representations of three-way three-mode data. In M. G. H. Jansen and W. H. van Schuur (eds.): *The many faces of multivariate data analysis*. RION, Groningen, 16-29.

DE SOETE, G., CARROLL, J. D. (1989): Ultrametric tree representations of three-way three-mode data. In R. Coppi and S. Bolasco (eds.): *Analysis of multiway data matrices*. North-Holland, Amsterdam, 415-426.

DE SOETE, G., CARROLL, J. D. (1994): K-means clustering in a low-dimensional Euclidean space. In E. Diday, Y. Lechevallier, M. Schader, P. Bertrand, P. Burtschy (eds.): *New approaches in classification and data analysis*, Springer-Verlag, Heidelberg.

DE SOETE, G., CARROLL, J. D., DESARBO, W. S. (1987): Least squares algorithms for constructing constrained ultrametric and additive tree representations of symmetric proximity data. *Journal of Classification*, 4, 155-173.

DE SOETE, G., DESARBO, W. S., CARROLL, J. D. (1985): Optimal variable weighting for hierarchical clustering: An alternating least-squares algorithm. *Journal of Classification*, 2, 173-192.

DEMING, W. E., STEPHAN, F. F. (1940): On a least squares adjustment of a sampled frequency table when the expected marginal totals are known. *Annals of Mathematical Statistics*, 11, 427-444.

DESARBO, W. S. (1982): GENNCLUS: New models for general nonhierarchical clustering analysis. *Psychometrika*, 47, 446-449.

DESARBO, W. S., CARROLL, J. D., CLARK, L. A., GREEN, P. E. (1984): Synthesized clustering: A method for amalgamating alternative clustering bases with differential weighting of variables. *Psychometrika*, 49, 57-78.

DESARBO, W. S., CRON, W. L. (1988): A conditional mixture maximum likelihood methodology for clusterwise linear regression. *Journal of Classification*, 5, 249-289.

DESARBO, W. S., DE SOETE, G. (1984): On the use of hierarchical clustering for the analysis of nonsymmetric proximities. *Journal of Consumer Research*, 11, 601-610.

DESARBO, W. S., JEDIDI, K., COOL, K., SCHENDEL, D. (1990): Simultaneous multidimensional unfolding and cluster analysis: An investigation of strategic groups. *Marketing Letters*, 2, 129-146.

DESARBO, W. S., MAHAJAN, V. (1984): Constrained classification: The use of a priori information in cluster analysis. *Psychometrika*, 49, 187-216.

DESARBO, W. S., MANRAI, A., BURKE, R. (1990): A nonspatial methodology incorporating the distance-density hypothesis. *Psychometrika*, 55, 229-253.

DESARBO, W. S., MANRAI, A. K., MANRAI, L. A. (1993): Non-spatial tree models for the assessment of competitive market structure: an integrated review of the marketing and psychometric literature. In J. Eliashberg and G. Lilien (eds.): *Handbook in operations research and management science: Marketing*. Elsevier, New York, 193-257.

DESARBO, W. S., OLIVER, R. L., RANGASWAMY, A. (1989): A simulated annealing methodology for clusterwise linear regression. *Psychometrika*, 4, 707-736.

DICKINSON, J. R. (1990): The bibliography of marketing research methods (3rd ed.). Lexington, Lexington, MA.

DILLON, W. R., MULANI, N., FREDERICK, D. G. (1989): On the use of component scores in the presence of group structure. *Journal of Consumer Research*, 16, 106-112.

DOYLE, P., SAUNDERS, J. (1985): Market segmentation and positioning in specialized industrial markets. *Journal of Marketing*, 49, 24-32.

DUBES, R., JAIN, A. K. (1979): Validity studies in clustering methodologies. *Pattern Recognition*, 11, 235-254.

ELROD, T., WINER, R. S. (1982): An empirical evaluation of aggregation approaches for developing market segments. *Journal of Marketing*, 46 (Fall), 65-74.

FIENBERG, S. E. (1970): An iterative procedure for estimation in contingency tables. *Annals of Mathematical Statistics*, 41, 907-17. (Erratum, p. 1778).

FINDEN, C. R., GORDON, A. D. (1985): Obtaining common pruned trees. *Journal of Classification*, 2, 255-276.

FOWLKES, E. B., GNANADESIKAN, R., KETTENRING, J. R. (1987):Variable selection in clustering and other contexts. In C. L. Mallows (ed.): *Design, data, and analysis*. Wiley, New York, 13-34.

FOWLKES, E. B., GNANADESIKAN, R., KETTENRING, J. R. (1988): Variable selection in clustering. *Journal of Classification*, 5, 205-228.

FURSE, D. H., PUNJ, G. N., STEWART, D. W. (1984): A typology of individual search strategies among purchasers of new automobiles. *Journal of Consumer Research*, 10, 417-431.

GAUL, W., BAIER, D. (1993): MARKtforschung und MARKeting MANagement [Market research and marketing management]. R. Oldenbourg, München.

GAUL, W., SCHADER, M. (1988a): Clusterwise aggregation of relations. *Applied Stochastic Models and Data Analysis*, 4, 273-282.

GAUL, W., SCHADER, M. (eds.). (1988b): Data, expert knowledge and decisions. Springer-Verlag, Berlin.

GLAZER, R., NAKAMOTO, K. (1991): Cognitive geometry: An analysis of structure underlying representations of similarity. *Marketing Science*, 10, 205-228.

GNANADESIKAN, R., KETTENRING, J. R. (1972): Robust estimates, residuals, and outlier detection with multiresponse data. *Biometrics*, 28, 81-124.

GORDON, A. D. (1980): Methods of constrained classification. In R. Tomassone (ed.): *Analyse de Donnes et Informatique*. INRIA, Le Chesnay, 161-171.

GORDON, A. D. (1981): Classification. Chapman and Hall, London.

GREEN, P. E., FRANK, R. E., ROBINSON, P. J. (1967): Cluster analysis in test market selection. *Management Science*, 13, B387-B400.

GREEN, P. E., HELSEN, K. (1989): Cross-validation assessment of alternatives to individual-level conjoint analysis: A case study. *Journal of Marketing Research*, 26, 346-350.

GREEN, P. E., KRIEGER, A. M. (1985): Buyer similarity measures in conjoint analysis: Some alternative proposals. *Journal of Classification*, 2, 41-61.

GREEN, P. E., KRIEGER, A. M. (1991): Segmenting markets with conjoint analysis. *Journal of Marketing*, 55 (October), 20-31.

GREEN, P. E., KRIEGER, A. M. (1993): An evaluation of alternative approaches to cluster-based market segmentation. Unpublished manuscript, Wharton School, University of Pennsylvania, Philadelphia.

GREEN, P. E., KRIEGER, A. M., SCHAFFER C. M. (1985): Quick and simple benefit segmentation. *Journal of Advertising Research*, 25, 9-17.

GREEN, P. E., SRINIVASAN, V. (1978): Conjoint analysis in consumer research: Issues and outlook. *Journal of Consumer Research*, 5, 103-123.

GREEN, P. E., SRINIVASAN, V. (1990): Conjoint analysis in marketing: New developments with implications for research and practice. *Journal of Marketing*, 54 (October), 3-19.

GROVER, R., SRINIVASAN, V. (1987): A simultaneous approach to market segmentation and market structuring. *Journal of Marketing Research*, 24, 139-153.

GROVER, R., SRINIVASAN, V. (1992): Evaluating the multiple effects of retail promotions on brand loyal and brand switching segments. *Journal of Marketing Research*, 29, 76-89.

HARTIGAN, J. A. (1975): Clustering algorithms. New York: Wiley. Translated into Japanese by H. Nishida, M. Yoshida, H. Hiramatsu, K. Tanaka, 1983. Micro Software, Tokyo.

HELSEN, K., GREEN, P. E. (1991): A computational study of replicated clustering with an application to market segmentation. *Decision Sciences*, 22, 1124-1141.

HUBERT, L. J., ARABIE, P. (1985): Comparing partitions. *Journal of Classification*, 2, 193-218.

HUTCHINSON, J. W., MUNGAL, A. (1992): Pairwise partitioning: A nonmetric algorithm for identifying feature-based similarity measures. Manuscript submitted for publication.

IACOBUCCI, D., HOPKINS, N. (1992): Modeling dyadic interactions and networks in marketing. *Journal of Marketing Research*, 29, 5-17.

JOHNSON, M. D., FORNELL, C. (1987): The nature and methodological implications of the cognitive representation of products. *Journal of Consumer Re-*

search, 14, 214-228.

KAMAKURA, W. A. (1988): A least squares procedure for benefit segmentation with conjoint experiments. *Journal of Marketing Research*, 25, 157-167.

KAMAKURA, W. A. (1992): A clusterwise multinomial logit model for multiple locally-independent choice sets. Unpublished manuscript, Owen Graduate School of Management, Vanderbilt University, Nashville.

KAMAKURA, W. A., AGRAWAL, J. (1990): A clusterwise multinomial logit model for benefit segmentation. Unpublished manuscript, Owen Graduate School of Management, Vanderbilt University, Nashville.

KAMAKURA, W. A., MAZZON, J. A. (1991): Value segmentation: A model for the measurement of values and value systems. *Journal of Consumer Research*, 18, 208-218.

KLASTORIN, T. D. (1985): The p-median problem for cluster analysis: A comparative test using the mixture model approach. *Management Science*, 31, 84-95.

LEGENDRE, P. (1987): Constrained clustering. In P. Legendre and L. Legendre (eds.): *Developments in numerical ecology* [NATO Advanced Study Institute Series G (Ecological Sciences)]. Springer-Verlag, Berlin, 289-307.

LEGENDRE, L., LEGENDRE, P. (1993): Numerical ecology (2nd ed.). Elsevier, Amsterdam.

LING, R. F. (1971): Cluster analysis. University Microfilms, Ann Arbor, MI. No. 71-22356.

MACQUEEN, J. (1967): Some methods for classification and analysis of multivariate observations. In L. M. Le Cam and J. Neyman (eds.): *Proceedings of the Fifth Berkeley Symposium on Mathematical Statistics and Probability*. University of California Press, Berkeley, (Vol. 1, 281-297).

MAHAJAN, V., JAIN, A. K. (1978): An approach to normative segmentation. *Journal of Marketing Research*, 15, 338-345.

MILLIGAN, G. W. (1994): Clustering validation: Results and implications for applied analyses. In P. Arabie, L. Hubert and G. De Soete (eds.): *Clustering and classification*. World Scientific, River Edge, New Jersey.

MILLIGAN, G. W., COOPER, M. C. (1985): An examination of procedures for determination of clusters in a data set. *Psychometrika*, 50, 159-179.

MILLIGAN, G. W., COOPER, M. C. (1986): A study of the comparability of external criteria for hierarchical cluster analysis. *Multivariate Behavioral Research*, 21, 441-458.

MILLIGAN, G. W., COOPER, M. C. (1988): A study of standardization of variables in cluster analysis. *Journal of Classification*, 5, 181-204.

MOORE, W. L. (1980): Levels of aggregation in conjoint analysis: An empirical comparison. *Journal of Marketing Research*, 17, 516-523.

MURPHY, R. A., TATHAM, R. L. (1979): Optimal construction of experimental clusters. *Management Science*, 25, 182-190.

MURTAGH, F. (1985): A survey of algorithms for contiguity-constrained clustering and related problems. *Computer Journal*, 28, 82-88.

NICOSIA, F. M., WIND, Y. (1977): Behavioral models of organizational buying processes. In F. M. Nicosia and Y. Wind (eds.): *Behavioral models for market analysis: Foundations for marketing action*. Dryden, Hinsdale IL, 96-120.

OGAWA, K. (1987): An approach to simultaneous estimation and segmentation in conjoint analysis. *Marketing Science*, 6, 66-81.

PRUZANSKY, S., TVERSKY, A., CARROLL, J. D. (1982): Spatial versus tree representations of proximity data. *Psychometrika*, 47, 3-24.

PUNJ, G., STEWART, D. W. (1983): Cluster analysis in marketing research: Review and suggestions for application. *Journal of Marketing Research*, 20, 134-148.

RAO, V. R., SABAVALA, D. J. (1981): Inference of hierarchical choice processes from panel data. *Journal of Consumer Research*, 8, 85-96.

RAO, V. R., SABAVALA, D. J., LANGFELD, P. A. (1977): Alternative measures for partitioning analysis based on brand switching data. Unpublished manuscript, Johnson Graduate School of Cornell University, Ithaca NY.

RAO, V. R., SABAVALA, D. J., ZAHORICK, A. J. (1982): Market structure analysis using brand switching data: A comparison of clustering techniques. In R. K. Srivastava and A. Shocker (eds.): *Analytical approaches to product and marketing planning: The second conference.* Marketing Science Institute, Cambridge MA, 17-25.

ROBLES, F., SARATHY, R. (1986): Segmenting the commuter aircraft market with cluster analysis. *Industrial Marketing Management*, 15, 1-12.

ROHLF, F. J. (1992): NTSYS-pc numerical taxonomy and multivariate analysis system (version 1.70). Exeter Software, Setauket, NY.

SCHADER, M. (ed.) (1992): Analyzing and modeling data and knowledge. Springer-Verlag, Berlin.

SCHADER, M., GAUL, W. (eds.) (1990): Knowledge, data and computer-assisted decisions. Springer-Verlag, Berlin.

SHEPARD, R. N., ARABIE, P. (1979): Additive clustering: Representation of similarities as combinations of discrete overlapping properties. *Psychological Review*, 86, 87-123.

SHOCKER, A. D., STEWART, D. W., ZAHORIK, A. J. (1990): Determining the competitive structure of product-markets: Practices, issues, and suggestions. *Journal of Managerial Issues*, 2, (2) 127-159.

SLATER, P. (1984): Tree representations of internal migration flows and related topics. Community and Organization Research Institute, University of California, Santa Barbara.

SOKAL, R. R., MICHENER, C. D. (1958): A statistical method for evaluating systematic relationships. *University of Kansas Science Bulletin*, 38, 1409-1438.

SRIVASTAVA, R. K. (1981): Usage-situational influences on perceptions of product-markets: Theoretical and empirical issues. In K. B. Monroe (ed.): *Advances in Consumer Research*, Vol. 8. Association for Consumer Research, Ann Arbor MI, 106-111.

SRIVASTAVA, R. K., ALPERT, M. I. (1982): A customer-oriented approach for determining market structures. In R. K. Srivastava and A. D. Shocker (eds.) *Proceedings of the Second Conference on Analytic Approaches to Product and Marketing Planning.* Marketing Science Institute, Cambridge MA, 26-57.

SRIVASTAVA, R. K., ALPERT, M. I., SHOCKER, A. D. (1984): A customer-oriented approach for determining market structures. *Journal of Marketing*, 48 (Spring), 32-45.

STEENKAMP, J.-B. E. M., WEDEL, M. (1991): Segmenting retail markets on store image using a consumer-based methodology. *Journal of Retailing*, 67, 300-320.

STEENKAMP, J.-B. E. M., WEDEL, M. (1993): Fuzzy clusterwise regression in benefit segmentation: Application and investigation into its validity. *Journal of Business Research*, 26, 237-249.

STEWART, D. W. (1981): The application and misapplication of factor analysis in marketing research. *Journal of Marketing Research*, 18, 51-62.

WARD, J. H., JR. (1963): Hierarchical grouping to optimize an objective function. *Journal of the American Statistical Association*, 58, 236-244.

WEDEL, M. (1993): Market segmentation research: A review of bases and methods with special emphasis on latent class models. In *Information based decision making in marketing*. European Society for Opinion and Marketing Research, Paris, 191-222.

WEDEL, M., STEENKAMP, J.-B. E. M. (1991): A clusterwise regression method for simultaneous fuzzy market structuring and benefit segmentation. *Journal of Marketing Research*, 28, 385-396.

WILKINSON, L. (1992): SYSTAT: The system for statistics. Systat, Inc., Evanston IL.

WIND, Y. (1978a): Introduction to special section on market segmentation research. *Journal of Marketing Research*, 15, 315-316.

WIND, Y. (1978b): Issues and advances in segmentation research. *Journal of Marketing Research*, 15, 317-337.

WIND, Y. J. (1982): Product policy: Concepts, methods, and strategy. Addison-Wesley, Reading MA.

WIND, Y., ROBERTSON, T. S. (1982): The linking pin role in organizational buying centers. *Journal of Business Research*, 10, 169-184.

Representation of Statistical Structures, Classification and Prediction Using Multidimensional Scaling

C. M. Cuadras[1], J. Fortiana, F. Oliva

Departament d'Estadistica. Universitat de Barcelona.
Diagonal 645, 08028–BARCELONA, Spain

Summary: We show that MDS and related methods provide techniques and solutions to a wide field of topics in statistics, classification and data analysis: distance–based regression; representation of Hoeffding's extremal correlations; examining a tree; representing parametric estimable functions in MANOVA; distance–based discrimination and classification; representation of a continuous random variable.

1. Introduction

The earliest books on Multivariate Analysis, e.g. those by M. G. Kendall, C. R. Rao, T. W. Anderson, H. L. Seal, D. F. Morrison, A. P. Dempster, A. M. Kshirsagar, published in the period 1957–1972, did not contain sections on Multidimensional Scaling (MDS). Later, due to the impact of works by R. N. Shepard, J. C. Gower, J. B. Kruskal, J. C. Lingoes, F. W. Young, J. D. Carroll, P. Arabie, J. de Leeuw and others as well as of the books by Shepard et al. and Romney et al. (1972), the interest on this method increased. Thus MDS is present in books on MA such as Bertier and Bouroche (1975), Mardia et al. (1979), Cuadras (1981b), Greenacre (1984), Seber (1984), and it is the main subject of e.g., Kruskal and Wish (1978) and Davison (1983).

Aplications of MDS contained in these books are generally restricted to representing and describing objects in a low–dimensional Euclidean space and interpreting the axes. But in Davison (1983) we find different applications, e.g. an MDS alternative to Thurstone's model of preferences.

Throughout this paper, we work with n objects labelled as $1, 2, \ldots, n$ and an $n \times n$ matrix $\boldsymbol{\Delta} = (\boldsymbol{\delta}_{ij})$ of distances between them. $\boldsymbol{X}$ will indicate the $n \times m$ matrix of Euclidean coordinates obtained via metric or nonmetric MDS. Hence, the Euclidean distances between points whose coordinates are rows of $\boldsymbol{X}$ reproduce (exactly or approximately) the matrix $\boldsymbol{\Delta}$.

2. The Distance–Based Regression Model

Regression analysis is possibly the most widely used, but also abused, statistical technique. A clear example of the latter is the linear regression of a

---

[1]Email: `carlesm@porthos.bio.ub.es`. The first author thanks H. H. Bock and W. Gaul for their kind invitation to attend the 18th Annual Conference, Gesellschaft für Klassifikation e.V. in Oldenburg. We also thank M. Greenacre for useful comments. Work supported in part by CGYCIT grant PB93–0784

continuous response variable y on a set $\boldsymbol{w}$ of p mixed (continuous, binary, nominal, ordinal) explanatory variables: even if theoretical difficulties are ignored, grossly divergent results can be obtained as a result of different ways of coding the categorical predictors. Arbitrariness can also cause trouble in the following regression framework: predictors are continuous, a linear model has a poor adjustment, and context knowledge of the data (e.g. a physical model) does not suggest a clear candidate f to build a nonlinear model $y = f(\boldsymbol{w}; \boldsymbol{\theta}) + e$.

The distance–based (DB) model for prediction is an MDS alternative in regression proposed and studied by Cuadras (1989) and Cuadras and Arenas (1990). Given n experimental units and a distance matrix $\boldsymbol{\Delta}$, computed on the basis of the (mixed) explanatory variables $\boldsymbol{w}$, y is regressed on the matrix of principal coordinates $\boldsymbol{X}$, i.e., the DB model is

$$\boldsymbol{y} = \beta_0 \mathbf{1} + \boldsymbol{X}_{(k)} \cdot \boldsymbol{\beta}_{(k)} + e_{(k)}$$

where $\boldsymbol{y}$ is the $n \times 1$ vector of observations on y, β_0 and $\boldsymbol{\beta}_{(k)}$ are parameters, and $\boldsymbol{X}_{(k)}$ is a matrix containing a suitable set of k columns extracted from $\boldsymbol{X}$.

The main properties of this model are:

- It reduces to the classic regression model when the Euclidean distance is used.
- The coefficient of determination and the prediction of y for a new case are readily obtained.
- It is quite general, as we can use any appropriate distance between observations. The real advantage appears with mixed variables. In this case, the distance function associated with Gower's similarity coefficient (Gower 1971a) is a good choice, and has the additional benefit of being able to handle missing values.
- The nonlinear regression model $y_i = f(w_{i1}, \dots, w_{ip}; \theta) + e_i, \quad i = 1, \dots, n$, for an unknown f, also admits the DB approach (Cuadras 1991) by using the distance $\boldsymbol{\delta}_{ij}^2 = |w_{i1} - w_{j1}| + \cdots + |w_{ip} - w_{jp}|$.

  For $p = 1$, this DB model is equivalent to an ordinary regression on Chebychev polynomials of the first kind (Cuadras and Fortiana 1993). Good performance for $p > 1$ leads us to conjecture that a similar result holds for higher dimensional predictors.

3. MDS of a Maximum Correlation Matrix

Hoeffding (1940) introduced the concept of maximum correlation, i.e. the maximum $\rho^+(F, G)$ of the set of correlation coefficients of bivariate distributions having F and G as marginals, which is reached for the upper Fréchet bound $H^+(x, y) = \min\{F(x), G(y)\}$.

This maximum correlation can serve as a measure of agreement between distribution functions, up to an affine transformation, since it equals 1 iff $F^* = G^*$, where F^* and G^* are the distribution functions of the variables standardized to mean 0 and variance 1.

The maximum correlation between standardized variables can be computed by means of

$$\rho^+(F,G) = \int_0^1 F^{-1}(u)\,G^{-1}(u)\,du. \tag{1}$$

With few exceptions, a closed formula for ρ^+ is unavailable. Cuadras (1992a, 1994) gives a lower approximation to ρ^+, when F and G are absolutely continuous. In general, a numerical evaluation of the improper integral (1) will be needed (Cuadras and Fortiana 1994a).

A set $\{F_1, \ldots, F_n\}$ of distribution functions has an associated $n \times n$ matrix $\boldsymbol{R}^+$ of maximum correlations. The fact that $\boldsymbol{R}^+$ is positive semi–definite ensures that performing a metric scaling on $\boldsymbol{R}^+$ results in a Euclidean representation of the given set.

Example 1 *Fig. 1 contains the matrix* $\boldsymbol{R}^+$ *for the distributions* U = *Uniform,* E = *Exponential,* N = *Normal,* G = *Gamma (with shape parameter* $p = 8$*),* L = *a log–normal distribution (the* log *of a standard normal distribution), as well as the empirical distribution* I *of the variable "sepal length" of* Iris setosa *in Fisher's Iris data. Omitted parameters are irrelevant to the results.*

Suppose now that we wish to ascertain the distribution of I *and we consider the candidates* U, E, N, G, L. $\rho^+(N, I)$ *is the largest correlation, revealing that possibly the sample* I *comes from a Normal distribution. Metric scaling on* $\boldsymbol{R}^+$ *and the resulting graph are useful tools to understand the proximity relations between* I *and the other distributions.*

U	E	N	G	L	I
1					
.866	1				
.977	.903	1			
.947	.977	.973	1		
.688	.893	.763	.872	1	
.974	.899	.987	.966	.745	1

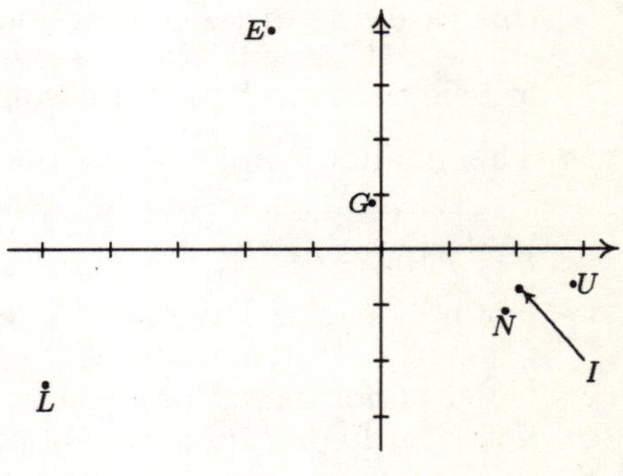

Fig. 1: Matrix R^+ of maximum correlations and its Euclidean representation.

4. MDS in MANOVA

In canonical variate analysis, given g random samples of the populations $\pi_j = N_p(\boldsymbol{\mu}_j, \boldsymbol{\Sigma})$, $j = 1, \ldots, g$, orthogonal coordinates are assigned to the estimated means $\widehat{\boldsymbol{\mu}}_1, \ldots, \widehat{\boldsymbol{\mu}}_g$. The populations, described by the means, are projected on the (orthogonal) canonical axes, giving a useful representation in a low–dimensional space.

Although the standard formulation is made in terms of the relative eigenstructure of the between–samples and within–samples cross–product matrices, similar results are reached if an MDS is performed on the Mahalanobis distance matrix between the means (Gower 1966).

A generalized canonical variate analysis can be defined for a multivariate linear model

$$\boldsymbol{Y} = \widetilde{\boldsymbol{X}} \cdot \boldsymbol{B} + \boldsymbol{E}, \tag{2}$$

where $\boldsymbol{Y}$ $(n \times p)$ and $\widetilde{\boldsymbol{X}}$ $(n \times q)$ are known, $\boldsymbol{B}$ $(q \times p)$ is the matrix of regression parameters and $\boldsymbol{E}$ $(n \times p)$ contains random errors. We suppose that the rows of $\boldsymbol{E}$ are iid as $N(\boldsymbol{0}, \boldsymbol{\Sigma})$.

Let us consider a set of parametric estimable functions

$$\boldsymbol{\psi}_i = \boldsymbol{p}_i \cdot \boldsymbol{B}, \quad i = 1, \ldots, k. \tag{3}$$

i.e., such that each row $\boldsymbol{p}_i$ is a linear combination of the rows of $\widetilde{\boldsymbol{X}}$. We define the squared distance between $\boldsymbol{\psi}_i$ and $\boldsymbol{\psi}_j$ as

$$m_{ij}^2 = (\boldsymbol{\psi}_i - \boldsymbol{\psi}_j) \cdot \boldsymbol{\Sigma}^{-1} \cdot (\boldsymbol{\psi}_i - \boldsymbol{\psi}_j)'.$$

If $\widehat{\boldsymbol{B}}$ is the OLS estimation of $\boldsymbol{B}$, this distance can be estimated replacing $\boldsymbol{\psi}_i$ (and similarly $\boldsymbol{\psi}_j$) by the Gauss–Markov estimator $\widehat{\boldsymbol{\psi}}_i = \boldsymbol{p}_i \cdot \widehat{\boldsymbol{B}}$ and $\boldsymbol{\Sigma}$ by $\widehat{\boldsymbol{\Sigma}} = \boldsymbol{Y}' \cdot [\boldsymbol{I} - \widetilde{\boldsymbol{X}} \cdot (\widetilde{\boldsymbol{X}}' \cdot \widetilde{\boldsymbol{X}})^{-} \cdot \widetilde{\boldsymbol{X}}'] \cdot \boldsymbol{Y}/(n-r)$, where $r = \operatorname{rank}(\widetilde{\boldsymbol{X}})$ (Rao 1973).

Performing a MDS on the distance matrix (m_{ij}), we have a geometric representation of (3).

Note that when (2) is the model for a multivariate one–way classification with g levels, then the $n = n_1 + \cdots + n_g$ rows of $\boldsymbol{Y}$ can be regarded as g samples, each consisting of n_j observations from π_j, and the classic canonical variate analysis is recovered as a particular case.

This extension was proposed by Cuadras (1974) and a list of applications can be found in Cuadras (1981b). See also Ahrens and Läntner (1974) and Coll et al. (1980).

Example 2 *Figures 2 and 3 show classic canonical variate representations of* 8 *populations of* Coleoptera *of the genus* Timarcha, *males and females, respectively, obtained from 5 biometrical variables (adapted from Petitpierre and Cuadras (1977), Cuadras (1981a)). Population 6 is closer to 4 and 5 for males but closer to 3 for females. Both relations seem contradictory from a taxonomic point of view.*

Using a crossed classification sex $\times$ population, *with interaction, and performing a generalized canonical analysis of the main effects (i.e. the parametric functions $\psi_i = \mu + \alpha_i$, $i = 1, \ldots, 8$), we obtain Fig. 4, in which population 6 appears as separated from 3, 4 and 5, independently of sex. Circles represent 95% confidence regions.*

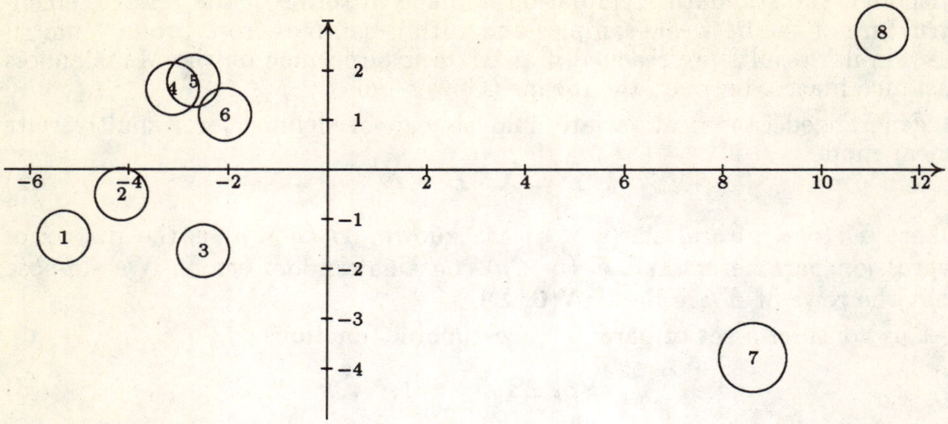

Fig. 2: Canonical variate analysis of 8 populations of males.

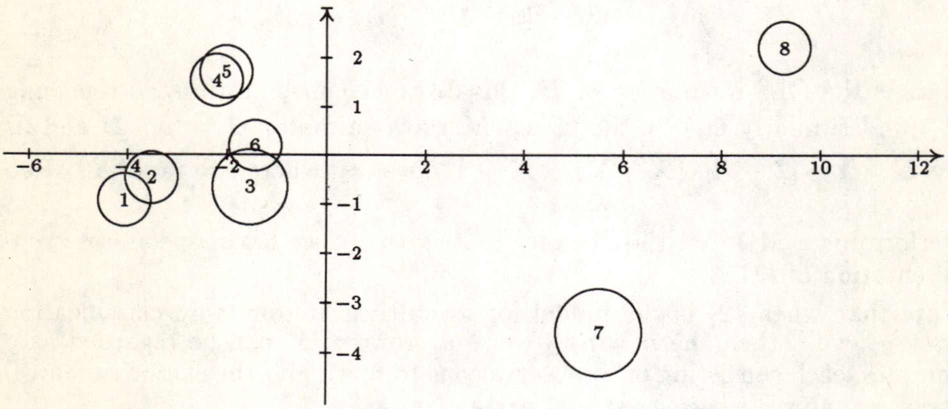

Fig. 3: Canonical variate analysis of 8 populations of females.

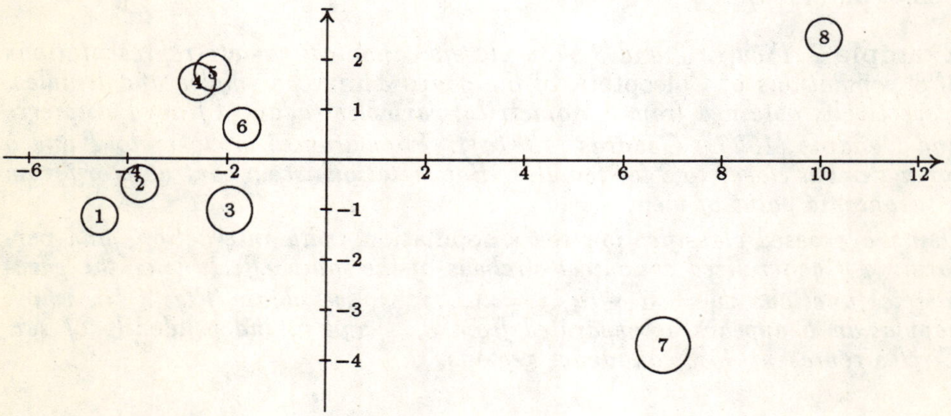

Fig. 4: Generalized canonical variate analysis of 8 populations (Males + Females, removing the influence of the sex effect).

5. MDS on a Tree

Hierarchical clustering and MDS have been widely used in combination for studying proximity data. Fig. 5, adapted from a comparative study of 13 European languages (Oliva et al. 1993), is a recent example. Several authors have studied the inter–relations between both models: Kruskal (1977), Pruzansky et al. (1982), Bock (1987) and Critchley (1988).

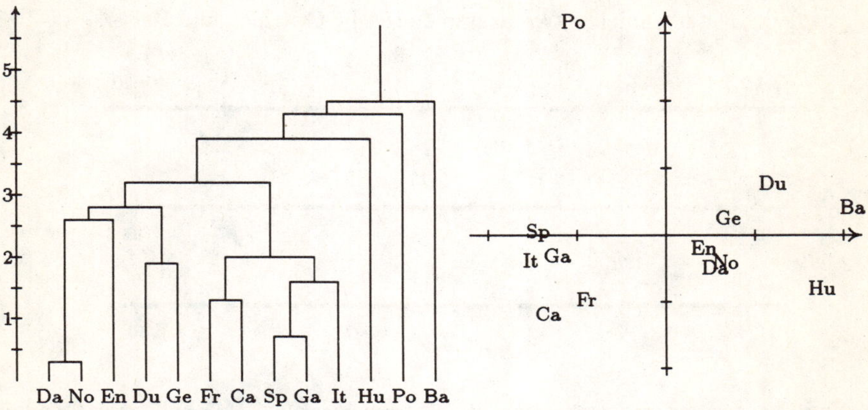

Fig. 5: Ultrametric tree and principal coordinate representation of 13 European languages: Basque, Catalan, Danish, Dutch, English, French, Galician, German, Hungarian, Italian, Norwegian, Polish and Spanish.

Holman (1972) showed that an ultrametric distance matrix $\boldsymbol{U} = (u_{ij})$ on a finite set $\mathcal{E} = \{1, \ldots, n\}$, $(n \geq 3)$ (the algebraic counterpart of a hierarchical clustering scheme) is a Euclidean distance matrix, proving a previous conjecture by Gower (1971b). More precisely, if $\boldsymbol{U}$ is definite ($i \neq j$ implies $u_{ij} > 0$), then $\mathcal{E}$ can be embedded in $\boldsymbol{R}^{n-1}$.

Experience with real data shows that two–dimensional Euclidean representations are often good enough to display $\mathcal{E}$, even when n is large, in apparent conflict with the full dimension $n-1$ needed according to Holman. Under this motivation the following explanations were given:

- Critchley and Heiser (1988) showed that $\mathcal{E}$ can be scaled in only one dimension, by finding a monotonic relation with $\boldsymbol{U}$.
- Cuadras and Oller (1987) explored the principal coordinates solution $\boldsymbol{X}$ obtained from $\boldsymbol{U}$ and showed that the first coordinates, $\boldsymbol{X}_0$, say, represent a partition of $\mathcal{E}$ in m maximal clusters $\mathcal{E}_1, \ldots, \mathcal{E}_m$, while each $\mathcal{E}_i$ is represented by a system of independent coordinates.
- Bock (1987) related a distance matrix $\boldsymbol{\Delta}$ with a given classification or partition $\mathcal{C} = (\mathcal{E}_1, \ldots, \mathcal{E}_m)$ of $\mathcal{E}$, finding a simultaneous Euclidean representation of the n objects of $\mathcal{E}$ and the m clusters of $\mathcal{C}$, described by the matrices of coordinates $\boldsymbol{X}^\star = \boldsymbol{Q}^{-1/2} \cdot \boldsymbol{X}$ and $\boldsymbol{Y}^\star = \boldsymbol{M}' \cdot \boldsymbol{X}^\star$, where $\boldsymbol{Q}$ and $\boldsymbol{M}$ are illustrated below.

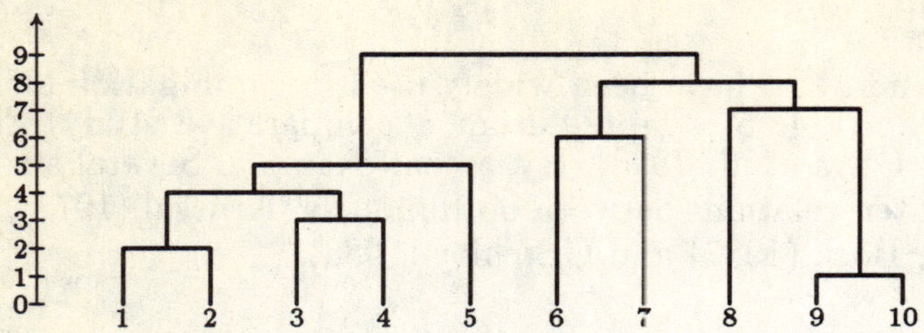

Fig. 6: The ultrametric tree example used by Critchley and Heiser.

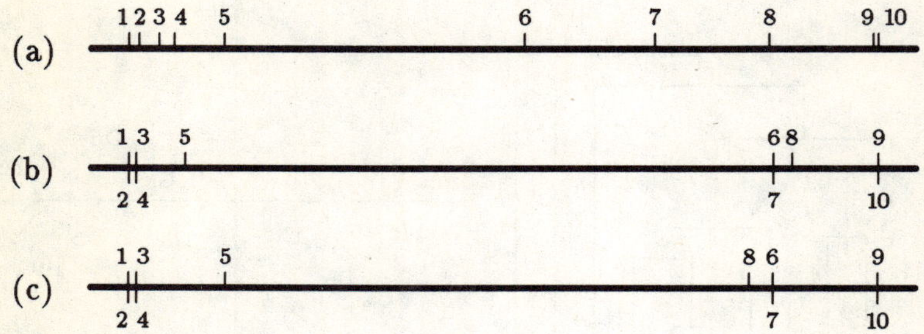

Fig. 7: One–dimensional representation of an hierarchical tree. (a) Following Critchley and Heiser, (b) Using MDS, (c) following Bock.

Example 3 *Fig. 6 represents an ultrametric matrix* $\boldsymbol{U}$ *of order* $n = 10$ *taken from Critchley and Heiser (1988). Here* $m = 6$ *and* $\mathcal{E}$ *has two isolated objects. The partition of* $\mathcal{E}$ *and the eigenvalues, computed using the formulas in Cuadras and Oller (1987), are*

$$\mathcal{E} = \{1,2\} + \{3,4\} + \{5\} + \{6,7\} + \{8\} + \{9,10\}, \tag{4}$$

and $\lambda_1 = \frac{1}{2} < \lambda_2 = 2 < \lambda_3 = \frac{9}{2} < \lambda_4 = 18 < \mu = 134$.
The link between the above approaches may be established noting that $\boldsymbol{X}_0$ *already separates the clusters in (4). The first principal dimension related to the greatest eigenvalue* μ *is enough to separate these clusters. Fig. 7 (a) gives the Critchley–Heiser scale for* $\mathcal{E}$*, 7 (b) is the Euclidean representation using MDS. Furthermore, applying Bock's results to the partition (4) and using standard weights, we have* $\boldsymbol{Q} = (\boldsymbol{I} + \boldsymbol{M} \cdot \boldsymbol{M}')$*, where* $\boldsymbol{M} = \boldsymbol{J} \cdot \boldsymbol{N}^{-1}$*,* $\boldsymbol{N} = \mathrm{diag}(2,2,1,2,1,2)$ *contains the sizes of the clusters, and* $\boldsymbol{J} = (c_{ij})$ *is an* $n \times m$ *matrix with entries* $c_{ij} = 0$ *iff* $i \notin \mathcal{E}_j$ *and* $c_{ij} = 1$ *iff* $i \in \mathcal{E}_j$*. Fig. 7 (c) gives the representation using the first column of* $\boldsymbol{Y}^\star$*.*

6. Distance–Based Discrimination

Let $\boldsymbol{X}$ be a random vector defined on a probability space $(\Omega, \mathcal{A}, P)$, with values on some $E \subset \boldsymbol{R}^p$ and let f be the probability density function of $\boldsymbol{X}$ with respect to a given measure λ. Assume that $\boldsymbol{\delta}(\cdot,\cdot)$ is a square integrable symmetric function on $E \times E$ (usually a dissimilarity), verifying that

$\boldsymbol{\delta}(\boldsymbol{x},\boldsymbol{y}) = \boldsymbol{\delta}(\boldsymbol{y},\boldsymbol{x}) \geq 0, \forall \boldsymbol{x},\boldsymbol{y} \in E$ and that

$$V_{\boldsymbol{\delta}}(\boldsymbol{X}) = \frac{1}{2}\int_{E\times E} \boldsymbol{\delta}^2(\boldsymbol{x},\boldsymbol{y})\, f(\boldsymbol{x})\, f(\boldsymbol{y})\, d\lambda(\boldsymbol{x})\, d\lambda(\boldsymbol{y}), \tag{5}$$

is finite. This quantity is a measure of dispersion of $\boldsymbol{X}$ with respect to $\boldsymbol{\delta}(\cdot,\cdot)$ and was called the *geometric variability* of $\boldsymbol{X}$ in Cuadras and Fortiana (1994b). When $\boldsymbol{\delta}(\cdot,\cdot)$ is the Euclidean distance and $p = 1$, $V_{\boldsymbol{\delta}}(\boldsymbol{X}) = \mathrm{Var}(\boldsymbol{X})$.
The following function of $\omega_0 \in \Omega$

$$\phi_{\boldsymbol{\delta},\boldsymbol{X}}(\omega_0) = \int_E \boldsymbol{\delta}^2(\boldsymbol{x}_0,\boldsymbol{x})\, f(\boldsymbol{x})\, d\lambda(\boldsymbol{x}) - V_{\boldsymbol{\delta}}(\boldsymbol{X}) \tag{6}$$

where $\boldsymbol{x}_0 = \boldsymbol{X}(\omega_0)$, can be understood as a *proximity function* of $\boldsymbol{x}_0$ to $\mathrm{E}(\boldsymbol{X})$. This interpretation can be justified on the following property (Cuadras 1989, 1991, 1992b).

> *If there is a representation of* $\boldsymbol{\delta}$*, i.e. a function* $\psi : \boldsymbol{R}^p \to L$*, where* L *is a Euclidean (or Hilbert) space, such that* $\boldsymbol{\delta}^2(\boldsymbol{x},\boldsymbol{y}) = \|\psi(\boldsymbol{x})-\psi(\boldsymbol{y})\|^2$*, then*
> $$\phi_{\boldsymbol{\delta},\boldsymbol{X}}(\omega_0) = \|\psi(\boldsymbol{x}_0) - \mathrm{E}(\psi(\boldsymbol{X}))\|^2, \tag{7}$$
> *assuming the existence of* $\mathrm{E}(\psi(\boldsymbol{X}))$ *and* $\mathrm{E}(\|\psi(\boldsymbol{X})\|^2)$.

If $\boldsymbol{x}_1,\ldots,\boldsymbol{x}_n$ are iid observations of $\boldsymbol{X}$, an estimation of (6) is given by

$$\widehat{\phi}(\omega_0) = \frac{1}{n}\sum_{i=1}^{n} \boldsymbol{\delta}^2(\boldsymbol{x}_0,\boldsymbol{x}_i) - \frac{1}{2n^2}\sum_{i,j=1}^{n} \boldsymbol{\delta}^2(\boldsymbol{x}_i,\boldsymbol{x}_j). \tag{8}$$

In this case, the representation space L is either $\boldsymbol{R}^s$ or $\boldsymbol{R}^s \oplus \boldsymbol{i}\boldsymbol{R}^t$, the map ψ is obtained by ordinary MDS, and the expected value in (7) reduces to the centroid $\overline{\psi(\boldsymbol{X})} = \frac{1}{n}\sum_{i=1}^{n} \psi(\boldsymbol{x}_i)$
The main interest of (8) is for ill–conditioned statistical problems, where building a probabilistic model would rely upon arbitrary or unverified hypotheses, while often a distance function $\boldsymbol{\delta}(\cdot,\cdot)$ between individuals appears naturally. For example, when one or more of the following conditions arises:

- $\boldsymbol{X}$ is a mixture of continuous and discrete variables.
- The data are incomplete.
- Observations are coded as character strings and/or the number of variables may change from one observation to another.
- The number of variables is greater than the number of observations.

Suppose that we have two populations π_1, π_2 (a generalization to $g \geq 2$ populations is easy). On the basis of a random vector $\boldsymbol{X}$ and a distance function $\boldsymbol{\delta}(\cdot,\cdot)$, we use (6) to construct ϕ_1, ϕ_2, related to π_1, π_2, respectively. The distance–based rule of discrimination is (Cuadras 1989):

$$\textit{Allocate } \omega_0 \textit{ to } \pi_\alpha \textit{ if } \phi_\alpha(\omega_0) = \min\{\phi_1(\omega_0), \phi_2(\omega_0)\}. \tag{9}$$

Some of its properties/advantages are:

- When $\pi_i = N(\boldsymbol{\mu}_i, \boldsymbol{\Sigma})$, $i = 1, 2$, and $\boldsymbol{\delta}$ is the Mahalanobis distance, (9) reduces to the classic linear discriminant. The quadratic discriminant for $\boldsymbol{\Sigma}_1 \neq \boldsymbol{\Sigma}_2$ is also obtained with a slightly modified $\boldsymbol{\delta}$.
- From (7), it turns out that (9) can be understood as a Matusita rule for discrimination, i.e. ω_0 is allocated to the *nearest population*. Note, however, that this fact follows from the underlying theory: only distances between observations are explicitly needed for computations.
- If observed variables are of mixed type, possibly with missing values, we can use the distance which is based on Gower's similarity coefficient (Gower 1971a) to construct the DB discriminant functions. See Krzanowski (1993), Ciampi et al. (1993) for additional discussion.
- DB discriminant is specially useful when information on individuals is constituted by (variable length) strings, e.g. words, DNA sequences.

Example 4 *Cuadras (1992a) reports an application to a problem in linguistics (Rosés 1990): to decide whether a diphthong whose first vowel is an atonic* i, *appearing immediately after a consonant in a Catalan word, should be pronounced as monosyllabic (π_1) or bisyllabic (π_2). A random sample of* 136 *and* 43 *words, respectively, whose pronounciation is known, were selected from a dictionary, and each word was coded in terms of five categorical variables. The linear (LDF) and quadratic (QDF) discriminant functions were used for comparison purposes. The log–linear model (LLM) could seem to be quite appropriate (Krzanowski 1988, p. 352), but one of the variables (the consonant preceding the* i*) has many states and LLM gives null likelihood in both groups for many words. In contrast, DB presents no problems and classifies words quite correctly. The leaving–one–out procedure yields 58 misclassifications for LDF, 38 for QDF and only 8 for the DB method, while LLM fails.*

7. MDS Solution for a Random Variable

Our aim is now to represent a continuous real random variable X following a MDS approach. Let us start with a uniform $(0,1)$ random variable U. In order to represent U with respect to the distance

$$\boldsymbol{\delta}(x,y) = \sqrt{|x-y|}, \qquad x, y \in (0,1),$$

we observe that the stochastic process $\mathbf{U} = \{U_t, 0 \leq t \leq 1\}$, where U_t is the indicator of the interval $(t, 1) \subset (0,1)$, following a Bernoulli distribution $B(1,t)$, can be interpreted as a continuous Euclidean configuration representing $(U, \boldsymbol{\delta})$, since $U = \int_0^1 U_t\, dt$, and the squared Euclidean distance between two trajectories x_t and y_t of $\mathbf{U}$ is given by $\int_0^1 (x_t - y_t)^2\, dt = |x-y|$. The duality between principal coordinates and principal components suggests us to find the principal components of the stochastic process $\mathbf{U}$. The covariance function is

$$K(s,t) = \mathrm{Cov}(U_s, U_t) = \min(s,t) - st \qquad 0 \leq s, t \leq 1, \tag{10}$$

and the principal components of $\mathbf{U}$ are the random variables

$$Z_j = \int_0^1 U_t f_j(t)\, dt = \frac{\sqrt{2}}{j\,\pi}\,[1 - \cos(j\,\pi\, U)].$$

Standardizing Z_j to mean 0 we have $C_j = \sqrt{2}\,\cos(j\,\pi\,U)/(j\,\pi)$.

The continuous metric scaling solution for $(U, \boldsymbol{\delta})$ can be expressed by noting that each value x of U is represented by the countable sequence

$$(C_1(x), C_2(x), \ldots, C_j(x), \ldots),$$

which satisfies the equality

$$\sum_{j=1}^{\infty}(C_j(x) - C_j(y))^2 = |x - y|.$$

The analogy with the finite MDS case is also supported by the fact that $C_1, C_2, \ldots$ are centered uncorrelated random variables, whose sum of variances equals 1/6, the trace of the kernel K.

This MDS solution can be used in prediction and goodness–of–fit assessment (Cuadras and Fortiana 1993), following a similar approach to distance–based regression (Section 2.). This representation can be generalized to any continuous second–order random variable X (Cuadras and Fortiana 1994b).

Again, this extension opens the way to obtaining new insights and applications of the MDS methodology.

References

Ahrens, H. and J. Läntner (1974). Mehrdimensionale Varianzanalyse. Akademic–Verlag, Berlin.

Bertier, P. and J. M. Bouroche (1975). Analyse des Données Multidimensionnelles. Presses Univ. de France, Paris.

Bock, H. H. (1987). On the interface between Cluster Analysis, Principal Component Analysis, and Multidimensional Scaling. In H. Bozgodan and A. K. Gupta (eds.), Multivariate Statistical Modeling and Data Analysis, pp. 17–33. D. Reidel Pub. Co.

Ciampi, A., L. Hendricks, and Z. Lou (1993). Discriminant analysis for mixed variables: Integrating trees and regression models. See Cuadras and Rao (1993), pp. 3–22.

Coll, M. D., C. M. Cuadras, and J. Egozcue (1980). Distribution of human chromosomes on the methaphase plate. Symmetrical arrangement in human male cells. *Genetical Research* 36, 219–234.

Critchley, F. (1988). The Euclidean structure of a dendrogram, the variance of a node and the question: "how many clusters really are there?". In H. H. Bock (ed.), Classification and Related Methods of Data Analysis, pp. 75–84. Elsevier Science Publishers B. V. (North–Holland).

Critchley, F. and W. Heiser (1988). Hierarchical trees can be perfectly scaled in one dimension. *Journal of Classification* 5, 5–20.

CUADRAS, C. M. (1974). Análisis discriminante de funciones paramétricas estimables. *Trabajos de Estadística e Investigación Operativa 25*(3), 3–31.

CUADRAS, C. M. (1981a). Análisis y representación multidimensional de la variabilidad. In Inter. Symp. Concpt. Meth. Paleo. Barcelona, pp. 287–297.

CUADRAS, C. M. (1981b). Métodos de Análisis Multivariante. EUNIBAR, Barcelona. 2ª edición, PPU, Barcelona (1991).

CUADRAS, C. M. (1989). Distance analysis in discrimination and classification using both continuous and categorical variables. In Y. Dodge (ed.), Statistical Data Analysis and Inference, pp. 459–473. North–Holland Publishing Co., Amsterdam.

CUADRAS, C. M. (1991). A distance based approach to discriminant analysis and its properties. Math. Preprint Series 90, Univ. de Barcelona.

CUADRAS, C. M. (1992a). Probability distributions with given multivariate marginals and given dependence structure. *Journal of Multivariate Analysis* 42, 51–66.

CUADRAS, C. M. (1992b). Some examples of distance based discrimination. *Biometrical Letters* 29, 3–20.

CUADRAS, C. M. (1994). A distribution with given marginals and given regression curve. In H. Sherwood and M. D. Taylor (eds.), Distributions with fixed marginals, Doubly Stochastic Measures and Markov Operators. Proceedings Univ. of Washington at Seattle, Aug. 1993. IMS Monograph Series. (to appear).

CUADRAS, C. M. AND C. ARENAS (1990). A distance based regression model for prediction with mixed data. *Communications in Statistics A. Theory and Methods* 19, 2261–2279.

CUADRAS, C. M. AND J. FORTIANA (1993). Continuous metric scaling and prediction. See Cuadras and Rao (1993), pp. 47–66.

CUADRAS, C. M. AND J. FORTIANA (1994a). Ascertaining the underlying distribution of a data set. In R. Gutiérrez and M. J. Valderrama (eds.), Selected topics on Stochastic Modelling. World Scientific, Singapore. (in press).

CUADRAS, C. M. AND J. FORTIANA (1994b). A continuous metric scaling solution for a random variable. *Journal of Multivariate Analysis* 50. (to appear).

CUADRAS, C. M. AND J. M. OLLER (1987). Eigenanalysis and metric multidimensional scaling on hierarchical structures. *Qüestiió* 11, 37–57.

CUADRAS, C. M. AND C. R. RAO (eds.) (1993). Multivariate Analysis, Future Directions 2. Elsevier, Amsterdam.

DAVISON, M. L. (1983). Multidimensional Scaling. John Wiley & Sons, New York.

GOWER, J. C. (1966). Some distance properties of latent root and vector methods used in multivariate analysis. *Biometrika* 53, 325–338.

GOWER, J. C. (1971a). A general coefficient of similarity and some of its properties. *Biometrics* 27, 857–874.

GOWER, J. C. (1971b). Statistical methods of comparing different multivariate analysis of the same data. In F. R. Hodson, D. G. Kendall, and P. Tautu (eds.), Mathematics in the Archaeological and Historical Sciences, pp. 138–149. Edinburgh University Press.

GREENACRE, M. J. (1984). Theory and Applications of Correspondence Analysis. Academic Press, London.

HOEFFDING, W. (1940). Masstabinvariante Korrelationtheorie. *Schriften des Math. Inst. Angew. Math. Univ. Berlin* 5, 181–233.

HOLMAN, E. W. (1972). The relation between hierarchical and Euclidean models for psychological distances. *Psychometrika* 37, 417–423.

KRUSKAL, J. B. (1977). The relationship between multidimensional scaling and clustering. In J. Van Ryzin (ed.), Classification and Clustering, pp. 17–44. Academic Press, New York.

KRUSKAL, J. B. AND M. WISH (1978). Multidimensional Scaling. Number 11 in Sage University Paper series on Quantitative Applications in the Social Sciences. Beverly Hills and London: Sage Publications.

KRZANOWSKI, W. J. (1988). Principles of Multivariate Analysis: a User's Perspective. Clarendon Press, Oxford.

KRZANOWSKI, W. J. (1993). The location model for mixtures of categorical and continuous variables. *Journal of Classification* 10, 25–49.

MARDIA, K. V., J. T. KENT, AND J. M. BIBBY (1979). Multivariate Analysis. Academic Press, London.

OLIVA, F., C. BOLANCE, AND L. DIAZ (1993). Aplicació de l'anàlisi multivariant a un estudi sobre les llengües europees. *Qüestiió* 17, 139–161.

PETITPIERRE, E. AND C. M. CUADRAS (1977). The canonical analysis applied to the taxonomy and evolution of the genus Timarcha Latr. (Coleoptera, Chrysomelidae). *Mediterranea* 1, 13–28.

PRUZANSKY, S., A. TVERSKY, AND J. D. CARROLL (1982). Spatial versus tree representations of proximity data. *Psychometrika* 47, 3–24.

RAO, C. R. (1973). Linear Statistical Inference and its Applications, 2nd Ed. John Wiley & Sons, New York.

ROMNEY, A. K., R. N. SHEPARD, AND S. NERLOVE (eds.) (1972). Multidimensional Scaling: Theory and Applications in the Behavioral Sciences. Vol. 2, Applications. Seminar Press, New York.

ROSÉS, F. (1990). Estudi de la "i" àtona en posició post–consonàntica. Univ. de Barcelona. Unpublished manuscript.

SEBER, G. A. F. (1984). Multivariate Observations. John Wiley & Sons, New York.

SHEPARD, R. N., A. K. ROMNEY, AND S. NERLOVE (eds.) (1972). Multidimensional Scaling: Theory and Applications in the Behavioral Sciences. Vol. I. Theory. Seminar Press, New York.

Null Models in Cluster Validation

A. D. Gordon

Mathematical Institute, University of St Andrews,
North Haugh, St Andrews KY16 9SS, Scotland

Dedicated to the memory of Richard Dubes

Summary : A brief overview is given of the problem of validation in classification studies. Attention is concentrated on the specification of appropriate null models for data, with respect to which one may assess some cluster structure that has been obtained as the output of a clustering algorithm. In addition to standard null models, a discussion is given of 'data-influenced' null models, in which the precise form of the null hypothesis is influenced by characteristics of the data set under investigation. To illustrate the importance of specifying relevant null models, the behaviour of U-statistics under these null models is used to assess individual clusters found when data were classified using some standard clustering criteria implemented in an agglomerative algorithm.

1. Introduction

In this paper, the word 'classification' is used to refer to the activity (and end result) of partitioning a set of objects into disjoint groups (or clusters) in such a way that objects in the same group are similar to one another and different from objects in other groups; on occasion, it is relevant to obtain a hierarchical classification, comprising a nested set of such partitions.

Many different clustering procedures for obtaining a classification have been proposed (see, for example, Hartigan (1975), Gordon (1981), and Jain and Dubes (1988)). However, the output of a clustering algorithm can provide a misleading indication of the group structure that is present in a data set, and increasing attention is being paid to the issue of validating the results of a classification study. General reviews of, and comments on, methods of cluster validation have been presented by Perruchet (1983), Jain and Dubes (1988, ch. 4), Bock (1989), and Gordon (1994a).

Four general classes of validation test can be identified (Jain and Dubes (1988, ch. 4)), comprising tests for (1) the complete absence of any group structure in the data, (2) the validity of an individual cluster, (3) the validity of a partition, and (4) the validity of a complete hierarchical classification. Logically, tests in the first class should precede use of a clustering algorithm (although some of the tests use the output of such an algorithm), and a classification should not be carried out if a null hypothesis of the absence of group structure cannot be rejected. However, this procedure rarely seems to have been followed, possibly because of a strong presumption by investigators that their data do contain group structure. It is also the writer's impression that investigators are rarely, if ever, concerned to validate a complete hierarchical classification, but rather are interested in assessing partitions or individual clusters (these possibly having been obtained as part of a hierarchical classification).

In statistical modelling, it is common practice for models to be formulated after a study of data, and for the adequacy of these models then to be assessed on a different data set that is believed to have the same properties as the original one. This rarely occurs in classification studies, possibly because interest often resides primarily in the (complete) data set under investigation and not in some larger collection of objects, from which the data are believed to be a sample. Attention has therefore been restricted to the use of so-called *internal* clustering indices to assess partitions or individual clusters produced by a clustering procedure. Such internal indices are based on the results of the cluster analysis, generally comparing them with the original data. Although not all validation research fits into this framework (*e.g.*, McIntyre and Blashfield (1980), Jain and Moreau (1987), Breckenridge (1989)), a common approach (and the one that is considered in this paper) is that information is sought about the distribution of the internal clustering index under a null model of the absence of structure in the data. This information is then used to assess the value of the index for the obtained cluster or partition. Care has to be taken to ensure that an appropriate comparison is made, since the output of a clustering algorithm will provide clusters that are more homogeneous than randomly-selected sets of objects.

The purpose of this paper is not to review the many different clustering indices that have been proposed, but rather to discuss the specification of the null model with respect to which the values taken by these indices are assessed. However, in order to illustrate the differences that can occur when different null models are used, and hence indicate the importance of giving careful consideration to the selection of appropriate null models, it seems relevant to describe some validation tests. Accordingly, the remainder of this section is devoted to defining two clustering indices that can be used in the assessment of the validity of individual clusters.

Several assumptions are made about the format of the data. First, that the n objects under investigation are described by p numeric variables, recorded in a pattern matrix $X \equiv (x_{ik})$, where x_{ik} denotes the value of the kth variable describing the ith object ($i = 1, \ldots, n; k = 1, \ldots, p$). Secondly, that consideration has been given to the selection, standardization, and weighting of these variables, so as to ensure that an appropriate measure of the pairwise dissimilarity d_{ij} between the ith and j^{th} objects is given by the Euclidean distance between them:

$$d_{ij} = \sqrt{\sum_{k=1}^{p} (x_{ik} - x_{jk})^2}.$$

The two clustering indices are based on comparing the ranks of within-cluster dissimilarities and selected between-cluster dissimilarities, using a

U-statistic (Mann and Whitney (1947)). Let

$$U_{ijkl} = \begin{cases} 0 & \text{if } d_{ij} < d_{kl} \\ 1/2 & \text{if } d_{ij} = d_{kl} \\ 1 & \text{if } d_{ij} > d_{kl}. \end{cases}$$

Let C denote the set of objects in the cluster (containing c objects) under investigation, and $\overline{C}$ denote the set of objects not belonging to C. The *local* U index is defined by :

$$U_L \equiv \sum_{i \in C} \sum_{j \in C \setminus \{i\}} \sum_{k \in \overline{C}} U_{ijik}.$$

Let $W \equiv \{(i,j)|i,j \in C, i < j\}$; and $B \equiv \{(k,l)|k \in C, l \in \overline{C}\}$. The *global* U index is defined by :

$$U_G \equiv \sum_{(i,j) \in W} \sum_{(k,l) \in B} U_{ijkl}.$$

The smaller the values of these indices, the more valid the cluster is deemed to be. Gordon (1994b) discusses how the indices are related to several proposals that have been made for assessing clusters and partitions, and describes a cluster validation test based on a U-statistic. After applying a relevant (partitioning or hierarchical) clustering algorithm A to one's data, the validity of an obtained cluster C (containing c objects) can be assessed as follows :

1. A random $(n \times p)$ pattern matrix is generated and clustered using algorithm A.
2. The value $U(c)$ for a (randomly-selected) cluster of size c is evaluated for each cluster size c that is found in the cluster output.
3. Steps 1 and 2 are repeated until $(m-1)$ values of $U(c)$ have been obtained for all desired values of c.
4. If the specified cluster C of size c in the original data set has a value of U that is less than the jth smallest simulated value of $U(c)$, the null hypothesis of randomness is rejected (and the cluster deemed to be valid) at the $\alpha = 100(j/m)\%$ level of significance.

The remainder of the paper addresses the problem of specifying appropriate null models.

2. Standard Null Models

Null models for the complete absence of structure in a data set can be formulated in terms of random pattern matrices or random dissimilarity matrices. Two broad classes of null model for pattern matrices can be distinguished (Bock (1985)).

Poisson model: The objects can be represented by points which are randomly located in some region R of p-dimensional space.

This model has also been referred to as the uniformity hypothesis (Bock (1985)) and the random position hypothesis (Jain and Dubes (1988, ch. 4)).

An alternative null model is that the data set comprises a single cluster of n objects, it being plausible that objects near the centre of the cluster are closer to one another than are points farther from the centre of the cluster.

Unimodal model: The variables describing the objects have been randomly selected from a unimodal distribution.

A null model based on the dissimilarity matrix is the following (Ling (1973)).

Random dissimilarity matrix model: The elements of the lower triangle of the dissimilarity matrix are ranked in random order, all $\binom{n}{2}!$ orderings being regarded as equally likely.

This model has also been referred to as the random graph hypothesis (Jain and Dubes (1988, ch. 4)). As it uses solely rank order information, its applicability would seem to be restricted to analyses based on monotone admissible (Fisher and Van Ness (1971)) clustering procedures, such as single link and complete link.

There are problems with each of these models.

1. In the Poisson model, it is necessary to specify the region R within which the points are to be uniformly distributed. Standard definitions of R that have been used are the p-dimensional unit hypercube and hypersphere (*e.g.*, Zeng and Dubes (1985), Dubes and Zeng (1987), Hartigan and Mohanty (1992)).
2. In the Unimodal model, it is necessary to specify the distribution used to generate the simulated data; a standard choice for numeric data has been the multivariate normal distribution with identity variance-covariance matrix (*e.g.*, Rohlf and Fisher (1968), Gower and Banfield (1975), Hartigan and Mohanty (1992)).
3. The random dissimilarity matrix model ignores second (and higher) order effects; *e.g.*, if d_{ij} has a low rank, one would expect d_{ik} and d_{jk} to have similar ranks. This property suggests that the model would be of dubious validity when dissimilarities are constructed from an underlying pattern matrix (but might be appropriate if dissimilarities are elicited directly from subjects).

Tab. 1 and 2 illustrate the effect that choice of null model can have on critical values of the U-statistics : Tab. 1 allows the assessment of single link clusters using U_L, and Tab. 2 allows the assessment of complete link clusters using U_G. In each case, $n = 50$ points were generated in $p = 2, 3, 5$, or 10 dimensions under the following null models :

- *P(□)* : Poisson in *p*-dimensional hypercube.
- *P(○)* : Poisson in *p*-dimensional hypersphere.
- *P(3-box)* : Poisson in *p*-dimensional hyperbox, in which one side is three times the size of all the other sides.
- *U(○)* : *p*-dimensional normal distribution with identity variance-covariance matrix.
- *U(3-ellipse)* : *p*-dimensional normal distribution with diagonal variance-covariance matrix, in which the largest variance is nine times the size of all the others.

In addition, elements of a dissimilarity matrix for 50 objects were generated under the Random dissimilarity matrix model *RDM.*

In each case, 99 simulated values of the U-statistic were obtained. Tab. 1 and 2 present five U-values for each combination (p, c), the first three under the Poisson model (in hypercube, hypersphere, and hyperbox, respectively), and the last two under the Unimodal model (spherical and ellipsoidal, respectively); the final column gives the U-value under the random dissimilarity matrix model. The Tables show the fifth smallest U-value that was found under each model; thus, a cluster whose U-value was strictly less than the appropriate tabulated value would be deemed to be valid at the $\alpha = 5\%$ level of significance. Particularly for small values of c, several simulated values of $U(c)$ can equal the observed value of U; this approach provides a larger value for the significance level than other approaches to the treatment of tied values.

c	$p=2$	$p=3$	$p=5$	$p=10$	*RDM*
4	0,0,0 0,0	0,0,0 0,0	1,1,1 1,0	3,7,6 3,2	63
5	1,1,0 2,0	1,2,3 3,3	5,8,9 7,5	17,26,13 27,9	154
6	4,2,4 3,6	6,4,9 9,10	21,23,15 22,15	48,45,28 52,27	307
8	16,8,19 16,17	43,32,34 37,28	71,70,53 90,54	197,189,111 173,86	690
10	58,48,46 57,44	77,115,49 121,67	197,173,116 181,155	344,471,165 345,198	1231
15	176,260,132 249,220	516,359,179 527,331	750,899,237 686,406	1228,1353,525 1018,678	2752
20	692,723,222 707,532	1176,1428,525 1273,624	1859,1785,589 1313,827	2455,3021,1190 1893,1305	4798
25	1439,1423,624 1199,875	2098,2065,602 1836,1108	2897,3274,1088 2216,1625	3927,4232,1595 2618,1798	6405
30	1937,2306,1155 1774,1098	2967,2826,1250 2186,1585	3945,4476,1483 2725,2233	4815,5662,1573 3345,2232	7640

Tab. 1: Results of simulation study of local U-statistic U_L for single link clusters when number of objects $n = 50$ and number of dimensions $p =$ 2,3,5,10. The fifth smallest (of 99) values of $U(c)$ is given for selected values of the cluster size c under the six standard null models described in the text.

The following features of Tab. 1 and 2 are noteworthy.

1. The random dissimilarity matrix model provides critical values that greatly exceed those provided by the other models; as mentioned earlier, this model seems inappropriate for the assessment of data that are described by a pattern matrix.
2. For larger values of the cluster size c, the models *P(3-box)* and *U(3-ellipse)* provide lower critical values than the other Poisson or Unimodal models; this result is to be expected, given the greater spread of points along a single direction under the first two models.
3. The Poisson models $P(\square)$ and $P(\bigcirc)$ tend to provide lower critical values than the Unimodal model $U(\bigcirc)$ for smaller values of c and p, and higher critical values for larger values of c and p; see also Tables 1 and 2 in Gordon (1994b).

c	$p=2$	$p=3$	$p=5$	$p=10$	*RDM*
4	0,0,0 0,0	0,0,0 0,1	2,5,2 3,7	15,14,8 26,12	75
5	3,1,4 0,5	3,4,5 13,5	15,18,19 26,16	76,58,47 59,50	302
6	14,1,6 29,18	12,25,18 38,24	93,72,66 101,51	217,161,137 220,112	769
8	74,46,36 58,86	158,138,176 250,199	468,375,182 526,238	813,680,430 816,589	2314
10	213,194,215 353,306	522,772,353 994,486	1318,1075,681 1343,746	1945,2185,1196 2500,1245	5428
15	1657,1864,1122 3134,2042	3789,3259,1708 4239,2636	7416,6366,2709 7145,3803	10708,10494,3526 9176,5615	19741
20	7367,6093,2708 8023,5907	14082,14321,4249 12551,9063	21749,18948,5776 19767,10831	27745,29404,6890 27373,14910	45568
25	16210,18492,6342 18398,11430	34623,30211,7192 31559,13811	41172,40389,11826 34821,20268	52170,51784,12158 44228,28687	78566
30	38238,38953,11846 35752,23180	50276,51972,16467 41758,22968	64756,63379,17559 53169,28799	76363,79223,20913 68323,37961	111138

Tab. 2: Results of simulation study of global U-statistic U_G for complete link clusters when number of objects $n = 50$ and number of dimensions $p = 2,3,5,10$. The format of the Table is the same as that for Tab. 1.

3. Data-Influenced Null Models

The results presented in Tab. 1 and 2 indicate the importance of giving careful consideration to the problem of selecting relevant null models. This section summarizes various ways in which the data may be used to influence the specification of the null model, the aim being to obtain tests that concentrate on assessing features of genuine interest and are not vitiated by less relevant mismatches between model and data.

3.1 Random Permutation of Pattern Matrix

In this null model, a random $(n \times p)$ pattern matrix is obtained by permuting the elements of each column of the original pattern matrix. Each column is independently modified, and so there are effectively $(n!)^{p-1}$ different pattern matrices. The null model assumes that each of these pattern matrices is equally likely, and the aim is to investigate if analysis of the one that contains the original data provides more homogeneous clusters than analysis of the permuted pattern matrices.

Variants of this model have been investigated for binary pattern matrices (*e.g.*, Harper (1978), Strauss (1982), Vassiliou et al. (1989)), some of this work restricting attention to generating matrices with the same row sums as the original pattern matrix.

This null model is conceptually different from the other ones described in this section : the relative homogeneity of clusters obtained in an analysis of the original pattern matrix (in comparison with those obtained in analyses of permuted pattern matrices) need not make any strong statement about the absolute homogeneity of clusters obtained in the analysis of the original data. The method also has the limitation that it ignores the correlation between variables : irrespective of this correlation structure, the procedure will generate 'pseudo-objects' within a hyperbox in attribute space. If the variables are strongly correlated (*e.g.*, height and weight of individuals), it is possible to generate pseudo-objects with impossible physical characteristics.

3.2 Convex Hull

The region R for the Poisson model can be chosen to be the convex hull of the points whose coordinates are given in the original pattern matrix (Smith and Jain (1984), Bock (1985)). Some justification for this choice is provided by the result that if data have been randomly generated in a convex region of their convex hull about its centroid provides an estimate of the boundary of the region (Ripley and Rasson (1977)); for statistics that depend only on the rank order of distances (such as the U-statistics, the dilation is unnecessary.

Algorithms for obtaining the convex hull of a set of points in p dimensions are given by Chand and Kapur (1970), Edelsbrunner (1987, ch. 8), and Preparata and Shamos (1988, ch.'s 3,4); heavy demands are made on computing resources for larger values of p. Hsuan (1979) describes an algorithm that randomly generates data within a polygonal region in the plane, by splitting it up into triangles; extensions of this idea to higher dimensions (Rubin (1984)) encounter the problem of the combinatorial explosion in the number of simplices as p increases. An alternative approach is to generate data within a larger region and use a rejection procedure (*e.g.*, Dobkin and Lipton (1976), Chazelle (1985)), but the set-up costs again appear heavy. An algorithm proposed by Smith and Jain (1984) for generating points within a region approximating the convex hull can provide poor approximations in regions in which there are no points close to the boundary of the convex hull.

3.3 Ellipsoidal Model

The covariance structure present in a data set can be imposed on the null model. A natural way of proceeding is to generate data randomly from a multivariate normal distribution whose variance-covariance matrix has been specified by the data, for example using the sample variance-covariance matrix. Alternatively (for scale-invariant indices, such as the U-statistics), one could obtain the relevant parameters from the covering ellipsoid of minimum volume that contains all the points (*e.g.*, Titterington (1975)). Robust procedures that allow a certain number of points to be disregarded in the process of estimation (*e.g.*, Cook et al. (1993)) seem to the writer to be less appropriate in this context.

3.4 Covering

The models described in Sections 3.2 and 3.3 are particularly relevant if the points in the data set fall into a convex or ellipsoidal region. Fig. 1(a) portrays a set of points for which an appropriate null model might be the Poisson model in a non-convex region. There has been a limited amount of work on identifying the 'shape' of a set of points (*e.g.*, Edelsbrunner et al. (1983)), concentrating on two-dimensional data and allowing 'holes' in the interior of the shape. It is tempting to seek to obtain a set of disks (or other basic regions), one centred on each point, whose union leaves no 'internal holes' and provides an accurate representation of the 'shape' of the set of points. Points could then be randomly generated within a larger region and rejected if they were found to belong to none of the disks; such a generating procedure would seem ideally suited to parallel computers (*e.g.*, Hwang and Briggs (1984)). However, a fundamental criticism of this approach is that it does not always seem relevant to follow the boundary of the data set too closely : *e.g.*, consider Fig. 1(b), for which there is a danger that the 'null' model could provide clustered (instead of unstructured) data.

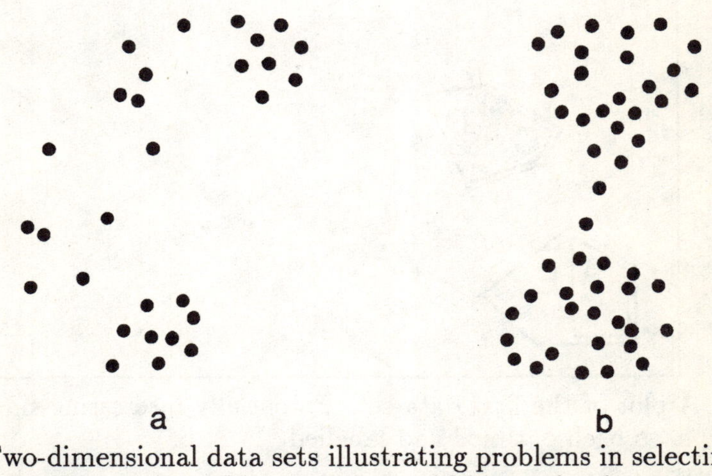

Fig. 1: Two-dimensional data sets illustrating problems in selecting a data-influenced region within which points should be randomly generated.

4. Examples and Discussion

This section summarizes the assessment of the clusters found in two artificial data sets, by simulating values of the U-statistics under several of the null models that have been described. Both data sets comprise $n = 50$ objects described by $p = 2$ numeric variables, allowing graphical representations of the data.

The first data set is portrayed in Fig. 2. These data were clustered using the single link method, and all clusters found were assessed using the local U-statistic, U_L, under the three standard null models $P(\square)$, $P(\bigcirc)$, and $U(\bigcirc)$, and the following data-influenced null models :

- *RP* : random permutation of columns of pattern matrix, as described in Section 3.1.
- *P(H)* : Poisson in convex hull of data set, as described in Section 3.2.
- $U(E_r)$: bivariate normal distribution with variance-covariance matrix specified by sample variance-covariance matrix ($r = 1$) or covering ellipse ($r = 2$), as described in Section 3.3.

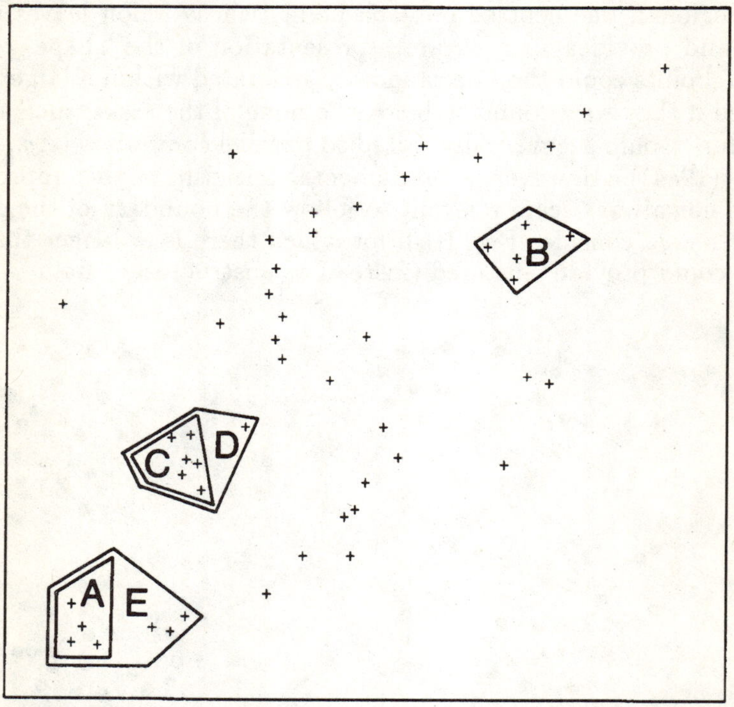

Fig. 2: A plot of the first data set. Potentially interesting single link clusters have been outlined and labelled.

Clusters that are deemed to be significant at the 10 % level of significance under at least one of these seven null models are outlined and labelled in Fig. 2. The levels of significance of these clusters under each of the null models are summarized in Tab. 3; a value of v in this Table means that the observed value of the local U-statistic was exceeded by $(100 - v)$ of the values simulated under the relevant null model. Bearing in mind the fact that these significance levels have been obtained using only 99 simulations, it is difficult to assign much importance to differences within each row of the Table. Although the cluster A of size 4 has an observed U-value of 0,

Cluster			Standard null models			Data-influenced null models			
	c	U_L	$P(\square)$	$P(\bigcirc)$	$U(\bigcirc)$	RP	$P(H)$	$U(E_1)$	$U(E_2)$
A	4	0	20	23	10	13	23	15	17
B	5	0	5	3	5	7	5	6	8
C	6	1	3	2	3	3	5	3	2
D	7	0	2	1	1	1	2	1	1
E	7	5	7	3	6	1	7	3	2

Tab. 3: Values of the local U-statistic U_L for the five single link clusters $(A-E)$ outlined in Fig. 2, together with the corresponding significance levels α (%) under seven null models.

it would not be deemed significant at conventional significance levels, but there is support for all of the other labelled clusters.

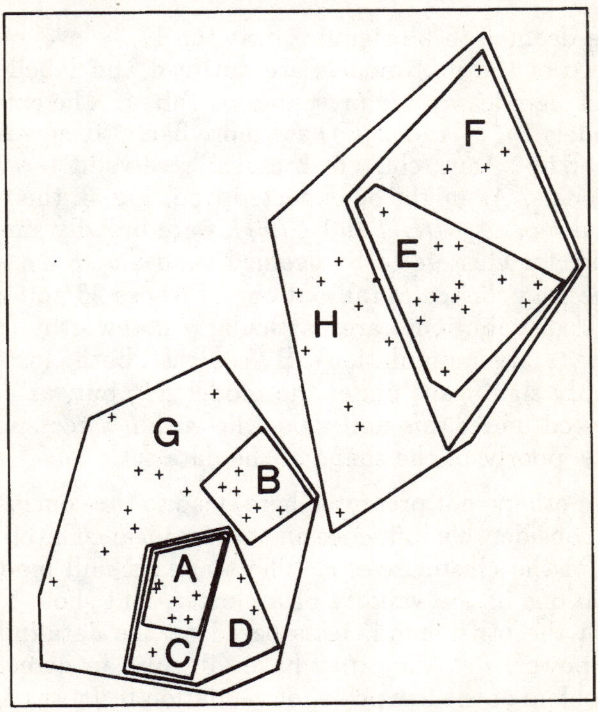

Fig. 3: A plot of the second data set. Potentially interesting sum-of-squares clusters have been outlined and labelled.

The second data set, portrayed in Fig. 3, comprises samples of 25 observations from each of two bivariate normal distributions, each having identity variance-covariance matrix, and whose mean vectors are a distance 5 apart; they are a rotated version of the data shown in Fig. 3 of Gordon (1994b). These data were clustered using Ward's (1963) agglomerative sum-of-squares algorithm, and all clusters found were assessed using the global U-statistic, U_G, under the same seven null models used with the first data set.

Cluster			Standard null models			Data-influenced null models			
	c	U_G	$P(\Box)$	$P(\bigcirc)$	$U(\bigcirc)$	RP	$P(H)$	$U(E_1)$	$U(E_2)$
A	6	4	5	3	4	4	2	3	3
B	6	13	10	7	6	7	12	6	6
C	7	22	5	3	2	8	6	3	2
D	9	190	15	6	5	17	9	8	9
E	13	1328	17	7	5	24	18	15	7
F	17	4683	20	10	7	28	35	21	16
G	24	4064	1	1	2	1	3	1	1
H	26	12654	1	1	2	1	20	8	3

Tab. 4: Values of the global U-statistic U_G for the eight sum-of-squares clusters $(A-H)$ outlined in Fig. 3, together with the corresponding significance levels α (%) under seven null models.

Clusters that are deemed to be significant at the 10 % level of significance under at least two of the null models are outlined and labelled in Fig. 3, and their levels of significance are presented in Tab. 4. The extent to which the spherical models $P(\bigcirc)$ and $U(\bigcirc)$ are more likely to suggest significant clusters is noteworthy : thus, clusters E and F receive little support under the other null models. As might be expected from Fig. 3, the results under the two ellipsoidal models, $U(E_1)$ and $U(E_2)$, were broadly similar. In general, it was harder for clusters to be deemed valid under the model $P(H)$: the results of assessing cluster H (all but one of whose 26 points came from one of the normal distributions) are particularly noteworthy (these results were verified using a larger simulation). By contrast, both clusters G and H were deemed highly significant under the model RP; but, as noted earlier, the points generated under this null model lie within a rectangular region, which corresponds poorly to the shape of the data set.

These results (and others, not presented here) lead to the conclusion that the null model has a considerable influence on the significance level, particularly for larger values of the cluster size, c. The standard null models can give misleading indications of the validity of a cluster, and should be deployed with caution. On the other hand, tests based on the data-influenced null models can lack power, *i.e.* they may have difficulty in detecting genuine clusters. This problem of the low power of validation tests was also noted by Hartigan and Mohanty (1992). It would be valuable to have available clustering indices that are powerful in detecting clusters, but it seems unlikely that any indices will be found that have good properties for all data sets.

The specification of a single null model, with respect to which clusters can be assessed, is not a straightforward matter. At present, it seems advisable to use a range of appropriate data-influenced null models, and synthesize the results provided by the significance levels. Thus, the conclusion reached from the analysis of the second data set would be that clusters A and G were strongly supported and that there was a lesser degree of support for the validity of clusters B, C, D and H.

References:

BOCK, H. H. (1985): On some significance tests in cluster analysis. *Journal of Classification, 2, 77–108.*

BOCK, H. H. (1989): Probabilistic aspects in cluster analysis. In : O. Opitz (ed.): *Conceptual and Numerical Analysis of Data.* Springer-Verlag, Berlin, 12–44.

BRECKENRIDGE, J. N. (1989): Replicating cluster analysis : Method, consistency, and validity. *Multivariate Behavioral Research, 24, 147–161.*

CHAND, D. R., and KAPUR, S. S. (1970): An algorithm for convex polytopes. *Journal of the Association for Computing Machinery, 17, 78–86.*

CHAZELLE, B. (1985): Fast searching in a real algebraic manifold with applications to geometric complexity. *Lecture Notes in Computer Science, 185, 145–156.*

COOK, R. D., HAWKINS, D. M., and WEISBERG, S. (1993): Exact iterative computation of the robust multivariate minimum volume ellipsoid estimator. *Statistics & Probability Letters, 16, 213–218.*

DOBKIN, D., and LIPTON, R. J. (1976): Multidimensional searching problems. *SIAM Journal on Computing, 5, 181–186.*

DUBES, R. C., and ZENG, G. (1987): A test for spatial homogeneity in cluster analysis. *Journal of Classification, 4, 33–56.*

EDELSBRUNNER, H. (1987): Algorithms in Combinatorial Geometry. Springer-Verlag, Berlin.

EDELSBRUNNER, H., KIRKPATRICK, D. G., and SEIDEL, R. (1983): On the shape of a set of points in the plane. *IEEE Trans. on Inform. Theory, IT–29, 551–559.*

FISHER, L., and VAN NESS, J. W. (1971): Admissible clustering procedures. *Biometrika, 58, 91–104.*

GORDON, A. D. (1981): Classification : Methods for the Exploratory Analysis of Multivariate Data. Chapman & Hall, London.

GORDON, A. D. (1994a): Clustering algorithms and cluster validation. In : P. Dirschedl and R. Ostermann (eds.): *Computational Statistics.* Physica-Verlag, Heidelberg, 503–518.

GORDON, A. D. (1994b): Identifying genuine clusters in a classification. *Computational Statistics & Data Analysis, 18, in press.*

GOWER, J. C., and BANFIELD, C. F. (1975): Goodness–of-fit criteria for hierarchical classification and their empirical distributions. In : L. C. A. Corsten and T. Postelnicu (eds.): *Proc. of the* 8th *Intern. Biometric Conference*, 347–361.

HARPER, C. W., Jr. (1978): Groupings by locality in community ecology and paleoecology : Tests of significance. *Lethaia, 11, 251–257.*

HARTIGAN, J. A. (1975): Clustering Algorithms. Wiley, New York.

HARTIGAN, J. A., and MOHANTY, S. (1992): The RUNT test for multimodality. *Journal of Classification, 9, 63–70.*

HSUAN, F. C. (1979): Generating uniform polygonal random pairs. *Applied Statistics, 28, 170–172.*

HWANG, K., and BRIGGS, F. A. (1984): Computer Architecture and Parallel Processing. McGraw-Hill, New York.

JAIN, A. K., and DUBES, R. C. (1988): Algorithms for Clustering Data. Prentice Hall, Englewood Cliffs, NJ.

JAIN, A. K., and MOREAU, J. V. (1987): Bootstrap techniques in cluster analysis. *Pattern Recognition, 20, 547–568.*

LING, R. F. (1973): A probability theory for cluster analysis. *Journal of the American Statistical Association, 68, 159–164.*

McINTYRE, R. M., and BLASHFIELD, R. K. (1980): A nearest-centroid technique for evaluating the minimum-variance clustering procedure. *Multivariate Behavioral Research, 15, 225–238.*

MANN, H. B., and WHITNEY, D. R. (1947): On a test of whether one of two random variables is stochastically larger than the other. *Annals of Mathematical Statistics, 18, 50–60.*

PERRUCHET, C. (1983): Une analyse bibliographique des épreuves de classifiabilité en analyse des données. *Statistiques et Analyse de Données, 8, 18–41.*

PREPARATA, F. P., and SHAMOS, M. I. (1988): Computational Geometry : An Introduction. Springer-Verlag, New York.

RIPLEY, B. D., and RASSON, J. P. (1977): Finding the edge of a Poisson forest. *Journal of Applied Probability, 14, 483–491.*

ROHLF, F. J., and FISHER, D. R. (1968): Tests for hierarchical structure in random data sets. *Systematic Zoology, 17, 407–412.*

RUBIN, P. A. (1984): Generating random points in a polytope. *Communications in Statistics : Simulation and Computation, B 13, 375–396.*

SMITH, S. P., and JAIN, A. K. (1984): Testing for uniformity in multidimensional data. *IEEE Trans. on Pattern Analysis and Mach. Intell. PAMI–6, 73–81.*

STRAUSS, R. E. (1982): Statistical significance of species clusters in association analysis. *Ecology, 63, 634–639.*

TITTERINGTON, D. M. (1975): Optimal design : Some geometrical aspects of D-optimality. *Biometrika, 62, 313–320.*

VASSILIOU, A., IGNATIADES, L., and KARYDIS, M. (1989): Clustering of transect phytoplankton collections with a quick randomization algorithm. *Journal of Experimental Marine Biology and Ecology, 130, 135–145.*

WARD, J. H., Jr. (1963): Hierarchical grouping to optimize an objective function. *Journal of the American Statistical Association, 58, 236–244.*

ZENG, G., and DUBES, R. C. (1985): A comparison of tests for randomness. *Pattern Recognition, 18, 191–198.*

Classifying Space and Analysing the Consequences: Spatial Analysis of Health Data

Robert Haining

Department of Geography,
University of Sheffield, Sheffield S10 2TN

Summary: The paper classifies the main forms of spatial analysis and briefly reviews advances in spatial analysis over the last twenty years. The paper then considers three problem areas using health data for illustration: (i) how to construct a spatial framework for analysis; (ii) how to construct reliable area rates that recognize the spatial distribution of areas and their variable populations; (iii) how to test for relationships between variables when rates vary in reliability and there may be spatial autocorrelation in the unexplained variation. The paper briefly considers the potentially important role of geographic information systems in the field of spatial data analysis.

1. Introduction and Classification of Spatial Data Analysis

Spatial data analysis (SDA) is a branch of data analysis where observations consist of attribute measurements taken at specified locations. SDA is distinguished from general forms of data analysis by the fact that the location of each data site is recorded. The position of each site is important first because some of the richness and information content of the data is bound up with associating data values to locations, second because in order to test hypotheses and model the data a knowledge of spatial relationships between observations may be essential and third because once the results of any analysis are known the researcher may need to be able to go back to specific sites to carry out follow up work.

Cressie (1991; p8-9) classifies spatial statistics into four different problem areas: geostatistical data, lattice data, point pattern data and finally object data. He distinguishes them through a common formalism which we simplify for present purposes. Let $\mathbf{s}$ be a data location in two dimensional space and let $\mathbf{Z}(\mathbf{s})$, a random quantity, constitute a data value at location $\mathbf{s}$. Let $\mathbf{s}$ vary over an index set or subset (D) of two dimensional space so as to generate the multivariate random field or random process $\{\mathbf{Z}(\mathbf{s}) : \mathbf{s} \in D\}$. We then have:

(a) Geostatistical data. D is a fixed subset and $\mathbf{Z}(\mathbf{s})$ is a random vector at location $\mathbf{s} \in D$. Data refer to a continuous spatial surface such as snow depth, soil pH at a specified depth or atmospheric pollution at ground level.
(b) Lattice data. D is a fixed (regular or irregular) collection of points in two

dimensional space. $\mathbf{Z}(\mathbf{s})$ is a random vector at location $\mathbf{s}$. Using D a graph can be specified which defines the neighbours of each point. Data refer to n fixed point locations (for example prices at a set of petrol retail outlets in a city) or to zones that partition an area into n regions (for example rates of mortality across a set of wards in a city).

(c) Point pattern data. D is a point process in two dimensional space (or subset). A point process is a process where $\mathbf{Z}(\mathbf{s}) = 1$ for all $\mathbf{s} \in D$. A marked point process is where $\mathbf{Z}(\mathbf{s})$ is a random vector at location $\mathbf{s} \in D$. Unlike case (b) the location of the points are not fixed but the outcome of some process. The data might refer, for example, to the distribution of trees in a forest or settlements across a region.

(d) Object data. D is a point process in two dimensional space and $\mathbf{Z}(\mathbf{s})$ is a random set. The data might refer to vegetation patches in an area or a cell culture.

2. Advances in Spatial Data Analysis

There has been a substantial and rapid growth in spatial data analysis in the last twenty years. This growth has taken a variety of forms:

(i) Development of theory

A review of the statistical literature in the early 1970s on how to analyse spatial data would have revealed a rather patchy development. There were papers dealing with problems of spatial sampling and the relative merits of different sampling strategies; a scattering of papers on specialist tests for pattern on two dimensional lattices and point distributions; some (rather difficult) papers on models for representing spatial variation that also emphasized the differences between time series modelling and spatial modelling. But there were no books that brought this literature together within a broader theoretical framework, as had been available for many years for time series analysis. "Spatial statistics" as a branch of statistics was relatively poorly developed. Bartlett's (1966) text for example devoted less than a few pages to spatial processes. Perhaps the main exception to this general comment was the systematic exploration by Matheron and others in France in the 1960s developing interpolation techniques for spatial data although for many years this was seen as a specialised set of techniques mainly of interest to geologists and mining engineers.

Since the early 1970s a number of major papers have appeared which have helped to bring a clearer structure to the field – for example the papers by Besag (1974) dealing with "lattice" data and by Ripley (1977) on point patterns both brought together material that had its origins in the pioneer work of Bartlett, Matern and Whittle. But not only important papers have appeared. The 1980s up to the present time have seen a growing number of texts that have sought to consolidate and bring an overview to the field or parts of it. Ripley's book (1981) was a 250 page monograph that dealt with a series of topics including sampling, interpolation, point process and lattice data analysis and the growing area of image analysis. Ten years later Cressie (1991) produced a 900 page volume entitled 'Statistics for Spatial Data' that covered a similar breadth of material. In between these two volumes Diggle's (1983) monograph focussed entirely on point pattern analysis and the book by Stoyan et al. (1989) dealt with point processes and stochastic geometry.

(ii) Development of specific techniques

Amongst the earliest spatial analysis techniques were those developed by Moran (1948), Geary (1954) and Krishna Iyer (1949) for testing for spatial pattern on regular lattices. With the development of models for spatial variation have come techniques for specifying and fitting such models, but a particularly important factor stimulating the development of new techniques has been the strong growth in interest in spatial analysis from the 'observational' sciences (as opposed to the experimental sciences) and with it an interest in techniques for analysing data that may not be recorded on a regular lattice, where data reliability is not high and where there is considerable uncertainty as to what exactly is being observed so that considerable resources have to be expended in identifying a good model. With these changes have come the requirements to develop analytical techniques more appropriate to such data. Cliff and Ord (1973;1981) pioneered the development of tests for pattern in the type of irregularly configured data frequently encountered in geography and the environmental sciences thereby extending the earlier work referenced above; Cressie (1984) introduced techniques for robust spatial analysis - techniques that were robust to data errors and outliers and provided a way of detecting patterns in such data.

(iii) New application fields

The first application areas for what has come to be known as "spatial statistics" were with respect to agricultural experiments, plant ecology and geological inference. There has been growing interest in spatial data analysis methods across a wide range of disciplines including the environmental sciences, earth sciences, engineering sciences, life sciences, observation sciences (remote sensing, stereology) and social sciences. In the same way that early developments in geostatistics were helpful to mining engineers needing to make decisions on potentially exploitable ore bodies so today the methodology of spatial statistics may help to underpin location decision making as well as decision making in identifying areas of social and economic deprivation or areas of health need. Such pressures have given added urgency and stimulus to the research field. A number of books have appeared over the last decade that have reviewed the growing theoretical and technical literature of spatial analysis in relation to various application areas. These books include Cliff and Ord (1981), Upton and Fingleton (1985), Anselin (1988) and Haining (1993).

(iv) Developments in computing technology

In common with many areas of data analysis the development of computing technology has been of considerable benefit to researchers wishing to analyse spatial data. The increase in computing speeds and data handling capacity have been important but also developments in interactive computing and computer graphics have had a significant impact on spatial data analysis. Advances in the speed and size of computers have been of benefit for spatial analysis because of the growth in large spatial data sets and because many of the techniques and model fitting procedures are (or were at one time) computationally quite demanding. Because the relative locations of sites may be important for data analysis, graphical display is an essential supplement to numerical analysis. Advances in computer graphics are therefore of special importance, particularly interactive graphics for the purposes of exploratory spatial data analysis (Haslett et al 1991). It is debatable which of these advances has had and will in the future have most impact on spatial analysis. With the growth of large spatial data sets there is clearly a need for computing power but because of the nature of spatial data and the philosophy underlying much contemporary data analysis the ability to interact with data, in ways that are highly visual and exploratory is also important.

Geographic information systems (GIS), computerised systems for the storage, retrieval, analysis and mapping of spatial data have developed over the last twenty years to provide potentially sophisticated platforms for performing spatial data analysis. There is currently much work exploring the extent to which GIS might facilitate a range of spatial analysis tasks (Goodchild et al 1992). If this research effort is successful it may contribute significantly to the wider adoption of spatial analysis methods because the software will then exist to interrogate spatial databases in a variety of different ways, carry out spatial analysis tasks and display results in graphical and cartographical form. This capability is not currently available and most of the results reported in spatial analysis are based on specialist software.

In this paper we shall review developments with respect to the second type of data type, lattice data, where D is a set of contiguous areas that partition a city and the random vector $\mathbf{Z}(\mathbf{s})$ includes disease incidence and socio-economic data. The aim is first to obtain a satisfactory description of the variation in disease incidence across a set of areas and second to seek an explanation of that variation in terms of the social and economic variables. The review will emphasize the special problems posed by spatial data analysis of this type and review some of the recent advances in spatial analysis that contribute to an effective and rigorous examination of the problem. To place the research in its context, the work reviewed here is part of a larger study looking at the incidence of colorectal cancer within Sheffield for the purpose of health needs assessment including the identification of 'problem regions' for the purpose of special targetting of health resources at geographically defined sub-groups.

In section three we consider the question of designing an appropriate areal framework for analysis, that is, the specification of D. In section four we will consider the problem of robust rate estimation on the chosen areal framework and in section five discuss the problem of testing for relationships between variables.

3. Constructing Areas for SDA

It is sometimes overlooked that in any area based study the definition of the boundaries of the study region and the way in which a region is partitioned can have a critical impact on the outcome of analysis. For example, in an analysis of crop productivity in England by county, Kendall (1939) observed: "a county which is recorded as moderate may in fact be a mixture of several very good and several very bad districts, or it may contain a large area on which no one of the (ten) crops is grown. The results can thus be taken to give only a general picture of the distribution of crop productivity over the country ... I hope that at some future time it will be possible to build up, from parish figures, estimates of yields and production in *natural farming type areas.*"

We need to distinguish two issues with respect to partioning the study area. First, the results of any analysis of a region will be dependent on the geographic scale of that analysis. The areal partition employed is analogous to a filter and pattern, and the processes responsible for pattern, are scale dependent. Second at any given scale several different sets of (internal) boundaries could be constructed. What should the internal partition be? In Kendall's study, there is the explicit wish to replace the 48 (heterogeneous) counties with 'natural farming areas'.

The design criteria for the construction of an areal framework for the spatial analysis of disease data are: (i) if rates are to be computed the areas should have large enough populations to yield reliable rates and moreover all the areas should have approximately similar population sizes so that area rates are of comparable reliability; (ii) if rates are to be compared with underlying socio-economic characteristics then the areas should be homogeneous with respect to those characteristics. Wards are a commonly employed spatial framework for area analysis in the UK but can vary greatly in size and are certainly not designed to be homogeneous so are usually inappropriate (Carstairs (1981)). On the other hand enumeration districts (EDs) are much smaller and could be aggregated to form regions socio-economically more uniform than wards.

The problem of region building shares common ground with the more general problem of scientific classification. A system of n spatial units is built from a system of k smaller spatial units (for example EDs). The usual objective is to minimize within region variation (of some set of attributes) and maximize between region variation. General classification routines can be used for this purpose but spatial classification procedures have been developed which incorporate locational information directly into the classification procedure (Cliff et al. (1975), Oliver and Webster (1989)). A problem with these spatial classifiers however is that they tend to trade off spatial compactness against intra-area homogeneity and may leave a residual scatter of small regions that are unallocated. Also, as the number (k) of smaller spatial units increases computing time may increase rapidly.

Haining et al. (1994) use a non spatial classification methodology based on information theory (Semple and Green (1984)) to classify and group EDs in Sheffield into Townsend deprivation regions. The original data (for all EDs) contains a certain amount of information that can be quantified according to the methods of information theory. Any grouping of the EDs into a smaller number of classes (each class represented by the attribute mean of those EDs falling into the class) will involve some loss of that original information. The aim, at the first stage, is to keep the number of classes to a minimum whilst ensuring that as little of the original information is lost as possible. Once the number of classes has been chosen the next step is to map the output. Dissolving boundaries between adjacent EDs in the same class generates an initial, and usually highly fragmented, regionalisation of the city. This can then be refined (or smoothed) manually by invoking additional rules to merge other adjacent regions that are in similar (adjacent) classes. The effect of this second stage can be tracked by the same information statistic used at the first stage. A GIS was used for this work.

This process created areas that were similar in size (although population differences still remained) and relatively homogeneous in terms of a single variable - the Townsend index of material deprivation. The method can, in principle, be generalized to more than one variable but the extent to which intra-area homogeneity could be preserved with respect to all the attributes must depend on the degree of spatial coincidence of those attributes and the desired size of the areas. In a multivariate context areas could be constructed that were homogeneous for all the attributes but the likely consequence is that the areas might be unacceptably small.

4. Rate Estimation for Geographic Areas

Computing area rates for disease occurences (which may be as incidence, prevalence or mortality statistics) raises a set of problems that have given rise to a distinctive methodology.

Let O_i denote the observed number of new cases (for example) of a disease in area i in a given period of time (say, one year). Let $E[x]_i$ denote the corresponding expected number of new cases given characteristic $[x]$ of the area. The maximum likelihood estimate of the relative risk (θ_i) in area i under the Poisson assumption (that each of a large number of individuals at risk have a small but equal probability of contracting the disease) is:

$$\hat{\theta}_i = O_i / E[x]_i. \tag{1}$$

It is usual to divide the population at risk into age cohorts and obtain age and sex specific rates for the disease for the appropriate reference group. This reference group might be all the people living in a city (if the contrasts are at the intra-urban scale) or all people living in the UK for example (if the contrasts are at the national scale). The expected number of cases in the denominator can be computed by multiplying age and sex specific rates by the numbers of individuals falling in each of these age/sex classes in region i

and summing over all the classes. This is referred to as the indirect method and (1) is referred to as the standardised relative risk.

If the aim is to highlight 'problem areas' where rates are unexpectedly high then the areas can be rank ordered and significance tests used to identify areas where the probability of obtaining an observed count as large or larger than O_i is computed under the Poisson assumption with parameter $E[x]_i$. This approach although often used has been criticised (Marshall (1991; p427)). More recently there have been new methods developed to improve relative risk estimation. These empirical Bayes estimators, as they are called, include estimators that take into account the variation in population sizes and the geographical distribution of the areas.

Empirical Bayes estimators (also known as James-Stein estimators) adjust for variation in the at risk population across the set of areas by assuming a mixture model for the distribution of relative risks. The gamma mixture model adjusts each relative risk so that areas with small populations are adjusted towards the mean of all the areas in the study region. This means that areas are not flagged as potential problem areas if their high rates are associated with small populations relative to the population sizes of the other areas.

The Bayes estimators described above treat each area independently ignoring the spatial distribution of the areas. Marshall (1991; p428) notes that "space characterizes the mapping problem and to ignore it seems unjustifiable." This is particularly true in the case of a city where the areas are contiguous. One form of spatial adjustment is to adjust rates to reflect neighbourhood means rather than the mean for the whole study area. For example if the areas are not socio-economically distinct or health events have been assigned to areas by a low accuracy data source so that there is some uncertainty as to which of several adjacent regions a disease event should be assigned or the disease is infectious, then an adjustment procedure in which rates are shrunk towards a neighbourhood mean may be more appropriate. Clayton and Kaldor (1987) use the log-normal model for this approach and Marshall (1991) has an extended discussion of Bayes estimators including under what circumstances the spatial dependence should be embodied in the prior or the likelihood. A recurrent problem however is the definition of neighbourhood. Although many neighbourhood schemes have been proposed (Haining (1993)) there is always a certain underlying 'ad hocery' in the selection which may be partially offset by exploring the sensitivity of results to alternative specifications of 'neighbourhood'.

5. Analysing Relationships

Spatial variation in relative risk is due to three sources (Pocock et al. 1981): (i) sampling variation (since the number of cases is a random variable subject to Poisson variation); (ii) explained variation (since certain socio-economic and environmental variables may be associated with variation in relative risk); (iii) unexplained variation (since areas differ in some respects that

cannot be accounted for by the variables in (ii)). A natural step is to use regression to test for relationships between city wide variation in these rates and variation in socio-economic and/or environmental characteristics. Two problems often arise in applying standard regression. These are:
(i) variation in sampling variances across the set of areas because the population sizes are different (heteroscedasticity);
(ii) spatial dependence in the errors(spatial autocorrelation).
The first is often handled by weighted regression, downweighting observations with small populations, or variance stabilising transformations (Cressie and Read (1989)). The model in Pocock et al. (1981) allows for both sampling variation and unexplained variation. This model is given by:

$$\boldsymbol{\theta} = \mathbf{X}\boldsymbol{\beta} + \boldsymbol{\eta}$$

where η_i is normally distributed error with mean 0 and variance $(\sigma^2 + (O_i/E[x]_i))$. The first term in the variance is associated with unexplained variation, the second with sampling variation. $\mathbf{X}$ is the matrix of observations on the independent variables (with ones in the first column) and $\boldsymbol{\beta}$ is the vector of regression parameters. Pocock et al. (1981) describe fitting procedures.

$\hat{\mathbf{Y}} = 4.57\,\mathbf{1} + 0.03\,\mathbf{X_1}$	s = 0.22	$R^2 \times 100 = 20.4\%$	$n = 48$
(0.03) (0.01)		Total SS= 2.85	Error SS= 2.26

Tab. 1: Results of least squares fitting of regression model with independent errors. (Values in brackets are standard errors; SS refers to sum of squares)

Tab. 1 displays the results of a bivariate regression of log Bayes adjusted rates (using the gamma-Poisson model) for the standardised relative risk (x100) of colorectal cancer within Sheffield 1979-83 ($\mathbf{Y}$) against the Townsend index of material deprivation ($\mathbf{X_1}$) for the corresponding areas. $\mathbf{1}$ is a vector of 1s and this model is related to the first model by setting $\mathbf{X} = [\mathbf{1}|\mathbf{X_1}]$ and $\boldsymbol{\eta}$ is normally distributed with mean zero and (constant) variance σ^2. In this case the adjustment for variation in population size (and hence reliability of rates) is bound up in the Bayes estimator of relative risk. The log adjustment was made in order to improve the linearity of the plot. The 48 areas were constructed as described in section 2 so that they are relatively homogeneous with respect to deprivation and this makes for a good spatial framework through which to explore the aggregate link between relative risk and deprivation.

The relationship is statistically significant but weak as might be expected given that we are seeking to describe disease incidence in terms of a single variable and given also the ambiguous relationship between colorectal cancer and deprivation through the influence of diet.

Area	Studentized Residual (>| 2.0 |)	Leverage ($> 2(p/n)$) ($>$ 0.083)	Cooks $\hat{D}$ ($> F(.5; p, n-p)$) ($>$ 0.70)
3	-2.28	0.04	0.13
4	-3.00	0.03	0.15
30	2.36	0.02	0.07
32	2.26	0.02	0.05
38	0.19	0.08	0.00

Tab. 2: Regression diagnostics from fit in Tab. 1

Tab. 2 reports diagnostics for the fit ($p = 2$, the number of model parameters; $n = 48$). Weisberg (1985) provides an explanation and discussion of these diagnostics. The values in brackets in Tab. 2 are the critical values for the data set. Four areas stand out as having large standardised residuals, two with rates that are large given their level of deprivation. There are no cases with high leverage and no cases with a large influence on the fit of the model. The hypothesis that the original 48 regression residuals are normally distributed was tested using a correlation test on the rankit plot which is essentially equivalent to the Shapiro-Wilk test. The correlation of 0.982 supports the hypothesis of a normal distribution.

When the areas are contiguous (as here) the assumption of independent errors needs checking. A number of such tests for spatial data are available. A generalized Moran coefficient (GMC) was used (see for example Haining (1993; p146)) in which areas are defined as neighbours if they share a common boundary. The row sums were not standardised to unity and the GMC as a standard normal variate was 2.50 indicating statistically significant spatial autocorrelation in the regression residuals. A good preliminary but quite effective check of whether there is residual autocorrelation can be obtained by simply plotting the value of each residual (horizontal axis) against the average of its neighbours (vertical axis) on a two dimensional scatterplot. This gives a visual indication as to whether there may be pattern in the residuals. A linear pattern suggests the presence of (positive) spatial autocorelation.

That the independence assumption on the residuals has been violated is a source of concern. The consequences are that the measure of fit (R^2) is inflated as are the parameter standard errors (Haining (1993)). So even the weak link with deprivation may not be statistically valid. With so few independent variables in the model the detection of spatial autocorrelation is not surprising. With the exception of infectious diseases, where residual autocorrelation may indicate the need for an autoregressive model on Y (Haining (1993; p340)), the presence of residual autocorrelation is usually a signal that important independent variables have been omitted from the model.

A regression model with autocorrelated errors was fit to the data. This model is one way to cope with the problem of omitted variables when no information is available on possible additional factors. The model used here is defined by:

$$\mathbf{Y} = \beta_0 \mathbf{1} + \beta_1 \mathbf{X_1} + \mathbf{u} \qquad \text{where} \qquad V(\mathbf{u}) = \sigma^2 \mathbf{V}$$

where again $\mathbf{Y}$ represents log transformed Bayes adjusted incidence rates for colorectal cancer and $\mathbf{X_1}$ represents the Townsend index of material deprivation. $V(\mathbf{u})$ denotes the variance covariance matrix for the errors with non zero off-diagonal values. A common form for $\mathbf{V}$ in these types of studies is:

$$\left((\mathbf{I} - \rho\mathbf{W})^T(\mathbf{I} - \rho\mathbf{W})\right)^{-1} \quad \text{and this arises by letting} \quad \mathbf{u} = \rho\mathbf{W}\mathbf{u} + \mathbf{e}$$

where $\mathbf{e}$ is $IN(\mathbf{0}, \sigma^2\mathbf{I})$, $\mathbf{I}$ is the identity matrix and $\mathbf{W}$ is a connectivity matrix that specifies the spatial relationships between the 48 areas. Each row and column of the matrix corresponds to one of the 48 areas. In the (i, j)th position in the matrix is a 1 if area i is a neighbour of j otherwise it is 0. The matrix is symmetric and diagonal values are zero. The parameter ρ is a measure of the strength of the neighbourhood dependence structure in the residuals and is estimated from the data. The value of the parameter ρ is constrained to lie between the reciprocals of the smallest and largest eigenvalues of the matrix $\mathbf{W}$ (Haining (1993)). In the case here the parameter range is the interval (-0.32,0.19). The spatial error model (on $\mathbf{u}$) which is known as a spatial autoregressive model represents a possible two dimensional analogue of the time series autoregressive model.

$$\begin{array}{llll} \mathbf{Y} = 4.49\,\mathbf{1} + 0.03\,\mathbf{X} + \tilde{\mathbf{u}} & \tilde{\mathbf{u}} = 0.12\,\mathbf{W}\tilde{\mathbf{u}} + \tilde{\mathbf{e}} & \\ \quad\;\; (0.05) \quad (0.01) & \qquad\quad (0.03) & \mathrm{n} = 48 \end{array}$$

Total SS= 2.85 Error(Regression) SS($\mathbf{u}$) = 2.52 Error(Final) SS($\mathbf{e}$) = 1.84

Tab. 3: Results of iterative maximum likelihood fitting of regression model with spatially autocorrelated errors. (Values in brackets are standard errors; SS refers to sum of squares).

Tab. 3 reports the result of fitting the regression model with correlated errors. The model is fitted by iterative maximum likelihood methods using the method in Ord (1975). Standard statistical packages will not fit the model so special routines have to be written. The estimates converged after 21 iterations. It is evident that the regression parameters have changed little.

As with the first model it is of interest to assess the model fit. A rankit plot and correlation test (0.98) indicates that the final set of residuals ($\mathbf{e}$) are normally distributed. No statistical test currently exists for spatial autocorrelation in these final residuals. An exploratory plot of the residuals against the average of their neighbours did however suggest that no strong spatial correlation remained. (Their correlation coefficient was 0.10.) However it would probably be more useful to look for other spatial patterns in the errors but unfortunately setting up alternative **W** matrices is still very tedious.

Area	Studentized residual (>| 2.0 |)	% change in the value of β_1, after deleting the case
3	-1.66	-9.24
4	-2.52	14.56
30	2.59	-15.44
32	2.08	12.39

Tab. 4: Regression diagnostics from fit in Tab. 3

Tab. 4 reports a limited number of diagnostics for the model with correlated errors (see Weisberg (1985) for a discussion of these diagnostics). The theory behind these diagnostics is presented in Martin (1992) where he obtains the expressions for leverage and Cooks D statistic. Leverage was not a problem in the original model and is still not a problem in the revised model and there are no influential cases. However it is interesting to note the new studentized residuals for the problem areas identified in the first model. All but one of these are smaller than before. The percentage shifts in the β_1 coefficient when individual cases are deleted are all small as well. Haining (1994) presents an extended discussion of the application of these spatial regression diagnostics many of which are awkward to implement and for which no standard software currently exists.

6. Conclusions

This paper has reviewed developments in the area of spatial statistics and spatial data analysis with particular reference to the analysis of medical data associated with geographical areas that are contiguous and of variable population size. A great deal of statistical theory and many techniques are now available for those working in this and related fields. An important need is now for software that will reduce the start up costs of working in this area. Analysing spatial data is currently a far more difficult and time consuming task than analysing time series data. Reference has already been made to the potentially valuable role that GIS can play as platforms for spatial analysis. However these systems are currently weak in terms of their analytical capabilities; but writing spatial analysis software for such systems could be a very effective way of bringing together visualisation and analysis capability.

Acknowledgements

Thanks to Marcus Blake and Stephen Wise for many helpful discussions on the topic of this paper. Marcus Blake obtained the data on which the regression work in section 5 is based.

References:

ANSELIN, L. (1988): Spatial Econometrics: Methods and Models. Dordrecht, Kluwer Academic Publishers.

BARTLETT, M. (1966): Stochastic Processes. Cambridge University Press.

BESAG, J.E. (1974): Spatial interaction and the statistical analysis of lattice systems. *Journal, Royal Statistical Society*, B36, 192-236.

CARSTAIRS, V. (1981): Small area analysis and health service research. *Community Medicine, 3(2): 131-139.*

CLAYTON, D. and KALDOR, J. (1987): Empirical Bayes estimates of age standardised relative risks for use in disease mapping. *Biometrics* 43, pp. 671-681.

CLIFF, A., HAGGETT, P., et al (1975): Elements of Spatial Structure. Cambridge University Press.

CLIFF, A.D. and ORD, J.K. (1973): Spatial Autocorrelation. Pion, London.

CLIFF,A.D and ORD, J.K. (1981): Spatial Processes: Models and Applications. London, Pion.

CRESSIE, N. (1984): Towards resistant geostatistics. In G. Verly et al (eds) Geostatistics for Natural Resources Characterization, 21-44. Dordrecht, Reidel.

CRESSIE, N. (1991): Statistics for Spatial Data, Wiley, New York.

CRESSIE, N. and READ, T.R.C. (1989): Spatial data analysis of regional counts. *Biometrical Journal*, 6, 699-719.

DIGGLE, P. (1983): Statistical Analysis of Spatial Point Patterns. Academic Press, London.

GEARY, R.C. (1954): The contiguity ratio and statistical mapping. *The Incorporated Statistician*, 5, 115-145.

GOODCHILD, M., HAINING, R.P. and WISE, S.M. (1992): Integrating Geographic Information Systems and Spatial Data Analysis: Problems and Possibilities. *International Journal of Geographical Information Systems*, 16, 407-24.

HAINING, R.P. (1993): Spatial data analysis in the social and environmental sciences. Cambridge University Press.

HAINING, R.P. (1994): Diagnostics for regression modelling in spatial econometrics. *Journal of Regional Science*, 34, 324-341.

HAINING, R.P., WISE, S.M. and BLAKE, M. (1994): Constructing regions for small area analysis: health service delivery and colorectal cancer (Unpublished ms).

HASLETT, J., BRADLEY, R., CRAIG, P.S. et al (1991): Dynamic graphics for exploring spatial data with application to locating global and local anomalies *American Statistician*, 45, 234-242.

KENDALL, M.G. (1939): The geographical distribution of crop productivity in England. *Journal, Royal Statistical Society*, 102, 21-48.

KRISHNA IYER, P.V.A. (1949): The first and second moments of some probability distributions on a lattice and their applications. *Biometrika*, 36, 135-41.

MARSHALL, R.J. (1991): A review of methods for the statistical analaysis of spatial patterns of disease. *Journal, Royal Statistical Society*, A, 154, 421-441.

MARTIN, R.J. (1992): Leverage, influence and residuals in regression models when observations are correlated. *Communications in Statistics: Theory and Methods*, 21, 1183-1212.

MORAN, P.A.P. (1948): The interpretation of statistical maps. *Journal, Royal Statistical Society*, B, 10, 243-51.

OLIVER, M.A. and WEBSTER, R. (1989): A geostatistical basis for spatial weighting in multivariate classification. *Mathematical Geology*, 21, 15-35.

ORD, J.K. (1975): Estimation methods for models of spatial interaction. *Journal, American Statistical Association*, 70, 120-6.

POCOCK, S.J., COOK, D.G. and BERESFORD, S.A.A. (1981): Regressions of area mortality rates on explanatory variables: what weighting is appropriate? *Applied Statistician* 30, 286-96.

RIPLEY, B. (1977): Modelling Spatial Patterns (with discussion). *Journal, Royal Statistical Society*,B, 39, 172-212.

RIPLEY, B. (1981): Spatial Statistics. New York, Wiley.

SEMPLE, R. and GREEN, M. (1984): Classification in Human Geography. In G.Gaile and C.Willmott (eds): Spatial Statistics and Models. Reidel, Dordrecht.

STOYAN, D., KENDALL, W.S. and MECKE, J. (1989): Stochastic Geometry and its Applications. New York, Wiley.

UPTON, G.J. and FINGLETON, B. (1985): Spatial Data Analysis by Example Volume 1: Point Pattern and Quantitative Data. New York, Wiley.

WEISBERG, S. (1985): Applied Linear Regression. New York, Wiley.

An Ordinal Model for Cluster Analysis – 15 Years in Retrospect

M. F. Janowitz

Department of Mathematics and Statistics
University of Massachusetts
Amherst, MA 01003, USA

Summary: A model for cluster analysis based on order theoretic techniques was introduced some 15 years ago. The paper seeks to provide a coherent introduction to this model, and to give an indication of some of the applications which it generated.

1. Introduction

Cluster analysis is generally regarded as a branch of exploratory data analysis whose goal is to suggest or discover the internal structure of a data set. It has the rather unique distinction of having been developed in parallel by researchers in a number of areas. This has led to many facts in the field being frequently "rediscovered" by various scientists. In its early days, the subject also suffered from a lack of any theoretical underpinnings, and indeed often seemed little more than a collection of various authors' favorite algorithms. What particularly seemed lacking was any sort of general mathematical model in which the various clustering algorithms could be compared and analyzed. In Jardine and Sibson (1968), several papers are mentioned that made early efforts to construct such a mathematical framework. No attempt will be made here to survey the vast literature involved in mathematical modeling of cluster algorithms. The model for cluster analysis that has come to be known as the *ordinal model* arose as a result of Jardine and Sibson (1968a), Sibson (1970) and the ensuing controversy that erupted between Jardine and Sibson and a team of researchers affiliated with CSIRO in Canberra, Australia (A list of references to this controversy appears in Jardine and Sibson (1971)). The essence of the argument seems to have been that Jardine and Sibson proposed a number of axioms for cluster algorithms from which it followed that if one wanted hierarchic clustering with no overlap between clusters at the same level, then the only allowable cluster algorithm was a technique called single-linkage clustering (see Jardine and Sibson ((1971a) p. 51). The Canberra researchers countered with an alternate set of axioms that they deemed important, and pointed out that they found techniques other than single-linkage clustering to produce more useful results. A number of authors then joined the fray. It seemed that what was really needed was a very general mathematical model which would include many of the useful cluster algorithms, so that one could analyze properties of competing cluster algorithms in a common framework. Thus was born the ordinal model (Janowitz (1978)). It is important to note, however, that though the model is order-theoretic in nature, it includes cluster algorithms that are not of that type. My goal in the present paper is to simply provide a coherent introduction to the ordinal model, and to outline some of the applications it generated. It will be assumed in what follows that the reader has a working knowledge of cluster analysis and the theory of ordered sets, so though all of the relevant notation will be introduced, the exposition will at times be terse.

2. The Ordinal Model

The input to a clustering problem is a finite nonempty set P together with finitely many attributes, or a dissimilarity coefficient (DC). The goal of a clustering algorithm is to produce some sort of nested sequence of classifications of P. Most (though not all) algorithms proceed by first constructing a DC, and then using the DC to produce the classifications. Our efforts will largely lie with that part of the procedure that goes from a DC to the output classifications. First we need some definitions. We agree to let $[0,\infty)$ denote the nonnegative real numbers, and $\Sigma(P)$ the set of reflexive, symmetric relations on P, ordered by inclusion.

Definition 2.1 *A dissimilarity coefficient* (DC) *is a mapping* $d : P \times P \mapsto [0,\infty)$ *that satisfies:*

(DC1) $d(x,x) = 0$ *for all* $x \in P$.

(DC2) $d(x,y) = d(y,x)$ *for all* $x, y \in P$.

It is called an ultrametric if it also satisfies

(DC3) $d(x,y) \leq \max\{d(x,z), d(y,z)\}$ *for all* $x,y,z \in P$.

Definition 2.2 *A numerically stratified clustering* (NSC) *is a mapping* $C : [0,\infty) \mapsto \Sigma(P)$ *that satisfies:*

(NSC1) $h \leq k$ *implies* $C(h) \subseteq C(k)$.

(NSC2) *There exists* $h \geq 0$ *such that* $C(h) = P \times P$.

(NSC3) *Corresponding to any* $h \geq 0$, *there is a number* $\delta = \delta(h) > 0$ *such that* $C(h) = C(h+\delta)$ *(continuity from the right).*

(NSC4) *If* $C(h)$ *is an equivalence relation for every* $h \in [0,\infty)$, *then* C *is called a dendrogram.*

It will be convenient to let $D(P)$ denote the set of dissimilarity coefficients on P, and $NSC(P)$ the set of numerically stratified clusterings. $D(P)$ is ordered by the rule $d_1 \leq d_2$ if and only if $d_1(x,y) \leq d_2(x,y)$ for all $x,y \in P$, and $NSC(P)$ is ordered in the analogous manner. A *cluster algorithm* can then be thought of as a transformation from $D(P)$ into $NSC(P)$, where the output is usually a dendrogram. But Jardine and Sibson (1971a) observe the following result:

Theorem 2.3 (i) *Corresponding to each* DC *d, there is an* NSC *Td given by*

$$Td(h) = \{(x, y) : d(x, y) \leq h\} \qquad (h \geq 0).$$

(ii) *Corresponding to each* NSC *C, there is a* DC d_C *given by*

$$d_C(x, y) = \min\{h \in [0, \infty) : (x, y) \in C(h)\} \qquad (x, y \in P).$$

(iii) *The correspondence* $d \mapsto Td$ *is a bijection between* $D(P)$ *and* $NSC(P)$ *whose inverse is the mapping* $C \mapsto d_C$.

(iv) *The* DC *d is an ultrametric if and only if Td is a dendrogram.*

Because of Theorem 2.3, a cluster method can either be thought of as a mapping $F : D(P) \mapsto D(P)$ or as a mapping $\overline{F} : NSC(P) \mapsto NSC(P)$. The two mappings are related by the fact that $[TF(d)](h) = [\overline{F}(Td)](h)$ for all $h \in [0, \infty)$ and all $d \in D(P)$. Jardine and Sibson (1971a) impose a number of axioms they feel any cluster method should satisfy. They are stated in somewhat different (but equivalent) terms in Janowitz (1978).

We are now ready to introduce the ordinal model. The first thing to notice is that by (DC1) and (DC2), we can view a DC as a mapping $d : P_2 \mapsto [0, \infty)$, where P_2 denotes the set of two element subsets of P. When this is done, we shall continue to write $d(x, y)$ in place of the more cumbersome notation $d(\{x, y\})$ since this can cause no confusion. Next we observe that reflexive, symmetric relations may naturally be identified with subsets of P_2. It follows that $\Sigma(P)$ forms a Boolean algebra that is naturally isomorphic to the lattice of all subsets of P_2. Once this is done, it is easily shown (Janowitz, (1978), Lemma 4.1, p. 60) that numerically stratified clusterings C are precisely the *residual* mappings from $M = [0, \infty)$ into $N = \Sigma(P)$. Note here that if M, N are ordered sets, then a mapping $C : M \mapsto N$ is called *residual* if there is a (necessarily unique) mapping $C^* : N \mapsto M$ having the property that

C, C^* are both isotone.

$CC^*(m) \geq m$ for all $m \in M$.

$C^*C(n) \leq n$ for all $n \in N$.

C^* is called the *residuated mapping* associated with C. It will be useful to let $\mathrm{Res}(N, M)$ denote the set of residuated mappings C^* from N into M, and $\mathrm{Res}^+(M, N)$ the associated set of residual mappings C from M into N. When $M = N$, the symbolism $\mathrm{Res}(M)$ and $\mathrm{Res}^+(M)$ will be used. In the special case where $C \in NSC(P)$, then $C^* : \Sigma(P) \mapsto [0, \infty)$ is given by

$$C^*(T) = \min\{h \in [0, \infty) : C(h) \supseteq T\} \qquad (T \in \Sigma(P)).$$

The associated DC d_C is simply the restriction of C^* to the singleton members $\{x, y\}$ of P_2.

In summary, here is the situation: Making use of the natural isomorphism between $\Sigma(P)$ and $\mathcal{P}(P_2)$, any DC $d : P_2 \mapsto [0, \infty)$ has a unique extension to a residuated mapping $Sd : \Sigma(P) \mapsto [0, \infty)$ given by $Sd(R) = \max\{d(x, y) :$

$xRy\}$, for any $R \in \Sigma(P)$. The residual map associated with Sd is none other than Td, where $Td(h) = \{\{x,y\} : d(x,y) \leq h\}$, for any $h \geq 0$. Thus we now have three versions of a cluster method: a mapping on $D(P)$, a mapping on $\text{Res}(\mathcal{P}(P_2), [0,\infty))$, or a mapping on $\text{Res}^+([0,\infty), \mathcal{P}(P_2))$. More generally, replacing $S = P_2$ by an arbitrary finite set S, we may extend these definitions as indicated below.

Definition 2.4 *Let S denote a finite set, and L a lattice with* 0. *An L-dissimilarity coefficient* (LDC) *is a mapping $d : S \mapsto L$, and an L-stratified clustering* (LSC) *is a member C of* $\text{Res}^+(L, \mathcal{P}(S))$. *It will be convenient to let $LD(S)$ denote the set of all* LDCs *on S, and $LSC(S)$ the set of all* LSCs.

Remark 2.5 *Let S and L be as in the above definition. Then:*

(i) *There is a natural bijection $d \mapsto Td$ between* LDCs *and* LSCs, *where $Td(h) = \{s \in S : d(s) \leq h\}$, $h \in L$. The inverse to this correspondence is given by $C \mapsto d_C$, where $d_C(s) = \bigwedge\{h \in L : s \in C(h)\}$, for each $s \in S$.*

(ii) *An L-cluster method may either be defined as a mapping F from $LD(S)$ into itself, or as a mapping $\overline{F} : LSC(S) \mapsto LSC(S)$, where $\overline{F}(Td) = T[F(d)]$ for all $d \in LD(S)$.*

(iii) *When $S = P_2$ for P a finite set, the condition for d to be an ultrametric becomes $d(x,y) \leq d(x,z) \vee d(y,z)$ for all $x,y,z \in P$.*

The above definition and remark represents all that is involved in the ordinal model. Following Janowitz (1978), it will be convenient to refer to it as Model L, where L is the underlying lattice in which dissimilarities are measured. In order to avoid trivial situations, it will always be assumed that L contains more than one member. When $S = P_2$ for some finite set P, and $L = [0,\infty)$, we are in the Jardine-Sibson setting, and for that reason will use the name Model JS. Note that if we just use a finite set S, then DCs can be thought of as measurements, and the resulting theory becomes a branch of abstract measurement theory. By allowing L to be a direct product of chains, we can think of dissimilarities taking values in the space of attributes associated with a set of objects, thus making contact with the ideas of *conceptual clustering* (Michalski and Stepp, 1983). Letting L be a family of distribution functions allows one to formulate a theory of clustering that is closely allied with the theory of probabilistic metric spaces (see Janowitz and Schweizer (1989) as well as Schweizer and Sklar (1983)). When S is just a set, there is some abuse of notation in discussing cluster methods, as we are not really doing cluster analysis except when $S = P_2$. Hopefully this will cause no confusion, as it costs nothing to phrase things in the slightly more general setting of allowing S to be an arbitrary nonempty finite set.

3. Monotone Equivariance

Unless otherwise specified, we shall be working in Model L as described above, with $L = [0, \infty)$. Recall that we are working with a finite nonempty set S, and that $d : S \mapsto [0, \infty)$. Recall further (Remark 2.5 that we view an L–cluster method (or simply a cluster method if there is no danger of confusion) as a mapping F of $LD(S)$ into itself. We may then follow Jardine and Sibson (1971a), and call a cluster method F *monotone equivariant* if:

For each order automorphism θ *of* $[0, \infty)$ *and each* LDC $d, F(\theta d) = \theta[F(d)]$.

Before doing anything else, we shall establish the exact nature of these cluster methods. Associated with each LDC d, we agree to call $h \in L$ a *splitting level* of d if $h = 0$ or $h = d(s)$ for some $s \in S$. This gives an increasing sequence:

$$0 = h_0 < h_1 < \ldots < h_i < \ldots h_t < h_{t+1}. \tag{1}$$

A set of the form $R_i = Td(h_i) = \{s \in S : d(s) \leq h_i\}$ is called a *splitting set* of d. Thus, corresponding to (1), we have a sequence of splitting sets:

$$R_0 \subset R_1 \subset \ldots \subset R_i \subset \ldots \subset R_t \subset S. \tag{2}$$

It is easy to see that the LDC d may be recaptured completely from knowledge of the sequences displayed in (1) and (2). To see this one need only note that for any $s \in S$, $d(s) = h_i$, where $i = \min\{R_j : s \in R_j\}$. Since a cluster method F transforms an LDC d into a second LDC $F(d)$, it produces a new pair of sequences of splitting levels and splitting sets:

$$0 = k_0 < k_1 < \ldots < k_i < \ldots < k_u < k_{u+1}. \tag{3}$$

$$T_0 \subset T_1 \subset \ldots \subset T_i \subset \ldots \subset T_u \subset S. \tag{4}$$

Because S necessarily occurs as the highest splitting set in both input and output, it (together with the corresponding splitting level) is often omitted from consideration. The resulting levels and subsets are called *proper* splitting levels and *proper* splitting sets. If d has t proper splitting sets, it will be convenient to say that d has *length* t, and write $\lambda(d) = t$. Thus if d is identically 0, then $\lambda(d) = 0$, and if d has a single proper splitting level, then $\lambda(d) = 1$, while the LDC represented by (3) has length $u + 1$, etc. We come now to a rather elegant characterization of monotone equivariance (see Janowitz (1978b) for background and a proof).

Theorem 3.6 *An L-cluster method F with* $L = [0, \infty)$ *is monotone equivariant if and only if the following two conditions hold for every LDC* d :

(i) *Every splitting level of* $F(d)$ *is a splitting level of* d.

(ii) *The splitting sets of* $F(d)$ *are completely determined by the proper splitting sets of* d. *Their construction ignores the values of the splitting levels of* d.

Let us examine the situation more closely. Suppose d is represented as in (1), (2), and $F(d)$ is represented as in (3), (4). Then every splitting level k_i of $F(d)$ is of the form h_j for some index j. It follows that $F(d)$ may be represented by the sequence (1) of splitting levels of d, and a sequence of subsets

$$W_0 \subseteq W_1 \subseteq \ldots \subseteq W_i \subseteq \ldots \subseteq W_t \subseteq S. \tag{5}$$

The idea is that the proper inclusions come at the places where h_i is of the form k_j. Thus a monotone equivariant cluster method is a transformation of a strictly increasing sequence of subsets of S

$$R_0 \subset R_1 \subset \ldots R_i \ldots \subset R_t \subset S$$

into an increasing sequence

$$W_0 \subseteq W_1 \subseteq \ldots W_i \ldots \subseteq W_t \subseteq S$$

When $S = P_2$ for some set P, splitting sets are called *splitting relations*, and there is a natural connection with graph-theoretic cluster methods. In fact, there is a bijection between $\Sigma(P)$ and undirected graphs with vertex set P. Corresponding to the reflexive, symmetric relation R on P, there is the graph G_R whose vertex set is P, and whose edges are the pairs $\{x, y\}$ such that xRy. Monotone equivariant cluster methods in Model JS can thus be seen as a natural vehicle for the study of the graph-theoretic cluster methods described by Matula (1971).

We next discuss the role of monotone equivariance and continuity in connection with the work of Jardine and Sibson. But first we pause to introduce some terminology that will be needed both for this and later material. Note that we are temporarily dropping the assumption that $L = [0, \infty)$.

Definition 3.7 *Let L be a lattice with 0, S a finite set, and F an L-cluster method.*

(i) *F is said to be a flat cluster method if there is a mapping γ : $\mathcal{P}(S) \mapsto \mathcal{P}(S)$ such that $\overline{F}(Td) = \gamma \circ Td$ for all LDCs d.*

(ii) *It is said to be semiflat if there is a family of mappings $\{\gamma_R\}_{R \in \Sigma(S)}$ on $\mathcal{P}(S)$ such that for every LDC d, $\overline{F}(Td) = \gamma_{Td(0)} \circ Td$*

Before we can understand these concepts, we need some more terminology.

Definition 3.8 *We agree to say that an L-cluster method is* compatible *with a mapping θ on L in case $F(\theta d) = \theta F(d)$ for every LDC d.*

The next two results appear in Janowitz (1978) as Lemma 7.1 and Theorem 7.4. For that reason, we state them without proof.

Lemma 3.9 *Let d be an LDC and $\theta \in \mathrm{Res}(L)$ with associated residual mapping θ^+. Then $F(\theta d) = \theta F(d)$ if and only if $\overline{F}[(Td) \circ \theta^+] = [\overline{F}(Td)] \circ \theta^+$.*

Theorem 3.10 *A necessary and sufficient condition for a given L-cluster method F to be flat is that F be compatible with all $\theta \in Res(L)$.*

The characterization of semiflat cluster methods in terms of compatibility is quite similar, and appears in Janowitz (1978a) as Theorem 3.8, p. 57. We state herewith the portion of that result that is pertinent to the present discussion.

Theorem 3.11 *A necessary and sufficient condition for a given L-cluster method F to be semiflat is that F be compatible with all $\theta \in Res(L)$ for which $\theta^+(0) = 0$.*

Supposing now that $L = [0, \infty)$, this implies that any flat or semiflat L-cluster method is monotone equivariant. It says even more than this. For if the input and output splitting sets are represented as in (2) and (5), then to say that F is flat is to say that every proper splitting set W_i of the output depends only on the corresponding input set R_i; to say that F is semiflat says that W_i depends on the pair of input sets $\{R_0, R_i\}$.

We next explain how all of this fits in with the model proposed by Jardine and Sibson (1971a). Recall that we are assuming that $L = [0, \infty)$, and let F be a cluster method in Model JS. We agree to say that F is *0-isotone* if $d_1 \leq d_2$ with $Td_1(0) = Td_2(0)$ implies that $F(d_1) \leq F(d_2)$. By Janowitz ((1981), Theorem 3, p. 109), the following conditions are equivalent:

(1) *F is semiflat.*

(2) *F is 0-isotone and monotone equivariant.*

(3) *F is left continuous and monotone equivariant.*

From Janowitz ((1981), Theorem 6, p. 111), comes the rather surprising result that the following conditions are equivalent:

(4) *F is flat.*

(5) *F is continuous and monotone equivariant.*

(6) *F is right continuous and monotone equivariant.*

Thus, though continuity would seem at first to be a rather strong assumption in the Jardine-Sibson model, it is actually the fact that all of their monotone equivariant cluster methods are already semiflat and hence left-continuous. The only role that continuity plays is to go from semiflat to flat cluster methods. Finally, it seems worth mentioning that to the author's knowledge, almost nothing is known about continuity or isotonicity for cluster methods that are not monotone equivariant.

We close this section by noting that considerations similar to those that led to the notions of *flat* and *semiflat* cluster methods can be used to produce a classification of ME cluster methods based on the assumptions they make about the validity of the values of the input LDC. Early work in this area was done by Baulieu (1985), (1991) and Powers (1988). More recently using the techniques of formal concept analysis (see Wille (1982)) a classification was constructed using compatibility with repect to 0-preserving isotone mappings. The objects in the underlying formal concept were taken to be the monotone equivariant cluster methods, the attributes the 0-preserving isotone mappings in $[0, \infty)$, and the relation $\perp$ between objects and attributies defined by $F \perp \theta$ iff F is compatible with θ. This work is contained in Janowitz and Wille (1993) and led to the seven classes of cluster methods indicated in Figure 1. If d is represented as in (1), (2) and $F(d)$ as in (1), (5), the classes may be represented as follows:

agglomerative: W_i depends on $R_0, \ldots, R_i$.
divisive: W_i depends on $R_i, \ldots, R_t$.
0-div: W_0 depends on R_0, and for $i > 0$, W_i depends on $R_0, R_i, \ldots, R_t$.
0-stable: W_0 depends on R_0, and for $i > 0$, W_i depends on all R_j.
Flat and **semiflat** are defined as before.

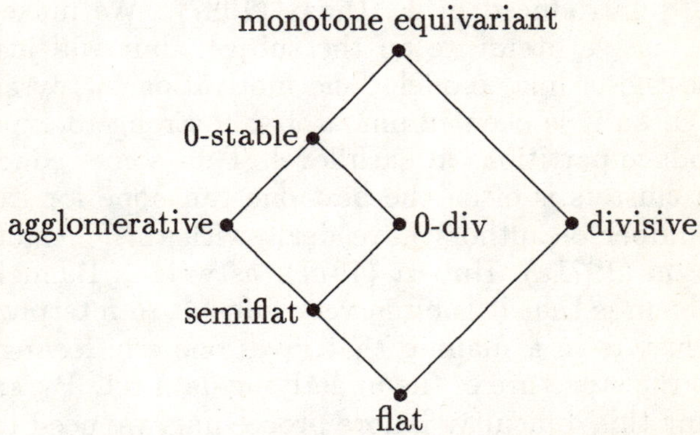

Fig. 1: Classes of monotone equivariant cluster methods.

Graph-theoretic classifications of cluster methods were introduced and studied in Janowitz and Stinebrickner (1993), (1993a).

4. Ordering of Input Data

The issue of how to deal with tied values in the input LDC is an important concern in the design of clustering algorithms. Jardine and Sibson ((1971a) p. 54 and p. 84) argue that standard clustering algorithms are ill-defined in that they depend on the order in which the input data is considered. In order to produce well-defined algorithms, they merge all pairs of clusters at level h that have dissimilarity h, and then take the transitive closure of the resulting relation in order to obtain an equivalence relation. But there is a very easy resolution to this issue, and this resolution also shows how to incorporate other structures into the ordinal model. *The extended definitions will only apply to material within the current section of the manuscript.* In what follows L will denote a lattice with 0 having at least two members, and $S = P_2$ for some finite set P.

Definition 4.1 *An* <u>*ordered*</u> LDC *is a pair* $(d, \preceq)$ *where* d *is an* LDC *on* S *and "*$\preceq$*" is an order on* P.

When $\preceq$ is a linear order on P, the standard clustering algorithms suddenly become well-defined, because part of the input data is the specification of a search order in case of ties. This is really a very modest assumption that only formalizes the reality of the fact that whenever data is entered into a computer, it is entered in some order, and this implicitly defines a linear order on P. The above definition only makes this implicit linear order explicit, and it seems to remove the objections raised by Jardine and Sibson. Thus there is a place in the ordinal model for both the Jardine–Sibson viewpoint, and the viewpoint adopted by many other authors, where in case of ties, the pair that the computer first comes upon is clustered. But if this were all that could be done, it would be a rather trivial observation. In fact, the notion of an ordered LDC allows concepts related to *pyramids* to be incorporated into the model. The idea of a pyramid appears to have been first formally introduced by E. Diday (1987). We make no attempt to survey the extensive literature on the subject, but will just show how pyramids fit into the ordinal model. The motivation for pyramids is easy to understand. Often it is deemed unreasonably strong to expect a cluster algorithm to produce partitions at each level. Thus some reduced degree of overlap between clusters is often the best one can hope for from a cluster algorithm. A number of authors have dealt with this. Among them are Jardine and Sibson (1971a), Hubert (1977), as well as Bandelt and Dress (1994). The problem is that it is often very difficult to interpret the output of overlapping clusters in a manner that gives one any feeling for what is being said about the structure of the underlying data set. Pyramids are one way of overcoming this difficulty. Before proceeding, we need to establish a way in which reflexive symmetric relations can produce clusters.

Definition 4.2 *Let R be a reflexive, symmetric relation on the finite set P. A linked set of R is a nonempty subset Q of P having the property that xRy for all $x, y \in Q$, and a maximal linked set* (ML-*set*) *of R is a linked set that is not properly contained in any other linked set of R.*

Remark 4.3 *The* ML-*sets of R are considered to be the clusters associated with R, and they have the following properties:*

(i) *Their union is all of P.*

(ii) *They are pairwise incomparable.*

Definition 4.4 *Let $(d, \preceq)$ be an ordered* LDC.

(i) *We say that d is convex if $a \preceq x \preceq b$ implies that $d(x, y) \leq d(a, b) \vee d(a, y) \vee d(b, y)$ for all $y \in S$.*

(ii) *A subset T of S is called convex if $a, b \in T$ with $a \preceq x \preceq b$ implies that $x \in T$.*

Theorem 4.5 *Let $(d, \preceq)$ be an ordered* LDC. *A necessary and sufficient condition for $(d, \preceq)$ to be convex is for every ML-set Q of any splitting relation of d to be a convex subset of S.*

Proof: Let $a \preceq x \preceq b$ with $a, b, y \in Q$, where Q is an ML-set of some $Td(h)$. If d is convex, $d(x,y) \leq h$ so x is linked with all members of Q. By maximality of Q, $x \in Q$. For the converse, note that if $h = d(a,b) \vee d(a,y) \vee d(b,y)$, then $\{a, b, y\}$ is contained in some ML-set Q of $Td(h)$. Convexity of Q then establishes that if $a \preceq x \preceq b$, then $d(x,y) \leq h$. ∎

A close contact with the theory of pyramids can now be extablished by noting that if d is ordered, with $(P, \preceq)$ a chain, then Theorem 4.5 implies that d is convex iff every ML-set of any $Td(h)$ is a closed interval of $(P, \preceq)$. If $\preceq$ is a linear order, then Condition (i) of Definition 4.4 can be replaced by $a \preceq x \preceq b$ implies that $d(a,b) \geq d(a,x) \vee d(b,x)$. A discussion of these and related issues is part of a manuscript currently in preparation.

5. Generalizations of the Ordinal Model

We present two rather different generalizations of the ordinal model. Though they differ in their approach, we shall note an interesting duality between them. They are both based on the notion that one does not need to assume that L is a lattice, and that cluster functions need not be directly related to residual mappings. Limitations of space preclude doing anything other than introducing the basic notions. The interested reader should go to the source material for further details. The first idea was due to Herden (1984), and was used to study a model for cluster analysis that was introduced by Hubert (1977) in connection with a careful analysis of clustering techniques that allow some restricted degree of overlap between clusters.

To introduce Herden's model, we let L and M be ordered sets, and assume that each has a smallest element 0. Think of M as taking on the role of $\Sigma(P)$, and L the role of $[0, \infty)$ in a standard clustering model. An *order ideal* of M is a nonempty subset I of M having the property that $m \in I$, $x \leq m$ implies that $x \in M$. We use the symbol $\mathcal{O}(M)$ to denote the order ideals of M, and order $\mathcal{O}(M)$ by set inclusion. For each $m \in M$, we define the *principal ideal* J_m generated by m by $J_m = \{x \in M : x \leq m\}$. There is a dual notion of principal filters defined by $F_m = \{x \in M : x \geq m\}$.

Definition 5.1 *A mapping $f : M \mapsto L$ is called a* <u>*preclustering function*</u> *if there exists a mapping $\overline{f} : \mathcal{O}(M) \mapsto L$ which satisfies the following conditions:*

(P0) $\overline{f}(\{0\}) = 0$.

(P1) *If $I \in \mathcal{O}(M)$ and if $\overline{f}(J_m) \leq t$ for all $m \in I$ then $\overline{f}(I) \leq t$.*

(P2) $\overline{f}(J_m) = f(m)$ *for all $m \in M$.*

The conditions of the above definition imply a number of things. First of all, it is immediate from (P0) and (P1) that for each $t \in L$, $\{I \in \mathcal{O}(M) : \overline{f}(I) \leq t\}$ is a principal ideal of $\mathcal{O}(M)$. This implies that $\overline{f} \in \text{Res}(\mathcal{O}(M), L)$, so $\overline{f}$ and hence f are each isotone. An application of (P2) now shows that

$\overline{f}$ is uniquely determined by f. Thus $f : M \mapsto L$ is a preclustering function iff there is a residuated mapping $\overline{f} : \mathcal{O}(M) \mapsto L$ such that $\overline{f}(J_m) = f(m)$ for all $m \in M$. This shows that there is a natural bijection between the set $\mathcal{PC}(M, L)$ of all preclustering functions from M into L and $Res(\mathcal{O}(M), L)$. For $f \in \mathcal{PC}(M, L)$ and $t \in L$, the analog of a level clustering in the sense of Hubert would be the set of maximal elements of $\{m \in M : f(m) \leq t\}$. If M is finite, this can be determined with no difficulty, but in general such elements simply may not exist. For the special case where L, M have the property that every chain has a least upper bound, the solution also is at hand. It is obtained by replacing $\mathcal{O}(M)$ with the order ideals of M that are closed under the formation of suprema of chains. We refer the reader to Herden (1984) for the details.

We now turn to a far reaching generalization of the fundamental bijections that seem to arise in connection with mathematical theories of classification. The research is due to Critchley and Van Cutsem (1993), (1993a). Our goal will be to describe their ideas, give an indication of how they are used, and finally to place the ideas in what we believe is the proper context in connection with the ordinal model for clustering. Their goal (as was the goal of Herden (1984)) was to investigate dissimilarity coefficients that take values in an arbitrary ordered set. We begin with the fundamental ideas.

Definition 5.2 *Let S be a nonempty set and L an ordered set. A mapping $m : L \mapsto \mathcal{P}(S)$ is called level if there is an associated mapping $f_m : S \mapsto L$ having the property that for each $t \in L$, $m(t) = \{s \in S : f_m(s) \leq t\}$.*

As pointed out in Critchley and van Cutsem (1993), the correspondence $m \mapsto f_m$ is a natural bijection between level mappings from L into $\mathcal{P}(S)$ and mappings $f : S$ into L. The next theorem is a restatement of Critchley and van Cutsem ((1993), Theorem 5.3). It contains the fundamental characterization of level mappings.

Theorem 5.3 *Let $m : L \mapsto \mathcal{P}(S)$. The following are then equivalent:*

(1) *m is level.*

(2) *m is isotone and for each $s \in S$, $\{t \in L : s \in m(t)\}$ has a minimum element.*

(3) *For each $s \in S$, $\{t \in L : s \in m(t)\}$ is a principal filter F_a of L; indeed, it is the principal filter generated by $a = f_m(s)$.*

The authors then investigate in some detail the fundamental bijections of the form $d \mapsto Td$ in situations where the underlying mappings d take values in an ordered set that is not a lattice. We shall not attempt to reproduce their results in detail here, but will instead show a parallel with the work of Herden.

We begin by letting M and L be ordered sets, and take $\overline{L}$ to be either the completion by cuts of L or the order ideals of L. It turns out that it will not matter which one is chosen. We then observe that any mapping $f : S \mapsto L$ may naturally be associated with a mapping $\overline{f} : S \mapsto \overline{L}$ by defining $\overline{f}(s) = J_{f(s)}$. But now $\overline{f}$ is a mapping from a set S into a complete lattice $\overline{L}$. It follows from standard properties of residuated mappings that $\overline{f}$ has a unique extension to a residuated mapping $g : \mathcal{P}(S) \mapsto \overline{L}$. Indeed, g is given by the formula $g(A) = \bigvee\{\overline{f}(s) : s \in A\}$. Now let g^+ be its associated residual mapping, and let $m_f : L \mapsto \mathcal{P}(S)$ be the level mapping associated with f. Then

$$\begin{aligned} g^+(J_t) &= \bigcup\{A \subseteq S : g(A) \subseteq J_t\} \\ &= \{s \in S : g(\{s\}) \subseteq J_t\} \\ &= \{s \in S : \overline{f}(\{s\}) \subseteq J_t\} \\ &= \{s \in S : f(s) \leq t\} = m_f(t). \end{aligned}$$

Suppose conversely that $m : L \mapsto \mathcal{P}(S)$, and that $g \in \mathrm{Res}(\mathcal{P}(S), \overline{L})$ has the property that:

(i) For each $s \in S$, $g(\{s\}) = J_t$ for some $t \in L$.

(ii) For each $t \in L$, $g^+(J_t) = m(t)$.

It then follows that if $f : S \mapsto L$ is defined by taking $f(s)$ to be the generator of the principal ideal formed by $g(\{s\})$, then $g^+(J_t) = \{s \in S : f(s) \leq t\}$, thus showing that m is level. Thus we have the dual of Herden's clustering functions. In short, we have proved:

Theorem 5.4 *Let $\overline{L}$ denote the completion by cuts or the lattice of order ideals of the ordered set L. A mapping $m : L \mapsto \mathcal{P}(S)$ is level iff there exists a residual mapping $\overline{m} : \overline{L} \mapsto \mathcal{P}(S)$ having the property that $\overline{m}(J_t) = m(t)$ for all $t \in L$, and that the associated residuated mapping sends singleton subsets of S into principal ideals of L.*

Example 5.5 We noted that in defining a level mapping, it does not matter whether $\overline{L}$ is taken to be the completion by cuts or the lattice of order ideals of L. In either case, there is an associated residual mapping $\overline{m} : \overline{L} \mapsto \mathcal{P}(S)$. This does not imply, however, that these residual mappings $\overline{m}$ are in any sense independent of the choice of $\overline{L}$. As an extreme case, suppose L is a set. The completion by cuts of L then only adjoins a unit element and a zero element to L. To say that $\overline{m}$ is a residual mapping from $\overline{L}$ into $\mathcal{P}(S)$ just requires that for all $s, t \in L$, $\overline{m}(J_s) \cap \overline{m}(J_t) = \overline{m}(0)$. On the other hand, the lattice of order ideals of L is naturally isomorphic to $\mathcal{P}(L)$, so much more is involved in defining $\overline{m}$.

Having made the observations contained in Theorem 5.4, we may now place level mappings in their proper order-theoretic framework.

Definition 5.6 *Let L, M be ordered sets with $\overline{L}$ denoting the completion by cuts or the lattice of order ideals of L. A mapping $m : L \mapsto M$ is called* *level in case there is a residual mapping* $\overline{m} : \overline{L} \mapsto \mathcal{O}(M)$ *having the property that* $\overline{m}(J_t) = J_{m(t)}$ *for all* $t \in L$, *and that the associated residuated mapping sends principal ideals of* M *into principal ideals of* L.

In order to apply this to form indexed hierarchies, we take M to be the ordered set of level clusterings of the set S, and L an ordered set. The point is that both Critchley and van Cutsem (1993) and Herden (1984) have similar goals. The tools for accomplishing the goals are in a sense dual to each other. It is quite likely that both approaches are useful. The goal of the discussion was to show that each approach may be subsumed by the ordinal model for clustering. It might be of some interest to see how level mappings apply to the model introduced by Hubert (1977).

6. Conclusion

In this paper we presented an overview of a model for cluster analysis that is order theoretic in nature. The underlying motivation behind the introduction of the model was presented, as well as some of the theory that the model has generated. Perhaps the most useful aspects of this "ordinal model" have involved a classification of hierarchical cluster algorithms according to the assumptions they make about the input data, as well as the ability to characterize algorithms within the model by means of their properties. By allowing dissimilarities to take values in arbitrary lattices, we briefly indicated a connection with a form of clustering based on distribution functions, and hinted at a connection with conceptual clustering. But some rather surprising connections also appeared. Among them were a connection to pyramids, and a connection with dissimilarities taking values in an arbitrary ordered set. This encompassed the work of both Herden and that of Critchley and Van Cutsem. In closing, we acknowledge that the research for this paper was supported in part by ONR Grant N00014-90-J-1008.

References:

BANDELT, H.-J. and DRESS, A. (1994): An order theoretic framework for overlapping clustering, preprint.

BAULIEU, F. (1985): Order theoretic classification of monotone equivariant cluster methods, *Algebra Universalsis, 20, 351–367.*

BAULIEU, F. (1991): Classification of normalized cluster methods in an order theoretic model, *Discrete Applied Mathematics, 32, (1991), 1–29.*

CRITCHLEY, F. and VAN CUTSEM, B. (1993): An order-theoretic unification and generalisation of certain fundanetal bijections in mathematical classification - I, preprint.

CRITCHLEY, F. and VAN CUTSEM, B. (1993a): An order-theoretic unification and generalisation of certain fundanetal bijections in mathematical classification - II, preprint.

DIDAY, E. (1987): Orders and overlapping clusters by pyramids, Rapports de

Recherche INRIA 730, INRIA-Rocqencourt.

HERDEN, G. (1984): Some aspects of clustering functions, *SIAM Journal on Algebraic and Discrete Methods, 5, 101–116.*

HUBERT, L. (1977): A set-theoretical approach to the problem of hierarchical clustering, *Journal of Math. Psychology, 15, 70–88.*

JANOWITZ, M. F. (1978): An order theoretic model for cluster analysis, *SIAM Journal on Applied Mathematics, 34, 55–72.*

JANOWITZ, M. F. (1978a): Semiflat L-cluster methods, *Discrete Mathematics, 21,, 47–60.*

JANOWITZ, M. F. (1978b): Monotone equivariant cluster methods, *SIAM Journal on Applied Mathematics, 34, 148–165.*

JANOWITZ, M. F. (1981): Continuous L-cluster methods, *Discrete Applied Mathematics, 3, 107-112.*

JANOWITZ, M. F. and STINEBRICKNER, R. (1993): Preservation of weak order equivalence, *Mathematical Social Sciences, 25, 181–197.*

JANOWITZ, M. F. and STINEBRICKNER, R. (1993a): Compatibility in a graph-theoretic setting, *Mathematical Social Sciences, 25, 251–279.*

JANOWITZ, M. F. SCHWEIZER, B. (1989): Ordinal and percentile clustering, *Mathematical Social Sciences, 18, 135–186.*

JANOWITZ, M. F. and WILLE, R. (1993): Compatibility with respect to isotone mappings, to appear.

JARDINE, N. and SIBSON, R. (1968): The construction of hierarchic and non-hierarchic classifications, *Computer Journal, 11, 177–184.*

JARDINE, N. and SIBSON, R. (1968a): A model for taxonomy, *Mathematical Biosciences, 2, 465–482.*

JARDINE, N. and SIBSON, R. (1971): Choice of methods for automatic classification, *Computer Journal, 14, 404–406.*

JARDINE, N. and SIBSON, R. (1971a): Mathematical Taxonomy, Wiley, New York.

MATULA, D. W. (1971): Graph theoretic techniques for cluster algorithms. In: J. van Ryzin, (ed.): *Classification and clustering*, Academic Press, New York, 95–129.

MICHALSKI, R. S., and STEPP, R. E., III (1983): Automated construction of classifications: conceptual clustering versus numerical taxonomy. *IEEE Transactions on Pattern Analysis and Machine Intelligence, 5, 396-410.*

POWERS, R. C. (1988): Order theoretic classification of percentile cluster methods, Ph.D. dissertation, University of Massachusetts, Amherst, MA, USA.

SCHWEIZER, B. and SKLAR, A. (1983): Probabilistic Metric Spaces, North-Holland, New York.

SIBSON, R. (1970): A model for taxonomy II, *Mathematical Biosciences, 6, 405–430.*

WILLE, R. (1982): Restructuring lattice theory: an approach based on hierarchies of concepts. In: I. Rival (ed.): *Ordered Sets*, NATO ASI Series 83, Reidel, Dordrecht.

An Overview and Recent Developments in Dual Scaling

Shizuhiko Nishisato[1]

The Ontario Institute for Studies in Education, and the University of Toronto, Toronto, Ontario, Canada M5S 1V6

Summary: Dual scaling quantifies such categorical data as contingency tables, multiple-choice data, sorting data, paired comparison data, rank-order data, and successive categories data. These data can be classified into two types, incidence data and dominance data. The present study is an overview of some key formulas and several conceptual problems, which require further investigations. Most of them are peculiar to data types, and some remedial procedures are suggested for them as interim measures. Awareness of these difficulties in dual scaling and other related methods seems to be the most notable recent development.

1. Introduction

Dual scaling, known also as correspondence analysis, homogeneity analysis and by many other names, is a method for quantifying categorical data, and as such it offers multidimensional scaling and principal component analysis of categorical data. This section contains a description of key formulas, properties and two types of categorical data.

1.1 Dual Scaling

For the purpose of illustration, let us consider the $n \times m$ contingency table, with f_{ij} being the typical element in row i and column j. Let us indicate by $f_{i.}$, $f_{.j}$, and $f_{..}$ the marginal frequency of row i, that of column j, and the total frequency in the table, respectively. Dual scaling decomposes f_{ij} as follows:

$$f_{ij} = \frac{f_{i.}f_{.j}}{f_{..}}[1 + \rho_1 y_{i1} x_{j1} + \rho_2 y_{i2} x_{j2} + \cdots + \rho_k y_{ik} x_{jk} + \cdots + \rho_T y_{iT} x_{jT}], \quad (1)$$

where ρ_k is the kth largest singular value, y_{ik} the ith element of the corresponding row singular vector, and x_{jk} the jth element of the corresponding column singular vector, which are called, in the context of dual scaling, the kth maximal correlation, the ith element of the corresponding optimal vector of row weights, and the jth element of that of column weights, respectively. The first term inside the square bracket, 1, indicates $\rho_0 y_{i_0} x_{j_0}$, where $\rho_0 = 1$, $y_{i_0} = 1$ $(i = 1, 2, \ldots, n)$, and $x_{j_0} = 1$ $(j = 1, 2, \ldots, m)$, and this is called a

[1]This study was supported by a grant from the Natural Sciences and Engineering Research Council of Canada to S. Nishisato.

trivial solution. The remaining T solutions are referred to as *proper solutions.* As may be obvious, the maximal value of T for the $n \times m$ contingency table is the smaller value of n and m minus 1, that is, $\min(n, m) - 1$.

From the above decomposition formula, therefore, one can say that dual scaling is nothing but singular value decomposition of f_{ij}. Another way of expressing it is in terms of the relation of f_{ij} to the three optimal quantities, ρ_k, y_{ik}, and x_{jk}:

$$y_{ik} = (\frac{1}{\rho_k})\frac{\sum_{j=1}^{m} f_{ij}x_{jk}}{f_{i.}}, \qquad x_{jk} = (\frac{1}{\rho_k})\frac{\sum_{i=1}^{n} f_{ij}y_{ik}}{f_{.j}}, \qquad k = 1, 2, \ldots, T. \tag{2}$$

These are called *substitution formulas* (Benzécri et al. (1973)), or *dual relations* (Nishisato (1980)). Noting that ρ_k is a maximal value, Equation (2) can be transformed into interesting formulas and properties:

(i) **Maximal product-moment correlation:** Assuming that the standard constraint of zero origins (i.e., the sum of responses weighted by y_{ik} and the sum of responses weighted by x_{jk} being both zero) is used, the product-moment (Pearsonian) correlation between responses weighted by y_{ik} and responses weighted by x_{jk} (3) is maximized.

$$\rho_k = \frac{\sum_{i=1}^{n}\sum_{j=1}^{m} f_{ij}y_{ik}x_{jk}}{\sqrt{\sum_{i=1}^{n} f_{i.}y_{ik}^2}\sqrt{\sum_{j=1}^{m} f_{.j}x_{jk}^2}} = \frac{y_k'Fx_k}{\sqrt{x_k'D_cx_k}\sqrt{y_k'D_ry_k}} \tag{3}$$

(ii) **Maximal correlation ratio:** The correlation ratios, (4), defined for responses weighted by y_{ik} and those weighted by x_{jk} are both maximized and identical.

$$\eta_k^2 = \frac{x_k'F'D_r^{-1}Fx_k}{x_k'D_cx_k} = \frac{y_k'FD_c^{-1}F'y_k}{y_k'D_ry_k} = \rho_k^2 \tag{4}$$

(iii) **Maximal average item-total correlation:** In dual scaling, the correlation ratio η^2 is equal to the average of the squared product-moment correlation between Item $j, j = 1, 2, \ldots, n$, and the total test score of optimally weighted n multiple-choice items, r_{jt}^2, hence dual scaling maximizes this average.

$$\eta_k^2 = \rho_k^2 = \frac{\sum_{j=1}^{n} r_{jt}^2}{n} \tag{5}$$

(iv) **Maximal internal consistency reliability:** As demonstrated by Lord (1958), the correlation ratio can be used to express the internal consistency reliability, often called Cronbach's alpha (Cronbach (1951)), or the generalized Kuder-Richardson reliability, which is under the standard constraints used in dual scaling,

$$\alpha = 1 - \frac{(1-\eta_k^2)}{(n-1)\eta_k^2}. \tag{6}$$

Thus maximization of η^2 means maximization of the reliability coefficient α. In other words, dual scaling, applied to multiple-choice data, produces scores for subjects which have maximal internal consistency reliability.

(v) **Principle of equivalent partitioning (PEP), or distributional equivalance (PDE):** When rows (columns) are partitioned into proportional rows (columns), or both rows and columns are partitioned likewise, the distribution of the correlation ratios remains the same as that of the original matrix. Nishisato (1984) introduced PEP for an algorithm, called *forced classification.* Benzécri et al. (1973) called it PDE.

(vi) **Principle of constant proportionality (PCP):** When an optimal solution is obtained, the quantified data satisfy that the between-option sum of squares of Item j to the item total sum of squares, $SS_j/SS_{t(j)}$, is equal to that of Item k, $SS_k/SS_{t(k)}$, for any two multiple-choice items j and k. PCP was derived by Nishisato and Sheu (1984) for the piecewise method of reciprocal averages, an algorithm for handling large data sets.

1.2 Two Types of Categorical Data

In quantifying categorical data, one must classify the data into two categories, incidence data and dominance data, for these two types follow the same rationale for quantification, but with different implications.

[1] **Incidence Data:** Responses are coded as 1 (presence) or 0 (absence), and data are either a set of 1's and 0's, or that of response frequencies. Typical data are the contingency table, multiple-choice data (expressed in terms of (1,0) response-patterns), and sorting data. In this type, each unit in the data is the object of quantification, and it is generally the case that row (column) marginals are not equal. Formula (1) applies to the decomposition of this data type.

[2] **Dominance Data:** Unlike incidence data, one response cannot be the object of quantification, but at least two responses are needed for quantification to be meaningful. Typical data are rank order data and paired comparison data. The judgment of stimuli X_j and X_k by Subject i is coded by (7).

$$_if_{jk} = \begin{cases} 1, & \text{if Subject } i \text{ prefers } X_j \text{ to } X_k, X_j \succ X_k. \\ 0, & \text{if Subject } i \text{ provides equality judgement}, X_j = X_k. \\ -1, & \text{if Subject } i \text{ prefers } X_k \text{ to } X_j, X_j \prec X_k. \end{cases} \tag{7}$$

In this type, the input data are transformed into a dominance data matrix, $\mathbf{E} = (e_{ij})$, with the typical element e_{ij} being the number of times Subject i preferred Stimulus j to the remaining $(n-1)$ stimuli minus the number of times Subject i preferred other stimuli to Stimulus j (see formula 8).

$$e_{ij} = \sum_{k=1(k \neq j)}^{n} {}_i f_{jk} \tag{8}$$

The dominance matrix **E** is subjected to singular value decomposition, with the understanding that each data unit is the outcome of the $(n-1)$ comparisons, that is, the same number of responses is involved across rows (columns).

Although these two types of data are both subjected to singular value decomposition, one should keep in mind the following differences between them:

(1) Incidence data have a trivial solution, while dominance data do not.

(2) Incidence data matrices have in general uneven marginals, while the total number of responses involved in column (row) p in the dominance matrix is equal to that in column (row) q.

With respect to (1) above, it will be shown that a trivial solution of incidence data is not trivial, but very influential on the outcome of quantification, and that the trivial solution is responsible for major problems that confront the current method of dual scaling. In addition to (1), the problem of uneven marginals in incidence data, (2), undermines usefulness of dual scaling as a routine method of data analysis. The main object of the current study is to present an overview of methodological problems with remedial ideas for them. Recognition of such problems seems to have been the most significant development in dual scaling over the past few years.

2. Problems and Suggested Alternatives

2.1 Perennial Problems of Joint Graphical Display

The point that weights for the rows, y_{ik}, and weights for the columns, x_{jk}, do not span the same space is well known. In spite of this logical difficulty, plotting both y_{ik} and x_{jk} in the same graph, called *joint* graphical display, has been very popular in practice, for dual scaling derives weights for rows and those for columns in such a way that the discrepancy between the two spaces is a minimum. To overcome this logical problem, Carroll et al. (1986, 1987) proposed an alternative, referred to as the CGS scaling. Greenacre (1989) criticized their approach, which led to a rebuttal (Carroll et al. (1989)) with no obvious conclusions.

As some researchers have done (e.g., Gifi (1990); Nishisato (1980)), let us rewrite Formula (2), and obtain (9).

$$\rho_k y_{ik} = \frac{\sum_{j=1}^{m} f_{ij} x_{jk}}{f_{i.}}, \qquad \rho_k x_{jk} = \frac{\sum_{i=1}^{n} f_{ij} y_{ik}}{f_{.j}}, \qquad k = 1, 2, \ldots, T. \tag{9}$$

Then we note that $\rho_k y_{ik}$ and $\rho_k x_{jk}$ are projections of y_{ik} and x_{jk} onto spaces of x_{jk} and y_{ik}, respectively: the pair $(\rho_k y_{ik}, x_{jk})$ spans the same space, and so does the pair $(y_{ik}, \rho_k x_{jk})$. Thus plotting either set of these pairs in one graph is logically sound. This, however, is not necessarily the best alternative, for the norms of the two variables in the joint graph can be quite different, especially when ρ_k is small. Then the graph can be very misleading.

The same problem of joint graphical display applies equally to multiple-choice data, another example of incidence data. To overcome the problem, one remedial approach of some promise is to plot only $\rho_k y_{ik}$ (e.g., the score of the respondent i on dimension k), label the point by his or her response pattern to the questions, and partition the space in terms of common patterns (Nishisato (1990)). Fig. 1 shows an example of this approach, using the data reported in Nishisato (1994, p. 153). This partitioning of the space in terms of options of Items 1 and 4 is only an example, and one can consider other combinations of common responses. This approach yields all graphs which are immediately interpretable and are logically sound. < **Suggestion 1** > Develop this type of graphical display. When the number of multiple-choice questions increases, one needs to develop a method for key item selection for multidimensional classification. This study is under way.

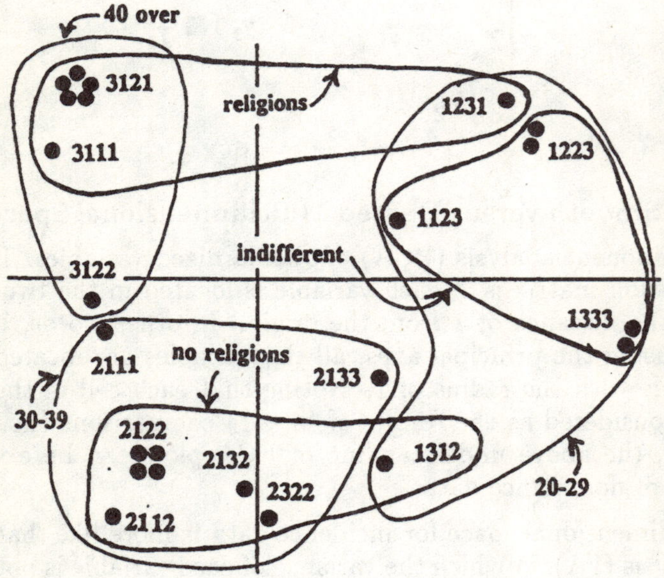

Q.1: "How old are your?"
(1) 20-29
(2) 30-39
(3) 40 or over
Q.2:"Children today are not as disciplined as I was a child"
(1) agree
(2) disagree
(3) I cannot tell
Q.3: "Children today are not as fortunate as I was a child"
(1) agree
(2) disagree
(3) I cannot tell
Q.4: "Religions should be taught at school"
(1) agree
(2) disagree
(3) indifferent

Fig. 1

One would expect that a similar graph of partitioned regions may be obtained by cluster analysis of Hamming distances between subjects' (1,0) response patterns. This, however, is not the case, for Hamming distances somehow lose the contributions of individual response options, imbedded in items, the most important information for interpretations.

Dominance data present an interesting situation. It is empirically known (Nishisato (1994)) that the rank order of the absolute value of the difference between y_{ik} and $\rho_k x_{jk}$ seems to correspond to the rank order in which Subject i ranked Stimulus j. Fig. 2 shows an example in which Subjects 1 and 2 ranked stimuli as X_1 first, then $X_2, X_3, X_4, X_5, X_6, X_7, X_8$, and X_9, while Subjects 3 and 4 ranked in order as $X_1, X_2, X_3, X_6, X_5, X_4, X_7, X_8$, and X_9. Thus it seems that joint graphical display of quantified dominance data offers a helpful means for data analysis. $<$ **Suggestion 2** $>$ For dominance data, plot $(y_{ik}, \rho_k x_{jk})$.

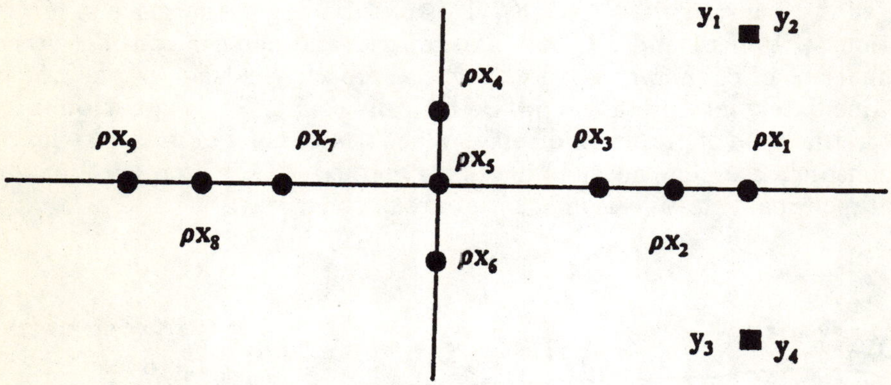

Fig. 2

2.2 Space Problem: Smooth versus Dented Multidimensional Space

Consider principal component analysis (PCA) of standardized variables. If the rank of the correlation matrix is 2, each variable is located in the two-dimensional space at the distance of 1 from the origin. In other words, if we plot the variables using the principal axes, all the variables are located somewhere on the circle with the radius of 1. Noting that each cell in the dominance matrix is considered as the results of $(n-1)$ comparisons (i.e., equal cell frequencies), the above understanding of PCA plots can be extended to dual scaling of dominance data.

In contrast, the multidimensional space for incidence data is more like that of common factor analysis (FA), in which the variance of each variable is not 1, but equal to the total variance (i.e., 1) minus the unique variance, that is, the communality h_j^2 for variable j, bounded between 0 and 1. As mentioned earlier, incidence data contain the trivial solution, and the removal of the trivial solution from standardized cross-product matrix creates non-uniform variances of individual variables. Thus if an incidence matrix contains two proper (i.e., nontrivial) solutions, the two dimensional plot of variables does not in general place the variables on a circle of a constant radius. The total space for proper dual scaling solutions may be considered as somewhat analogous to the common factor space in FA. In dual scaling, categories and multiple-choice options with very few responses tend to produce exceedingly large variances (see "outliers" in Section 2.4). Can we then standardize the space for proper solutions of incidence data? Keep in mind, however, that the standardized incidence data will no longer maximize η^2. < **Suggestion 3** > Watch out those categories or multiple-choice options with very few responses, and if they can be merged with others, do so; otherwise, assess their effects on the scaling results.

2.3 Accountability Problem

When PCA and FA are used, one of the popular statistics is the total variance explained by component (factor) k, say δ_k. Although PCA and FA are viewed as representing two distinct cases in the previous section, these methods provide easily interpretable statistics δ_k. Whether this is the case with dual scaling depends on the data types. The statistic δ is defined for a proper solution as

$$\delta_k = \frac{100\eta_k^2}{\sum_{t=1}^{T}\eta_t^2}(\%), \qquad t = 1, 2, \ldots, T \tag{10}$$

As far as this formula is concerned, there does not seem to be any problem of interpretation. It is simply the percentage of each solution to the total set of solutions. Nevertheless, this statistic is totally uninterpretable when it is used for dual scaling of incidence data.

[I-1] $n \times m$ **Contingency Table:** The denominator of (10) can be expressed in terms of the chi-square statistic.

$$\sum_{t=1}^{T}\eta_t^2 = \frac{\chi^2}{f_{..}} \tag{11}$$

Thus the denominator of (10) is a function of the row-column association in the table, and, more importantly, so is the numerator. This leads to a strange case in which the larger the row-column association the smaller the δ_1 (see a few examples in Tab. 1). To overcome this difficulty, alternatives to δ have been proposed, that is, percentage homogeneity ν_k (Nishisato and Nishisato (1984)), percentage to the total space γ_k, and percentage to the absolute maximum θ_k (Nishisato (1993)).

$$\nu_k = \delta_k \sum_{t=1}^{T}\eta_t^2 = \frac{100\eta_k^2}{\max(\eta^2)} = 100\eta_k^2, \quad \gamma_k = \frac{100\eta_k^2}{1+\sum\eta_t^2}, \quad \theta_k = \frac{100\eta_k^2}{\min(n,m)-1} \tag{12}$$

Statistics on these measures from four contingency tables, $\mathbf{F_1}, \mathbf{F_2}, \mathbf{F_3}, \mathbf{F_4}$, are listed in Tab. 1. Compare the four values of η_1^2 with the corresponding values of the other four measures. Although ν is not a measure of the total variance explained, it is easy to interpret. γ is a measure of the total variance of a given data set, including the trivial solution, explained by Solution k, and θ a measure of the maximal variance for the given data size, accounted for by the solution. These three measures appear to be more reasonable than

δ . < **Suggestion 4** > Use ν_k, and report the other measures, together with tr(C) and η_k^2.

$$\mathbf{F_1}=\begin{bmatrix}20 & 0 & 0\\ 0 & 10 & 0\\ 0 & 0 & 8\end{bmatrix}, \mathbf{F_2}=\begin{bmatrix}18 & 1 & 1\\ 1 & 8 & 1\\ 1 & 1 & 6\end{bmatrix}, \mathbf{F_3}=\begin{bmatrix}16 & 3 & 1\\ 3 & 4 & 3\\ 1 & 3 & 4\end{bmatrix}, \mathbf{F_4}=\begin{bmatrix}14 & 3 & 3\\ 3 & 4 & 3\\ 3 & 3 & 2\end{bmatrix},$$

Case	tr(C)	η_1^2	δ_1	γ_1	θ_1	ν_1
1	2.0000	1.0000	50	33	50	100
2	1.0600	0.6259	59	30	31	63
3	0.3795	0.3685	98	27	18	37
4	0.1400	0.1395	100	12	7	14

Tab. 1: Measures of Information Explained by Solution 1

[I-2] **Multiple-Choice Data:** The sum of correlation ratios of nontrivial solutions is equal to the average number of response options minus 1.

$$\sum_{t=1}^{T} \eta_t^2 = \overline{m} - 1 \tag{13}$$

Thus, for a questionnaire with six response alternatives per question, δ cannot exceed 20%, for the total variance is 5 and the absolute maximum of the correlation ratio is 1. Benzécri (1979) and Greenacre (1987) proposed other measures that involve transformations of data for decomposition, scaling of which provide larger values than δ. Our object here, however, is to find a measure for the results of standard dual scaling/correspondence analysis. In this context, one measure of some promise is to redefine proper solutions: when the correlation ratio becomes smaller than $1/n$, where n is the number of multiple-choice questions, the internal consistency reliability, α, becomes negative (Nishisato (1980)). Thus one can define proper solutions as those, of which correlation ratios are greater than $1/n$, that is, the average of all η^2 's (Nishisato (1994)). Then the proposed alternative to δ is δ^*, which is obtained by substituting the sum of correlation ratios that are greater than $1/n$ for the denominator of (10). < **Suggestion 5** > Use ν_k and α, and report $\delta_k^*, \delta_k, \eta_k^2$ and tr(C), that is, (13).

As for dominance data, there does not seem to be any interpretation problem of δ , but the correlation ratio is generally much smaller than that of incidence data.

[D-1] **Rank-Order Data** and **Paired Comparison Data**: For n variables and N subjects, these two types of data provide the same sums of correlation ratios.

$$\sum_{t=1}^{n-1} \eta_t^2 = \frac{n+1}{3(n-1)} \tag{14}$$

Unlike incidence data in which one response is the unit of analysis, dominance data require a minimum of two responses to be meaningful because of the comparative nature. Then the total variance (i.e., the sum of correlation ratios) attains its maximum of 1 when $n = 2$, and its minimum of 1/3 as n approaches infinity. This range also applies to each correlation ratio, the reason why the correlation ratio for dominance data is typically much smaller than that of incidence data. When n is greater than 2, the correlation ratio cannot attain 1 even if the data can be explained exhaustively by one solution, but δ_1 is then 100%. Thus δ means what we expect it to be, and the correlation ratio is somewhat misleading. < **Suggestion 6** > Use δ_k.

[D-2] **Successive Categories Data:** Another example of dominance data is successive categories data with $(m+1)$ successive rating categories, collected from N subjects on n stimuli. The sum of correlation ratios is

$$\sum_{t=1}^{n+m-1} \eta_t^2 = \frac{n+m+1}{3(n+m-1)} \tag{15}$$

Since the minimum number of $n + m$ for the data to be meaningful is two, the conclusion on δ and the correlation ratio is exactly the same as in [D-1]. < **Suggestion 7** > Use δ_k.

2.4 Problems Caused by Too many Response Options

Out of all the data types so far discussed, quantification of multiple-choice data seems most problematic. The following are some of them.

(i) **Over-quantification:** Suppose that we categorize continuous variates, each into m categories. As the value of m increases, dual scaling solutions would show that the inter-variable correlation between any two variables approaches 1, independently of the correlation between the two original continuous variables. What does that increased correlation mean? There seem to be two implications: (1) dual scaling is not a method to approximate the underlying data structure, and if the recovery of the original data structure is one's major concern, one must devise another method; (2) extreme high correlations can be effectively used to classify subjects or objects, for the resultant quantifications provide clearcut discriminative information. < **Suggestion 8** > Do not use too many response options per question.

(ii) **Expanded dimensionality:** The total number of solutions is generally equal to the total number of response options of all the items, m, minus the number of items, n, that is, $m - n$. Thus, if each of twenty items has five options, dual scaling of the data collected from many subjects can yield 80 solutions, while PCA of the data can produce at most 20 solutions. In this example, each PCA component is associated with four DS solutions (see Fig. 3 for the plots of the first 3 and 9 solutions).

What do these extra solutions mean? Although this extra detailed analysis by DS may be useful in some cases, one should question if such patterns as (extremely serious, very serious) and (very serious, extremely serious) should indeed be treated as distinct in data analysis. < **Suggestion 9** > Do not

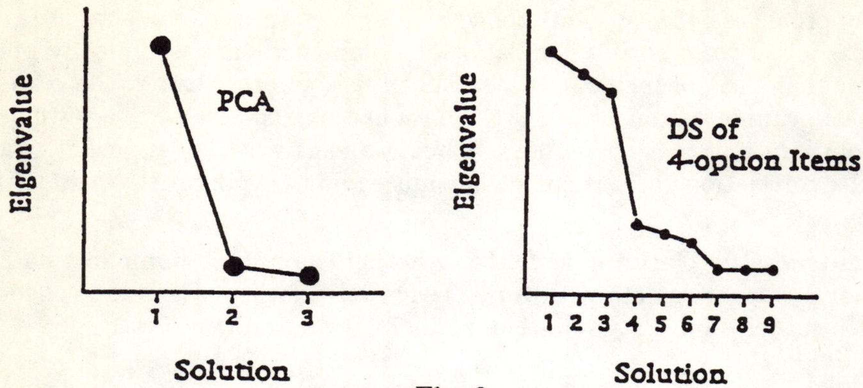

Fig. 3

use too many options per question.

(iii) **Contributions of items and options:** For N subjects and m_j options of Item j, the total sum of squares of weighted responses of Option p, SS_{jp}, and that of Item j, SS_j, are given by (16).

$$SS_{jp} = \sum_{k=1}^{T} f_{jp}x_{jpk}^2 = n(N - f_{jp}), SS_j = \sum_{p=1}^{m_j} SS_{jp} = f_{..}(m_j - 1). \qquad (16)$$

These two formulas indicate that those unpopular response options and those items with many options contribute more information, as reflected by SS, to the total set of solutions. That SS_{jp} is a linearly decreasing function of f_{jp} is a direct consequence of discarding the trivial solution. These two aspects have an important implication for designs of questionnaires. Methodologically, standardizing the quantification in terms of options or items is a way to eliminate those artificial effects, which, however, will result in the sacrifice of optimality of quantification. < **Suggestion 10** > Use the same number of options for as many items as possible, and do not use too many options per item.

(iv) **Outlier responses:** A single response that affects the quantification drastically is defined as an outlier response. Formula (16) may not appear serious, but if one response option is chosen only by one person out of 100 the item which contains this option is likely to show an extremely large value of r_{jt} as a consequence of an exceptionally large absolute value of the weight for the option. See an extreme case of such an example and a number of resistant or robust methods of quantification in Nishisato (1994, pp. 293-300). < **Suggestion 11** > Follow Suggestion 3, or use the method of reciprocal trimmed averages, discussed in the above reference for robust methods.

2.5 Missing Responses

Data analysis has been constantly plagued by missing responses. In the exploratory type of categorical data analysis such as dual scaling, there seems to be neither a model for missing responses, nor a model for data generation available. This latter aspect makes it difficult to carry out Monte Carlo investigations on different methods for handling missing responses. Empirically speaking, the *passive method*, in which missing responses are treated as unavailable responses (thus ignored), seems reasonable as long as missing responses occur in a random fashion. Under this condition, analysis of multiple-choice data, paired comparision, rank-order and successive categories data can be handled relatively easily.

Another approach to missing responses is to impute responses for the missing cells. For multiple-choice data, one can introduce an extra option for each item to record missing responses, thus transforming incomplete data to complete data. This idea is appealing under a situation in which subjects omit responses for some reasons. When missing responses are abundant, this approach may work. Otherwise, a few missing responses may result in outlier responses.

One may be tempted to impute responses for missing responses in such a way that η^2 be a maximum. This approach, however, is absurd, for under the scheme the more missing responses the higher the value of η^2. The hot-deck method of imputation appears reasonable, but again as the number of missing responses increases η^2 can easily get spuriously high, which one can infer from the experience with forced classification (see Nishisato (1994, pp. 241-242)). < **Suggestion 12** > In the absence of a satisfactory method of imputation, use the passive method, described above. See also an interesting paper by van Buuren and van Rijckevorsel (1992).

From a different perspective, some work has been done to answer "How many is too many?." In particular,

Question 1: What is the maximal number of missing responses that a data set can contain before it is judged unworthy of analysis?

Question 2: What is the maximal number of missing responses that a single item can tolerate?

Question 3: What is the maximal number of missing responses that a single respondent is allowed to make while maintaining the validity of his/her score?

For a given set of response patterns with missing responses, different imputations produce different values of η^2. Retain the largest value, η^2_{max}, and the smallest value, η^2_{min}. Question 1, for example, was investigated by empirically constructing the confidence interval for η^2_{max} and that for η^2_{min} as a function of the number of missing responses. The decision rule is to give up data analysis when the difference between the two extreme values becomes statistically significant, for then different imputations may produce significantly different results, as reflected on η^2. Because there is no way to simulate realistic categorical data, the results are tentative: the critical point for question 1 ranges between 11% to 25%, that for question 2 between 5% to 20%, and that for question 3 between 11% to 50% (Ahn (1991); Nishisato and Ahn (1994)).

3. Concluding Remarks

These problems discussed above represent some of the difficulties that confront dual scaling today. There are many other problems of interest that one can investigate, and such problems become visible once the method is applied to a variety of real data. Other than those numerical problems, there exists a major practical problem of how to interpret the scaling outcome. Admitting that the interpretation involves a lot of subjective judgment, hence experience, we must provide the users of the method with as many helpful statistics and as much guidance as possible. Since the results must be interpreted in reference to the input data, an efficient and objective interpretive system needs to be developed for relating the outcome to the input data. Computer programs should have many help files for clarification of technical terms and for decision making, and should be intelligent enough to cope with minor errors and missing responses in the data set. Much work is still needed in order to make dual scaling a routine method for data analysis.

References:

AHN, H. (1991): Effects of missing responses in multiple choice data on dual scaling. Unpublished doctoral dissertation, University of Toronto.

BENZCRI, J.-P. (1979): Sur le calcul des taux d'inertie dans l'analyse d'un questionnaire. *Cahiers de l'Analyse des Donnes*, 4, 377-378.

BENZCRI, J.-P. et al. (1973): L'Analyse des donnes: II. L'Analyse des correspondances. Paris: Dunod.

CARROLL, J.D., GREEN, P.E. and SCHAFFER, C.M. (1986): Interpoint distance comparisons in correspondence analysis. *Journal of Marketing Research*, 23, 271-280.

CARROLL, J.D., GREEN, P.E. and SCHAFFER, C.M. (1987): Comparing interpoint distances in correspondence analysis: A clarification. *Journal of Marketing Research*, 24, 445-450.

CARROLL, J.D., GREEN, P.E. and SCHAFFER, C.M. (1989): Reply to Greenacre's commentary on the Carroll-Green-Schaffer scaling of two-way correspondence analysis solutions. *Journal of Marketing Research*, 26, 366- 368.

CRONBACH, L.J. (1951): Coefficient alpha and internal structure of tests. *Psychometrika*, 16, 297-334.

GREENACRE, M.J. (1987): Measuring total variation and its components in multiple correspondence analysis. Murray Hill, NJ, AT&T Bell Laboratories Statistical Research Report.

GREENACRE, M.J. (1989): The Carroll-Green-Schaffer scaling in correspondence analysis: A theoretical and empirical appraisal. *Journal of Marketing Research*, 26, 358-365.

LORD, F.M. (1958): Some relations between Guttman's principal components of scale analysis and other psychometric theory. *Psychometrika*, 23, 291-296.

NISHISATO, S. (1980): Analysis of categorical data: Dual scaling and its applications. Toronto: University of Toronto Press.

NISHISATO, S. (1984): Forced classification: A simple application of a quantification technique. *Psychometrika*, 49, 25-36.

NISHISATO, S. (1987): Robust techniques for quantifying categorical data. In: I.B. McNeil and G.J. Umphrey (eds.): *Foundations of Statistical Inference.* Dordrecht: D. Reidel Publishing, 209-217.

NISHISATO, S. (1990): Dual scaling of designed experiments. In M. Schader and W. Gaul (eds.): *Knowledge, Data and Computer-Assisted Decisions.* Berlin: Springer-Verlag, 115-125.

NISHISATO, S. (1993): On quantifying different types of categorical data. *Psychometrika*, 58, 617-629.

NISHISATO, S. (1994): Elements of dual scaling: An introduction to practical data analysis. Hillsdale, NJ: Lawrence Erlbaum Associates.

NISHISATO, S. and AHN, H. (1994): When not to analyze data: Decision making on missing responses in dual scaling. *Annals of Operations Research (in press).*

NISHISATO, S. and NISHISATO, I. (1984): An introduction to dual scaling. Toronto: MicroStats.

NISHISATO, S. and SHEU, W.J. (1980): Piecewise method of reciprocal averages for dual scaling of multiple-choice data. *Psychometrika*, 45, 467- 478.

VAN BUUREN, S. and VAN RIJCKEVORSEL, J.L.A. (1992): Imputation of missing categorical data by maximizing internal consistency. *Psychometrika*, 57, 567-580.

Gibbs Sampling in AR Models with Random Walk Priors

Wolfgang Polasek, Song Jin

Institute for Statistics and Econometrics,
University of Basel, CH-4051 Basel

Summary: The paper analyses univariate autoregressive AR(p) models with tightness prior. The framework of the model is the conjugate normal linear model where the prior distribution is assumed to be a random walk process. The deviation from the prior distribution is measured by the tightness (hyper-) parameter λ. It is shown how the estimation of the starting values can be incorporated into the Gibbs sampling scheme. We demonstrate this approach with simulated and economic time series. It is found that for typical economic sample size the sampling fluctuations influence the posterior distribution considerably and informative prior distributions seem to be useful, especially for prediction.

1. Introduction

Bayesian VAR models have been used increasingly for macroeconomic forecasting in the last decade, because the forecasting properties have been found to outperform the corresponding classical models. The assumption of a tightness or smoothness structure for the lag coefficients implies a so-called hierarchy of prior parameters, also called hyperparameters. In the approach used by Litterman(1986) these hyperparameters for the tightness model are fixed by the modeler. Using the Gibbs sampler and a proper prior distribution for all parameters (Gelfand and Smith (1990)), we can find the posterior distribution of all parameters by simulation which utilises all so-called full conditional distributions (f.c.d.). In a conjugate Normal-Wishart framework the full conditional distributions can be derived in closed form which also allows an efficient generation of random numbers.

Smoothness models have been proposed by Shiller (1973), where the specification of the hyperparameters was left open. We will show that the Gibbs sampler can be used in a similar way as in the tightness models to find the posterior distribution of all parameters. The Gibbs sampler has been applied for time series model by e.g. Chib(1993), and Marriot et al. (1992). Gibbs sampling has the big advantage that one can impose many additional complication to a basic model, like outliers, heteroscedasdicity, or errors in variables (Polasek (1994)).

In the next section we introduce the univariate tightness model for the AR(p) model (Polasek 1993). Then we demonstrate with simulated data the behavior of the starting values and the importance of the prior values for these parameters. As a real example, we estimated the Swiss consumption time series.

2. The Tightness Autoregression Model: B-AR(p)

2.1 Introduction to the Tightness Autoregression Model: B-AR(p)

Let $\mathbf{y}$ be a vector of a univariate time series of length T, and we want to estimate an autoregressive process of lag p:

$$\begin{aligned}\mathbf{y} &= \alpha\mathbf{1}_n + \mathbf{y}_{-1}b_1 + \mathbf{y}_{-2}b_2 + \ldots + \mathbf{y}_{-p}b_p + \epsilon \\ &= \mathbf{X}\mathbf{b} + \epsilon \end{aligned} \tag{1}$$

where $\mathbf{b} = (\alpha, b_1, \ldots, b_p)$ and the n×(p+1) regressor matrix $\mathbf{X} = [\mathbf{1}_n : \mathcal{L}_p\mathbf{y}]$ consists of the constant and the past of the left hand variable $\mathbf{y}$. $\mathcal{L}_p$ is the matrix lag operator $\mathcal{L}_p\mathbf{y} = [\mathbf{y}_{-1} : \ldots : \mathbf{y}_{-p}]$.

The univariate autoregressive tightness model has the following hierarchical linear model structure

$$\mathbf{y} \sim N(\mathbf{Xb}, \sigma^2\mathbf{I}_T), \tag{2}$$
$$(\mathbf{b}, \sigma^{-2}) \sim NoGa(\mathbf{b}_*, \lambda\mathbf{H}_*, \sigma_*^2, n_*), \tag{3}$$
$$\lambda^{-1} \sim Ga(\lambda_*, l_*), \tag{4}$$
$$\mathbf{y}_- \sim N[\mathbf{y}_-^*, \mathbf{\Psi}_*]. \tag{5}$$

λ is the tightness parameter and the (p+1)×(p+1) prior covariance matrix $\mathbf{H}_*$=diag(d_0^*,$\mathbf{D}_*$) includes the unknown variance component for the intercept and the known tightness structure for the p past lags $\mathbf{D}_*$= diag(1,1/2,...,1/p). The diagonal structure is adopted for simplicity for the precision matrix $\mathbf{H}_*^{-1}$=diag(1/d_0^*,1,2,...,p). The p×1 vector $\mathbf{y}_-$ contains the p starting values and we assume that prior information about the starting values is available by the normal distribution N[$\mathbf{y}_-^*$,$\mathbf{\Psi}^*$]. By introducing the starting values as parameters which are estimated, we can now use all n=T observations of the observed time series instead the n=T-p values, if we condition on the p first observed time series values.

The joint distribution for the data $\mathbf{Y}$=($\mathbf{y}$,$\mathbf{X}$) and the parameters $\theta = (\mathbf{b}, \sigma^2, \lambda, \mathbf{y}_-)$ is given by the product

$$\begin{aligned} p(\theta, \mathbf{Y}) \propto\ & N[\mathbf{y}|\mathbf{Xb}, \sigma^2\mathbf{I}_n]N[\mathbf{b}|\mathbf{b}_*, \lambda\mathbf{D}_*]Ga[\lambda^{-1}|\lambda_*, l_*] \\ & \cdot Ga[\sigma^{-2}|\sigma_*^2, n_*]N[\mathbf{y}_-|\mathbf{y}_-^*, \mathbf{\Psi}_*]. \end{aligned} \tag{6}$$

Let θ^c be the symbol for the complementary parameters in a full conditional distribution. Then the full conditional distributions for θ^c can be derived in blocks as follows:

2.1.1 For the regression coefficients:

$$p(\mathbf{b}|\mathbf{Y}, \theta^c) \propto (\sigma^{-2})^{n/2} exp\{-\frac{1}{2\sigma^2}(\mathbf{y} - \mathbf{Xb})'(\mathbf{y} - \mathbf{Xb})\}$$

$$\cdot exp\{-\frac{1}{2}(\mathbf{b}-\mathbf{b}_*)'\lambda^{-1}\mathbf{H}_*^{-1}(\mathbf{b}-\mathbf{b}_*)\}$$
$$\propto N[\mathbf{b}|\mathbf{b}_{**}, \mathbf{H}_{**}] \qquad (7)$$

a normal distribution with the moments

$$\mathbf{H}_{**}^{-1} = \mathbf{H}_*^{-1}\lambda^{-1} + \sigma^{-2}\mathbf{X}'\mathbf{X}, \qquad (8)$$
$$\mathbf{b}_{**} = \mathbf{H}_{**}[\lambda^{-1}\mathbf{H}_*^{-1}\mathbf{b}_* + \sigma^{-2}\mathbf{X}'\mathbf{y}]. \qquad (9)$$

2.1.2 The f.c.d. for the residual precision:

$$p(\sigma^{-2}|\mathbf{Y},\theta^c) \propto (\sigma^{-2})^{n/2} exp\{-\frac{1}{2\sigma^2}(\mathbf{y}-\mathbf{X}\mathbf{b})'(\mathbf{y}-\mathbf{X}\mathbf{b})\}$$
$$\cdot(\sigma^{-2})^{n_*/2-1} exp\{-\frac{1}{2}n_*\sigma_*^2/\sigma^2\}$$
$$\propto Ga[\sigma^{-2}|\sigma_{**}^2, n_{**}] \qquad (10)$$

a gamma distribution with the parameters $n_{**}=n_*+n$, and

$$n_{**}\sigma_{**}^2 = n_*\sigma_*^2 + (\mathbf{y}-\mathbf{X}\mathbf{b})'(\mathbf{y}-\mathbf{X}\mathbf{b}) \qquad (11)$$

2.1.3 The full conditional distribution for the tightness parameter:

$$p(\lambda|\mathbf{Y},\theta^c) \propto |\lambda\mathbf{H}_*|^{-1/2} exp\{-\frac{1}{2}(\mathbf{b}-\mathbf{b}_*)'\lambda^{-1}\mathbf{H}_*^{-1}(\mathbf{b}-\mathbf{b}_*)\}$$
$$\cdot\lambda^{-l_*/2+1} exp\{-\frac{1}{2}\lambda^{-1}l_*\lambda_*\}$$
$$\propto Ga[\lambda^{-1}|\lambda_{**}, l_{**}] \qquad (12)$$

is a gamma distribution with the parameters $l_{**}=l_*+p+1$, and

$$l_{**}\lambda_{**} = l_*\lambda_* + (\mathbf{b}-\mathbf{b}_*)'\mathbf{H}_*^{-1}(\mathbf{b}-\mathbf{b}_*) \qquad (13)$$

2.1.4 The distribution of the starting values $\mathbf{y}_-$*:*

The first elements of an AR(p) process suffer from the 'starting value problem'. In Bayesian terms starting values can be simply viewed as unknown parameters which can be estimated from the data. The convenient feature of the Gibbs sampler is, that the full conditional distribution of the starting values can be also expressed in closed form as a normal distribution where we can draw samples. So the 'starting value problem' adds just another iteration step in the Gibbs sampler.

For an AR(p) process we have p starting values which we collect in the p×1 vector $\mathbf{y}'_-=(y_{-1},\ldots,y_{-p})$. The first p elements of the AR(p) regression model are denoted as $\mathbf{y}_0=(y_1,\ldots,y_p)$ and

$$\mathbf{y}_0 = \mathbf{X}_0\mathbf{b} + \epsilon_0.$$
$$= \alpha\mathbf{1}_p + \begin{pmatrix} y_{-1} & y_{-2} & y_{-3} & \cdots & y_{-p} \\ y_0 & y_{-1} & y_{-2} & \cdots & y_{-p+1} \\ \cdots & \cdots & \cdots & \cdots & \cdots \\ y_{p-2} & y_{p-3} & \cdots & y_0 & y_{-1} \end{pmatrix} \begin{pmatrix} b_1 \\ b_2 \\ \cdots \\ b_p \end{pmatrix} + \epsilon_0 \qquad (14)$$

This can be written also as

$$
\begin{aligned}
y_0 &= \alpha \mathbf{1}_p + \begin{pmatrix} 0 & 0 & \dots & 0 & 0 \\ y_0 & 0 & \dots & 0 & 0 \\ \dots & \dots & \dots & \dots & \dots \\ y_{p-2} & y_{p-3} & \dots & y_0 & 0 \end{pmatrix} \mathbf{b} \\
&\quad + \mathbf{b} y_{-1} + \dots + \begin{pmatrix} b_p \\ 0 \\ \dots \\ 0 \end{pmatrix} y_{-p} + \epsilon_0 \\
&= \alpha \mathbf{1}_p + \mathbf{Y}^\circ \mathbf{b} + \mathbf{A}\mathbf{y}_- + \epsilon_0 \qquad (15)
\end{aligned}
$$

where $\mathbf{a}=\alpha\mathbf{1}_p+\mathbf{Y}^\circ\mathbf{b}$ does not depend on the starting values, and the $\mathbf{Y}^\circ$ and the matrix $\mathbf{A}$ contain known data and regression coefficients, respectively:

$$
\mathbf{Y}^\circ = \begin{pmatrix} 0 & 0 & \dots & 0 & 0 \\ y_0 & 0 & \dots & 0 & 0 \\ \dots & \dots & \dots & \dots & \dots \\ y_{p-2} & y_{p-3} & \dots & y_0 & 0 \end{pmatrix}, \qquad (16)
$$

$$
\mathbf{A} = \begin{pmatrix} b_1 & b_2 & \dots & b_{p-1} & b_p \\ b_2 & b_3 & \dots & b_p & 0 \\ \dots & \dots & \dots & \dots & \dots \\ b_p & 0 & \dots & 0 & 0 \end{pmatrix} \qquad (17)
$$

Note that $\mathbf{A}$ and $\mathbf{Y}^\circ$ are constructed in such a way that they have complementary roles for the decomposition of observed data ($\mathbf{Y}^\circ$), regression coefficients ($\mathbf{A}$) and starting values ($\mathbf{y}_-$). Using the prior information about the starting values $\mathbf{y}_- \sim \mathrm{N}[\, \mathbf{y}_-^*, \mathbf{\Psi}_* \,]$, we find the f.c.d. posterior distribution for $\mathbf{y}_-$ to be

$$
\begin{aligned}
p(\mathbf{y}_-|\mathbf{Y},\theta^c) &\propto (\sigma^{-2})^{n/2} exp\{-\frac{1}{2\sigma^2}(\mathbf{y}_0 - \mathbf{X}_0\mathbf{b})'(\mathbf{y}_0 - \mathbf{X}_0\mathbf{b})\} \\
&\quad \cdot exp\{-\frac{1}{2}(\mathbf{y}_- - \mathbf{y}_-^*)'\mathbf{\Psi}_*^{-1}(\mathbf{y}_- - \mathbf{y}_-^*)\}. \qquad (18)
\end{aligned}
$$

Now the first quadratic form can be written by the above transformation as $(\mathbf{y}_0 - \mathbf{X}_0\mathbf{b})'(\mathbf{y}_0 - \mathbf{X}_0\mathbf{b})$ with $\mathbf{a}=\mathbf{Y}^\circ\mathbf{b}+\alpha\mathbf{1}_n$. Therefore the full conditional can be obtained as

$$
p(\mathbf{y}_-|\mathbf{Y},\theta^c) = N[\mathbf{y}_-^{**}, \mathbf{\Psi}_{**}] \qquad (19)
$$

a normal distribution with the parameters

$$
\begin{aligned}
\mathbf{\Psi}_{**}^{-1} &= \mathbf{\Psi}_*^{-1} + \sigma^{-2}\mathbf{A}'\mathbf{A}, \qquad (20) \\
\mathbf{y}_-^{**} &= \mathbf{\Psi}_{**}[\mathbf{\Psi}_*^{-1}\mathbf{y}_-^* + \sigma^{-2}\mathbf{A}'(\mathbf{y}_0 - \mathbf{a})]. \qquad (21)
\end{aligned}
$$

The posterior mean for the starting values is the matrix weighted average between the prior location $\mathbf{y}_-^*$ and the first p time series values $\mathbf{y}_0$ which are adjusted by the vector **a** containing the information ('back-casting') of the AR model. The starting values will now be used to construct the n×(p+1) **X** matrix: $\mathbf{X}=[\mathbf{1}_T:\mathcal{L}_p\mathbf{y}^\circ]$ consists of the constant and the past of the left hand variable **y**.

The Gibbs sampler can be invoked with the starting values which are based on simple OLS estimates: $\mathbf{b}^{(0)}=\mathbf{b}_{OLS}=(\mathbf{X}'\mathbf{X})^{-1}\mathbf{X}'\mathbf{y}$, $\sigma^{2(0)}=\sigma^2_{OLS}$, and $\lambda^{(0)}=(\mathbf{b}_{OLS}-\mathbf{b}_*)'\mathbf{H}_*^{-1}(\mathbf{b}_{OLS}-\mathbf{b}_*)$, where the X matrix contains no starting values. As prior values for the starting values $\mathbf{y}_-$ we suggest to take the first observation, i.e. $\mathbf{y}_-^* = y_1\mathbf{1}_p$, and a tightness variance as well: $\mathrm{Var}(\mathbf{y}_-) = \mathrm{Var}(\mathbf{y})\mathbf{D}_*$, where $\mathbf{D}_* = \mathrm{diag}\ (1,1/2,\ldots,1/p)$. Note that a small $\mathbf{\Psi}_*$ puts more weight on the prior mean $\mathbf{y}_-^*$ and reduces sampling fluctuations for large p.

In case of centered time series, i.e. $\tilde{y}_t = y_t - \bar{y}$, t=1,...,n, which improves the convergence of the Gibbs sampler in time series models, we suggest to take as prior mean for the starting values $\mathbf{y}_-^* = (y_1 - \bar{y})/p\ \mathbf{1}_p$. We call this shrinkage prior for the starting values. This prior puts more weight on the prior starting values which should be close to zero. This approach has been found to produce posterior distributions for the starting values which are less variance inflated if the lag length becomes large.

2.2 Simulation Results

We used simulated time series to test the inference procedure. The function is a simple random walk series, i.e. $y_t=y_{t-1}+\epsilon_t$, t=1,...,n, where $\epsilon_t \sim N(0,0.01)$, with the starting values $y_0=1$. This random walk series with n=200 observations is called 'RW-200'. In Fig. 1, besides the Gibbs sampling output of the tightness model we show in the histograms of the marginal posterior distribution, the analytically calculated posterior distribution of an usual normal-gamma model - as solid line overlay plot - where we take a conjugate normal distribution $(\mathbf{b}, \sigma^{-2}) \sim \mathrm{NoGa}(\mathbf{b}_*,\lambda_*\mathbf{H}_*,\sigma_*^2,n_*)$, i.e. the interim step of a tightness prior distribution is omitted.

We can see that for the short time series, e.g. n=200, the prior information for λ and σ^2 is important for the convergence behavior. The Gibbs sampler produces an inflated posterior distribution if the prior information is not close to the true values, especially for short series. One should carefully choose the prior parameters. If the process converges to the 'wrong posterior distribution', the variances of the regression coefficients, the prediction and the starting values are larger than in the correct case. Thus we suggest a prior sensitivity analysis: We search over a range of prior values and take the one where the posterior variance is the smallest. If a time series is long enough, e.g. 1000 observations, the prior information is less important. The Gibbs sampler in a long series converges easily to the right parameters.

2.3 An Example for Swiss Macro-Economic Time Series

A simple B-AR(p) model for quarterly consumption (in real prices) is analysed for Switzerland for the period 1966.4-1988.4. We start the analysis with a univariate AR(4) process. The prior parameters are set to l_*=l,

λ_*=0.04, n_*=1, and σ_*^2=0.01. The convergence of the Gibbs sampler could be achieved very fast. The posterior distribution for the parameters of the univariate AR(4) model for real Swiss consumption is shown in the Figure 2. The first and third columns of the figure contain the descriptive statistics of the marginal distribution of the parameters (including the OLS values if available) while the second and fourth ones show the histograms.

3. Conclusions

The estimation of hyperparameters in econometric models like in the Bayesian VAR model is an unsolved problem, because it requires heavy computations in non-standard form. The paper has demonstrated that the Gibbs sampler solves the estimation problem of the hyperparameters quite elegantly if the tightness prior information can be specified by conjugate distributions. Furthermore, it is possible to extend the approach to hierarchical tightness and smoothness models, which will be reported in a separate paper. A further advantage is that all small sample parameter distribution and predictive distributions can be simulated.

References:

Chib S. (1993): Bayes regression with autoregressive errors: A Gibbs sampling approach. *Journal of Econometrics, 58, 275-294.*

Gelfand A.E., Smith A.F.M. (1990): Sampling based approaches to calculating marginal densities. *Journal of American Statistcal Association, 85, 398-409.*

Littermann R.B. (1986): A statistical approach to economic forecasting. *Journal of Business and Economic Statistics, 4, 1-24.*

Marriott J., Ravishanker N., Gelfand A.E., Pai J. (1992): Bayesian analysis of ARMA processes: Complete sampling based inference under full likelihood. mimeo, University of Connecticut.

Polasek W. (1993): Gibbs sampling in VAR models with tightness priors. mimeo, University of Basel.

Polasek W. (1994): Gibbs sampling in B-VAR models with latent variables. WWZ-discussion papers Nr. 9415, University of Basel.

Shiller R.J. (1973): A distributed lag estimator derived from smoothness priors. *Econometrica, 41, 775-788*

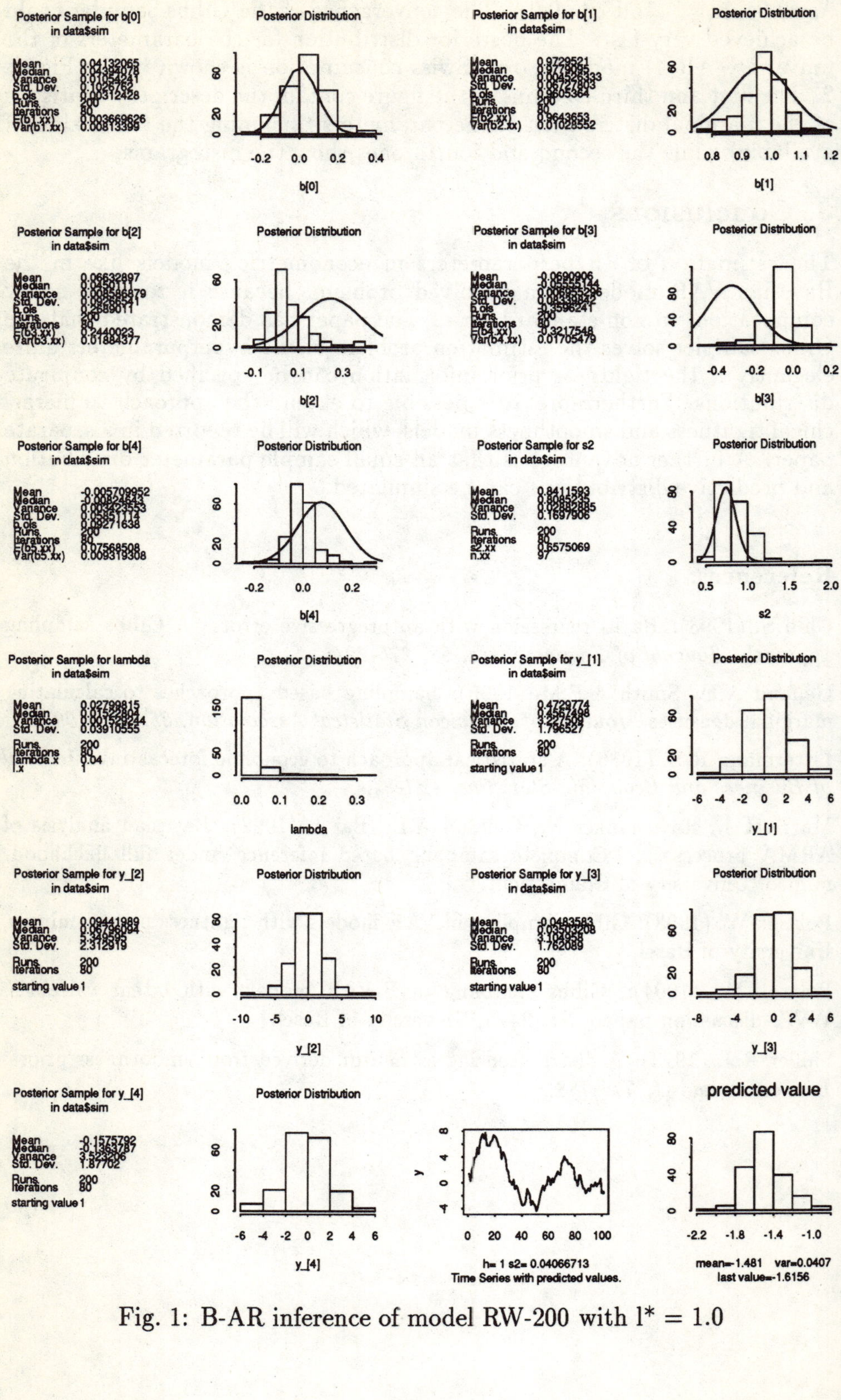

Fig. 1: B-AR inference of model RW-200 with $l^* = 1.0$

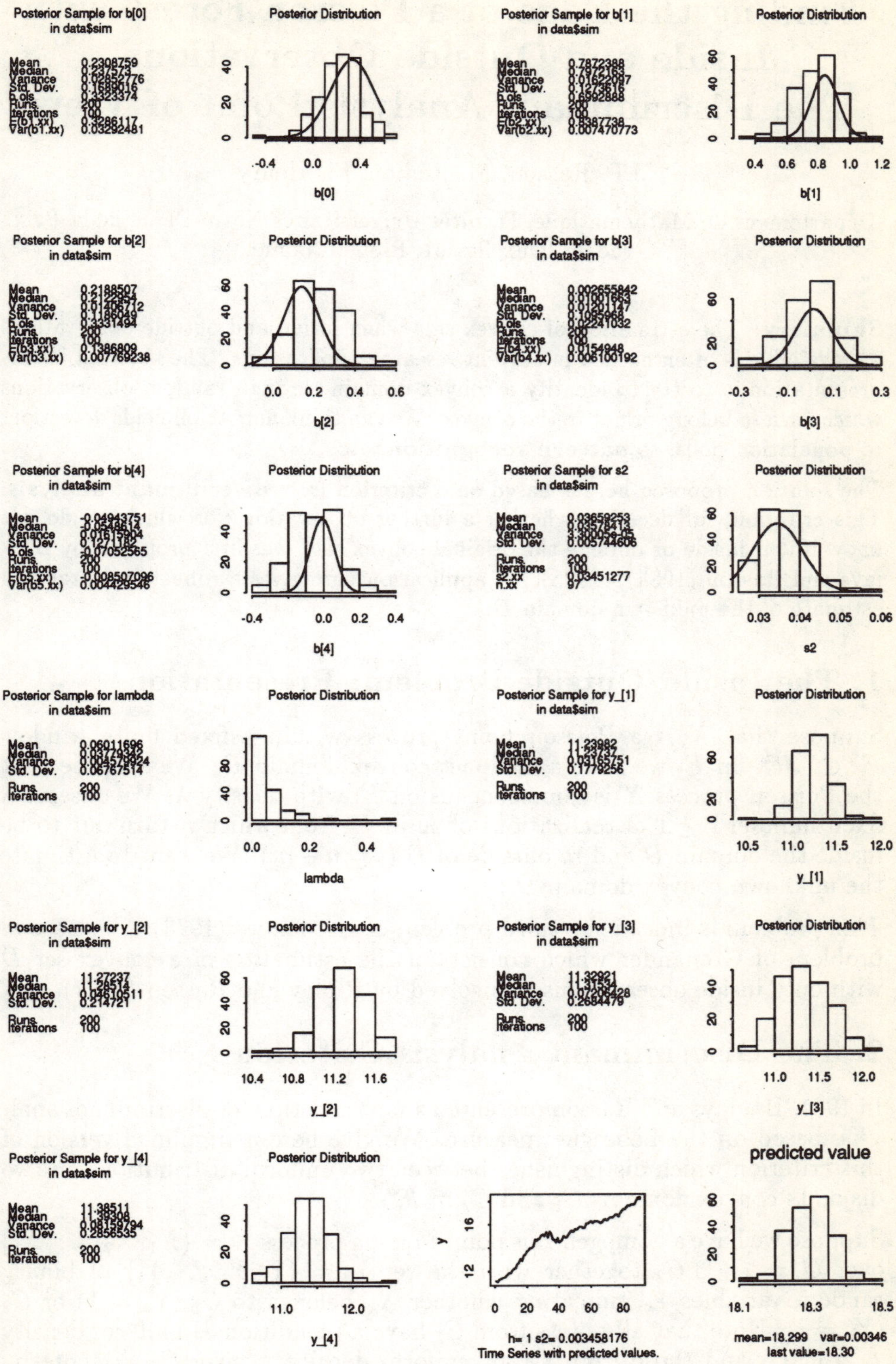

Fig. 2: B-AR inference of real Swiss consumption

Finding the Edge of a Poisson Forest with Inside and Outside Observations: The Discriminant Analysis Point of View

J.P. Rasson, M. Rémon, Fl. Henry

Département de Mathématique, Facultés Universitaires Notre-Dame de la Paix, rue de Bruxelles 61, B-5000 Namur

Summary: The estimation of convex sets when inside and outside observations are available is often needed in current research applications. The key idea in this presentation is to try to identify a convex domain through random observations which turn to belong or not to the convex. We can think here to oil fields detection, to population polls, to **pattern recognition**, etc.

The solution proposed here is based on a criterion from **discriminant analysis**. This criterion, for deciding whether a further observation [for which we do not know if it is inside or outside the original convex set], was first proposed by Baufays and Rasson (1984), (1985). Its application here gives a robust and practical estimate of the unknown domain D.

1. The Inside/Outside Problem: Presentation

Suppose that X is a Poisson point process within a fixed finite window $F \subset \mathbb{R}^d$. In F, we have a compact convex domain D. We suppose that the Poisson process X is homogeneous on F, with density λ. We observe a fixed number $t \geq 1$ of realizations of X in F, from which n turn out to be inside the domain D and m outside of D ($t = n + m$). We want to estimate the unknown convex domain D.

This problem is indeed the third problem of Grenander (1973). The second problem of Grenander which consists of the estimation of a convex set D with only inside observations was solved by Ripley and Rasson (1977).

2. The Discriminant Analysis Criterion

In 1984, Baufays and Rasson presented a new criterion for discriminant analysis, based on the Lebesgue measure. We give here a simplified version of this criterion which distinguishes between two uniform distributions on two disjoints convex domains C_1 and C_2 in $\mathbb{R}^d$.

Suppose we have a homogeneous point Poisson process $X = \{X_1, X_2, \ldots, X_t\}$ over $C = C_1 \cup C_2$ together with the set $Y = \{Y_1, Y_2, \ldots, Y_t\}$ of binary random variables Y_k indicating whether X_k belongs to C_1 ($Y_k = 1$) or C_2 ($Y_k = 2$). Note that all points from C_1 have a (conditional) uniform density f_1 over C_1 and those from C_2 an uniform density f_2 over C_2. Denote by $p_1 = \frac{m(C_1)}{m(C)}$ and $p_2 = \frac{m(C_2)}{m(C)}$ the probabilities that $Y_k = 1$ and $Y_k = 2$. Let

$(x, y) = (x_1, y_1, ..., x_t, y_t)$ be a training sample with a fixed number t of points $x_1, \ldots, x_t$ from $C_1 \cup C_2$ with indicators $y_1, \ldots, y_t$ We suppose, without any loss of generality, that $x_1, \ldots, x_n$ belong to C_1 and $x_{n+1}, \ldots, x_t$ to C_2. The problem is to decide if the realization x_0 of a new uniform point process X_0 over $C_1 \cup C_2$ belongs to C_1 or C_2.

Conditionally on n and m fixed, the likelihood for (x, y) is

$$\begin{aligned}
L(x, y) &= \prod_{i=1}^{n+m} \frac{1}{m(C)} \mathbb{1}_{[x_i \in C]} \\
&= \left(\frac{1}{m(C)}\right)^{n+m} \prod_{i:y_i=1} \mathbb{1}_{[x_i \in C_1]} \prod_{i:y_i=2} \mathbb{1}_{[x_i \in C_2]} \\
&= \left(\frac{1}{m(C)}\right)^{n+m} \prod_{i=1}^{n} \mathbb{1}_{[x_i \in C_1]} \prod_{i=n+1}^{n+m} \mathbb{1}_{[x_i \in C_2]} \\
&= \left(\frac{1}{m(C_1) + m(C_2)}\right)^{n+m} \mathbb{1}_{[H(x_1,...,x_n) \subseteq C_1]} \mathbb{1}_{[H(x_{n+1},...,x_{n+m}) \subseteq C_2]}
\end{aligned}$$

where $m(C)$ is the Lebesgue measure of C, $H(x_1, ..., x_n)$ is the convex hull based on $x_1, ..., x_n$, and $\mathbb{1}_{[A]} = 1$ if A is true and 0 otherwise.

The maximum likelihood estimators for C_1 and C_2 are $H(x_1, ..., x_n)$ and $H(x_{n+1}, ..., x_{n+m})$ respectively. Suppose a new point x_0 is observed, without knowing to which set C_1 or C_2 it belongs. Its likelihood is:

$$\begin{aligned}
L(x_0) &= \frac{1}{m(C)} \mathbb{1}_{[x_0 \in C]} \\
&= \frac{1}{m(C_1) + m(C_2)} \left(\mathbb{1}_{[x_0 \in C_1]} + \mathbb{1}_{[x_0 \in C_2]}\right) \\
&= \frac{p_1}{m(C_1)} \mathbb{1}_{[x_0 \in C_1]} + \frac{p_2}{m(C_2)} \mathbb{1}_{[x_0 \in C_2]} \\
&= p_1 f_1(x_0) + p_2 f_2(x_0)
\end{aligned}$$

This likelihood is the mixing of two uniform densities f_1 and f_2 over C_1 and C_2, with respective weights p_1 and p_2.

If a sample point x_0 belongs to C_1 [respectively to C_2], its likelihood becomes $p_1 f_1(x_0)$ [resp. $p_2 f_2(x_0)$]. Assuming equal costs of misclassification, the allocation rule which minimizes the Bayesian risk is (Hand, 1982, p. 88):

"allocate x_0 to C_1 if $p_1 f_1(x_0) > p_2 f_2(x_0)$
and to C_2 if $p_1 f_1(x_0) < p_2 f_2(x_0)$".

Unfortunately, the distribution densities f_1 and f_2 depend on the unknown convex sets C_1 and C_2. Baufays and Rasson (1984) suggested to use the maximum likelihood estimate of C_1 and C_2 with x_0 considered as a member of C_1 [res-pectively C_2] to compute $p_1 f_1(x_0)$ [resp. $p_2 f_2(x_0)$]. The proposed allocation rule is:

"allocate the new observation x_0 to C_1 if $p_1\widehat{f_1(x_0)} > p_2\widehat{f_2(x_0)}$ and to C_2 otherwise".

Therefore, x_0 will be allocated to C_1 if

$$\frac{1}{m(H(x_1,...,x_n,x_0))+m(H(x_{n+1},...,x_{n+m}))} > \frac{1}{m(H(x_1,...,x_n))+m(H(x_{n+1},...,x_{n+m},x_0))}$$

i.e. if

$$S_1(x_0) < S_2(x_0)$$

where

$$S_1(x_0) \equiv m(H(x_1,...,x_n,x_0)) - m(H(x_1,...,x_n))$$

and

$$S_2(x_0) \equiv m(H(x_{n+1},...,x_{n+m},x_0)) - m(H(x_{n+1},...,x_{n+m})).$$

$S_i (i = 1, 2)$ represents the increase in the Lebesgue measure of the maximum likelihood estimators of C_i when x_0 is considered as a member of C_i. See Fig. 1.

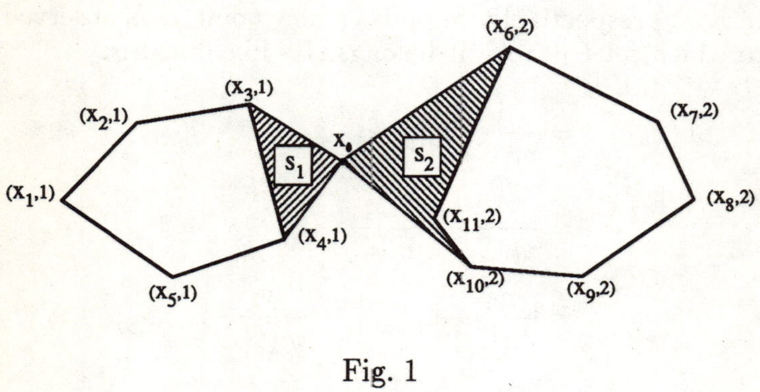

Fig. 1

As C_1 and C_2 are known to be disjoint domains, we have to exclude situations where x_0 is such that $H(x_1,...,x_n,x_0) \cap H(x_{n+1},...,x_{n+m}) \neq \emptyset$ [or respectively $H(x_1,...,x_n) \cap H(x_{n+1},...,x_{n+m},x_0) \neq \emptyset$]. When such a situation occurs, we assign to $S_1(x_0)$ [resp. $S_2(x_0)$] an arbitrary infinite value. The locus of points x_0 such that $p_1\widehat{f_1(x_0)} = p_2\widehat{f_2(x_0)}$, i.e. $S_1(x_0) = S_2(x_0)$, gives us the boundary of the allocation regions.

3. Application to the Inside/Outside Problem

The same idea is applied to the inside/outside problem, with D and $\bar{D}$ playing the role of C_1 and C_2. The major difference here is that $\bar{D}$ is no longer a convex set.

Here again we note $(x, y) = \{x_1, y_1, \ldots, x_{n+m}, y_{n+m}\}$ the realizations of the homogeneous Poisson process X and the labelling variable Y : $y_i = 1$ if $x_i \in D$ and $y_i = 2$ otherwise. Let $y_i = 1$ for $i = 1, ..., n$ and $y_i = 2$ for $i = n+1, ..., n+m$, without restricting generality.

Conditionaly on n and m fixed, the likelihood function for (x, y) is

$$
\begin{aligned}
L_D(x,y) &= \left(\frac{1}{m(F)}\right)^{n+m} \prod_{i=1}^{n} \mathbb{1}_{[x_i \in D]} \prod_{i=n+1}^{n+m} \mathbb{1}_{[x_i \in \bar{D}]} \\
&= \left(\frac{1}{m(D)+m(\bar{D})}\right)^{n+m} \mathbb{1}_{[H(x_1,\ldots,x_n) \subseteq D]} \, \mathbb{1}_{[J(x_{n+1},\ldots,x_{n+m}|x_1,\ldots,x_n) \subseteq \bar{D}]}
\end{aligned}
$$

where $J(x_{n+1}, ..., x_{n+m}|x_1, ..., x_n)$ is the "shadow" statistic defined in Hatchel, Meilijson and Nadas (1981). It is defined as:

$$
J(x_{n+1}, ..., x_{n+m}|x_1, ..., x_n) = \bigcup_{i:y_i=2} \{x_i + \lambda(x_i - b) \in I\!R^d | \lambda \geq 0, b \in H(x_1, ..., x_n)\}.
$$

$(H(.), J(.|.))$ is a minimal sufficient statistic for the estimation of D. See Fig. 2.

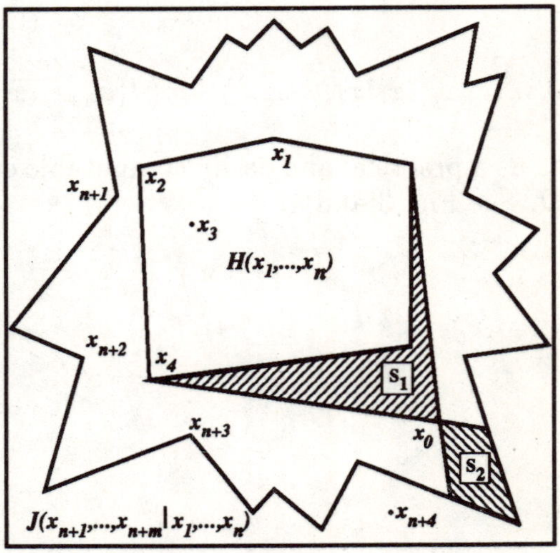

Fig. 2

It is known that $J(x_{n+1}, ..., x_{n+m}|x_1, ..., x_n)$ has similar properties as the convex hull statistic $H(x_1, ..., x_n)$. It is a consistent estimate of $\bar{D}$. It is robust with respect to small changes in the location of the data points. It satisfies the equivariance requirement and underestimates the Lebesgue measure of $\bar{D}$ in the same way as $m(H(x_1, ..., x_n))$ does for $m(D)$, ie

$$E[m(J(x_{n+1}, ..., x_{n+m}|x_1, ..., x_n))] = \left(1 - \frac{E[V_{m+1}]}{m+1}\right) m(\bar{D})$$

with $E[V_{m+1}]$ being the expected number of extreme points of $J(.|.)$ for $m+1$ observations in $\bar{D}$. See Ripley and Rasson (1977) and Rémon (1993).

The use of $H(x_1, ..., x_n)$ and $J(x_{n+1}, ..., x_{n+m}|x_1, ..., x_n)$ as estimates of the two unknown domains [here D and $\bar{D}$] in the criterion proposed by Baufays and Rasson is the key idea of this paper.

One gets then the following boundary of the regions allocating to D and $\bar{D}$. This is the set of points x_0 such that:

$$p_D \widehat{f_D}(x_0) = p_{\bar{D}} \widehat{f_{\bar{D}}}(x_0)$$

i.e.

$$\frac{1}{m(H(x_1,...,x_n,x_0))+m(J(x_{n+1},...,x_{n+m}|x_1,...,x_n))} = \frac{1}{m(H(x_1,...,x_n))+m(J(x_{n+1},...,x_{n+m},x_0|x_1,...,x_n))}$$

i.e.

$$S_1(x_0) = S_2(x_0)$$

where

$$S_1(x_0) \equiv m(H(x_1, ..., x_n, x_0)) - m(H(x_1, ..., x_n))$$

and

$$S_2(x_0) \equiv m(J(x_{n+1}, ..., x_{n+m}, x_0|x_1, ..., x_n)) - m(J(x_{n+1}, ..., x_{n+m}|x_1, ..., x_n)).$$

This boundary gives us a practical and easily computable estimate $\hat{D}$ for the unknown domain D. See Fig. 3 and 4.

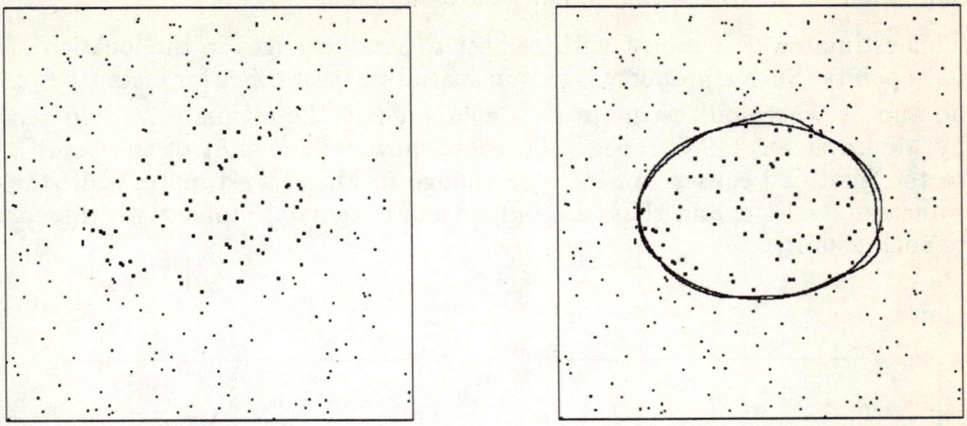

Fig. 3: Ellipsoidal D with $m(D) = 0.20$: $t = 300$ observations with $n = 68$ from D yield $\hat{D}$ with $m(\hat{D}) = 0.20$ and $m(D \cup \hat{D} \setminus D \cap \hat{D}) = 0.011$.

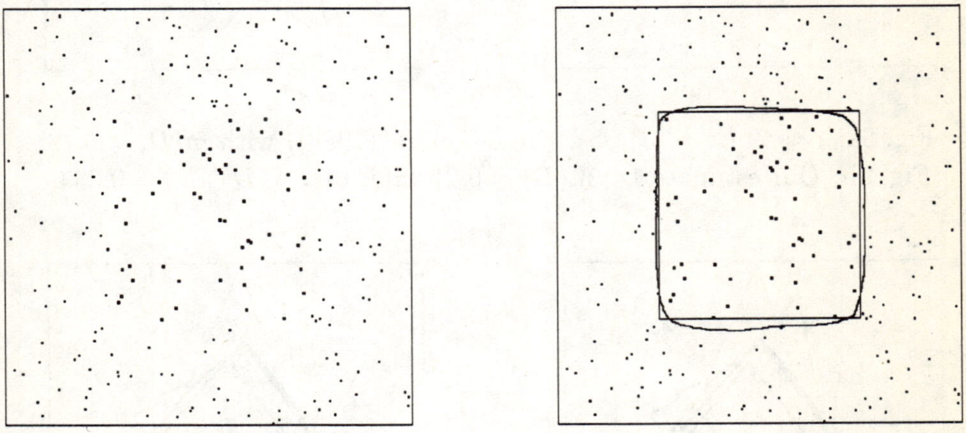

Fig. 4: Quadratic D with $m(D) = 0.25$: $t = 250$ observations with $n = 59$ from D yield $\hat{D}$ with $m(\hat{D}) = 0.26$ and $m(D \cup \hat{D} \setminus D \cap \hat{D}) = 0.018$.

4. Some Properties of $\hat{D}$

The estimator $\hat{D}$ yields a consistent estimation of D, as it is bound by two consistent estimators $H(.)$ and $J(.|.)$ of D. It has a piecewise continuous boundary. Unfortunately it happens not to be a convex set. On the other hand, this last feature turns out to be an advantage when a similar reasoning is applied to the estimation of non-convex domains.

This estimator $\hat{D}$ is robust with respect to small changes in the location of data points. Such a property is rare in spatial statistics. See for instance Fig. 5b and 5c where our estimate $\hat{D}$ is compared to the estimate $\tilde{D}$ proposed by Moore et al. (1988). Fig. 5d shows how a little shift of the point z to the right (z') causes considerable change to Moore's estimate, while the sufficient statistic, and thus also our estimate, remains stable with respect to such a shift.

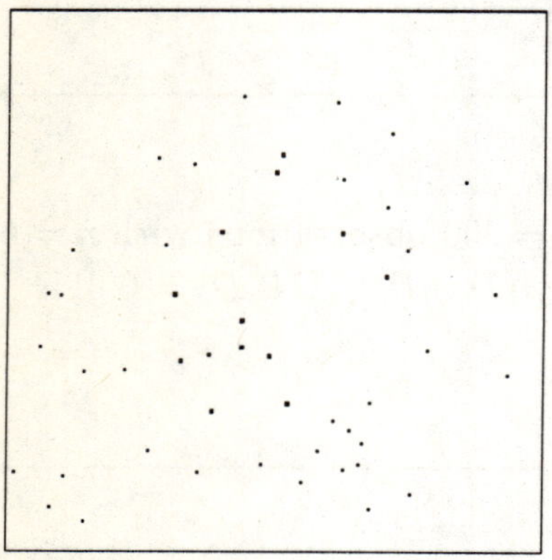

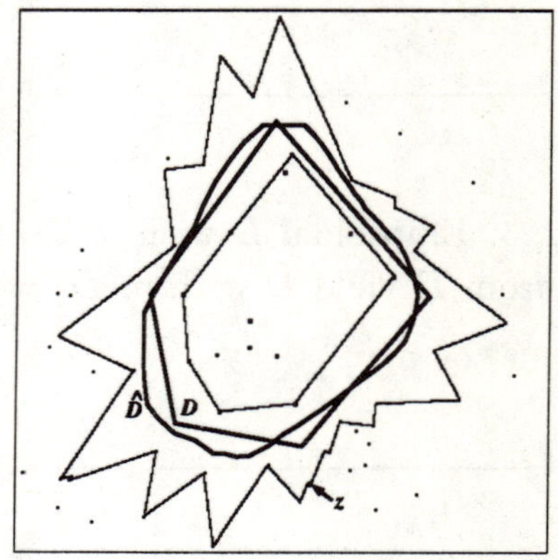

Fig. 5a: $t = 49$ ($n = 13$) obs. (Moore et al. (1988)) with $m(D) = 0.19$
Fig. 5b: Our estimate $\hat{D} : m(\hat{D}) = 0.21, m(D \cup \hat{D} \setminus D \cap \hat{D}) = 0.033$.

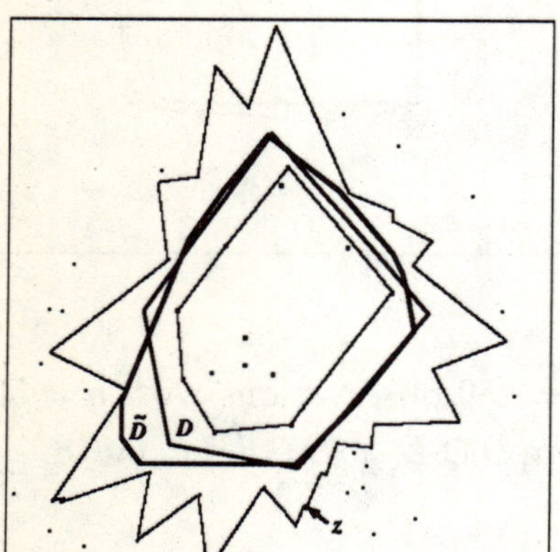

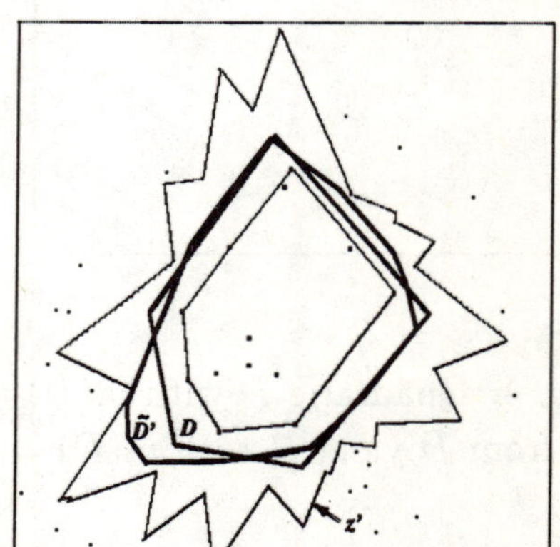

Fig. 5c: Moore's estimate $\tilde{D} : m(\tilde{D}) = 0.21, m(D \cup \tilde{D} \setminus D \cap \tilde{D}) = 0.028$.
Fig. 5d: Idem with z' : $m(\tilde{D}') = 0.20, m(D \cup \tilde{D}' \setminus D \cap \tilde{D}') = 0.029$.

A more theoretical solution to the original problem - with a convex domain - was worked out by Rasson, Rémon and Kubushishi (1993) and should be published soon. This theoretical solution is linked to Ripley and Rasson (1977). It requires, however, a large computer time and is less robust to small changes of the data.

5. Conclusion

Our estimate for D, based on a well known discriminant analysis criterion, seems to be a fruitfull tool for pattern recognition. It is indeed quite straightforward to generalize it to a non-homogeneous Poisson process, to non convex domains D or to higher dimensions.

References:

BAUFAYS, P., and RASSON, J.P. (1985): A new geometric discriminant rule. *Computational Statistics Quaterly, vol. 2, issue 1, 15-30.*

BAUFAYS, P., and RASSON, J.P. (1984): Une nouvelle règle de classement utilisant l'enveloppe convexe et la mesure de Lebesgue. *Statistique et Analyse des Données, vol. 9, no 2, 31-47.*

GRENANDER, U. (1973): Statistical geometry: a tool for pattern analysis. *Bulletin of the American Mathematical Society, vol. 79, 829-856.*

HACHTEL, G.D., MEILIJSON, I. and NADAS, A. (1981): The estimation of a convex subset of $\mathbb{R}^k$ and its probability content. IBM research report, Yorktown Heights N.Y.

HAND, D.J. (1982): Kernel Discriminant Analysis. Research studies press, England.

MOORE, M., LEMAY, Y. and ARCHAMBAULT, S. (1988): Algorithms to reconstruct a convex set from sample points. Computing Science and Statistics : Proceedings of the 20th Symposium on the Interface, Eds. E.J. Wegman, D.T. Gantz and J.J. Miller, ASA, Virginia, 553-558.

RASSON, J.P., REMON, M. and KUBUSHISHI, T. (1993): Finding the edge of a Poisson forest with inside and outside observations. Internal report 93/14, Dept. of Mathematics, FUNDP, Namur.

REMON, M. (1994): The estimation of a convex domain when inside and outside observations are available. *Supplemento ai Rendiconti del Circolo Matematico di Palermo, serie II, no 35, 227-235.*

RIPLEY, B.D., and RASSON, J.P. (1977): Finding the edge of a Poisson forest. *Journal of Applied Probability, 14, 483-491.*

Spatial Fibre and Surface Processes – Stereological Estimations and Applications

Konrad Sandau

Fachbereich Mathematik & Naturwissenschaften der FH Darmstadt,
Schöfferstr. 3 ,D-64295 Darmstadt, Germany

Summary: The aim of this paper is to give an introduction to the treatment of spatial fibre and surface processes especially under the aspects of modelling and stereology. In the last two decades mathematical tools were developed for modelling spatial fibres and surfaces as they occur in biology or other scientific disciplines. These tools also allow to quantify characteristics of the objects like length or surface area. In quantitative microscopy actually the objects are sectioned. Therefore the spatial characteristics have to be derived from plane sections. This is the fundamental problem treated in stereology, a mathematical discipline, collecting methods to relate the measurements on sections to the characteristics of three-dimensional structures. Several examples demonstrate in which way theory and applications of this kind can be combined passing the three stages 'modelling, stereology, statistics'.

1. Introduction

In a wide variety of scientific disciplines it is of interest to quantify spatial structures and in particular fibres and surfaces. In biology surfaces are borders of biological systems as organs or borders between different systems. The area of a surface may for example give relevant information about the physiological capacity. To enlarge the area, biological surfaces often have a very complex shape. A very famous example is the alveolar surface of the lung (Weibel, 1979). Other surfaces of the lung were also considered (Michel & Cruz-Orive, 1988), recently with very modern techniques and methods (Pache et al., 1993). Depending on objects and techniques the use of relative characteristics as 'surface density', meaning the surface area per unit volume, is preferred. Mayhew (1981) even investigates different kinds of surface densities of cell membranes. A very complex surface in the epitheliochorial placenta was considered by Hasselager (1986). However, the border between different phases is also of interest for geophysicists looking at the porous structure of rock material (Pape et al., 1987) or rock fracture surfaces (Riss & Gentier, 1989). Furthermore other material scientists (e.g. Wright & Karlsson, 1983) as well as soil scientists (Bodziony et al., 1993, Vogel et. al., 1993) are looking at surface densities. And it was an engineer in metallurgy, Saltykov (1945) who derived the well known stereological formula $S_V = 2 \cdot I_L$ even about 50 years ago. This formula was discovered independently by another five scientists of different disciplines in the following eleven years until 1956 (cf. Mattfeldt, 1990, p. 24).

In corresponding quantitative investigations fibres are characterized by length and length density. The objects considered in biological applications are e.g. transport systems like capillary networks (Mathieu et al., 1983; Mattfeldt & Mall, 1984) or dendrites of neurons (Howard et al., 1992). In agricultural sciences channels formed by roots of plants (Lang & Melhuish, 1970; Sandau, 1987a) or by animals (Vogel et al., 1993; Monestiez et al., 1993) are some objects of interest. Scientists, looking at metallurgical microstructures investigate e.g. dislocation lines ore grain edges (Gokhale, 1990).

Common to most of the cited objects is that they can not be observed spatially. Only sections or sometimes projections of thick slices are available, which are to be interpreted as two-dimensional images. In a lot of cases these images have been obtained by microscopes of different kind. In recent years also images have been got by other technical systems as e.g. tomographical systems. So we also consider a special aspect of quantitative microscopy, where measurements of the three-dimensional structure have to be taken from two-dimensional images. This is a mathematical problem – the problem of stereology. A more precise formulation of the definition of Stereology is given by Stoyan et al. (1987, p. 279) and shall be cited here:

"The objective of stereology is to draw inferences about the geometrical properties of three-dimensional structure when information is only available in some lower-dimensional form via linear probes, planar sections, or projections of thick slices."

The sciences using microscopes for quantitative information had a strong formative influence on the historical development of stereology (Miles, 1987 pp. 5-78). The cited definition above also reflects this. However, regarded as a mathematical discipline, stereology has its place between stochastical geometry and spatial statistics. In stochastic geometry the first important results concerning stereology of fibres and surfaces were Buffon's needle experiment described 1777 (cf. Kendall & Moran, 1963) and the theorem of Cauchy (cf. Santaló, 1976). The result of Buffon gave an answer to the question how to estimate the length of a plane curve, using a grid of parallel lines (Steinhaus, 1930). Moran (1966) calculated bounds of the variance of a corresponding estimator. The three-dimensional extensions to fibres and surfaces are obvious.

A fundamental work about stereological estimators and the correct randomization of test sets (test lines, test planes) was given by Miles & Davy (1976). The mathematical progress concerning spatial point processes (cf. Ripley, 1981) and of the random set theory (Matheron, 1975) initiated a new development in stereology. The objects 'fibres' or 'surfaces' are no longer considered as fixed objects. They are interpreted as realizations of a random variable. Then the object under study is imagined as having been generated according to a stochastic model (The mathematical definition of such a model is given below). These two approaches are called design-based approach and model-based approach, described and discussed in Jensen, et al. (1985). In the present paper we follow the model-based approach for fibres and surfaces. The definitions of models and a thorough treatment of the mathematical tools and derivations can be found in Stoyan et al. (1987).

In the next two sections the random variables 'fibre process' and 'surface process' are defined. Furthermore these processes are extended to weighted processes. The weight allows to characterize each point of a fibre or surface by an additional value. If e.g. a fibre system represents a plant root, then the weight might be the thickness of the plant root at each point. With these definitions most of the above cited applications can be modelled adequately. Length density and surface density are natural intensity parameters of the model. The modelling part and the statistical analysis are connected by an existence theorem, guaranteeing that a process has a joint distribution of direction and weight. To show stereological thinking we first realize that spatial processes induce plane processes on test lines and test planes. Then two aspects are picked out. In section 4. we derive some relations between the intensities of the spatial processes and the induced intensities of the processes on the test sets, following some classical methods of modern stereology. To this aim randomization schemes for the direction of the test sets are introduced. The second aspect deals with distributions of direction and weight. Section 5. illuminates the development of this aspect citing three investigations of this topic. In the last section a recently finished project is described in more detail. It should be emphasized that this last application especially was taken up to demonstrate that all such investigations pass the three stages 'modelling, stereology, statistics'.

2. Spatial Fibre Processes

Throughout the paper $\Re$ denotes the set of real numbers, $\mathbf{R}_n$ the system of Borel sets in $\Re^n$, $\mathbf{B}_n$ the system of bounded Borel sets in $\Re^n$ and ν_n shall be the Lebesgue measure on $[\Re^n, \mathbf{R}_n]$.

Mecke & Nagel (1980) define a fibre $\gamma \subseteq \Re^3$ by a biunique function $\gamma(\cdot) = \big(\gamma_1(\cdot), \gamma_2(\cdot), \gamma_3(\cdot)\big) : [0,1] \to \Re^3$ with continuously differentiable components and the norm of the vector $\dot{\gamma}(t)$, meaning the derivation of $\gamma(t)$, is positive for all $t \in (0,1)$. Then the set γ is given by $\gamma = \gamma([0,1])$. A fibre system F consists of an at most countable number of fibres where each set $B \in \mathbf{B}_3$ intersects only a finite number of fibres. The fibres of one fibre system shall have no inner points in common; only endpoints of different fibres may coincide. The fibre system F implies a Radon measure μ_F on $[\Re^3, \mathbf{R}_3]$, which measures the length $\mu_F(B)$ of the fibre system F in a set $B \in \mathbf{B}_3$ by adding up the lengths of all the intersections of the fibres of F with B. The set of all such fibre systems is denoted by $\mathcal{F}$ and $\mathbf{F}$ is the σ-algebra generated by all sets $\{F \in \mathcal{F} \mid \mu_F(B) < x\}$, where $x \in \Re$ and $B \in \mathbf{B}_3$.

A random variable Φ defined on the space $[\mathcal{F}, \mathbf{F}]$ with the distribution P_Φ is called fibre process. Then P_Φ is a probability measure on $[\mathcal{F}, \mathbf{F}]$. P_Φ is invariant concerning to shifts in $\Re^3$ the fibre process is called stationary. For a stationary fibre process Φ the expected length in a set $B \in \mathbf{B}_3$ is equal to the expected length in a set B_x, which is the set B translated with $x \in \Re^3$, that means

$$\mathbf{E}\mu_\Phi(B) = \int \mu_F(B) P_\Phi(dF) = \mathbf{E}\mu_\Phi(B_x). \tag{1}$$

This property implies, that the expectation is proportional to the 'volume' of B, i.e.

$$\mathbf{E}\mu_\Phi(B) = L_V \cdot \nu_3(B) \tag{2}$$

where L_V is the intensity of the process, which is called length density, and the equation

$$L_V = \int \mu_F([0,1]^3) P_\Phi(dF) \tag{3}$$

shows, that the mean length per unit volume is a good characterization of L_V.

At almost all points of a fibre system F well defined tangent directions exist, because the components of $\gamma(.)$ are differentiable. A direction is an element of the set $\mathcal{H}$, the set of all spatial directions of lines h through the origin (or equivalently the set of points of the (upper) hemisphere of the unit sphere). A related σ-algebra $\mathbf{H}$ is generated by the sets $\{H \subseteq \mathcal{H} \,|\, \cup_{h\in H} h \in \mathbf{R}_3\}$. The directions of the tangents along all F$\in \mathcal{F}$ weighted by P_Φ can be described by a distribution R of the directions on $[\mathcal{H}, \mathbf{H}]$, sometimes called the rose of directions. To get R by an integral representation a function $h_F : \Re^3 \to \mathcal{H}$ is defined giving the direction $h_F(x)$ of the tangent if $x \in F$ and the tangent exists and $h_F(x) = v$ otherwise, where v is the vertical direction. Following Mecke & Stoyan (1980) and Mecke & Nagel (1980) there exists a uniquely defined distribution R, so that

$$\iint u(x, h_F(x)) \mu_F(dx) P_\Phi(dF) = L_V \iint u(x,h) R(dh) dx \tag{4}$$

holds for all measurable functions u:$\Re^3 \times \mathcal{H} \to [0, \infty)$.

It should be noticed, that a stationary fibre process is called isotropic, if the distribution P_Φ is invariant to rotations. Then the rose of directions R must be invariant to rotations and therefore R must be the uniform distribution on $[\mathcal{H}, \mathbf{H}]$.

A natural extension of a fibre process is a weighted fibre process. The weight allows to characterize each point of a fibre by an additional value. If e.g. a fibre system represents a plant root, then the weight might be the thickness of the plant root at each point. Another example of an interesting weight is the curvature of the fibre at each point (if it exists). For simplicity, we only consider real-valued weights here. But generalizations are possible. By the way, even the direction of the tangent in a point of the fibre could be interpreted as a weight.

The formal way to extend fibre processes to weighted fibre processes starts with the definition of a weighted fibre by $\gamma_w(.) = \big(\gamma_1(.), \gamma_2(.), \gamma_3(.), w(.)\big)$: [0,1]$\to \Re^4$, where $w(.)$ is the weight function and $\gamma_i(.)$, i$= 1,2,3$ are defined as above. The fibre system is constructed as in the unweighted case and also the length measure μ_{F_w} is not affected by the weight. But finer measures $\mu_{F_w(W)}$ can be defined, looking only to that part of F having weights in $W \in \mathbf{R}_1$. Then a set B$\in \mathbf{B}_3$ is measured by assigning to B the measure $\mu_{F_w(W)}(B)$, which gives the sum of the lengths of all the intersections of B with fibres of this part of F. This requires a finer σ-algebra $\mathbf{F}_w$, but it can be defined in the same way as $\mathbf{F}$ above using measures $\mu_{F_w(W)}$, $W \in \mathbf{R}_1$. A random variable Φ_w defined on the space $[\mathcal{F}_w, \mathbf{F}_w]$ with the distribution P_{Φ_w} is called now a weighted fibre process and P_{Φ_w} is a probability measure on $[\mathcal{F}_w, \mathbf{F}_w]$.

Furthermore we define similarly to the function h_F a function $w_F : \Re^3 \to \Re$, assigning to each $x \in \mathcal{R}^3$ the weight of F_w at the point x if $x \in F_w$ and $w_F(x) = 0$ otherwise. Then there exists a uniquely defined distribution M, which is a joint distribution of h and w, so that

$$\iint u(x, w_F(x), h_F(x))\mu_{F_w}(dx)P_{\Phi_w}(dF_w) = L_V \iint u(x, w, h)M(d(w, h))dx \tag{5}$$

holds for all measurable functions u:$\Re^3 \times \Re \times \mathcal{H} \to [0, \infty)$.

To illustrate the importance of a joint distribution M in an example we consider again weights describing the thickness of plant roots. We can imagine that roots more vertically orientated tend to be thicker than those more horizontally orientated. Such dependencies can also be quantified by a joint distribution M.

3. Spatial Surface Processes

In the same way as the spatial fibre processes spatial surface processes can be defined, starting with a surface $\rho \subseteq \Re^3$ defined by a biunique function $\rho(.) = \big(\rho_1(.), \rho_2(.), \rho_3(.)\big) : [0, 1]^2 \to \Re^3$ fulfilling conditions corresponding to those of the spatial fibres as given above. Then a surface system S consists of an at most countable number of surfaces where each set B$\in \mathbf{B}_3$ intersects only a finite number of surfaces and the surfaces of one surface system shall have no inner points in common. A different definition is often used (Pohlmann et al., 1981; Stoyan et al., 1987) by considering $\mathbf{K}$, the class of all finite unions of convex compact subsets of $\Re^3$ (cf. Matheron, 1975). Then a surface system is a set S which can be represented as the surface of a set A$\subseteq \Re^3$, having the property that A$\cap K \in \mathbf{K}$ for any $K \in \mathbf{K}$.

The surface system S implies a Radon measure μ_S on $[\mathcal{R}^3, \mathbf{R}_3]$, which measures the area $\mu_S(B)$ of the surface system S in a set B$\in \mathbf{B}_3$. The set of all such surface systems is denoted by $\mathcal{S}$ and $\mathbf{S}$ is the σ-algebra generated by all sets $\{S \in \mathcal{S} \mid \mu_S(B) < x\}$, where $x \in \Re$ and B$\in \mathbf{B}_3$. A random variable Ψ defined on the space $[\mathcal{S}, \mathbf{S}]$ with the distribution P_Ψ is called surface process and P_Ψ is an accompanying probability measure on $[\mathcal{S}, \mathbf{S}]$. If the distribution P_Ψ is invariant concerning to shifts in $\Re^3$ the surface process is called stationary. A stationary surface process Ψ also has an intensity S_V, which is called here surface density and which is given by

$$S_V = \int \mu_S([0,1]^3) P_\Psi(dS), \tag{6}$$

and gives the factor of the proportionality

$$\mathbf{E}\mu_\Psi(B) = S_V \cdot \nu_3(B). \tag{7}$$

Because of the assumptions we have made, the surface system S has at almost all points a well defined direction described by the normal direction of the surface. This implies the existence of a uniquely defined distribution R of directions on $[\mathcal{H}, \mathbf{H}]$ (rose of directions). Pohlmann et al. (1981) proved, that

$$\iint u(x, h_S(x))\mu_S(dx)P_\Psi(dS) = S_V \iint u(x,h)R(dh)dx \tag{8}$$

holds for all measurable functions u:$\Re^3 \times \mathcal{H} \to [0,\infty)$, where the function $h_S : \mathcal{R}^3 \to \mathcal{H}$ gives the direction $h_S(x)$ of the normal if $x \in S$ and the normal exists and $h_S(x) = v$ otherwise. For an isotropic stationary surface process again the rose of direction R is a uniform distribution on $[\mathcal{H}, \mathbf{H}]$.

The extension from surface processes Ψ to weighted surface processes Ψ_w shall not be pointed out here in detail. The concept of weighted fibre processes has demonstrated the principles. An example of a weighted surface process will be given in sections 5. and 6. But it should be stressed, that a theorem and formula corresponding to eq.(5), also given in Pohlmann et al. (1981), guarantees the existence of a uniquely defined distribution M, which is a joint distribution of direction h and weight w. The importance of this theorem is caused in linking the part of modelling the spatial structure with the world of statistics.

4. Intensities on Test Planes and Test Lines

Stationary spatial processes induce stationary processes on intersecting test planes. A spatial fibre process induces a point process, a surface process induces a fibre process. Furthermore, a surface process also induces a point process on a test line. Stationarity guarantees the existence of intensities of the induced processes. The induced intensities can be observed and measured on the test planes and lines. This shows the necessity to know a relationship between the intensities of the spatial process and its induced process. Such a relationship shall be derived here for the spatial fibre process Φ with finite length density L_V. The test plane $T(g)$ is orientated by their normal direction $g \in \mathcal{H}$, but the location of $T(g)$ is not considered because Φ is stationary. The induced point process $\Xi(g)$ on $T(g)$ has the intensity $P_A(g)$, which can be related with L_V by

$$P_A(g) = L_V \cdot \mathcal{F}_R(g). \tag{9}$$

$\mathcal{F}_R : \mathcal{H} \to \Re$ means the Buffon transformation (Beneš, 1993) given by

$$\mathcal{F}_R(g) = \int_{\mathcal{H}} |\cos(g,h)| R(dh), \tag{10}$$

where (g, h) denotes the angle between the two directions and R is the rose of direction of the process Φ. This formal concept concentrates the results about Buffon's needle experiment (cf. Kendall & Moran, 1963). In particular an isotropic process implies that $\mathcal{F}_R(g) = \frac{1}{2}$ is constant and we get the well known stereological formula

$$L_V = 2 \cdot P_A. \tag{11}$$

If the process Φ is not isotropic this can be compensated by randomizing the direction g. That means a random direction G is taken uniformly distributed with distribution U on $[\mathcal{H}, \mathbf{H}]$ and the test plane $T(g)$ is a realization of the randomization. The mean intensity can be calculated via

$$P_A = \mathbf{E}(P_A(G)) = \int_{\mathcal{H}} P_A(g)U(dg) = L_V \int_{\mathcal{H}} \mathcal{F}_R(g)U(dg) = \frac{1}{2}L_V. \tag{12}$$

We get the same formula as in eq. (11),

$$L_V = 2 \cdot P_A \tag{13}$$

but in a different, more general meaning. This randomization procedure of the test plane is often called IUR-sampling, where IUR means *Isotropic Uniform Random*, but it should be stressed, that IUR-sampling also considers the randomization of location in the case of bounded reference spaces (cf. Miles & Davy, 1976). It is difficult to realize an IUR-sampled intersection plane e.g. in microscopical applications and even a special procedure was developed for this (Mattfeldt et al., 1990).

It is obvious that methods, where the randomization of the intersection plane is more simple to handle are more popular. Such a method was proposed by Baddeley (1984; Baddeley et al., 1986), for estimating the (finite) surface density S_V of a spatial surface process Ψ. A test line $T_1(g)$ with direction $g \in \mathcal{H}$ intersects the surface process. The intersection $\Psi \cap T_1(g)$ is a point process Ξ_1 on $T_1(g)$ having the point intensity $I_L(g)$. Again the fixed direction g is replaced by a random direction G uniformly distributed on $[\mathcal{H}, \mathbf{H}]$ and the same consideration as above using the Buffon transformation lead to another very famous stereological equation

$$S_V = 2 \cdot \mathbf{E}(I_L(G)) = 2 \cdot I_L. \tag{14}$$

The considerations yielding eq.(13) and eq.(14) demonstrate the often discernible duality between $fibre - plane$ and $surface - line$ very clearly. The practical relevance of this method is among others the procedure to get a randomized realization $T(g)$ of the line. In a first step a plane is taken by VUR-sampling (VUR means *Vertical Uniform Random*). To do this a vertical direction $v \in \mathcal{H}$ is fixed and the direction h of a test plane $T(h)$ has to be orthogonal to v. Each of the remaining horizontal directions h is uniquely defined by an angle $\varphi \in [0, \pi)$ and $T_v(\varphi)$ is the corresponding vertical plane. The randomization is done by using a uniformly distributed angle variable φ on $[0,\pi)$. The intersection $\Psi \cap T_v(\varphi)$ shows a fibre process on the plane. Now in a second step a line $T_1(\vartheta)$ is taken in $T_v(\varphi)$, defined by the angle ϑ between the line and the vertical direction v. Again the angle ϑ is replaced by a random angle variable ϑ on $[0,\pi)$, but this variable has now a sine-weighted distribution. The pair (ϑ, φ) defines a uniformly distributed random variable on $[\mathcal{H}, \mathbf{H}]$. A further idea is, to replace the randomized line on $T_v(\varphi)$ by a curve, of which tangents are sine-weighted distributed. Cycloids have this property (Baddeley et al., 1986).

A third method shall be cited (Gokhale, 1990), where the intensity L_V of a spatial fibre process Φ can be determined by the orthogonal projections of vertical slices. That means, the VUR-sampling of planes is done as described above, but now the planes are slices $T_v(\varphi, d)$ and have the constant thickness d in all directions. The orthogonal projection of $\Phi \cap T_v(\varphi, d)$ induces a fibre process $\Phi_p(\varphi, d)$ on the projection plane with intensity $L_A(\varphi, d)$. The intersection of $\Phi_p(\varphi, d)$ with cycloids (turned 90°) yields a point process with intensity I_L on the cycloids. These cycloids can be imagined as the projections of cycloidal shaped surfaces in the slice. Then again we see the above mentioned duality between fibre and surface and the result

$$L_V = \frac{2}{d} \cdot I_L \tag{15}$$

becomes comprehensible.

Obviously estimators of intensities can be developed based on formulae like eqs. (13), (14), (15). Then the estimators have the advantage to be unbiased. But a second important characteristic of an estimator is its variance. Because the variance generally depends on the shape of fibres and surfaces it is difficult to get an estimate of the variance or other second order characteristics. Nevertheless some results were found in recent years especially for the variance of estimators of L_V (Vedel-Jensen & Kieu, 1992; Beneš, 1994; Chadœuf & Beneš, 1994).

Finally it seems worth to remark, that the described methods also can be used in a variant way to quantify the size of spatial objects absolutely. Two methods shall be mentioned here. Cruz-Orive & Howard (1991) estimated the length of a spatial curve by vertical projections. The area of a spatial surface can be estimated by intersections with lines of randomized direction, but systematically orientated in two or three orthogonal directions, where the last method is called 'spatial grid' (Sandau, 1987b; Hahn & Sandau, 1989; Sandau & Hahn, 1994).

5. Distributions of Directions and Weights

In sections 2. and 3. we have seen that weighted fibre and surface processes can be characterized by their uniquely existing joint distributions of direction and weight. These distributions induce observable distributions on test sets and the essential question is: is it possible to reconstruct the joint or at least the marginal distributions of direction and weight from the distributions on the test sets? A classical result for unweighted fibre processes was proved by Mecke & Nagel (1980) who consider the rose of direction of a spatial fibre process. We saw in section 4. that a spatial fibre process induces point processes on planes $T(g)$, having intensities $P_A(g)$, $g \in \mathcal{H}$. Therefore $P_A(g)$ represents a distribution on $[\mathcal{H}, \mathbf{H}]$, not necessarily a probability distribution, which is obviously determined by the Buffon transformation (eq. (10)). Mecke & Nagel (1980) have demonstrated a calculation to get the rose of direction R if $P_A(g)$ is known.

Assuming that direction and weight are stochastically independent random variables the distributions can be treated separately. One of the first considerations of this kind was given by Jensen et al. (1979). The distribution of a biological membrane (human glomerular basement membrane) was derived from orthogonal intercepts on IUR-sampled test planes. The membrane is modelled by a stationary spatial surface process with the weight 'thickness'. The IUR-sampling scheme allows to compensate the lack of information due to the fact that orthogonal intercepts in the plane do not show the true thickness of the membrane. The derivation culminates in an integral equation determining the distribution of the thickness. Two numerical solutions are proposed. An unfolding procedure using grouped data (cf. Weibel, 1980) and a parametric approach where the distribution of the weight is assumed to be in the class of the generalized gamma distributions.

Another weight was considered by Stoyan & Gerlach (1987). They looked at the weight 'curvature radius' of an isotropic stationary spatial fibre process. Isotropy guarantees the stochastical independence between direction and weight and uniformly distributed directions. It remains to determine the distribution R_3 of the curvature radius. The orthogonal projection of the intersection of the process with slices (thick planes) of a constant thickness induces a plane fibre process. The distribution R_2 of the curvature radius of this plane process is observable. Stoyan & Gerlach (1987) demonstrate how to relate the distributions R_3 and R_2 by an integral equation. The integral equation was numerical solved by an unfolding procedure. It should be noticed, that, owing to the isotropy, the directions of the slices must not be randomized.

The next section deals with an example where the joint distribution of direction and weight is considered, and where the dependency is implemented in a special way using a parametric approach.

6. Stereological Treatment of Width and Direction of Fissures in Soil

In pedobiology fissures of soil are considered in vertical sections. To get information about the properties of the soil the joint distribution of spatial direction and width is of interest. The fissures are modelled as a stationary weighted surface process Ψ_w with the weight 'width'. Stationarity guarantees a joint distribution. This distribution shall have a probability density function (pdf for short) $M(\tau, \vartheta, \varphi)$ of the spatial direction (in polar coordinates (ϑ, φ), where ϑ is the colatitude) and the width τ. The soil scientists usually sample vertical sections, where the vertical direction means vertical to the ground. This restriction initiated our investigations and the results are described in Sandau (1993), Sandau & Vogel (1993) and Vogel et al. (1993). A parametric approach is used and as a further assumption, adequate to this application, the rose of direction shall be rotational symmetric around the vertical direction. The test set T_1 is a vertical line because this choice do not require any further assumptions concerning the randomization of the vertical planes taken by the pedobiologists. The surface process induces a weighted point process on T_1 with the two weights 'length ℓ of the orthogonal intercept' through the intersection of the fissure with the section plane and 'inclination angle ϑ'' between the vertical direction and the normal of the intersection of the fissure in the section plane. The joint distribution of these two weights is observable given that Ψ_w hits T_1 ($\Psi_w \uparrow T_1$ for short) and shall have the pdf $N(\ell, \vartheta' | \Psi_w \uparrow T_1)$. Now we have to derive a relation between M and N. In a first step we get the conditional pdf $M(\tau, \vartheta, \varphi \mid \Psi_w \uparrow T_1)$. It is easy to realize that

$$M(\tau, \vartheta, \varphi | \Psi_w \uparrow T_1) = c \cdot M(\tau, \vartheta, \varphi) \cdot \cos \vartheta, \tag{16}$$

where c is a norming constant. In a second step this result is to transform to the observable variables ℓ and ϑ' which can be done by geometrical considerations introducing a (blind) third component ξ. This transformation maps $(\tau, \vartheta, \varphi)$ biuniquely to (ℓ, ϑ', ξ) and with the help of this map we get the pdf $N(\ell, \vartheta', \xi \mid \Psi_w \uparrow T_1)$ from $M(\tau, \vartheta, \varphi \mid \Psi_w \uparrow T_1)$. In a third step we take the marginal pdf

$$N(\ell, \vartheta' | \Psi_w \uparrow T_1) = \int N(\ell, \vartheta', \xi | \Psi_w \uparrow T_1) d\xi, \tag{17}$$

which determines the observable distribution on the test line. Now we introduce a parametric family for the distribution of the weights. To this aim we use the product representation

$$M(\tau, \vartheta, \varphi) = M_b(\vartheta) \cdot M_w(\tau | \vartheta). \tag{18}$$

The pdfs are independent of φ because M is rotational symmetric. For the rose of direction we propose an axial version of the Bingham-Mardia distribution (Bingham & Mardia, 1978), which is a small circle distribution and has the pdf

$$M_b(\vartheta) = c \cdot \mathbf{e}^{-\kappa(\cos(2\vartheta) - \cos(2\vartheta_0))^2} \cdot \sin \vartheta, \tag{19}$$

where $\vartheta_0 \in [0, \frac{\pi}{2}]$ is the preferred colatitude, $\kappa \in \Re$ a parameter of concentration and c$\in \Re^+$ a norming constant. The conditional distribution of the width shall be taken from the family of the lognormal distributions with the pdf

$$M_w(\tau | \vartheta) = \frac{1}{\sqrt{2\pi} \cdot \sigma\tau} \cdot \mathbf{e}^{\frac{1}{2\sigma^2}(\ln \tau - a(\vartheta))^2}, \tag{20}$$

having the moments $\mathbf{E}(T|\vartheta) = \mathbf{e}^{a(\vartheta)} \cdot \mathbf{e}^{\frac{\sigma^2}{2}}$ and $Var(T|\vartheta) = \mathbf{E}^2(T|\vartheta) \cdot (\mathbf{e}^{\sigma^2} - 1)$. The dependency between τ and ϑ is realized by $\mathrm{a}(\vartheta) = \mathrm{b} + \mathrm{a} \cdot \cos\vartheta$, where a and b are real constants with $\mathrm{b} \in \Re^{+} and$ $\mathrm{a} < b$. This relation includes as a special case the independence (a= 0). The relation emerges from the idea that in soil sciences a standard width b of a fissure becomes smaller by the pressure of the material above (a negative) and the resulting power along the normal of the fissure is therefore proportional to $\cos\vartheta$.

Combining M_b and M_w the distribution M is determined by five parameters. These parameters also determine the observable pdf $N(\ell, \vartheta' \mid \Psi_w \uparrow T_1)$. The Chi-square method of goodness of fit is proposed to estimate these five parameters. In the paper of Sandau & Vogel (1993) special problems as e.g. the problem of the starting values are discussed in detail. To improve the estimation procedure the same considerations are repeated for a horizontal test line. This is especially important if the surface process is vertical orientated as e.g. the surface of gorges or ravines. The results of the measurements on vertical and horizontal test lines are then combined in one estimation procedure, taking into account the structure of the surface process.

References:

BADDELEY, A.J. (1984): Vertical sections. In: Ambartzumian & Weil (Eds.): *Stochastic Geometry, Geometric Statistics, Stereology.* Teubner, Leipzig, 43-52.

BADDELEY, A.J., GUNDERSEN, H.J.G., CRUZ-ORIVE, L.M. (1986): Estimation of surface area from vertical sections. *J. Microsc., 142: 259-276.*

BENEŠ, V. (1993): On anisotropic sampling in stereology. *Acta Stereol., 12/2: 185-190.*

BENEŠ, V. (1994): On second-order formulas in anisotropic stereology. *Adv. in Appl. Prob.*, submitted.

BINGHAM, C., MARDIA, K.V. (1978): A small-circle distribution on the sphere. *Biometrika, 65: 379-389.*

BODZIONY, J., KONSTANKIEWICZ, K., MLYNARCZUK, M., RATAJCZAK, T. (1993): Evaluation of the pore structure of soil specimens. *Acta Stereol., 12/2: 243-248.*

CHADŒUF, J., BENES, V. (1994): On some estimation variances in spatial statistics. *Kybernetika, 3,* to appear.

CRUZ-ORIVE, L.M., HOWARD, C.V. (1991): Estimating the length of a bounded curve in three dimensions using total vertical projections. *J. Microsc., 163: 101-113.*

GOKHALE, A.M. (1990): Unbiased estimation of curve length in 3-D using vertical slices. *J. Microsc., 159: 133-141.*

HAHN, U., SANDAU, K. (1989): Precision of surface area estimation using spatial grids. *Acta Stereol, 8/2: 425-430.*

HASSELAGER, E. (1986): Surface exchange area of the porcine placenta: morphometry of anisotropic interdigitating microvilli. *J. Microsc., 141: 91-100.*

HOWARD, C.V., CRUZ-ORIVE, L.M., YAEGASHI H., (1992): Estimating neuron dendritic length in 3D from total vertical projections and from vertical slices. *Acta Neurol Scand. Suppl., 137: 14-19.*

JENSEN, E.B., BADDELEY, A.J., GUNDERSEN, H.J.G. SUNDBERG, R. (1985): Recent trends in stereology. *Int. Stat. Review, 53: 99-108.*

JENSEN, E.B., GUNDERSEN, H.J.G., OSTERBY, R. (1979): Determination of membrane thickness distribution from orthogonal intercepts. *J. Microsc., 115: 19-33.*

KENDALL, M.G. & MORAN, P.A.P. (1963): *Geometrical Probability.* Griffin, London.

LANG, A.R.G. & MELHUISH, F.M. (1970): Lengths and diameters of plant roots in non-random populations by analysis of plane surfaces. *Biometrics, 26: 421-431.*

MATHERON, G. (1975): *Random Sets and Integral Geometry.* J. Wiley & Sons, New York.

MATHIEU, O., CRUZ-ORIVE, L.M., HOPPELER, H., WEIBEL, E.R. (1983): Estimating length density and quantifying anisotropy in skeletal muscle capillaries. *J. Microsc., 131: 131-146.*

MATTFELDT, T. (1990): Stereologische Methoden in der Pathologie. G. Thieme Verlag, Stuttgart.

MATTFELDT, T., MALL, G. (1984): Estimation of length and surface of anisotropic capillaries. *J. Microsc., 135: 181-190.*

MATTFELDT, T., MALL, G., GHAREHBAGHI, H. (1990): Estimation of surface area and length with the orientator. *J. Microsc., 159: 301-317.*

MAYHEW, T.M. (1981): On the relative efficiencies of alternative ratio estimators for morphometric analysis of cell membrane surface features. *J. Microsc., 122: 7-14.*

MECKE, J., NAGEL, W. (1980): Stationäre räumliche Faserprozesse und ihre Schnittzahlrosen. *J. Information Processing and Cybernetics EIK, 16: 475-483.*

MECKE, J., STOYAN, D. (1980): Formulas for stationary planar fibre processes I - general theory. *Statistics, Math. Operationsforschung und Statistik, 11: 267-279.*

MICHEL, R.P., CRUZ-ORIVE, L.M. (1988): Application of the Cavalieri principle and vertical sections method to lung: estimation of volume and pleural surface area. *J. Microsc., 150: 117-136.*

MILES, R.M. (Ed.) (1987): Twenty-five years of stereology. *Acta Stereol., 6/II,* Ljubljana.

MILES, R.E., DAVY, P. (1976): Precise and general conditions for the validity of a comprehensive set of stereological fundamental formulae. *J. Microsc., 107:*

211-226.

MONESTIEZ, P., KRETZSCHMAR, A., CHADŒUF, J. (1993): Modelling natural burrow systems in soil by fibre process: Monte-Carlo test on independence of fibre characteristics. *Acta Stereol. 12/2, 237-242.*

MORAN, P.A.P. (1966): Measuring the length of a curve. *Biometrika, 53: 359-364.*

PACHE, J.-C., ROBERTS, N., VOCK, P., ZIMMERMANN, A., CRUZ-ORIVE, L.M. (1993): Vertical LM sectioning and parallel CT scanning designs for stereology: application to human lung. *J. Microsc., 170: 9-24.*

PAPE, H., RIEPE, L., SCHOPPER, J. R. (1987): Theory of self-similar network structures in sedimentary and igneous rocks and their investigation with microscopical and physical methods. *J. Microsc., 148: 121-147.*

POHLMANN, S., MECKE, J., STOYAN, D. (1981): Stereological formulas for stationary surface processes. *Statistics, Math. Operationsforschung und Statistik, 12/3: 429-440.*

RIPLEY, B.D. (1981): *Spatial Statistics.* J. Wiley & Sons, New York.

RISS, J., GENTIER, S. (1989): Linear and areal roughness of non planar rock surfaces of fracture. *Acta Stereol., 8/2: 677-682.*

SALTYKOV, S. A. (1945): Stereometric Metallography. 1st edition. State Publishing House for Metals Sciences, Moscow. (cited from Weibel, 1980).

SANDAU, K. (1987a): Estimating the length density of roots using vertical sections. *Acta Stereol., 6/1: 43-60.*

SANDAU, K. (1987b): How to estimate the area of a surface using a spatial grid. *Acta Stereol., 6/3: 31-36.*

SANDAU, K. (1993): An estimation procedure for the joint distribution of spatial direction and thickness of flat bodies using vertical sections. Part I: Theoretical considerations. *Biom. J., 35: 649-660.*

SANDAU, K., VOGEL, H.J. (1993): An estimation procedure for the joint distribution of spatial direction and thickness of flat bodies using vertical sections. Part II: An application in soil micromorphology. *Biom. J., 35: 661-675.*

SANDAU, K., HAHN, U. (1994): Some remarks on the accuracy of surface area estimation using the spatial grid. *J. Microsc., 173: 67-72.*

SANTALÓ, L. A. (1976): *Integral Geometry and Geometric Probability.* Addison-Wesley.

STEINHAUS, H. (1930): Zur Praxis der Rektifikation und zum Längenbegriff. *Ber. Sächs. Akad. Wiss. Math. Nat. Kl., 82: 120-130.*

STOYAN, D., GERLACH, W. (1987): Stereological determination of curvature distributions of spatial fibre systems. *J. Microsc., 148: 297-305.*

STOYAN D., KENDALL, W. S., MECKE, J. (1987): *Stochastic Geometry and its Applications.* Akademie Verlag, Berlin.

VEDEL JENSEN E.B., KIEU, K. (1992): A note on recent research in second-order stereology. *Acta Stereol. 11/1, 569-579.*

VOGEL, H.J., WELLER, U., BABEL, U. (1993): Estimating orientation and width of channels and cracks at soil polished blocks - a stereological approach. *Geoderma, 56: 301-316.*

WEIBEL E.R. (1979): *Stereological Methods*, Vol. I. Academic Press, London.

WEIBEL E.R. (1980): *Stereological Methods*, Vol. II. Academic Press, London.

WRIGHT, K., KARLSSON, B. (1983): Topographic quantification of non-planar localized surfaces. *J. Microsc., 130: 37-51.*

Robustizing Mixture Analysis Using Model Weighting

Michael P. Windham

Mathematics and Statistics
Utah State University
Logan, UT 84322–3900, USA

Summary: This paper presents two methods for modifying a statistical estimation or model fitting procedure. The first shows how to extend the procedure to mixture distributions. The second shows how to make the procedure more robust using weighting. The two modifications are then combined to produce robust methods for mixture analysis.

1. Introduction

Mixture analysis is used to measure characteristics of subpopulations in a population, from data on individuals that does *not* include information on the subpopulations to which the individuals belong. Mixture analysis in the statistical sense refers to fitting a mixture model to the empirical distribution of the observed data. In what follows I will characterize the probability structure of random variable in terms of what I will call its "density function." This function is the usual probability density function for continuous random variables, and the probability distribution function for discrete; or in general it is the Radon-Nikodym derivative of the cumulative distribution function, with respect to a suitable measure on the support of the random variable. For example, for a data set $\{x_1, \ldots, x_n\}$ the "empirical density" would be $f(x) = \frac{1}{n} \sum_j \epsilon(x - x_j)$, where $\epsilon(x) = 1$, if $x = 0$ and is zero otherwise.

A random variable X has a mixture distribution, if its density is given by

$$g(x) = \sum_{i=1}^{k} \pi_i g_i(x),$$

where $k \geq 2$ is the number of subpopulations, g_i is the density for the i-th subpopulation and π_i is the i-th mixing parameter satisfying $\pi_i \geq 0$ and $\sum_i \pi_i = 1$. Implicit in this structure is another random variable, call it Z, with values in $\{1, \ldots, k\}$, that identifies the subpopulation to which an individual belongs. The parameter $\pi_i = P(Z = i)$ is the probability of belonging to the i-th subpopulation, g_i is the conditional density for X knowing $Z = i$, and the density g is the marginal for X. From this point of view, one can say that mixture analysis deals with modeling the joint probability structure of X and Z from observations of X alone. It will be assumed that each g_i is a member of a specified family of densities, $\mathcal{G}$, and that we are seeking, for a

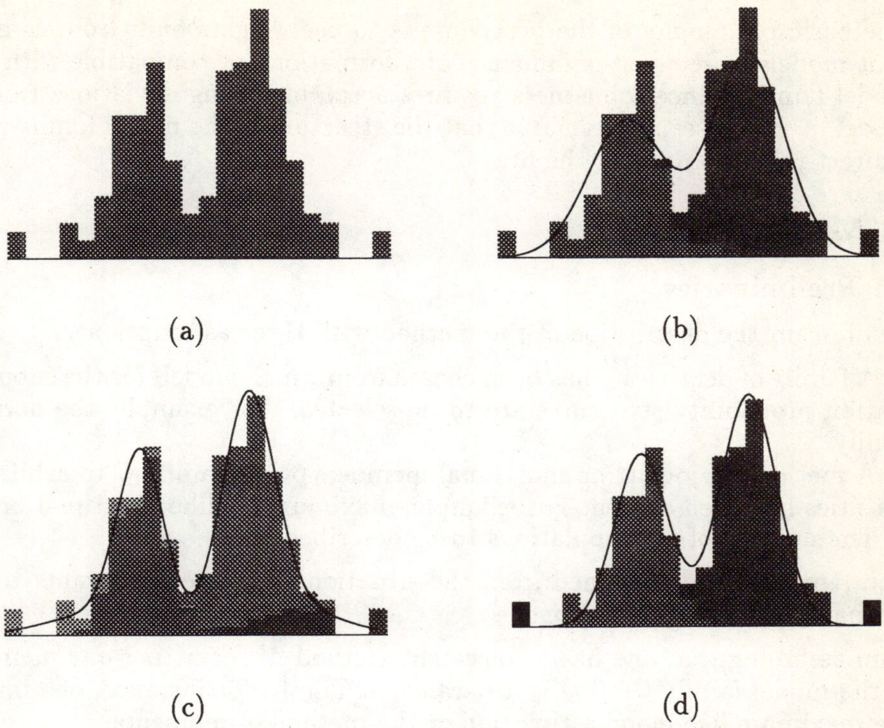

Fig. 1: Several approaches to mixture modeling

given k, the k-component mixture of members of $\mathcal{G}$ to model the distribution of the data.

There are many methods for fitting mixture models, but the most commonly used are not necessarily robust. That is, they are sensitive to outliers, contamination or a poor choice for the family of subpopulation models, $\mathcal{G}$.

Fig. 1 illustrates the problems caused by, for example, outliers. It also shows the result of applying the method I am proposing. Fig. 1(a) shows a histogram of univariate data. Fig. 1(b) shows the result of fitting a mixture of two normals with maximum likelihood estimation. The fit looks reasonable at first glance, but a closer examination suggests that the tails are not fit well and the modes are too low. The problem is caused by outliers. The data are a sample from a mixture of two normals with outliers added on both ends. One approach to dealing with the possibility of outliers is to look for more subpopulations, hoping that the outliers will form their own groups. The result of fitting a mixture of four normals is shown in Fig. 1(c) and it is clear from this example that the approach may not work well. I am proposing here the procedure that produced the result in Fig. 1(d), fitting a two normal mixture. The fit to the bulk of the data is much better than the other attempts, and the outliers have been identified as the two outside bars in the histogram. We will see how the method was applied to these data in more detail later.

The basic philosophy of the procedure is to use weights built from component models to reduce the influence of information not compatible with the model family. Since robustness requires accommodating deviations from a model, it would seem reasonable that the structure of the model family play a direct role in obtaining the fit.

2. Methodology

2.1 Preliminaries

I will begin the description of the method with three assumptions.

1. A family of densities $\mathcal{G}$ has been chosen from which models for the subpopulation probability structures are to be selected. For example, the normal family.
2. A method M for fitting individual members of the family $\mathcal{G}$ to arbitrary densities has been chosen. For example, maximum likelihood estimation.
3. The number of subpopulations to be described is k.

The family $\mathcal{G}$ is determined from the situation, e.g. the user wants to fit normals to the subpopulations.

I am assuming that one has a consistent method M for fitting *one* member of the model family $\mathcal{G}$. If $\mathcal{G}$ is a parametric family, for instance, one might use maximum likelihood estimation or the method of moments.

It will be convenient in what follows to view the method as one that assigns to a given *density* a model g from the family $\mathcal{G}$, rather than fitting a model to data sets. From this more general perspective, a data set would be "represented" by its empirical density. Formally, I will say that we have a function, M, that assigns to each density f a density g in $\mathcal{G}$. For example maximum likelihood estimation for normal models can be defined by assigning to a density f

$$M(f) = \phi(\,\cdot\,; E_f[X], \mathrm{var}_f[X]), \tag{1}$$

where $\phi(\,\cdot\,; 0, I)$ is the standard normal density. For data, the expected values with respect to the empirical density, would simply be sample averages. Shortly, I will describe a technique for extending M to fit mixtures of members of $\mathcal{G}$ and a method for robustizing M. Combining the two yields the robustized method for mixtures.

2.2 Extending to Mixtures

(a) Virtually any method M for fitting models from a given family of densities $\mathcal{G}$ to an arbitrary density can be extended in a natural way to a method for fitting finite mixtures of members of $\mathcal{G}$. If f is the mixture $f = \sum_i \pi_i f_i$, then a reasonable model fit using members of $\mathcal{G}$ for f would be provided by $g = \sum_i \pi_i M(f_i)$.

(b) Suppose, on the other hand, we have a mixture of models from $\mathcal{G}$, $g = \sum_i \pi_i g_i$, then, in the notation of the Introduction, $p_i(x) = \pi_i g_i(x)/g(x)$

is $P(Z=i|X=x)$ the conditional density of Z knowing X for data sampled from g. For any density f, let $\pi_i^+ = E_f[p_i(X)]$ and $f_i(x) = p_i(x)f(x)/\pi_i^+$, then f_i is a density and $f(x) = \sum_i \pi_i^+ f_i(x)$. In other words, we can induce a mixture structure on any density f using a mixture g of models from $\mathcal{G}$, with, in fact, the same conditional structure for $Z|X$, that is, $p_i(x) = \pi_i g_i(x)/g(x) = \pi_i^+ f_i(x)/f(x)$.

Putting these two together, if we have a mixture g of members of $\mathcal{G}$ we can induce a mixture structure on any f, and if f is a mixture we can model it using a mixture of members of $\mathcal{G}$. The situation is a natural one for iteration, that is, given an arbitrary f, pick a mixture of models, $g = \sum_i \pi_i g_i$, with g_i in $\mathcal{G}$, induce a mixture structure on f, $\sum_i \pi_i^+ f_i$, as in (b), then obtain the model $g^+ = \sum_i \pi_i^+ M(f_i)$ for it determined by method M as in (a), and repeat to convergence. The resulting density g^* *models* the structure on f that it *induces.*

The extension process may or may not produce a new model fitting method. For example, if M is maximum likelihood estimation, then the extension is also, and the iterative procedure is simply the EM algorithm (Dempster et al. 1977). On the other hand, this technique allows one to extend such simple procedures as method of moments to mixtures.

2.3 Robustizing a Model Fitting Method

Any density f can be *weighted* by a member of the model family $\mathcal{G}$ as follows. For a nonnegative number c and a density g in $\mathcal{G}$, the weighted density $f_{c,g}$ is defined by $f_{c,g}(x) = a\, g^c(x) f(x)$, where the constant a is chosen so that $f_{c,g}$ is a density. Data that is compatible with both is emphasized, up-weighted, and data compatible with neither is down-weighted. The degree to which g influences the resulting density is regulated by the exponent c. In particular, for $c = 0$ no weighting occurs, but as c is increased the weighting density g has greater effect.

The goal is to fit f with a member of $\mathcal{G}$ that is the most compatible with f in some sense. The compatibility is achieved by using a known model fitting method, M, and choosing the model g^* so that, applying M to f weighted in an appropriate way, and applying M to g^*, weighted in the same way, produce the same results. The weighting is also is done using g^*. In particular, *robustized model fit* $M_c(f)$ is defined to be the density g^* in $\mathcal{G}$ that satisfies

$$M(f_{c,g^*}) = M(g^*_{c,g^*}). \tag{2}$$

For example, maximum likelihood for normal models, $\mathcal{G}$, the function M given in (1) is applied to a data set $\{x_1, \ldots, x_n\}$ with empirical density $\hat{f}$. For $g = \phi(\,\cdot\,; m, V)$ a normal density, $M(\hat{f}_{c,g}) = \phi(\,\cdot\,; \hat{m}, \hat{V})$, where for $w_c(x; m, V)$ being the c-th power of the normal density g normalized so that $\sum_j w(x_j; m, V) = 1$, we have

$$\hat{m} \;=\; \sum_j w_c(x_j; m, V) x_j$$

$$\hat{V} = \sum_j w_c(x_j; m, V)(x_j - \hat{m})(x_j - \hat{m})'. \tag{3}$$

On the other hand, for the same normal density g, the density $g_{c,g}$ is also normal, with parameters m and $V/(c+1)$, that is, $M(g_{c,g}) = \phi(\,\cdot\,; m, V/(c+1))$. Therefore, $M_c(\hat{f}) = \phi(\,\cdot\,; m_c, V_c)$ is the normal density, where m_c and V_c satisfy

$$\begin{aligned} \sum_j w(x_j; m_c, V_c)x_j &= m_c \\ \sum_j w(x_j; m_c, V_c)(x_j - m_c)(x_j - m_c)' &= \frac{V_c}{c+1}. \end{aligned} \tag{4}$$

In the general case, the solution to (2) is obtained by an iterative procedure. Beginning with a convenient model g^0, for example, $g^0 = M(f)$, the iteration from g^{N-1} to g^N is given by solving

$$M(g^N_{c,g^N}) = M(f_{c,g^{N-1}}), \tag{5}$$

that is, compute $M(f_{c,g^{N-1}}) = \hat{g}$, and find the model, g^N, to which when weighted by itself, M assigns $\hat{g}$. The latter is often easy to do. Continuing the normal example, if the right hand side of (5) produced parameters m and V, the parameters of the model g^N would be m and $(1+c)V$.

The last problem to solve is to decide which exponent c to use. The exponent c is called a tuning constant in the jargon of robust statistics, and many methods have them. An advantage of this one is that it can usually be chosen automatically, by optimizing an appropriate criterion. The criterion I have used is related to the asymptotic efficiency of the model fitting process. The details of the relationship are given in Windham (1994b). Let me simply say two things about it. First, the criterion is a function of c that is an upper bound for the asymptotic efficiency of the model fitting procedure when using c. It takes the value one, if f is actually a member of the model family, but can be much smaller than one, if f is not very compatible with $\mathcal{G}$. In that sense it measures goodness of fit. Secondly, the criterion can often be estimated easily using the convergence rate of the iterative procedure in (5). In particular, if the model fitting procedure is a parameter estimation method, then the model at each iteration corresponds to a parameter θ^N that identifies it. The convergence rate is approximately $r = |\theta^L - \theta^{L-1}|/|\theta^{L-1} - \theta^{L-2}|$, where L is the iteration at which the process is stopped. The criterion is then given by $\rho(c) = (c/r - c)^2$. The final robustized model corresponds to the value of c that maximizes ρ.

2.4 Robustizing Mixture Analysis

The extension of M to mixtures and the robustizing are easily combined. The iterative procedure for the extension can be described as in 2.2 by

$$g = \sum_i \pi_i g_i \longrightarrow g^+ = \sum_i \pi_i^+ M(f_i),$$

where π_i^+ and f_i are given in **2.2**. All that is needed to robustize the extension is to replace M with M_c. That is, a complete iteration of the model fitting algorithm from $g = \sum_i \pi_i g_i$ to g^+ is given by

$$g \to p_i = \frac{\pi_i g_i}{g} \begin{matrix} \nearrow & \pi_i^+ = E_f(p_i) & \searrow \\ \searrow & f_i = \frac{p_i}{\pi_i^+} f & \nearrow \end{matrix} f = \sum_i \pi_i^+ f_i \to g^+ = \sum_i \pi_i^+ M_c(f_i)$$

The criterion to choose c is $\rho(c) = \min_i \rho_i(c)$, where ρ_i is the criterion for applying M_c in the i-th component. In effect, the goodness of fit of the mixture is measured by the worst of the fits of its components.

3. Example Revisited

The data pictured in Fig. 1 are a random sample of size 200 from a mixture of two normals, $.4\,\phi(x;0,1) + .6\,\phi(x;4,1)$, to which six outliers were added, three near -4 and three near 8. The maximum likelihood estimates for the parameters of the 200 uncontaminated data are $\pi_1 = .41$, $m_1 = .07$, $v_1 = .97$, $m_2 = 4.06$, and $v_2 = 1.04$. Adding the outliers gives maximum likelihood estimates of $\pi_1 = .42$, $m_1 = -.01$, $v_1 = 1.73$, $m_2 = 4.15$, and $v_2 = 1.54$. The outliers have moved the mean estimates slightly in their direction, but have seriously inflated the variances. Applying my robustized maximum likelihood produced a maximum for the criterion ρ at $c = .73$ and estimates $\pi_1 = .41$, $m_1 = .16$, $v_1 = .88$, $m_2 = 3.97$, and $v_2 = 1.00$, which correspond more closely to the parameters for the data without outliers and the model from which they were obtained. The outliers were identified using the weights obtained in the robustizing within components. If the weights were less than a cutoff value for a data point, it was considered an outlier. The cutoff was $.01^c$.

The effect of the robustizing can be seen clearly in Fig. 2. The top diagram is a plot of the exponent $0 \le c \le 1$ versus the parameter estimates of the parameters obtained for each c. This plot suggests that as the exponent is increased the effect of the outliers is reduced, finally disappears, and the parameter estimates stabilize. This phenomenon is typical of the experiments I have performed. The lower plot shows qualitatively the criterion ρ and the percent outliers detected. These plots also show the disappearance of the effects of the outliers and the identification of their presence.

4. Discussion

I have described two modifications of statistical parameter estimation or model fitting. The extension to mixtures is essentially the same as the one I presented at the IFCS Conference in Paris in 1993 (Windham 1994a). At that conference I also presented a robustization procedure. The one presented here is somewhat different, and I believe better. One important difference is the way the exponent is chosen. In the previous presentation the choice was made using characteristic functions and was time consuming,

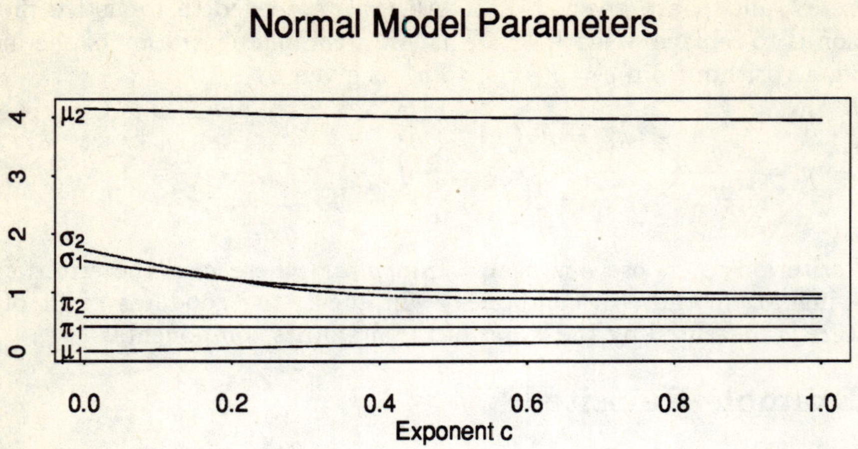

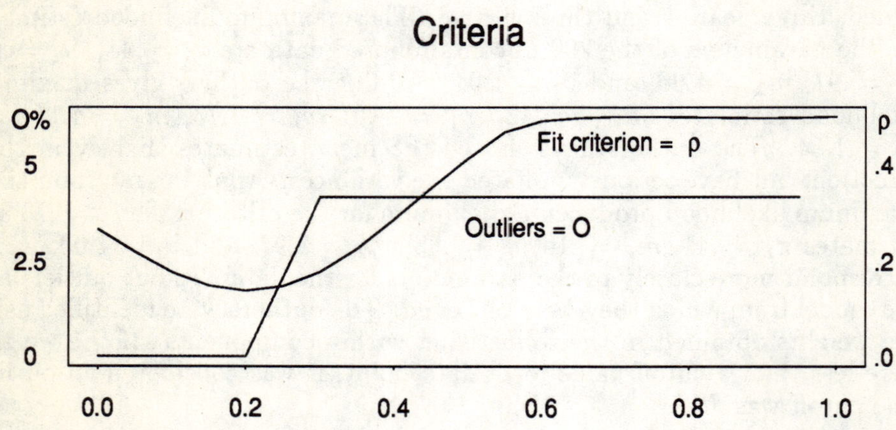

Fig. 2: Univariate normal mixture with outliers

if not difficult, to compute, and not as reliable as the criterion based on convergence rates.

It may seem strange that the convergence rate could have a statistical interpretation and be useful in assessing model fit, but I have used similar ideas with some success in the past (Windham and Cutler, 1992).

References:

DEMPSTER, A.P., LAIRD, N.M., and RUBIN, D.B. (1977): Maximum-likelihood from Incomplete Data Via the EM Algorithm. *Journal of the Royal Statistical Society, Series B, 39, pp. 1–38.*

WINDHAM, M.P., (1994a) Mixture Analysis with Noisy Data. In *New Approaches in Classification and Data Analysis. Proceedings of the Fourth Con-*

ference of the International Federation of Classification Societies (IFCS-93). E. Diday, Y. Lechavllier, M. Schader, P. Bertrand, B. Burtschy (eds.) Heidelberg: Springer-Verlag, 155–160.

WINDHAM, M.P., (1994b) Robustizing Model Fitting. *Journal of the Royal Statistical Society,* submitted.

WINDHAM M.P. and CUTLER A. (1992): Information Ratios for Validating Mixture Analyses. *Journal of the American Statistical Association*, 87, 1188–1192.

Evaluation of the First Life Table Published in 1888 in Japan

Keiji YAJIMA

School of Management
Science University of Tokyo
Kuki-City, Saitama 346 Japan

Summary: Rikitaro Fujisawa received his doctor's degree in mathematics at Strassburg Kaiser-Wilhelm University in 1886 and in the proceeding year he was appointed as a professor of Tokyo Imperial University. He published a paper on a life table in 1888 in a Tokyo Mathematics-Physics Journal intending to promote the life insurance business in the country. According to an article written by the president of one insurance company in the memorial book of Fujisawa the professor had intended to settle the insurance premium rates a little higher than the real figures in order to make secure a profit for life insurance enterprises. He also requested that companies make promises of refundment to the insured when a surplus had become clear. Data analysis of the life table seems to prove negative characteristics contrary to his intention and it likely resembles European tables at that time.

1. Introduction

Rikitaro Fujisawa was born in 1861 and in 1878 entered the present University of Tokyo, Faculty of Science, Department of Physics. He graduated in 1882. Then, in 1883 he was ordered to England and Germany to complete his education. In 1886 he obtained a doctorate by presenting a thesis on the theory of series applying function theory to Strassburg Kaiser-Wilhelm University. He returned from abroad in May, 1887, and in the next month he was appointed as a professor of the newly built Faculty of Mathematics, Tokyo University. He held his seat till 1921, having reached the traditional retirement age, and afterwards he was appointed as a member of the House of Peers in 1924. He finished his distinguished life in 1933.

In the era of rapid westernization of Japanese culture he conducted himself as a leader in many fields, including mathematics education, Wasan, that is Japanese arithmetic, statistical theory and others. He was confident of the importance of the life insurance business "to prevent the invasion of destructionism which prevails in Europe". He published a paper on a life table in a Tokyo Mathematics-Physics Journal in 1888. His table was based on a compilation and adjustment of four mortality tables: a 1881-86 report, a 1886 report, a 1886 population survey, a 1887 mortality report and a 1887 population survey. In 1911 he was appointed as a member of the investigation project team for post-office life insurance annuities. Post-office life insurance activities, which he helped to develop, started in 1916.

He left many English texts, most of which were delivered as speeches to guests from abroad. In these documents he had a very clear mind to establish a social funding system against hazards such as earthquakes, typhoons, crop failures, and infectious diseases. He backed up the theoretical issues of insurance problems dealing with statistical and mathematical aspects.

The first life table which included tables issued by the government before 1920 is recognized as almost negligible by the specialists of today regarding the preparation procedure of the table. On the other hand it is interesting to review the table from the facts described below.

2. Life Table for the Insurance Company

The biggest problem in 1889 for the Nippon Life Insurance Corp. to begin business was how to settle the rates of life insurance. Mr. S. Hikone, the first president afterwards, contacted certain medical doctors but his efforts were in vain. Accidentally he found a new book 'Introduction to Life Insurance Establishment', the author of which was Fujisawa, and he sent members of his staff to meet the author.

After a couple of visits the members could hear from Fujisawa that 'In Japan the business of life insurance is not yet in the primitive stage and so it is better to settle the insurance premium rates a little higher, not less than the actual ones, and after observing the state of matters the insurance company is recommended to refund to the insured when the appearance of surplus would occur. Therefore if you would like my help you should make promise of the refundment principle'. Hikone and other founders of the company agreed to his proposal. Fujisawa then hired two mathematicians during the period from August 28 to September 3 and the revisions of the table published in 1888 were made. The company was established on September 20.

According to Fujisawa the life table which had been prepared by Hikone at that time was adjusted to the other tables of two Japanese life insurance companies, Meiji Corp. and Teikoku Corp. They had actually adopted the tables issued in western countries. On the other hand, he made a second table for the company revising his first table upon consideration of census registration material received from the Census Office of the Department of Home Affairs. Naruse wrote that the company refunded to the insured a profit for the first time in December, 1897, but the details are unclear.

3. Assessment of the Fujisawa's Tables

There are two tables prepared by Fujisawa, one of which was issued in 1888 in the journal mentioned, the other one was made in 1889. The first life table consisted of life expectancies for ages ranging from 10 to 90. The second table covered data ranging from 0 to 94 without mention of gender. The expectancy values of the second table exceeded the first table values in the interval of ages from 10 to 47 by less than 0.4 years. From an age of 47 there is a small fluctuation between the first and second table. The absolute values of differences are less than 0.3 below the age of 86. Over this age there are larger differences due to the size of samples (see the curve of 1889–1888 in Fig. 5). Both tables consisted of 1,000 people at the time of birth. Recent life tables are dealing with 100,000 people as standard.

Age	*Death*	*Survival*
10	6	1000
11	6	994
12	6	988
...	...	...
89	4	14
90	0	10

Tab. 1: Life Table

Fujisawa wrote in his paper that the English table which he referred to as a basis consisted of 5,000 people and that he decided to adopt the number of 1,000 for his table as an acceptable size. In his second table, the number started from 1,416 because the original table started from an age of 10 for a number of 1,000 people.

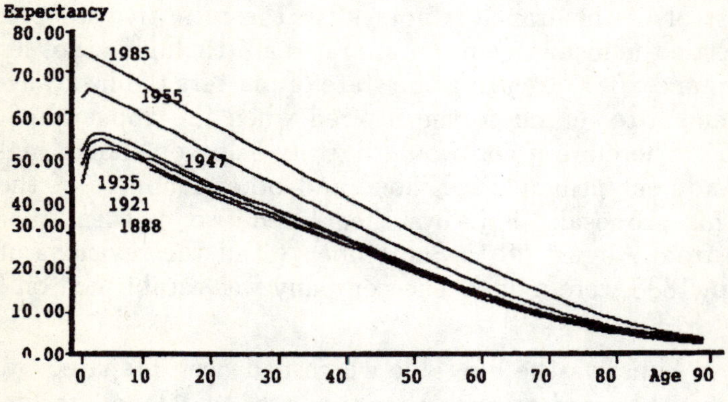

Fig. 1: Life expectancies(Japan)

Fig. 1 shows a general view of the life tables for such years as 1888, 1921, 1935, 1947, 1955 and 1985 where the horizontal axis corresponds to ages from 0 to 90 and the vertical axis indicates the life expectancy value. The 1888 curve in the figure reflects the first table and starts from the age of 10 but is almost the same as the second table. Those survey years are adopted for convenience and in fact the authorized or reliable life tables start from the year 1921, which is labeled as the 4th life table figured out from the 1921-1925 data. The 5th table was issued in 1926 for data from 1926-1930 and the 6th in 1935 for data from 1935-1936. There is no 7th table and the 8th table corresponds to the year 1947. After the 9th table a regular time cycle of five years was established starting from the year 1950.

The table of 1888, that is the first Fujisawa table, and the tables of 1891, 1899, 1909 which were issued by the government had faults as the expectancy data did not reach the suitability of former values. The revised values of life expectancy for the period of years between 1895 and 1923 are shown in Fig. 2 in the curve indicated as Japan together with the original values curve. From the sense of continuity the revised curve looks more reliable. In Fig. 2, the two curves for England/Wales and Germany also indicate discontinuity in the years before 1895. Furthermore, values of England and Germany in 1895 are equal to the values of Japan 30 to 35 years later. Fujisawa made a table by using the mortality data of one thousand people. But the amount of data was insufficient and he was unable to grasp the geographical distribution of the Japanese prefectural life expectancies (Fig. 8 will show life expectancies distribution and prefectures using correspondence analysis).

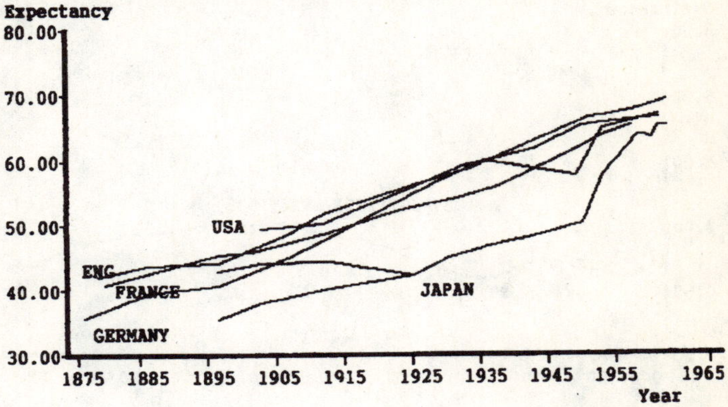

Fig. 2: Life expectancies(Germany, Japan)

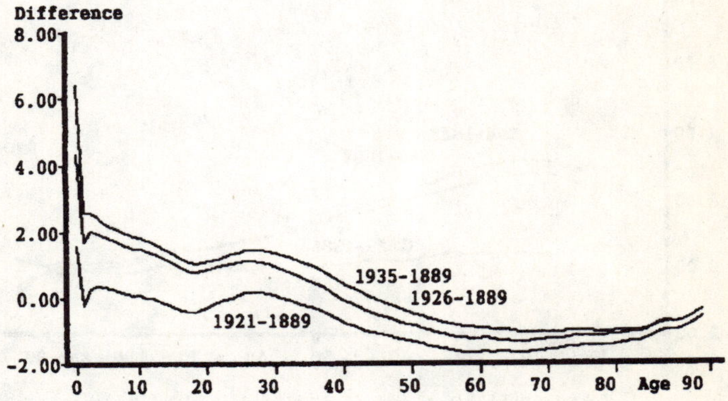

Fig. 3: Feature of Fujisawa's table(1)

In Fig. 3 there are three curves, 1921-1889, 1926-1889 and 1935-1889, showing differences between, for example values of 1921 expectancy and of 1888, where the 1888 table corresponds to the second Fujisawa table. The large values located at zero age indicate the three different values of mean expectancy, 1.55, 4.31 and 6.41. There are some fluctuations at the age of 2 due to sample errors caused by high infant death rates. Authorized tables of 1921, 1926 and 1935 are divided into male and female tables and in Fig. 3 the male table is chosen for convenience to show the characteristics of the differences. These three curves show local minimum values near the age of 17 and local maximum values as being near the age of 27.

Fig. 4 shows characteristics of Fujisawa's table compared with the authorized tables, 1921, 1926 and 1935, where the estimated values for each age are calculated from the three tables. Estimation values are calculated for comparison by using the simple minimum squares extrapolation method with equal distances between two subsequent four survey years. The indicated ages 17 and 27 also appear as in Fig. 3.

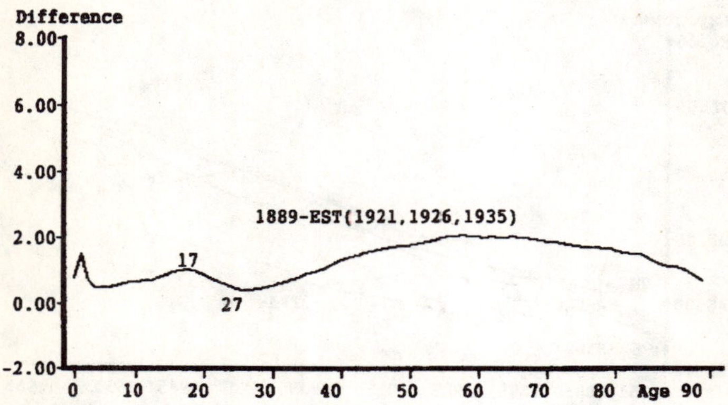

Fig. 4: Feature of Fujisawa's table(2)

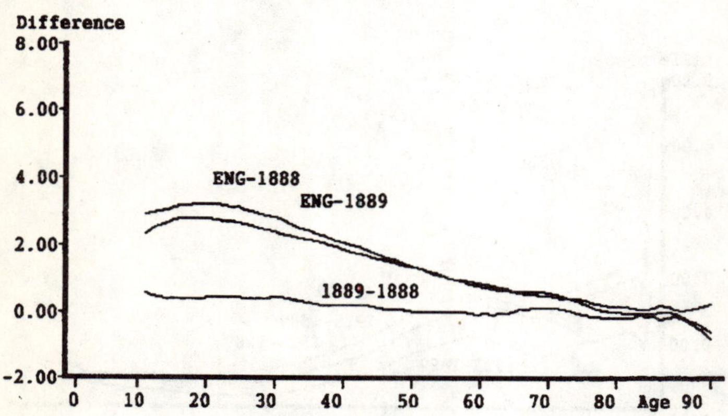

Fig. 5: Feature of Fujisawa's table(3)

Fig. 5 shows the differences between the England tables and two Fujisawa tables, and the differences between two Fujisawa's tables, the first and second. As the England table and the first Fujisawa table consist of a starting age of 10 all curves begin from an age of 10. The England table was made public in 1843 and was used by 17 life insurance companies. The two difference curves in Fig. 5 are simpler than the ones of Fig. 3, which are the differences from later survey tables, in the sense that there is no up and down behaviour.

Instead, it implies that the basic curve selected in the England table was similar to the curve utilized by two pre-existing Japanese companies with some technical modifications.

Fig. 6 shows details of death rate curves where a curve of 1889 has a slightly unnatural behaviour when compared to the curves of 1921 and 1926, which are also very far from the curves of later years.

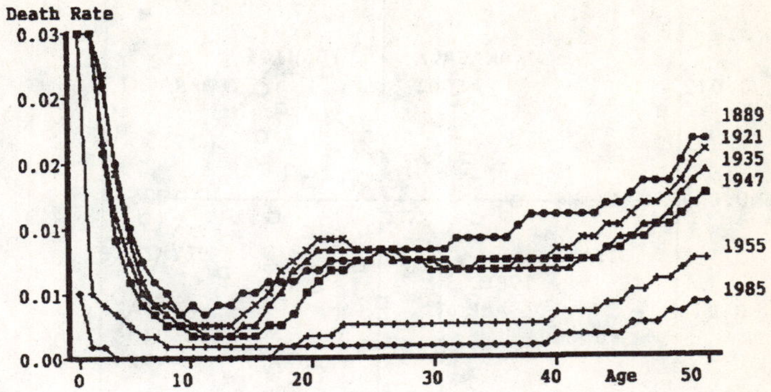

Fig. 6: Death rates(Japan)

Fig. 7 shows the cluster analysis of twelve expectancy tables, the England table of 1843, Fujisawa's table, and tables of 1891, 1899, 1909, 1921, 1926, 1935, 1947, 1948, 1949, 1950 where ages are taken from 10 to 70 with a 5-year interval. The dendrogram shows that a resemblance exists between the England table of 1843 and the Japanese 1948, 1949 and 1950 tables.

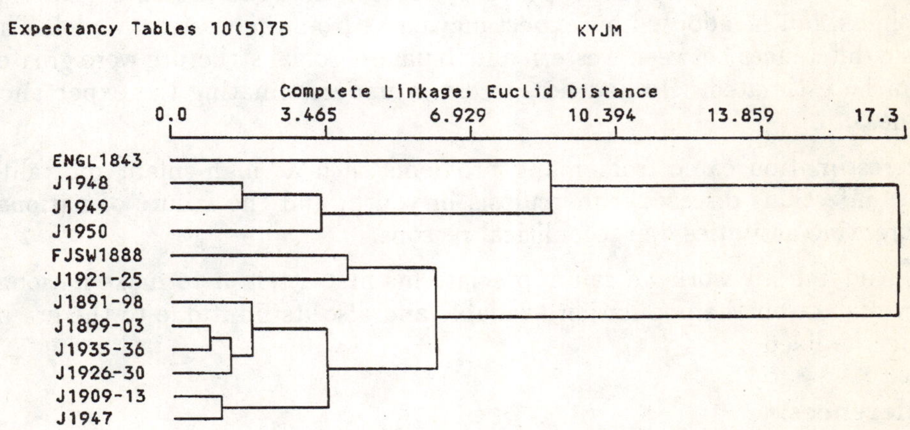

Fig. 7: Cluster analysis of expectancy tables

Fig. 8 shows the geographical characteristics of the expectancies when combining data from prefectures and survey years. There are groups composed of north and south areas and the Japanese sea region and Pacific sea region. The characteristics of the geographical position are related to sampling strategies and if samples are taken from restricted areas with small sample size then bias could not be unavoidable.

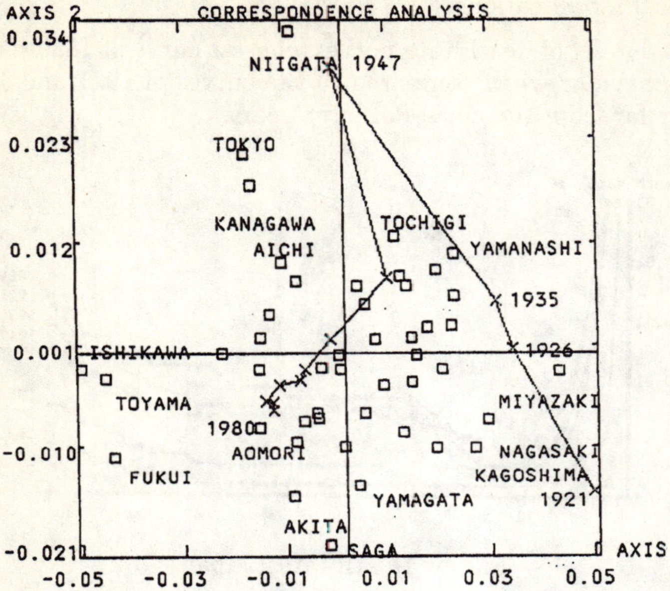

Fig. 8: Expectancies and prefectures

4. Conclusion

Fujisawa intended to set his expectancy table in a way to promote insurance business, but he adopted his expectancy curve from the western model. The basic differences between western and Japanese social structure were greater than he estimated. He had no success in underestimating the expectancy value.

Overestimation came from many problems such as high infant mortality rate, infectious diseases, tuberculosis in youth, and the failure of national quarantine activities due to political reasons.

By studying his work we can appreciate his first attempt to make national survey data on the population available and also its limit due to the era in which he lived.

References:

[1] *Commemoration of Dr. Fujisawa, (in Japanese)*, Dr. Fujisawa Commemoration Committee (1928).

[2] *Collected Works of Dr. Fujisawa, vol. 1 (in Japanese), vol. 2(in English), vol. 3(in Japanese)*, Dr. Fujisawa Commemoration Committee (1934).

Classification Related Results and Other Aspects of Data Analysis

Incomplete Data Matrices and Tests on Randomly Missing Data

U. Bankhofer

Institut für Statistik und Mathematische Wirtschaftstheorie, Universität Ausgburg, D-86135 Ausburg

Summary: In a practical analysis of data, the problem of incomplete data matrices is usually solved by estimating and substituting the missing values. Nevertheless, the imputation of missing values is only appropriate if the data are randomly missing. In general, the appropriate use of any missing-data method requires fundamental knowledge of the reasons and the underlying missing-data mechanism. With an analysis of the structure of the incomplete data matrix, the effects of the missing-data mechanism to the data under consideration can be investigated. Regarding the possible relations of dependence concerning the missing data, there are a few methods to test the existence of a non-systematic missing-data mechanism. The results of these tests can be sufficient conditions to reject the acceptance of randomly missing data or necessary conditions to accept a non-systematic missing-data mechanism.

1. Introduction

To analyze the similarity or dissimilarity of the objects of a finite **set of objects** $N = \{1, ..., n\}$, the variables of a **set of variables** $M = \{1, ..., m\}$ are usually taken to characterize these objects. The recorded values are collected in a **data matrix** A. In the event of missing data this matrix is incomplete and has the following form:

$$A = (a_{ik})_{n,m} = \begin{pmatrix} a_{11} & \cdots & \cdots & a_{1m} \\ \vdots & \circ & & \vdots \\ & & & \circ \\ \vdots & & \circ & \vdots \\ a_{n1} & \cdots & \cdots & a_{nm} \end{pmatrix} = \left(A^{obs}, A^{mis}\right). \qquad (1)$$

The circles in (1) denote the missing values and the whole matrix can be splitted in an observed part A^{obs} of the data with the corresponding present values and a missing part A^{mis} of the data with the unknown realizations of the missing values.

Additionally a **missing-data-indicator matrix** $V = (v_{ik})_{n,m}$ is introduced. In this matrix the value one indicates that the corresponding element of A is present, while a zero represents a corresponding missing value of A.

2. Reasons of Incomplete Data Matrices

2.1 General Reasons of Missing Data

Regarding the process of an analysis of data, the following general reasons of incomplete data matrices can be noted (Lösel and Wüstendörfer (1974)):

In the course of the **discussion of the data base**, an incorrect or inadequate design of the investigation can lead to missing values. For example, the question about the age of the children causes an error in the design of the investigation as soon as a person without children is interviewed. An inadequate design of the investigation, for example, is given in the event of misunderstanding questions.

When doing the **inquiry of data**, among other things, missing values can be caused by the refusal of a response, a lacking knowledge of the respondent, time problems in a survey, problems of motivation and understanding, the inattention of the observer and, in the event of a secondary analysis, the existing incompleteness of secondary data.

Finally, in the course of the **processing of data**, coding and transfer errors and the deletion of impossible or incorrect data can lead to missing values.

2.2 Non-Systematic Missing-Data Mechanisms

With regard to an adequate treatment of an incomplete data matrix, only the effects of the mentioned reasons of missing values to the present data are significant. This fact is taken into consideration when defining the following non-systematic missing-data mechanisms (Rubin (1976)):

Suppose that the data matrix A has a parametric distribution whose specification is not relevant in the following consideration and the indicator matrix V has a distribution with parameter Θ. Then the data are **missing at random (MAR)**, if

$$f(V \mid A^{obs}, A^{mis}, \Theta) = f(V \mid A^{obs}, \Theta) \quad \forall\ A^{mis}, \tag{2}$$

i.e. the probability of response is independent of the missing data. Analogously the data are **observed at random (OAR)**, if

$$f(V \mid A^{obs}, A^{mis}, \Theta) = f(V \mid A^{mis}, \Theta) \quad \forall\ A^{obs}, \tag{3}$$

i.e. the probability of response is independent of the observed data. Finally the data are **missing completely at random (MCAR)**, if

$$f(V \mid A^{obs}, A^{mis}, \Theta) = f(V \mid \Theta) \quad \forall\ A^{mis}, A^{obs}, \tag{4}$$

i.e. the qualities MAR and OAR are given simultaneously and therefore the probability of response is independent of the missing and the observed data.

3. Analysis of the Structure of Incomplete Data Matrices

3.1 Motivation and Overview

The consequences of the underlying missing-data mechanism concerning the evaluation of the data are shown in Fig. 1.

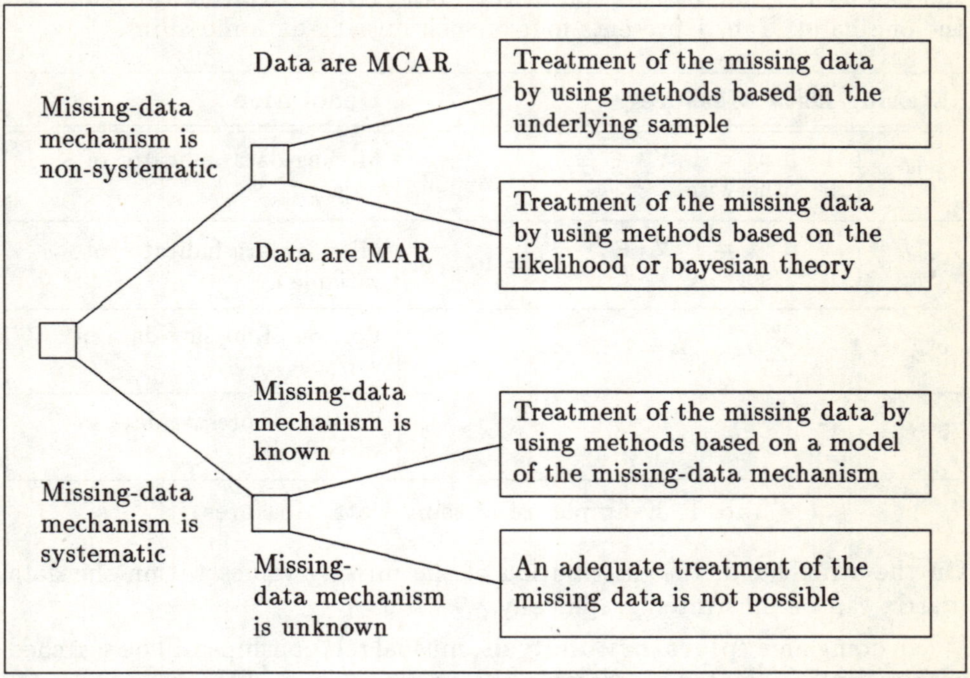

Fig. 1: Consequences of the Missing-Data Mechanism

If the missing-data mechanism is non-systematic the missing data can be taken into consideration in a proportional simple and appropriate manner. The use of methods based on the likelihood or bayesian theory, for example the EM-algorithm or multiple regression methods, is possible if only the quality MAR for the data is given. If the data are MCAR, in addition, the generally simplier methods based on the underlying sample, for example the mean imputation or the complete-case analysis, can be used. In the event of a systematic missing-data mechanism, an adequate treatment of the missing values is only possible if the missing-data mechanism is known. In this case a model of the underlying missing-data mechanism must be used.

Fig. 1 shows that an adequate treatment of the missing values is only possible if there is knowledge about the underlying missing-data mechnism. With that the necessity of an **analysis of structure of the incomplete data matrix** results. In the event of such an analysis of structure, only the underlying data and not all factors of influence concerning the missing data can be analyzed. Therefore the results of an analysis of structure can be sufficient to reject the acceptance of randomly missing data, but only necessary to accept a non-systematic missing-data mechanism. In general, the analysis of the structure of an incomplete data matrix, which should be extensive as possible, can be distinguished in a descriptive, an exploratory, and an inductive analysis.

In the case of a **descriptive analysis**, missing-data measures which describe the extent and the distribution of the missing values can be calculated on the one hand. Tab. 1 presents four of such missing-data measures.

Missing-Data Measure	**Denotation**
$v_{i\bullet}^{ind} = \begin{cases} 1 & \text{if } v_{ik} = 1 \quad \forall\, k \in M \\ 0 & \text{otherwise} \end{cases} \quad ; i = 1, ..., n$	Missing-data indicator of object i
$v_{\bullet k}^{ind} = \begin{cases} 1 & \text{if } v_{ik} = 1 \quad \forall\, i \in N \\ 0 & \text{otherwise} \end{cases} \quad ; k = 1, ..., m$	Missing-data indicator of variable k
$\tilde{v}_{\bullet k}^{mis} = 1 - \frac{1}{n} \sum_{i=1}^{n} v_{ik} \quad ; k = 1, ..., m$	Portion of missing data in variable k
$\tilde{v}_{\bullet k}^{obs} = \frac{1}{n} \sum_{i=1}^{n} v_{ik} \quad ; k = 1, ..., m$	Portion of present data in variable k

Tab. 1: Examples of Missing-Data Measures

On the other hand, the distribution of the missing values within the data matrix can be illustrated graphically.

When doing an **exploratory analysis**, mutual relationships will be searched within the data matrix concerning the missing values. For example, correlations based on the indicator matrix can be calculated or the similarities of the rows respectively columns of the indicator matrix can be analyzed.

Finally, in the case of an **inductive analysis**, tests on object-specific, variable-specific and general accumulations of missing values on the one hand and, on the other hand, tests on non-systematic missing-data mechanisms can be done. These latter test methods are introduced in the following subsection.

3.2 Tests on Non-Systematic Missing-Data Mechanisms

The basic problem is to test the null hypothesis H_0 that the data are randomly missing versus the alternative hypothesis H_1 that the data are systematically missing. According to the analyzed relation of dependence of the missing data Tab. 2 gives a survey of the possible test methods.

When doing an analysis of the dependence of the missing data on the realizations of these values, the quality MAR and implicitly the quality OAR are tested. The starting-point is the data matrix. In addition, information about the starting sample is required, so that a practical use of this investigation is restricted. The application of a **goodness-of-fit test** implies a comparison of the distribution of the present data and the distribution of the starting sample. If the starting sample is a random sample, the distribution of the population can be used. Possible test methods are the χ^2-goodness-of-fit test and the Kolmogoroff-Smirnoff test. In the case of the **parametric one-sample tests**, the measures of position and dispersion of the present data are compared with the corresponding measures of the starting sample. Depending on the scales of measurement and the distribution of the data, the one-sample sign test, the one-sample Gauss test, the one-sample t-test, the approximate Gauss test, and the χ^2-test of the variance can be used.

Problem	**Direction**	**Starting-Point**	**Test Method**
Analysis of the dependence of the missing data on the realizations of these values	MAR, OAR	Data matrix	Goodness-of-fit tests, parametric one-sample tests
Analysis of the dependence of the missing data on the absence of the data in other variables or objects	MAR, OAR	Indicator matrix	Test of *Kim and Curry*
Analysis of the dependence of the missing data on present values in other variables or objects	MAR	Data and indicator matrix	Tests on difference of location or independence, test of *Little*

Tab. 2: Survey of Test Methods

An analysis of the dependence of the missing data on the absence of the data in other variables or objects is appropriate to test the MAR and OAR assumption. The starting-point is the indicator matrix, so that the corresponding **test of Kim and Curry** (1977) is particularly proper for practical use. In the following only the case of a variable-specific analysis will be regarded. An object-specific investigation can be done analogously. The idea of the test of *Kim and Curry*, which is represented formally in Fig. 2, is to compare the real number of objects with no missing values, one missing value, and more than one missing value respectively with the corresponding expected frequencies. In detail, h^{nom} is the real number of objects with no missing values, h^{mom} is the real number of objects with more than one missing value and h_k is the real number of objects which have a missing value exclusively in variable k. The corresponding expected values characterized by a tilde are calculated over the relative portions of missing or present data in the single variables. The test statistic results as the sum of the relative quadratic deviations of the real and the expected values. The null hypothesis H_0 is rejected if the value of the test statistic is greater than the corresponding quantile of the χ^2-distribution. Fundamentally, a refinement of the test statistic would be possible to the effect that the number of objects with more than one missing value is splitted into the number of objects with two missing values, three missing values etc.. Nevertheless, in the event of large data matrices this is not necessary and in the event of small data matrices the problem arises that the expected frequencies are possibly too small.

Step 1:	Statement of the hypothesis H_0: the data are randomly missing in the sense, that they do not depend on the missingness in other variables H_1: the data are systematically missing
Step 2:	Selection of the significance level α
Step 3:	Calculation of the test statistic z $z = \frac{(h^{nom} - \tilde{h}^{nom})^2}{\tilde{h}^{nom}} + \frac{(h^{mom} - \tilde{h}^{mom})^2}{\tilde{h}^{mom}} + \sum_{k \in M^{mis}} \frac{(h_k - \tilde{h}_k)^2}{\tilde{h}_k}$ with $h^{nom} = \sum_{i=1}^{n} v_{i\bullet}^{ind}, \; \tilde{h}^{nom} = n \cdot \prod_{k \in M^{mis}} \tilde{v}_{\bullet k}^{obs},$ $h^{mom} = n - h^{nom} - \sum_{k \in M^{mis}} h_k, \; \tilde{h}^{mom} = n - \tilde{h}^{nom} - \sum_{k \in M^{mis}} \tilde{h}_k,$ $M^{mis} = \left\{k \in M : v_{\bullet k}^{ind} = 0\right\}, \; \tilde{h}_k = n \cdot \tilde{v}_{\bullet k}^{mis} \cdot \prod_{l \in M^{mis}, l \neq k} \tilde{v}_{\bullet l}^{obs},$ $h_k = \lvert\{i \in N : v_{ik} = 0 \wedge v_{il} = 1 \; \forall \, l \in M, l \neq k\}\rvert$
Step 4:	Rejection of H_0, if $z \in \langle x_{1-\alpha}; \infty\rangle$, where the quantile is from the χ^2-distribution with $(\mid M^{mis} \mid +1)$ degrees of freedom

Fig. 2: Test of Kim and Curry

Finally, the analysis of the dependence of the missing data on present values in other variables or objects is only appropriate to analyze the MAR assumption and requires both, the data and the indicator matrix. In the case of the **tests on difference of location**, the present data of an ordinal or quantitative variable is partitioned into two groups by the indicator variable and then the measures of position in these two groups are compared. Given nominal variables, the **independence** of the indicator variables is analyzed. According to the scales of measurement and the distribution of the data the two-sample sign test, the two-sample Gauss test, the two-sample t-test, the approximate two-sample Gauss test or the χ^2-test can be used. In comparison with the tests on difference of location, which must be applied on each variable, the **test of Little** (1988) only uses one test statistic for the whole data matrix. Nevertheless, this test has two disadvantages. First, a quantitative data matrix is required and second, estimates of the vector of means and the covariance matrix are used. Since the quality OAR for the data is investigated, only methods based on the MAR assumption can be used for estimating. With that, this test is restricted in its practical use and a further representation is therefore omitted.

4. Example

Finally, an example shall illustrate the presented test methods. The data matrix of 15 statistical software packages for PC's is shown in Tab. 3. The variables price level, user interface and programing interface are self-explaining and the values of the other variables are scoring values which indicate a percentage efficiency of the software packages in the single topics. Fundamentally, the data in the shaded fields are regarded as missing.

Objects \ Variables	Price Level	User Interface	Programing Interface	Descriptive Statistics	Tests	Multivariate Methods	Special Topics	Business Graphics	Statistical Graphics
BMDP	high	M/K	yes	92.66	56.00	59.28	48.66	0.00	20.00
CRUNCH	medium	M/K	no	85.33	62.00	15.71	0.00	20.00	11.25
CSS	medium	M/K	yes	80.33	64.00	56.85	77.00	100.00	83.00
MICROSTAT II	low	M/K	no	61.66	79.00	14.28	2.66	0.00	9.00
MINITAB	medium	K	yes	62.33	64.00	14.85	31.00	0.00	28.24
NCSS	medium	M	no	76.66	67.00	49.00	80.33	92.00	69.75
P-STAT	high	M/K	yes	92.66	100.00	24.57	55.66	46.00	22.25
RS/1	high	M/K	yes	54.33	73.00	10.28	27.66	49.00	51.00
SAS	high	M/K	yes	89.66	78.00	71.14	94.00	100.00	51.25
SPSS	high	M/K	yes	89.33	86.00	60.85	52.00	86.00	20.75
STATA	medium	M/K	yes	65.66	68.00	27.42	42.66	89.00	36.25
STATGRAPHICS	senior	M/K	no	86.66	78.00	33.71	56.33	97.00	42.00
STATISTIX	low	M	no	67.33	81.00	15.14	22.00	0.00	16.00
STATPAC GOLD	medium	M/K	no	77.66	68.00	17.71	39.66	43.00	21.25
SYSTAT	senior	M/K	yes	86.66	64.00	61.00	45.00	89.00	85.25

Legend:
M = Menu driven system
K = Command driven system
M/K = Combination of menu and command driven system

Tab. 3: Data Matrix of 15 Statistical Software Packages for PC's

The dependence of the missing data on the realizations of these values is considered in the variable *Multivariate Methods*, where the five largest values are regarded as missing. To run a χ^2-goodness-of-fit test, a normal distribution with $\mu = 35.45$ and $\sigma = 20.25$ is assumed. These parameters are calculated from the starting data of all 15 objects. With a p-value of 0.011 the constructed dependence can be verified. Additionally assuming a known variance the one-sample Gauss test gives the same results. In this case the mean of the present data is compared with the hypothetical mean computed from the starting data. The p-value of this two-sided test is 0.039.

The dependence of the missing data on the absence in other variables is considered in the variables *Business Graphics* (8) and *Statistical Graphics* (9). Either the values of both variables are present or missing. In this case the test of *Kim and Curry* can be used. Regarding only these two variables the following real and expected frequencies are calculated: $h^{nom} = 10$, $\tilde{h}^{nom} = 6.66$, $h^{mom} = 5$, $\tilde{h}^{mom} = 1.66$, $h_8 = 0$, $\tilde{h}_8 = 3.33$, $h_9 = 0$, $\tilde{h}_9 = 3.33$. The test statistic is 15.06 and the p-value of this test is 0.002, i.e. when using a significance level greater than 0.002, the constructed dependence can be verified.

Finally, a dependence of the missing values in the variable *Price Level* on the present values in the variable *Descriptive Statistics* is constructed in this way that the price levels of the objects with the five smallest values of the variable *Descriptive Statistics* are missing. Assuming a normal distribution for the variable *Descriptive Statistics* and unknown, but identical variances within the two groups which are formed by the missing-data indicator of the variable *Price Level*, the two-sample t-test can be used. Regarding the p-value of 0.000 (two-sides alternatives) the significant deviation of the group means is obvious.

5. Conclusion

In order to enable an adequate treatment of an incomplete data matrix, an analysis of the structure is indispensable. The results of such an analysis can either justify the assumption of randomly missing data or make a systematical missing-data mechanism more transparent. The introduced test methods show some possibilities of an inductive analysis of the structure of an incomplete data matrix.

References:

KIM, J.O., CURRY, J. (1977): The Treatment of Missing Data in Multivariate Analysis. *Sociological Methods and Research, 6, 215–239.*

LITTLE, R.J.A. (1988): A Test of Missing Completely at Random for Multivariate Data with Missing Values. *Journal of the American Statistical Association, 83, 1198–1202.*

LÖSEL, F., WÜSTENDÖRFER, W. (1974): Zum Problem unvollständiger Datenmatrizen in der empirischen Sozialforschung. *Kölner Zeitschrift für Soziologie und Sozialpsychologie, 26, 342–357.*

RUBIN, D.B. (1976): Inference and Missing Data. *Biometrika, 63, 581–592.*

Valuations and Hierarchizations

Klaus Biedermann, Rudolf Wille

Forschungsgruppe Begriffsanalyse, Fachbereich Mathematik der Technischen Hochschule Darmstadt, Schloßgartenstr. 7, D-64289 Darmstadt

Summary: In data analysis, the interplay of *valuations*, mapping objects into an ordered set of values, and *hierarchizations*, assigning to the values suitable subsets of the object set, is of central interest. Their relationship can be appropriately described and analyzed by methods for *formal concept analysis*. The essential step is to form suitable formal contexts with valuations as formal objects and hierarchizations as attributes. The corresponding concept lattices which contain all structural information of the interplay between valuations and hierarchizations can be completely determined.

1. Bijections in Classification

The general view of this paper is: Data yield information about objects - conceptual orders make objects understandable. Often, data are given by *valuations* of objects from which *hierarchizations* are deduced to order those objects conceptually. The relationship between such valuations and hierarchization is of central interest in data analyis. In formal concept analysis, valuations are understood as attributes of a many-valued context which are conceptually scaled to reach a conceptual hierarchization (cf. Ganter and Wille (1989)). In cluster analysis, valuations are basic to describe similarity between objects so that they usually evaluate pairs of objects. F. Critchley and B. Van Cutsem (1992) have given an order-theoretic unification and generalization of fundamental *bijections* between valuations and hierarchizations which occur in *mathematical classification*. In this paper we transform their approach into the setting of *formal concept analysis* to give a new understanding of their investigations and to open new perspectives. Concerning basic notions and results of mathematical order theory and formal concept analysis, the reader is referred to Davey and Priestley (1990), Wille (1984), Wille (1992), Ganter and Wille (1994).

The general approach starts by fixing a set G whose elements are called *objects* (objects are not necessarily singletons but might also be subsets, n-tuples, functions etc.). A *valuation* on G is defined to be a map v from G into an ordered set $\underline{P} := (P, \leq)$ whose elements are called *values*. $\underline{P}$ might be a chain, an antichain, or anything between those extreme types of (partially) ordered sets. The set of all valuations from G into $\underline{P}$ is denoted by $V(G, \underline{P})$ or shortly by V. A *hierarchization* on G *indexed* by $\underline{P}$ is defined as an order-preserving map from $\underline{P}$ into the power set of G ordered by inclusion. For a hierarchization h, the images $h(p)$ ($p \in P$) are called the *classes* of h. The set of all hierarchizations from $\underline{P}$ into the power set of G is denoted by $H(G,\underline{P})$ or shortly by H. A hierarchization $h \in H$ is called *covering* if $G = \bigcup_{p\in P} h(p)$; the subset of H consisting of all covering hierarchization is denoted by H_c or more explicitly by $H_c(G,\underline{P})$. A hierarchization $h \in H$ is called *disjoining* if $\bigcap_{p\in P} h(p) = \emptyset$; the subset of H consisting of all disjoining hierarchizations is denoted by H_d or more explicitly by $H_d(G,\underline{P})$. The valuations and hierarchizations are naturally ordered by $v \leq w :\Leftrightarrow \forall g \in G : v(g) \leq w(g)$ and $h \leq k :\Leftrightarrow \forall p \in P : h(p) \subseteq k(p)$. Basically, the general order-theoretic setting was created by M. F. Janowitz (1978) when he observed that it is meaningful and fruitful to replace the reals by an arbitrary ordered set of values. F. Critchley and B. Van Cutsem (1992) have used the order-theoretic setting to unify the description and explanation of bijections between certain valuations and hierarchizations.

The basic question how to assign to a given valuation an appropriate hierarchization is intensively discussed in formal concept analysis and conceptually answered by the method of *conceptual scaling* (cf. Ganter and Wille (1989)). For an ordered set $\underline{P} := (P \leq)$ of values, there are two basic methods to derive hierarchizations: The first follows the *principle of confirmation* in defining, for any valuation $v \in V$, the corresponding (covering) hierarchization h_v by

$$h_v(p) := \{g \in G | v(g) \leq p\} \text{ for all } p \in P;$$

the second expresses the *principle of discrimination* in defining, for any valuation $v \in V$, the corresponding (disjoining) hierarchization $\overline{h}_v$ by

$$\overline{h}_v(p) := \{g \in G | v(g) \not\geq p\} \text{ for all } p \in P.$$

Both methods yield a bijection between the valuations and specific hierarchizations which is stated more explicitly by the following proposition:

Proposition 1.1: *Let G be a set (of objects) and let $\underline{P} := (P, \leq)$ be an ordered set (of values). Then $v \mapsto h_v$ and $v \mapsto \overline{h}_v$ describe injective maps from $V(G,\underline{P})$ into $H(G,\underline{P})$ satisfying $v \leq w \Leftrightarrow h_v \geq h_w$ and $v \leq w \Leftrightarrow \overline{h}_v \geq \overline{h}_w$ for all $v, w \in V(G,\underline{P})$.*

Proof: Obviously, $v \leq w$ implies $h_v \geq h_w$ and $\overline{h}_v \geq \overline{h}_w$. Conversely, $h_v \geq h_w$ yields $g \in h_w(w(g)) \subseteq h_v(w(g))$ for each $g \in G$, hence $v \leq w$. Similarly, $\overline{h}_v \geq \overline{h}_w$ yields $g \notin \overline{h}_v(v(g)) \supseteq \overline{h}_w(v(g))$ for each $g \in G$, hence $w \geq v$. Now, $h_v = h_w$ or $\overline{h}_v = \overline{h}_w$ implies $v \leq w$ and $w \leq v$, i.e. $v = w$. This means that $v \mapsto h_v$ and $v \mapsto \overline{h}_v$ describe injective maps. □

A natural question is how the specific hierarchizations h_v and $\overline{h}_v$ can be characterized within $H(G,\underline{P})$. This may be answered as follows (recall that, in an ordered set $(S, \leq)$, the *principal filter* generated by an element a of S is defined by $[a) := \{x \in S | a \leq x\}$; furthermore, one defines the *principal ideal* $(a] := \{x \in S | x \leq a\}$):

Proposition 1.2: *Let h be a hierarchization. There exists a valuation v with $h = h_v$ if and only if $F_{g,h} := \{p \in P \mid g \in h(p)\}$ is a principal filter for all $g \in G$. There exists a valuation v with $h = \overline{h}_v$ if and only if*

$I_{g,h} := \{p \in P \mid g \notin h(p)\}$ is a principal ideal for all $g \in G$.

Proof: Let $h{=}h_v$ for some valuation v. Then $g{\in}h_v(v(g))$ implies $v(g){\in}F_{g,h}$. If $g{\in}h(p)$ for $p{\in}P$, by the definition of h_v, we have $v(g) \leq p$. This proves $F_{g,h} = [v(g))$. Conversely, let $F_{g,h}$ be a principal filter for all $g \in G$. Then we define $v(g) := q$ if $F_{g,h} := [q)$. By the definition of $F_{g,h}$ and of v, $g \in h(p)$ is equivalent to $v(g) \leq p$ for all $g \in G$ and $p \in P$, hence $h = h_v$. Secondly, let $h = \overline{h}_v$ for some valuation v. Then $g \in \overline{h}_v(v(g))$ implies $v(g) \in I_{g,h}$. If $g \notin h(p)$ for $p \in P$, by the definition of $\overline{h}_v$, we have $p \leq v(g)$. This proves $I_{g,h} = (v(g)]$. Conversely, let $I_{g,h}$ be a principal ideal for all $g \in G$. Then we define $v(g) := q$ if $I_{g,h} := (q]$. By the definition of $I_{g,h}$ and of $v, g \notin h(p)$ is equivalent to $p \leq v(g)$ for all $g \in G$ and $p \in P$, hence $h = \overline{h}_v$. □

Often, one is interested in restrictions of the bijections $v \mapsto h_v$ and $v \mapsto \overline{h}_v$ to suitable subsets of V and H specified by interesting properties for valuations and hierarchizations. A general scheme to establish such restricted bijections is given by the following proposition:

Proposition 1.3: *If $\underline{P}$ is a $\vee$–semilattice, then $v(g) \leq v(g_1) \vee \ldots \vee v(g_n)$ is equivalent to $\forall p \in P : g_1, \ldots, g_n \in h_v(p) \Rightarrow g \in h_v(p)$. If $\underline{P}$ is a $\wedge$–semilattice, then $v(g) \geq v(g_1) \wedge \ldots \wedge v(g_n)$ is equivalent to $\forall p \in P$: $g_1, \ldots, g_n \notin \overline{h}_v(p) \Rightarrow g \notin \overline{h}_v(p)$.*

Proof: Obviously, there are the following equivalences:

$$\begin{aligned} & v(g) \leq v(g_1) \vee \ldots \vee v(g_n) \\ \Longleftrightarrow\ & \forall p \in P : v(g_1), \ldots, v(g_n) \leq p \Rightarrow v(g) \leq p \\ \Longleftrightarrow\ & \forall p \in P : g_1, \ldots g_n \in h_v(p) \Rightarrow g \in h_v(p); \end{aligned}$$

$$\begin{aligned} & v(g) \geq v(g_1) \wedge \ldots \wedge v(g_n) \\ \Longleftrightarrow\ & \forall p \in P : v(g_1), \ldots, v(g_n) \geq p \Rightarrow v(g) \geq p \\ \Longleftrightarrow\ & \forall p \in P : g_1, \ldots, g_n \notin \overline{h}_v(p) \Rightarrow g \in \overline{h}_v(p). \end{aligned}$$ □

Since F. Critchley and B. Van Cutsem (1992) have been concerned with the *paradigm of similarity* in mathematical classification, they primarily consider the case of $G := S \times S$ where $v(x_1, x_2) \leq v(y_1, y_2)$ has the interpretation that the degree of dissimilarity of x_1 to x_2 is less than or equal to the degree of dissimilarity of y_1 to y_2. In particular, Critchley and Van Cutsem assume that $\underline{P}$ has a smallest element 0 which indicates the complete absense of dissimilarity. To apply Proposition 1.3, one has additionally to assume that $\underline{P}$ is a $\vee$- or $\wedge$-semilattice. Then Proposition 1.3 yields, for instance, that $v \mapsto h_v$ ($v \in V$ with $v(x,x) = 0$ and hence $(x,x) \in h_v(0) \subseteq h_v(p)$) restricts to the bijection between *ultrametrics* and *dendrograms* (cf. Janowitz (1978)):

$$v(x,y) \leq v(y,x) \Longleftrightarrow \forall p{\in}P : (y,x){\in}h_v(p) \Rightarrow (x,y){\in}h_v(p)$$
$$v(x,z) \leq v(x,y) \vee v(y,z) \Longleftrightarrow \forall p{\in}P : (x,y),(y,z){\in}h_v(p) \Rightarrow (x,z){\in}h_v(p)$$

2. The Confirmation Context

The relationship between the valuation v and the hierarchization h_v can be understood as consequence of a general relation between valuations and covering hierarchizations. This relation, named C, is defined by

$$vCh :\Longleftrightarrow \forall g \in G : g \in h(v(g))$$

and interpreted by saying: the valuation v *confirms* the hierarchization h. In other words, the relation C indicates when an object g is always an element of the class $h(p)$ of its value $p = v(g)$. The relation C gives rise to a formal context $\mathbb{C} := (V, H_c, C)$ which might be described more explicitly by $\mathbb{C}(G, \underline{P}) := (V(G, \underline{P}), H_c(G, \underline{P}), C)$; we call $\mathbb{C}$ the *confirmation context* with respect to G and $\underline{P}$. Establishing this formal context is motivated by the idea that its formal concepts may give essential insights into the interplay of valuations and covering hierarchizations according to the principle of confirmation. By all means, the bijection $v \mapsto h_v$ can be seen in $\mathbb{C}$ if one determines the concepts (v^{CC}, v^C) generated in $\mathbb{C}$ by the valuations v. This is performed by the following proposition:

Proposition 2.1: *For any valuation v, the formal concept $([v), [h_v))$ is the smallest concept of $\mathbb{C}$ having v in its extent.*

Proof: We have the following equivalences: $vCh \Leftrightarrow \forall g \in G : g \in h(v(g)) \Leftrightarrow \forall g \in G\ \forall p \in P : (v(g) \leq p \Rightarrow g \in h(p)) \Leftrightarrow \forall p \in P : h_v(p) \subseteq h(p) \Leftrightarrow h_v \leq h$; therefore $v^C = [h_v)$. Furthermore, $w \in [h_v)^C \Leftrightarrow \forall h \in h_v\ \forall g \in G : g \in h(w(g)) \Leftrightarrow \forall g \in G : g \in h_v(w(g)) \Leftrightarrow \forall g \in G : v(g) \leq w(g) \Leftrightarrow v \leq w$; therefore, $[h_v)^C = [v)$. This proves that $([v), [h_v))$ is the concept generated by v. □

Since $\bigcap_{i \in I}[h_i) = [h)$ if $h(p) = \bigcup_{i \in I} h_i(p)$ $(p \in P)$ for arbitrary $h_i \in H_c$ $(i \in I)$, Proposition 2.1 yields:

Corollary 2.2: *The intents of $\mathbb{C}$ different from H_c are the principal filters of $(H_c, \leq)$ different from H_c.*

For a better understanding of the information coded in the confirmation context, it is helpful to determine its concept lattice. Usually, this becomes easier if one reduces the context if possible. There are quite special hierarchizations from which all others can be derived conceptually: We define the *elementary hierarchization* $h_{g,p}$ for an object g in G and a value p in P with $(p] \neq P$ by

$$h_{g,p}(q) := \begin{cases} G \setminus \{g\} & \text{if } q \leq p, \\ G & \text{if } q \not\leq p, \end{cases}$$

for all $q \in P$. The set of all elementary hierarchizations denoted by H_e or more explicitly by $H_e(\underline{P}, G)$ is the desired set of hierarchizations which is shown by the following proposition (notice that $H_e \subseteq H_c$):

Proposition 2.3: *For any elementary hierarchization, the formal concept $(\{v \in V | v(g) \not\leq p\}, [h_{g,p}))$ is the largest concept of $\mathbb{C}$ having $h_{g,p}$ in its intent; in particular, these concepts are exactly the $\bigwedge$-irreducible elements of the concept lattice $\mathfrak{B}(\mathbb{C})$. The $\bigvee$-irreducible elements of $\mathfrak{B}(\mathbb{C})$ are precisely the concepts $([v), [h_v))$ for which there is a $g \in G$ such that $(v(g)] \neq P$ and $v(f)$ is maximal in $\underline{P}$ for all $f \in G \setminus \{g\}$.*

Proof: First we show that $vCh_{g,p} \Leftrightarrow v(g) \not\leq p$. By definition of C and $h_{g,p}$, $vCh_{g,p}$ is equivalent to $g \in h_{g,p}(v(g))$ which is equivalent to $v(g) \not\leq p$. Thus $h_{g,p}^C = \{v \in V \mid v(g) \not\leq p\}$. Now, let $h \in \{v \in V \mid v(g) \not\leq p\}^C$. For $q \not\leq p$ in $\underline{P}$ and $f \in G$, we choose a $v \in V$ with $v(f) = v(g) = q$. Since $v(g) \not\leq p$ it follows that vCh and so $f \in h(v(f)) = h(q)$. Thus, $h(q) = G$. For $q \leq p$ and $f \in G\backslash\{g\}$, we choose a $v \in V$ with $v(f) = q$ and $v(g) \not\leq p$ (recall that p is not the greatest element of $\underline{P}$). Then it follows that vCh and so $f \in h(v(f)) = h(q)$. Thus, $h(q) \supseteq G\backslash\{g\}$. This proves that $h \in [h_{g,p})$. Conversely, we start with an $h \in [h_{g,p})$. Let $v \in V$ with $v(g) \not\leq p$ and let $f \in G$. If $v(f) \not\leq p$ then $h(v(f)) = G \ni f$. If $v(f) \leq p$ then $f \neq g$ and $h(v(f)) \supseteq G\backslash\{g\}$ because $h \geq h_{g,p}$; hence $f \in h(v(f))$. This proves vCh and therefore $h \in \{v \in V \mid v(g) \not\leq p\}^C$. Thus, $(\{v \in V \mid v(g) \not\leq p\}, [h_{g,p}))$ is the concept generated by $h_{g,p}$. The $\bigwedge$-irreducibility is proved by using the $\nearrow$-relation in $\mathbb{C}$. Let $v \nearrow h$. Then there is a $g \in G$ with $g \notin h(v(g))$. For $p := v(g)$, we consider $h_{g,p}$. It follows that $h \leq h_{g,p}$ and $(v, h_{g,p}) \notin C$. Therefore $h = h_{g,p}$. In general, let $h_{g,p} \in H_e$. We choose a $v \in V$ with $v(g) = p$ and $v(f) = q \not\leq p$ for all $f \in G\backslash\{g\}$. Then $(v, h_{g,p}) \notin C$. Moreover, let $h \in H_c$ with $h_{g,p}^C \subset h^C$ or, equivalently, $[h_{g,p}) \supset h^{CC} \ni h$. Thus $h_{g,p} < h$ which implies $h(p) = G$ and so $h(v(G)) = \{G\}$. It follows vCh. Hence $v \nearrow h_{g,p}$. Therefore the concepts generated by the elementary hierarchizations are exactly the $\bigwedge$-irreducible elements of $\underline{\mathfrak{B}}(\mathbb{C})$. The $\bigvee$-irreducibility is proved by using the $\swarrow$-relation in $\mathbb{C}$. Let $v \in V$ with $(v(g)] \neq P$ for some $g \in G$ such that $v(f)$ is maximal for all $f \in G\backslash\{g\}$; furthermore, let $v(g) = p$. Then $h_{g,p} \in H_e$ and $(v, h_{g,p}) \notin C$. Let $w \in V$ with $v^C \subset w^C$, i.e. $[h_v) \subset [h_w)$. It follows $h_w < h_v$ and $v < w$. By assumption, $v(f) = w(f)$ for all $f \in G\backslash\{g\}$. But $v \neq w$ implies $v(g) < w(g)$. Therefore $h_{g,p}(w(G)) = \{G\}$ and hence $wCh_{g,p}$. Thus $v \swarrow h_{g,p}$. Conversely, if $(v(g)] = P$ for all $g \in G$ then, for all $h \in H_c$, vCh and hence $v \not\swarrow h$. If $(v(g)] \neq P$ for some $g \in G$ but $v(\tilde{f})$ is not maximal for some $\tilde{f} \in G\backslash\{g\}$, then there is a $w \in V$ with $w(\tilde{f}) > v(\tilde{f})$, $w(g) = v(g)$, and $w(f) = v(f)$ for all $f \in G\backslash\{g, \tilde{f}\}$. Then $v^C \subset w^C$ and $(w, h_{g,w(g)}) \notin C$. Hence $v \not\swarrow h_{g,v(g)}$. This completes the proof. □

Now, we can summarize that the confirmation context $\mathbb{C}$ has a unique *reduced subcontext* if P has enough maximal elements. Structurally, the always existing attribute-reduced subcontext $\mathbb{C}_r := (V, H_e, C_r)$ with $C_r := C \cap (V \times H_e)$ is even more interesting. By the preceding proposition, $\mathbb{C}_r$ is isomorphic to the context $\tilde{\mathbb{C}}_r := (V, G \times \hat{P}, \tilde{C}_r)$ where $\hat{P} := P\backslash\{\bigvee P\}$ and $v\tilde{C}_r(g,p) :\Leftrightarrow v(g) \not\leq p$ (note that $\{\bigvee P\}$ might be empty). Thus, it follows:

Corollary 2.4: $\underline{\mathfrak{B}}(\mathbb{C}) \cong \underline{\mathfrak{B}}(\mathbb{C}_r) \cong \underline{\mathfrak{B}}(\tilde{\mathbb{C}}_r)$

The corollary indicates how to understand the concept lattice of the confirmation context structurally: The key is to view the context $\tilde{\mathbb{C}}_r$ as the *semiproduct* of the contexts $(\{g\} \times P, \{g\} \times \hat{P}, \not\leq_g)$ $(g \in G)$ with $(g,p) \not\leq_g (g,q) :\Leftrightarrow p \not\leq q$, i. e. $\tilde{\mathbb{C}}_r \cong (\bigtimes_{g \in G} \{g\} \times P, \bigcup_{g \in G}\{g\} \times \hat{P}, \nabla)$ where

$(g, p_g)_{g\in G}\nabla(f,q) :\Leftrightarrow (f,p_f) \not\leq_f (f,q)$. It is known how the concept lattice of a semiproduct is determined by the concept lattices of its factors (cf. Ganter (1988)) and how the concept lattice of $(P, \hat{P}, \not\leq)$ is formed (cf. Wille (1985)). This yields the following theorem:

Theorem 2.5: *Let G be a set, let $\underline{P} := (P, \leq)$ be an ordered set, and let $\underline{\mathfrak{F}} := (\mathfrak{F}, \subseteq)$ be the ordered set of all non-empty order filters of $\underline{P}$. Then*

$$\underline{\mathfrak{B}}(\mathbb{C}) \cong \begin{cases} \underline{\mathfrak{F}}^{|G|} & \text{if } \bigvee P \text{ exists,} \\ (\mathfrak{F}^{|G|} \cup \{\emptyset\}^{|G|}, \subseteq) & \text{otherwise,} \end{cases}$$

where $(X,Y) \mapsto (\{v(g) \mid v \in X\})_{g\in G}$ describes the natural isomorphism.

By Theorem 2.5, $\underline{\mathfrak{B}}(\mathbb{C})$ can be constructed in the following way: for $(F_g)_{g\in G} \in \underline{\mathfrak{F}}^{|G|}$, the extent of the correspondig concept in $\underline{\mathfrak{B}}(\mathbb{C})$ consists of all valuations v with $v(g) \in F_g$ for all $g \in G$ and the intent consists of all hierarchizations h for which $p \in F_g$ implies $g \in h(p)$ for all $g \in G$.

3. The Discrimination Context

The relationship between the valuation v and the hierarchization $\overline{h}_v$ can be understood as consequence of a general relation between valuations and disjoining hierarchizations. This relation, named D, is defined by

$$vDh :\Leftrightarrow \forall g \in G : g \notin h(v(g))$$

and interpreted by saying: the valuation v *discriminates* the hierarchization h. In other words, the relation D indicates when an object is never an element of the class of its value. The relation D gives rise to a formal context $\mathbb{D} := (V, H_d, D)$ which might be described more explicitly by $\mathbb{D}(G,\underline{P}) := (V(G,\underline{P}), H_d(G,\underline{P}), D)$; we call $\mathbb{D}$ the *discrimination context* with respect to G and $\underline{P}$. Establishing this context is motivated by the idea that its formal concepts may give essential insights into the interplay of valuations and disjoining hierarchizations according to the principle of discrimination. By all means, the bijection $v \mapsto \overline{h}_v$ can also be seen in $\mathbb{D}$ if one determines the concepts (v^{DD}, v^D) generated in $\mathbb{D}$ by the valuations v. This is performed by the following proposition:

Proposition 3.1: *For any valuation v, the formal concept $((v], (\overline{h}_v])$ is the smallest concept of $\mathbb{D}$ having v in its extent.*

Corollary 3.2: *The intents of $\mathbb{D}$ unequal H_d are the principal ideals of $(H_d, \leq)$ unequal H_d.*

This proposition and the following assertions are direct consequences of the corresponding results in Section 2 by the following *duality principle*: $h \mapsto h^\perp$ with $h^\perp(p) := G\backslash h(p)$ for $p \in P$ describes an isomorphism between the ordered sets $H_c(G,(P,\leq))$ and $H_d(G,(P,\geq))$; obviously, $(h_v)^\perp = \overline{h}_v$ if h_v is formed with respect to the ordered set $(P,\leq)$ and $\overline{h}_v$ is formed with respect to the dually ordered set $(P, \geq)$.

The concept lattice of the discrimination context can also be determined completely if one uses its attribute-reduced context. To obtain the hierarchizations which are reduced in $\mathbb{D}$, we define the *anti–elementary hierarchization* $\overline{h}_{g,p}$ for an object g in G and a value p in P with $[p) \neq P$ by

$$\overline{h}_{g,p}(q) := \begin{cases} \{g\} & \text{if } q \geq p, \\ \emptyset & \text{if } q \not\geq p, \end{cases}$$

for all $q \in P$. The set of all anti-elementary hierarchizations, denoted by $\overline{H}_e$ or more explicitly by $\overline{H}_e(\underline{P}, G)$, is the unique reduced set of hierarchizations which is shown by the following proposition:

Proposition 3.3: *For any anti–elementary hierarchization, the formal concept* $(\{v \in V | v(g) \not\geq p\}, (\overline{h}_{g,p}])$ *is the largest concept of* $\mathbb{D}$ *having* $\overline{h}_{g,p}$ *in its intent; in particular, these concepts are exactly the* $\bigwedge$*-irreducible elements of the concept lattice* $\underline{\mathfrak{B}}(\mathbb{D})$*. The* $\bigvee$*-irreducible elements of* $\underline{\mathfrak{B}}(\mathbb{D})$ *are precisely the concepts* $((v], (\overline{h}_v])$ *for which there is a* $g \in G$ *such that* $[v(g)) \neq P$ *and* $v(f)$ *is minimal in* $\underline{P}$ *for all* $f \in G \setminus \{g\}$.

Now, we can summarize that the discrimination context $\mathbb{D}$ has a unique *reduced subcontext* if $\underline{P}$ has enough minimal elements. Structurally, the always existing attribute-reduced context $\mathbb{D}_r := (V, \overline{H}_e, D_r)$ with $D_r := D \cap (V \times \overline{H}_e)$ is even more interesting. By the preceding proposition, $\mathbb{D}_r$ is isomorphic to the context $\widetilde{\mathbb{D}}_r := (V, G \times (P \backslash \{\bigwedge P\}), \widetilde{D}_r)$ where $v\widetilde{D}_r(g,p) :\Leftrightarrow v(g) \not\geq p$. Thus, it follows:

Corollary 3.4: $\underline{\mathfrak{B}}(\mathbb{D}) \cong \underline{\mathfrak{B}}(\mathbb{D}_r) \cong \underline{\mathfrak{B}}(\widetilde{\mathbb{D}}_r)$

From the corollary we obtain as in section 2 the following theorem:

Theorem 3.5: *Let* G *be a set, let* $\underline{P} := (P, \leq)$ *be an ordered set, and let* $\underline{\mathfrak{I}} := (\mathfrak{I}, \subseteq)$ *be the ordered set of all non-empty order ideals of* $\underline{P}$*. Then*

$$\underline{\mathfrak{B}}(\mathbb{C}) \cong \begin{cases} \underline{\mathfrak{I}}^{|G|} & \text{if } \bigwedge P \text{ exists,} \\ (\mathfrak{I}^{|G|} \cup \{\emptyset\}^{|G|}, \subseteq) & \text{otherwise,} \end{cases}$$

where $(X, Y) \mapsto (\{v(g) \mid v \in X\})_{g \in G}$ *describes the natural isomorphism.*

By Theorem 3.5, $\underline{\mathfrak{B}}(\mathbb{D})$ can be contructed in the following way: for $(I_g)_{g \in G} \in \mathfrak{I}^{|G|}$, the extent of the correspondig concept in $\underline{\mathfrak{B}}(\mathbb{D})$ consists of all valuations v with $v(g) \in I_g$ for all $g \in G$ and the intent consists of all hierarchizations h for which $p \in I_g$ implies $g \notin h(p)$ for all $g \in G$.

4. Examples

Let G:={1,2} and $\underline{P}$: 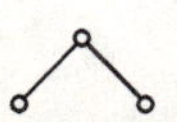 . 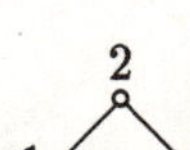

and 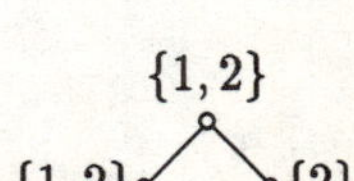

are instances of how we shortly denote valuations and hierarchizations (empty sets occuring in the range of hierarchizations are omitted).

By Proposition 2.3, we obtain the reduced confirmation context

	{1, 2} {1, 2} {2}	{1, 2} {1, 2} {1}	{1, 2} {1} {1, 2}	{1, 2} {2} {1, 2}
2 1	×	×	×	
1 2	×	×		×
1 2	×		×	×
2 1		×	×	×

This context is a rather familiar one. Its concept lattice is isomorphic to $\underline{B_4}$, the 4-dimensional Boolean lattice:

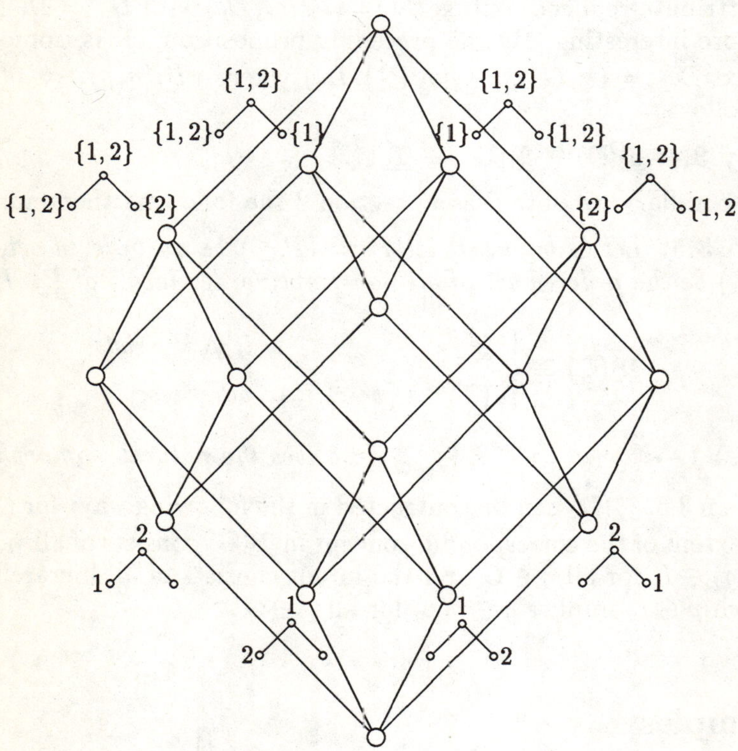

Since $(\mathcal{F}\backslash\emptyset, \subseteq)$ is , Theorem 2.5 yields the same result.

By Proposition 3.3, we obtain the reduced discrimination context

	top {1}, right {1}	top {1}, left {1}	top {1}	top {2}, right {2}	top {2}, left {2}	top {2}
right 1,2		×	×		×	×
left 2, right 1		×	×	×		×
top 2, right 1		×	×			
left 1, right 2	×		×		×	×
left 1,2	×		×	×		×
top 2, left 1	×		×			
top 1, right 2					×	×
top 1, left 2				×		×

In this case it is easier to determine the corresponding concept lattice by Theorem 3.5: $(\mathcal{I}, \subseteq)$ is 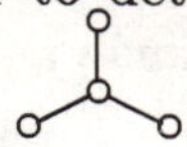 and thus we obtain:

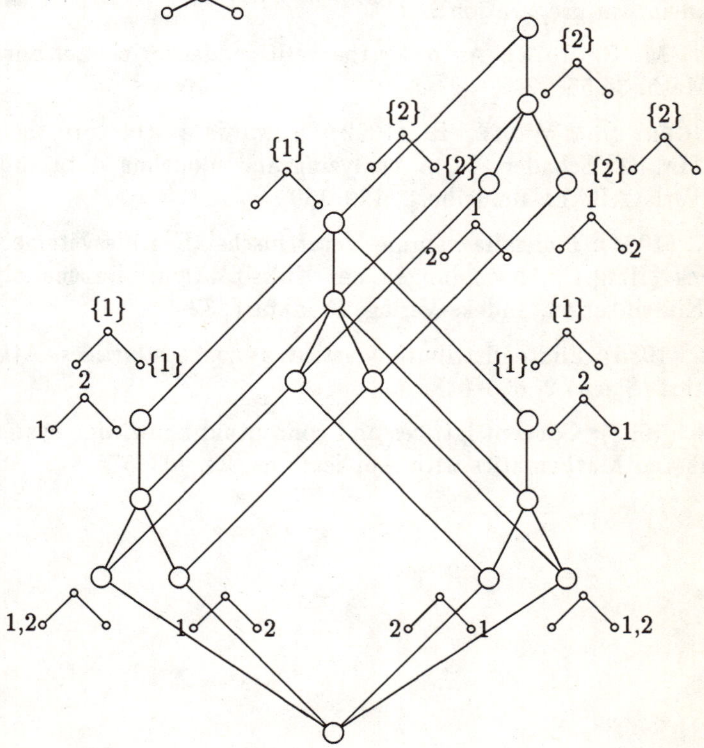

For both cases there is a simple method how to determine the missing valuations and hierarchizations. This calculation, further results and illustrating examples can be found in Biedermann (1994).

References:

BIEDERMANN, K. (1994): Bewertungen und Hierarchisierungen. Staatsexamenarbeit, FB Mathematik, TH Darmstadt.

CRITCHLEY, F. and VAN CUTSEM, B. (1992): An order-theoretic unification of certain fundamental bijections in mathematical classification I,II. Research Reports, Dept. Stat. Univ. Warwick. To appear in: B. Van Cutsem (ed.): Classification and Dissimilarity Analysis. Lecture Notes in Statistics. Springer-Verlag, New York.

DAVEY, B. A. and PRIESTLEY, H. A. (1990): Introduction to lattices and order. Cambridge University Press, Cambridge.

GANTER, B. (1988): Composition and decomposition in formal concept analysis. In: H. H. Bock (ed.): Classification and related methods of data analysis. North–Holland, Amsterdam, 561–566.

GANTER, B. and WILLE, R. (1989): Conceptual scaling. In: F. Roberts (ed.): Applications of combinatorics and graph theory to the biological and social sciences. Springer-Verlag, New York, 139–167.

GANTER, B. and WILLE, R.: Formale Begriffsanalyse. B. I. –Wissenschaftsverlag, Mannheim (in preparation).

JANOWITZ, M. F. (1978): An order theoretic model for cluster analysis. SIAM J. Appl. Math. 34,55–72.

STRAHRINGER, S. , WILLE, R. (1992): Towards a structure theory for ordinal data. In: M. Schader (ed.): Analyzing and modeling data and knowledge. Springer–Verlag, Berlin-Heidelberg, 129–139.

WILLE, R. (1984): Liniendiagramme hierarchischer Begriffssysteme. In: H. H. Bock (Hrsg.): Anwendungen der Klassifikation: Datenanalyse und numerische Klassifikation. Indeks-Verlag, Frankfurt, 32–51.

WILLE, R. (1985): Finite distributive lattices as concept lattices. Atti Inc. Logica Mathematica (Siena) 2, 635–648.

WILLE, R. (1992): Concept lattices and conceptual knowledge systems. Computers and Mathematics with Applications, 23, 493–515.

Recent Developments in Multimode Clustering

Thomas Eckes

Fachbereich Gesellschaftswissenschaften, Bergische Universität Wuppertal, Gaußstr. 20, D-42097 Wuppertal, Germany

Summary: In recent years several models and corresponding algorithms for clustering two- or higher-mode data have been developed, including the additive-clustering approach (e.g., DeSarbo, 1982), the tree-fitting approach (e.g., De Soete & Carroll, 1989), and the error-variance approach (e.g., Eckes & Orlik, 1993). The present paper relates various types of data frequently collected in the behavioral and social sciences to prominent models of multimode clustering and demonstrates the versatility of three-mode clustering using a real data set drawn from social-psychological research.

1. Introduction

Discrete models and corresponding algorithms allowing the simultaneous classification of two or more modes or sets of entities (e.g., stimuli, responses, subjects, features, situations) are currently attracting a lot of attention in the behavioral and social sciences. Since Carroll and Arabie's (1980) review of the field, much progress in multimode data analysis has been made. This is most clearly evidenced by the large variety of sophisticated models and substantial applications that have appeared over the last ten years or so (see, for recent reviews, Arabie & Hubert, 1992, 1994).

The purpose of the present paper is to provide an overview of two- and higher-mode cluster-analytic models with special emphasis given to three-mode clustering.

2. Data Types

Using the terminology of Carroll and Arabie (1980), a given data matrix has a particular number of *modes* and a particular number of *ways*. The number of modes is the number of distinct sets of entities; the number of ways is the number of sets of entities, including the repetition of one or more of these sets. Modes and ways, respectively, are denoted by capital letters A, B, and C. Depending on the respective number of modes and ways, four basic types of data relevant in the context of multimode clustering can be distinguished.

• **A × A data**. These are *two-way one-mode* data since there is only one mode that is repeated. Typical examples are similarity or dissimilarity rating data, co-occurrence data resulting from usage of the free-sort technique, or euclidean distances computed between row or column elements of a two-way two-mode data matrix described next.

• **A × B data**. Since there are two distinct sets of entities involved, these data are called *two-way two-mode* data. Prominent examples are (a) *profile data* resulting from ratings of n stimuli on m attribute scales or from measurements of n persons on m variables, (b) *confusions data* showing how often a given stimulus has been identified as (or confused with) some other stimulus, and (c) *preference* or *dominance data* occurring when various subjects render preference judgments over a number of different stimuli.

• **A × A × B data**. In this case, two of the three ways correspond to the same set of entities. Usually, such *three-way two-mode* data are gathered when the focus of the research is on some kind of individual or group differences in the perception or conception of a given set of stimuli. For instance, one-mode two-way similarity data may be collected under different experimental conditions, with the set of conditions constituting the second mode.

• **A × B × C data**. This data type is characterized by three ways corresponding to three distinct sets of entities; that is, none of the ways is repeated. Data of the *three-way three-mode* kind typically result from ratings of n stimuli on m attribute scales by p subjects or from measurements of n persons on m variables under p experimental conditions or at p moments in time.

Of course, these four data types do not encompass all that is empirically possible and/or potentially relevant in a given study. For instance, Cattell's (1966) data box conceptualization takes into account as many as ten modes. It should be noted, however, that empirical studies focusing on more than three modes simultaneously are fairly rare.

3. Models and Methods of Multimode Clustering

Generally speaking, multimode clustering aims at a simultaneous, hierarchic or nonhierarchic classification of at least two distinct sets of entities into internally homogeneous and externally isolated clusters. Two- or three-mode hierarchical clustering (or ultrametric tree) representations are only one example out of a large variety of currently available models for multimode data. In the following, a short overview of some of the most prominent cluster analytic models and methods suited to the representation of structure in two- and higher-mode data is given. Tab. 1 portrays the relationships between the four basic data types and corresponding models. Since the focus of this paper is on recent developments in two- and three-mode representations, only these will be dealt with.

• **LS tree-fitting**. Using a least squares (LS) approach, an ultrametric or additive tree representation can be constructed whose distances correspond maximally to the observed dissimilarities, where correspondence is measured by a least squares criterion (De Soete et al., 1984). Such an approach has a high degree of methodological flexibility since it (a) can deal with incomplete data sets, (b) is applicable to the two-way two-mode case, as well as to the three-way three-mode case, and (c) provides additional features such as allowing for constraints on the topology of the tree or incorporating optimal variable weighting (see, for reviews, Carroll & De Soete, 1994; De Soete, 1988).

Of special interest is a set of procedures for representing three-way three-mode data by one or more ultrametric trees (De Soete & Carroll, 1989). In the first, distances are computed from the complete data matrix weighting the levels of two modes (say, the last two) optimally; then, using these distances, the levels of the first mode are represented by the terminal nodes of an ultrametric tree. In the second, the data are optimally aggregated over one mode (say, the third); then the first two modes are jointly represented by the terminal nodes of an ultrametric tree. In the third, the first two modes are jointly represented for each level of the third mode by an ultrametric tree subject to the constraint that all p trees have the same topology (but possibly different height values).

• **BEA**. The bond energy algorithm (BEA) originally proposed by McCormick et al. (1972) aims at permuting the rows and columns of a data matrix in such a way as to push the numerically larger matrix elements together. This is accomplished by maximizing the summed "bond strengths" over row and column permutations of the input matrix, where the bond strength between two nearest neighbor elements is defined as their product. In recent years this algorithm has been substantially improved (Arabie & Hubert, 1990; Arabie et al., 1988) and investigated in the context of simulated annealing (Schleutermann et al., 1990) and blockmodeling (Arabie et al., 1990).

• **EVA**. The error-variance approach (EVA) proposed by Eckes and Orlik (1993, 1994) yields a two- or higher-mode ultrametric tree representation and optionally allows constructing an overlapping cluster solution given a preselected number of disjoint clusters. In contrast to De Soete and Carroll's third procedure that represents three-mode three-way data by a series of two-mode ultrametric trees, Eckes and Orlik's approach jointly portrays the three modes in a *single* ultrametric tree. At each step of the agglomerative process, the basic algorithm merges those clusters whose fusion results in the smallest possible increase in an internal heterogeneity measure.

• **Additive box clustering**. A model-based alternative to the error-variance approach and several corresponding algorithms have been advanced by Mirkin et al. (1994). The technique developed for fitting this kind of clustering model, which is closely related to Shepard and Arabie's (1979) ADCLUS (for ADditive CLUStering) model, utilizes a doubly greedy optimization strategy: first, clusters are obtained sequentially rather than simultaneously, and second, each cluster is formed incrementally, with an element-by-element addition/deletion strategy; each greedy step maximizes the explained part of the total sum of squares.

• **GENNCLUS**. The GENNCLUS (for GENeral Nonhierarchical CLUStering) methodology is a generalization of the ADCLUS model to the case of nonsymmetric or two-mode proximity data (DeSarbo, 1982). It makes use of a series of gradient-based techniques and combinatorial optimization methods in an alternating least squares framework.

• **HICLAS**. De Boeck and Rosenberg's (1988) hierarchical classes (HICLAS) model is restricted to object-by-attribute data matrices with binary entries of 0, 1. In this model, objects with identical sets of attributes are grouped into hierarchically ordered object classes; analogously, attributes are grouped into hierarchically ordered attribute classes. The hierarchy of object classes is linked to the hierarchy of attribute classes by means of a (symmetric) association relation.

	Models		
Data	One-mode	Two-mode	Three-mode
A × A	hierarchical clustering, LS tree-fitting, ADCLUS	—	—
A × B	Ward's (1963) method, LS tree-fitting, partitioning methods	BEA, EVA, LS tree-fitting, additive box clustering, GENNCLUS, HICLAS	—
A × A × B	INDCLUS, INDTREES	—	—
A × B × C	LS tree-fitting	LS tree-fitting	BEA, EVA

Tab. 1: Data Types and Clustering Models

Note. A × A = two-way one-mode data. A × B = two-way two-mode data. A × A × B = three-way two-mode data. A × B × C = three-way three-mode data. The listing of procedures is far from being complete; it is only intended to highlight important classes of currently available procedures pertaining to the respective data type-model combinations.

There are two models specifically designed to handling A × A × B data, both yielding one-mode individual differences representations. Carroll and Arabie's (1983) INDCLUS (for INdividual Differences CLUStering) model represents the data in terms of common features that are differentially weighted depending on the level of the third way (see also Arabie et al., 1987). Similarly, in the INDTREES (for INdividual Differences in TREE Structures) model (Carroll et al., 1984) the levels of the mode corresponding to the first two ways are represented as terminal nodes of tree structures (ultrametric or additive trees) that have a common topology but different height values or branch lengths per level of the third way. As a matter of fact, De Soete and Carroll's two-mode procedure for representing three-way three-mode data outlined above is analogous to the INDTREES model.

4. An Illustrative Application

Applications of two- and higher-mode clustering can be found in such diverse fields as sociometric data analysis (e.g., Arabie et al., 1990), marketing research (e.g., DeSarbo & De Soete, 1984; Espejo & Gaul, 1986), social stereotype research (Eckes, 1994), structural analysis of perceived situations (Eckes, 1993, in press), and physical attractiveness research (Eckes & Hassebrauck, 1993), to mention but a few. In the following, an illustrative example of a three-mode clustering representation of three-way three-mode data is given.

Twenty subjects were asked to rate on seven-point scales the appropriateness of acting in a way corresponding to each of 5 traits in any of 10 situations. Subjects were students from an introductory psychology course, traits were items loading most highly on each of the Big Five personality dimensions (John, 1990), and everyday situations were selected from a former study of situation cognition (Eckes, 1993). The rating data were arranged in a 5 (traits) $\times$ 10 (situations) $\times$ 20 (persons) matrix. In a preparatory step of the analysis, a rowwise reflection of the data within each column and block was performed; that is, entries were first duplicated for each row and then the duplicated entries were rescored by multiplying with -1. This reflection procedure ensured that each pole of a given bipolar rating scale constituted a feature of its own which could be clustered separately from the respective opposite pole. Using the three-mode error-variance approach (Eckes & Orlik, 1994), four disjoint clusters were identified in a first step and then, in a second step, these clusters were augmented by an overlapping procedure. Due to space limitations, only the final four cluster solution is presented here (see Tab. 2).

Quite obviously, the majority of subjects consider it inappropriate to appear tense at a swimming pool, when going by bus, when going window shopping, at a bar, or when meeting a friend (Cluster A). Clusters C and D are also highly consensual clusters in that many subjects agree in judging a given trait as inappropriate or appropriate in a given situation. Cluster B is special in this respect since there is only one person element belonging to it (i.e., S12). Such an idiosyncratic cluster clearly demonstrates that there are distinctive patterns of construing the (in-)appropriateness of traits in situations. Hence, homogeneity of the subject sample is an assumption whose validity should not be taken for granted, otherwise the risk is high to miss out on interesting questions. For example, it could be asked what made this subject judge the appropriateness of being talkative or sympathetic and not being tense in situations like meeting a friend of a family get together so much higher than any other subject. Surely, further data are needed to answer questions like this, but the point is that one should be in a position to ask them in the first place. Three-mode or even four-mode clustering is the method of choice in such cases.

	Mode		
Cluster	Traits	Situations	Persons
A	−tense	swimming pool, going by bus, going window shopping; *bar, meeting a friend*	S11, S15, S6, S19, S10, S8, S18, S4; *S12, S5, S9, S7, S16, S17, S13*
B	talkative, sympathetic; *−tense*	university cafeteria, bar, meeting a friend, family get together, lecture; *going window shopping*	S12
C	−imaginative	in an elevator	S14, S7, S20, S2, S13; *S12, S5, S6, S9, S4*
D	thorough	supermarket; *lecture*	S5, S9, S17, S16; *S20, S15, S8, S13*

Tab. 2: Three-Mode Clustering Solution.

Note. Trait elements with a minus sign refer to traits considered inappropriate in the respective situation(s). Elements added to any of the disjoint clusters in the second step of the analysis appear in italics. Isolated elements (i.e., elements not belonging to any of the four clusters) are: −thorough, −sympathetic, tense, imaginative, −talkative; S3, S1.

References:

ARABIE, P., CARROLL, J.D., and DESARBO, W.S. (1987): *Three-Way Scaling and Clustering*. Sage, Newbury Park.

ARABIE, P., and HUBERT, L.J. (1990): The Bond Energy Algorithm Revisited. *IEEE Transactions on Systems, Man, and Cybernetics, 20, 268–274.*

ARABIE, P., and HUBERT, L.J. (1992): Combinatorial Data Analysis. *Annual Review of Psychology, 43, 169–203.*

ARABIE, P., and HUBERT, L.J. (1994): An Overview of Combinatorial Data Analysis. In: P. Arabie, L. Hubert and G. De Soete (eds.): *Clustering and Classification.* World Scientific, River Edge.

ARABIE, P., HUBERT, L.J., and SCHLEUTERMANN, S. (1990): Blockmodels from the Bond Energy Approach. *Social Networks, 12, 99–126.*

ARABIE, P., SCHLEUTERMANN, S., DAWS, J., and HUBERT, L. (1988): Marketing Applications of Sequencing and Partitioning of Nonsymmetric and/or Two-Mode Matrices. In: W. Gaul and M. Schader (eds.): *Data, Expert Knowledge and Decisions.* Springer-Verlag, Berlin, 215–224.

CARROLL, J.D., and ARABIE, P. (1980): Multidimensional Scaling. *Annual Review of Psychology, 31, 607–649.*

CARROLL, J.D., and ARABIE, P. (1983): INDCLUS: An Individual Differences Generalization of the ADCLUS Model and the MAPCLUS Algorithm. *Psychometrika, 48, 157–169.*

CARROLL, J.D., CLARK, L.A., and DESARBO, W.S. (1984): The Representation of Three-Way Proximity Data By Single and Multiple Tree Structure Models. *Journal of Classification, 1, 25–74.*

CARROLL, J.D., and DE SOETE, G. (1994): Spatial, Non-spatial and Hybrid Models for Scaling. In: P. Arabie, L. Hubert and G. De Soete (eds.): *Clustering and Classification.* World Scientific, River Edge.

CATTELL, R.B. (1966): The Data Box: Its Ordering of Total Resources in Terms of Possible Relational Systems. In: R.B. Cattell (ed.): *Handbook of Multivariate Experimental Psychology.* Rand McNally, Chicago, 67–128.

DE BOECK, P., and ROSENBERG, S. (1988): Hierarchical Classes: Model and Data Analysis. *Psychometrika, 53, 361–381.*

DESARBO, W.S. (1982): GENNCLUS: New Models for General Nonhierarchical Clustering Analysis. *Psychometrika, 47, 449–475.*

DESARBO, W.S., and DE SOETE, G. (1984): On the Use of Hierarchical Clustering for the Analysis of Nonsymmetric Proximities. *Journal of Consumer Research, 11, 601–610.*

DE SOETE, G. (1988): Tree Representations of Proximity Data by Least Squares Methods. In: H.H. Bock (ed.): *Classification and Related Methods of Data Analysis.* North-Holland, Amsterdam, 147–156.

DE SOETE, G., and CARROLL, J.D. (1989): Ultrametric Tree Representations of Three-Way Three-Mode Data. In: R. Coppi and S. Bolasco (eds.): *Multiway Data Analysis.* North-Holland, New York, 415–426.

DE SOETE, G., DESARBO, W.S., FURNAS, G.W., and CARROLL, J.D. (1984): The Estimation of Ultrametric and Path Length Trees From Rectangular Proximity Data. *Psychometrika, 49, 289–310.*

ECKES, T. (1993): A Two-Mode Clustering Study of Situations and Their Features. In: O. Opitz, B. Lausen and R. Klar (eds.): *Information and Classification.*

Springer-Verlag, Berlin, 510–517.

ECKES, T. (1994): Features of Men, Features of Women: Assessing Stereotypic Beliefs about Gender Subtypes. *British Journal of Social Psychology, 33, 107-123.*

ECKES, T. (in press): Features of Situations: A Two-Mode Clustering Study of Situation Prototypes. *Personality and Social Psychology Bulletin.*

ECKES, T., and HASSEBRAUCK, M. (1993): Multimodale Analysen in der physischen Attraktivitäts-Forschung. In: M. Hassebrauck and R. Niketta (eds.): *Physische Attraktivität.* Hogrefe, Göttingen, 95–121.

ECKES, T., and ORLIK, P. (1993): An Error Variance Approach to Two-Mode Hierarchical Clustering. *Journal of Classification, 10, 51–74..*

ECKES, T., and ORLIK, P. (1994): Three-Mode Hierarchical Cluster Analysis of Three-Way Three-Mode Data. In: H.H. Bock, W. Lenski and M.M. Richter (eds.): *Information Systems and Data Analysis.* Springer-Verlag, Berlin, 219–227.

ESPEJO, E., and GAUL, W. (1986): Two-Mode Hierarchical Clustering as an Instrumemt for Marketing Research. In: W. Gaul and M. Schader (eds.): *Classification as a Tool of Research.* North-Holland, Amsterdam, 121–128.

JOHN, O.P. (1990): The "Big Five" Factor Taxonomy: Dimensions of Personality in the Natural Language and in Questionnaires. In: L.A. Pervin (ed.): *Handbook of Personality: Theory and Research.* Guilford Press, New York, 66–100.

McCORMICK, W.T., SCHWEITZER, P.J., and WHITE, T.W. (1972): Problem Decomposition and Data Reorganization by a Clustering Technique. *Operations Research, 20, 993–1009.*

MIRKIN, B., ARABIE, P., and HUBERT, L.J. (1994): Additive Two-Mode Clustering: The Error-Variance Approach Revisited. Paper submitted.

SCHLEUTERMANN, S., ARABIE, P., HUBERT, L.J., and BRONSARD, F. (1990): Some Algorithms for "Bond Energy" Data Analysis, Including Simulated Annealing. In: M. Schader and W. Gaul (eds.): *Knowledge, Data and Computer-Assisted Decisions.* Springer-Verlag, Berlin, 139–152.

SHEPARD, R.N., and ARABIE, P. (1979): Additive Clustering: Representation of Similarities as Combinations of Discrete Overlapping Properties. *Psychological Review, 86, 87–123.*

WARD, J.H. (1963): Hierarchical Grouping to Optimize an Objective Function. *Journal of the American Statistical Association, 58, 236–244.*

Gaussian Neural Networks Applied to the Cluster Analysis Problem

Christian Firmin, Denis Hamad

Centre d'Automatique de Lille, Bâtiment P2,
Université des Sciences et Technologies de Lille
F-59655 Villeneuve d'Ascq, Cedex, France

Summary: This paper describes a Gaussian neural network (GNN) applied to the cluster analysis problem. The GNN architecture is constituted by one layer of Gaussian units and one output unit which provides an estimation of the probability density function of the mixture. During the training of the network, a weighted competitive learning approach is used to estimate both the mean vector and the covariance matrix for each Gaussian function of the hidden units. The key problem with the GNN networks is the determination of the number of units in the hidden layer. This problem is solved by means of three information criteria. The interest of this approach lies in the adjusting of the number of units in an unsupervised context. Some results are reported and the performance of this approach is evaluated.

1. Introduction

In this paper, we propose to use a GNN network for the problem of classification in an unsupervised environment. The architecture of the Gaussian network is devoted to the mixture identification problem. It is made of a three feed-forward layers : the input layer simply passes the input values to the hidden layer which is composed of Gaussian units and the output layer constituted of one unit which provides an estimation of the probability density function.

The training of this network consists in adjusting its parameters, namely the mean vector and the non-diagonal covariance matrix for each Gaussian unit and a set of weights that relates the Gaussian units to the output unit. In our network the Gaussian functions are hyperellipsoidally shaped while they are hyperspherically shaped when using the radial basis function network (RBF), Moody and Darken (1989), or the probabilistic neural network (PNN), Specht (1990). During the training phase of the network, a weighted competitive learning approach is used to estimate both the mean vector, and the non-diagonal covariance matrix for each Gaussian function of the hidden units. This learning scheme may avoid the convergence towards non significant local minima of the parameters of the Gaussian functions, due to the initialisation problem, Ahalt et al. (1990), Delsert et al. (1993).

The number of units in the hidden layer, which is also the number of components in the mixture, is estimated by means of three informational criteria:

the logarithm of the likelihood criterion (LLC), the Akaike information criterion (AIC), Akaike (1974), and the minimum description length criterion (MDL), Rissanen (1978). The two last criteria are known as penalized log-likelihood criteria. The penalty term depends upon the number of hidden units, in the Gaussian finite mixture context, Bozdogan (1992), Culter and Windham (1994). The interest of these criteria lies in the adjusting of the number of hidden units in an unsupervised context and in the fact that it does not require any threshold to be fixed.

2. The Finite Mixture Identification Problem

Let us consider a set of unlabelled samples drawn from a finite mixture of Gaussian densities. If we assume that each Gaussian component represents one cluster, the pattern classification problem can then be viewed as a parameter estimation problem. The key problem is then to find the number of components and to estimate the parameters of the components, namely the mean vectors and covariance matrices and their *a priori* probabilities. When the mixture is identified, a Bayes classifier is constructed by means of classical discriminant functions, Duda and Hart (1973), Postaire and Vasseur (1981).

Let the given data set consists of Q unlabelled N-dimensional samples $X_q \in R^N$, each denoted by a vector $X_q = [X_{q,1}; X_{q,2}; \ldots; X_{q,N}]^T$, where $q = 1, 2, \ldots, Q$ and let us assume that the underlying probability density function, $f(X)$, actually follows the model:

$$f(X) = \sum_{k=1}^{K} \pi_k g_k (X; M_k, \Sigma_k) \tag{1}$$

- where the $K-1$ independent mixing parameters π_k, $k = 1, \ldots, K$, are the *a priori* probabilities, and must satisfy the constraints:

$$\pi_k \geq 0; \qquad \sum_{k=1}^{K} \pi_k = 1 \tag{2}$$

- and where $g_k (X; M_k, \Sigma_k)$ is the k-th component multivariate normal density function associated to the cluster C_k, with mean vector M_k and covariance matrix Σ_k:

$$g_k (X; M_k, \Sigma_k) = (2\pi)^{-N/2} | \Sigma_k |^{-1/2} \exp \left[-\frac{1}{2} (X - M_k)^T \Sigma_k^{-1} (X - M_k) \right] \tag{3}$$

Σ_k^{-1} and $| \Sigma_k |$ are the inverse and the determinant of the covariance matrix Σ_k respectively. The subscript T denotes the transpose.

Let $\theta = (\pi_k, M_k, \Sigma_k)$ be the vector constituted by the $1 + N + N(N+1)/2$ unknown parameters of the k-th component and $\Theta = (\theta_1, \ldots, \theta_k, \ldots, \theta_K)$ the vector of all unknown parameters.

The problem is to estimate these unknown parameters constituting the vector Θ and the number K of components in the mixture. Two approaches are usually used: the maximum likelihood estimator and the K-means procedure, Duda and Hart (1973). However, the results of these algorithms depend upon the initialisation of the centres of the clusters which frequently leads to a non significant local minima. In the following we treat this problem by the weighted competitive learning technique.

3. Architecture and Training of the Gaussian Neural Network

The GNN is a feed-forward type of network constituted of three layers: the input, the hidden and the output layers. The input units simply send the input values through connecting arcs to the hidden units. Thus, each hidden unit receives unaltered input values. The transfer functions of the hidden units are Gaussian functions. The arcs connecting the output of the hidden units to the output unit are weighted. The output layer is constituted of one unit which provides the probability density function of the mixture provided the parameters of the Gaussian functions in the hidden layer as well as the number of units in this layer are correctly adjusted.

In RBF networks, the centre of the radial functions are estimated by means of a batch K-means clustering algorithm. The radial function of each hidden unit is isotropic and its width is calculated by the distance of the centre of this unit to its P-nearest neighbour centre, Moody and Darken (1989). Another approach, known as generalised radial basis functions, uses the regularization technique to estimate the parameters of these functions, Poggio and Girosi (1990).

In this paper, we use another approach to simultaneously adjust the means and the covariance matrices of the Gaussian functions. The principle of this approach combines the use of the adaptive Mahalanobis distance and the weighted competitive learning concepts.

The weighted competitive learning technique consists in incorporating a coefficient in each neural unit which indicates the number of times this unit is the winner during the learning stage, Ahalt et al. (1990), Delsert et al. (1993). If a given unit wins the competition frequently the corresponding coefficient and, consequently, its distance to the data sample, increases. This phenomenon reduces the likelihood that the unit should be the winner in the next steps. Then, other units, with lower coefficients, have more chance to win the competition.

3.1 Estimation of the Gaussian Parameters

The learning technique consists in presenting sequentially the Q unlabelled observations and in adjusting the parameters of the winner unit after each presentation according to the following scheme:

$$M_i(t+1) = M_i(t) + \alpha(t)\left[X(t) - M_i(t)\right] \tag{4}$$

$$\Sigma_i(t+1) = \Sigma_i(t) + \alpha(t)\left[(X(t) - M_i(t))^T (X(t) - M_i(t)) - \Sigma_i(t)\right] \quad (5)$$

- i is the index of the winner unit, i.e. the unit whose mean vector is the closest to the observation $X(t)$ presented at time t, according to the modified Mahalanobis distance:

$$D_i(X(t), M_i(t)) = a_i(t)\left[(X(t) - M_i(t))^T \Sigma_i^{-1}(t)(X(t) - M_i(t))\right] \quad (6)$$

where $a_i(t)$ is the number of times the i-th unit wins the competition. Note that, for singular solutions of Σ_i, we use the Euclidean distance instead the Mahalanobis distance in equation (6).

- $\alpha(t)$ is the learning rate defined by (7):

$$\alpha(t) = \alpha_0(1 + t/t_0)^{-1} \quad (7)$$

where parameters α_0 and t_0 are adjusted by the analyst. This type of learning schemes are known as search-then-converge schedule and its propertie has been studied by Darken and Moody (1991). Note that, up to now, it has only be applied to search the centres or means in the K-means clustering procedure. In this paper, we apply it in order to find the mean vectors and the non-diagonal covariance matrices.

3.2 Estimation of the Weights of the Connections Between the Hidden Units and the Output Unit

In the supervised RBF network the weights of the arcs connecting the hidden units to the output unit are calculated by minimising the error between the desired and the estimated outputs, in the least square sense. However, in an unsupervised context, the weights of these connections are computed such that the output of the output unit is an estimation of the probability density function of the mixture. Under the hypothesis that each Gaussian function represents one cluster, these weights are the *a priori* probabilities of the different clusters. Then, at the end of the learning process, they can be estimated by the ratio between the number of samples associated to the kth cluster and the size of the available data set:

$$\pi_k = \frac{1}{Q}\sum_{q=1}^{Q} I_k(X_q) \quad (8)$$

where $I_k(X_q)$ is equal to 1 if X_q belongs to the kth cluster and to 0 otherwise.

4. Determination of the Number of Gaussian Units

The only unknown parameter which remains to be adjusted is the number K of Gaussian units constituting the hidden layer. Three informational criteria of validity functionals are considered: the log-likelihood, the penalized log-likelihood, Akaike (1974), and the minimum description length, Rissanen (1978). For most details concerning validity functionals for mixture models, see (Bozdogan 1992), Culter and Windham (1994).

The first validity functional which we consider is the log-likelihood criterion, (LLC):

$$\mathrm{LLC} = -2\sum_{q=1}^{Q} \log \left[\sum_{k=1}^{K} \hat{\pi}_k g_k \left(X_q, \hat{M}_k, \hat{\Sigma}_k \right) \right] \tag{9}$$

where $\hat{M}_k$ and $\hat{\Sigma}_k$ are computed by equations (4) and (5) of the weighted competitive learning algorithm and $\hat{\pi}_k$ is estimated at the end of the clustering procedure by equation (8).

Akaike suggested to use a penalized log-likelihood criterion known as Akaike's information criterion (AIC), Akaike (1974). The penalty term is proportional to the number of independent parameters of the mixture. According to Bozdogan, the minimum value of the AIC is given by the following relationship to, Bozdogan (1992):

$$\mathrm{AIC} = \mathrm{LLC} + 3n_p(K) \tag{10}$$

$n_p(K)$ is the number of independent parameters which is given by :

$$n_p(K) = (K-1) + (KN) + (KN(N+1)/2) \tag{11}$$

where the first term in the right side of equation (11) indicates the number of independent mixing proportions, the second term the number of elements of the mean vectors and the third term the number of independent elements of covariance matrices of the components of the mixture.

In the coding theory concept, Rissanen, Rissanen (1978), introduced the idea of stochastic complexity and proposed the criterion of minimum description length (MDL) defined by Rissanen (1978):

$$\mathrm{MDL} = \mathrm{LLC} + \log(Q)\, n_p(K) \tag{12}$$

In the following, the three criteria (10), (11) and (12) are used in order to adjust the number of Gaussian units.

5. Experimental Results

Many experiments have been carried out in order to evaluate the performance of this unsupervised classification approach for different shapes of clusters, degrees of overlapping between clusters, dimensionalities of data and numbers of samples in each cluster, Firmin and Hamad (1994). However, for the sake of clarity, only two experiments are presented in what following in which. For the two experiments the vector centres of the Gaussian units have been randomly initialized and the covariance matrices have been initialized to identity matrices.

Experiment 1 - The data set used for the first experiment is shown in Fig. 1 (a). There are five clusters of 300 samples each. The clusters are drawn from five Gaussian distributions centred at (0,0), (1,1), (1,-1), (-1,-1) and (-1,1) in the data space. The covariance matrix of the first cluster is a diagonal one with $\Sigma_1 = \text{diag}\{0.2, 0.2\}$. The covariance diagonal matrices of the other four clusters are taken to $\text{diag}\{0.05, 0.3\}$ with rotational angle $\pm\pi/4$.

The number of units in the hidden layer of the Gaussian network varies from

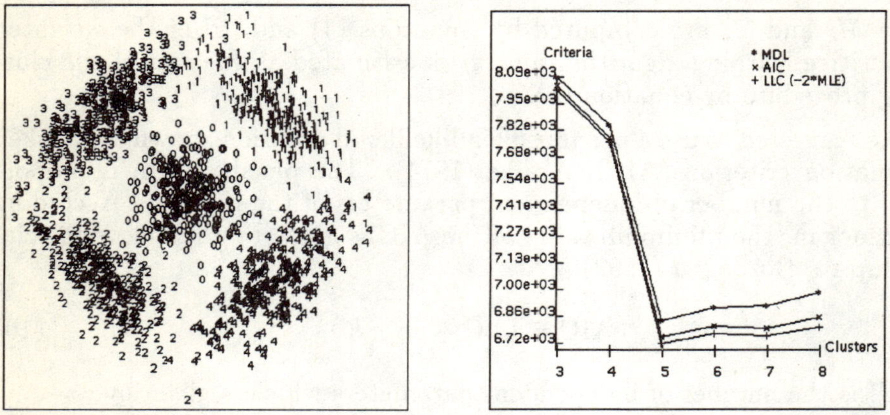

(a) (b)

	unit 1	unit 2	unit 3	unit 4	unit 5	
cluster 1	**286**	0	0	1	0	287
cluster 2	4	**292**	1	0	3	300
cluster 3	2	4	**298**	2	0	306
cluster 4	4	0	1	**293**	2	300
cluster 5	4	4	0	4	**295**	307
	300	300	300	300	300	1500

(c)

Fig. 1:(a) The data set used for experiment 1. Each cluster is composed of 300 samples drawn from five different Gaussian distributions. (b) Variations of the LLC, AIC and MDL criteria as a function of the number of clusters, the actual number of clusters is equal to 5. (c) The confusion matrix obtained by the Gaussian neural network when K=5.

3 to 8, and for each value, the parameters of the network are estimated and the three preceeding criteria are computed. The actual number of units, which is also the number of clusters, corresponds to the minimum of the three criteria, as it is clearly indicated in Fig. 1 (b). The results of the unsupervised pattern classification are evaluated by the confusion matrix in Fig. 1 (c). The error rate is equal to 2.4 % which corresponds to 36 misclassified samples over a total of 1500 samples.

Experiment 2 - In the second experiment, we use the Fisher's IRIS data set as an experimental test. This data set, consisting of four features measured on 150 flowers, 50 in each of three species (iris setosa, iris versicolor and iris virginia), has been used in many papers to illustrate various unsupervised and supervised classifier designs. Note that, the three clusters are not spherically-shaped. Typical error rates for supervised designs are 0 to 5 "mistakes"; and for unsupervised designs, around 15 "mistakes", Pal et al. (1993).

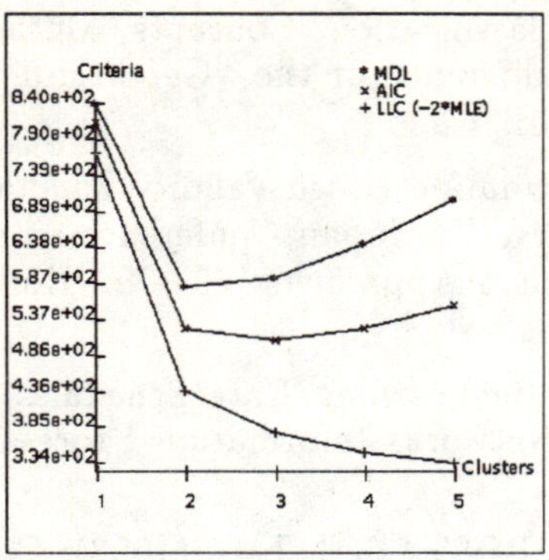

(a)

	unit 1	unit 2	unit 3	
cluster 1	**50**	0	0	50
cluster 2	0	**45**	2	47
cluster 3	0	5	**48**	53
	50	50	50	150

(b)

Fig. 2: (a) Variation of LLC, AIC and MDL criteria as functions of the number of clusters. (b) The confusion matrix for the Fisher's IRIS obtained by the Gaussian neural network with K=3.

In order to choose the number of components, the three criteria have been represented in Fig. 2 (a). The log-likelihood criterion LLC decreases when the number of clusters increases and is useless to detect the true number of clusters. The minimum of the MDL criterion occurs for a number of clusters equal to two. Only the AIC criterion had successfully determined the true number of clusters which is equal to three. The classification approach produces only 7 "mistakes" , as it is shown in the confusion matrix represented in Fig. 2 (b). We notice that, the K-means algorithm misclassifies 16 samples.

6. Conclusion

The unsupervised classification problem is investigated using a Gaussian neural network. A weighted competitive learning scheme was used to simultaneously estimate the mean vector and the non-diagonal covariance matrix of each component. This learning scheme is able to avoid the convergence towards non significant local minima of the parameters of Gaussian functions due to the initialisation problem. The determination of the number of hidden units of the network, and, therefore, the number of unknown components, is achieved by minimizing an informational criteria.

References:

AHALT, S., KRISHNAMURTHY, A., CHEN, P., and MELTON, D. (1990): Competitive Leaning Algorithm for Vector Quantization. Neural Networks, 3, 277-290.

AKAIKE, H., (1974): A New Look at the Statistical Model Identification. IEEE Trans. on Automatic Control, AC-19, 6, 716-722.

BOZDOGAN, H. (1992): Choosing the Number of Component Clusters in the Mixture- Model Using a New Informational Complexity Criterion of the Inverse-Fisher Information Matrix. Information and Classification, Concepts, Methods and Applications. Proceeding of the 16-th Conference of the "Gesellshaft für Klassification e.V.", University of Dormund, April 1-3.

CULTER, A., and WINDHAM, M. (1994): Information-Based Validity Functionals for Mixture Analysis. Proceeding of the First US/Japan Conference on the Frontier of Statistical Modeling: An Informational Approach, 149-170. Kluwer Academic Publishers. Printed in the Netherlands.

DARKEN, C., and MOODY, J. (1991): Note on Learning Rate Schedules for Stochastic Optimization. Advances in Neural Networks Information Processing Systems 3, Morgan Kauffmann Publishers, 1991.

DELSERT, S., HAMAD, D., DAOUDI, M., and POSTAIRE, J.-G. (1993): Competitive Learning Neural Networks Applied to Multivariate Data Set Reduction. IEEE Int. Conf. on SMC, 4, 496-500, Le Touquet, France.

DUDA, R. and HART, P. (1973): Pattern Classification and Scene Analysis. New York: John Wiley & Sons.

FIRMIN, C. and HAMAD, D., (1994): Gaussian Based Neural Networks Applied to Pattern Classification and Multivariate Probability Density Estimation. WCCI '94, IEEE International Conference on Neural Networks, Orlando, Florida, June 26-July 2.

KOHONEN, T. (1990): The Self Organizing Map. Proceedings of IEEE (9), 1464-1479.

MOODY, J., and DARKEN, C. (1989): Fast Learning in Networks of Locally-Tuned Processing Units. Neural Computation, 1, 281-294.

PAL, N., BEZDEK, C., and TSAO, E. (1993): Generalized Clustering Networks and Kohonen's Self-Organizing Scheme, IEEE Trans. on Neural Networks, 4, 549-557.

POSTAIRE, J.-G., and VASSEUR, C. (1981): An Approximate Solution to Normal Mixture Identification with Application to Unsupervised Pattern Classification. IEEE Trans. on Pattern Analysis and Machine Intelligence, PAMI-3, 2, 163-179.

SPECHT, D. (1990): Probabilistic Neural Networks. Neural Networks, 3, 109-118.

Graph-Theoretic Models for Testing the Homogeneity of Data

E. Godehardt[1], A. Horsch[1]

[1] AG Biometrie der Klinik für Thorax- und Kardiovaskular-Chirurgie, Heinrich Heine-Universität, Postfach 10 10 07, D-40001 Düsseldorf

Summary: In cluster analysis, the random graph model $\mathcal{G}_{n,p}$ and $\mathcal{G}_{n,p}$-based multigraph models have been used for purposes of statistical modelling of data and testing the randomness of outlined clusters. While being appropriate for non-metric data, such models supposing independence of all edges do not take into account the triangle inequality which is valid for metric data. We will introduce graph models $\mathcal{I}_{n,d}$ and $\mathcal{I}_{t,n,(d_1,\ldots,d_t)}$ for random intersection graphs in $\mathcal{R}^1$ and multigraphs in $\mathcal{R}^t$ under which the triangle inequality holds. We derive limit theorems for the distribution of random variables which describe important properties of these random intersection graphs. While being asymptotically equivalent for some properties like the limit distribution of the number of isolated points, the $\mathcal{G}_{n,p}$-model and the $\mathcal{I}_{n,d}$-model differ in numerous aspects.

1. Introduction

Graph-theoretical models are useful when the structures of data sets have to be explored. In cluster analysis, the similarity matrix induces a graph structure on the n objects to be clustered (*vertices* $1,\ldots,n$): two vertices are connected by an *edge* if and only if the corresponding objects are similar enough. Using the concept of random graphs, we can derive test statistics to test the randomness of the clusters found. This means that we look whether properties found in the graph obtained from the data, are likely under the assumption of randomness in the process of edge drawing or not. The following two probability models for random graphs are usually used.

Uniform model (Erdős (1960)). *A random graph $\mathcal{G}_{n,N}$ on n vertices is obtained by choosing N edges at random among the $\binom{n}{2}$ possible edges, so that all possible choices are supposed to be equiprobable.* •

Binomial model (Gilbert (1959)). *For each pair of n labelled vertices make a random choice, independent of each other, whether or not to join the vertices of the pair by an edge with the common probability p. This gives a random graph $\mathcal{G}_{n,p}$.* •

By considering the asymptotic behaviour of sequences of random graphs, we can describe the "typical stucture" of random graphs. For deriving asymptotic results, the numbers N in $\mathcal{G}_{n,N}$ and the probabilities p in $\mathcal{G}_{n,p}$ are usually considered as specific functions of n, additionally assuming $N/n^2 \to 0$ or $p(n) \to 0$ for $n \to \infty$. If we put $\binom{n}{2}\,p(n) = N(n)$, then sequences of random graphs $(\mathcal{G}_{n,N})_{n\to\infty}$ and of random graphs $(\mathcal{G}_{n,p})_{n\to\infty}$ show the same asymptotic behaviour. This has been proved in Bollobás (1985) and Łuczak (1990). From Erdős (1960), the threshold probability functions for Poisson approximations are well known for various random variables in the $\mathcal{G}_{n,p}$-model, including the number V of edges, the degree $U_{i,\cdot}$ of a vertex i, the number X of isolated vertices and the number Z of components. We state these results as Theorem 1.

Theorem 1. *For sequences $(\mathcal{G}_{n,p})_{n\to\infty}$ of random graphs with edge probabilities $p = p(n)$ we get:*

1. *For $p(n) = (c + o(1))/n^2$, the number V of edges tends to a Poisson distribution, $V \to \mathrm{Po}(\lambda)$ where $\lambda = c/2$.*

2. *For $p(n) = (c + o(1))/n$, the degree $U_{i,\cdot}$ of any vertex i tends to a Poisson distribution, $U_{i,\cdot} \to \mathrm{Po}(\lambda)$ where $\lambda = c$.*

3. *For $p(n) = (\log n + c + o(1))/n$, the number X of isolated points and the number Z of components, diminished by 1, both tend to the same Poisson distribution, $X \to \mathrm{Po}(\lambda)$ and $Z - 1 \to \mathrm{Po}(\lambda)$ where $\lambda = e^{-c}$.* •

Note that we have the same threshold functions for the number of isolated vertices and the number of components, decremented by 1. That means that for $p(n)$ as given in part (3), we expect only one non-trivial component for large numbers of vertices; the other components are isolated vertices.

If we can assume that the edges are drawn at random under the hypothesis of homogeneity of the data, then we can use these results for testing the hypothesis of homogeneity in a data set as follows. We can choose, for example, the number of isolated points as a test statistic. For n not too small, and p given by part (3) of Theorem 1, the typical structure of a random graph is that of a graph having one "giant" component and additionally a small number of isolated points. Therefore, the data set is assumed to be inhomogeneous—and thus the detected cluster structure is said to be real—if the number of isolated vertices found in the graph generated from the data differs too much from the one we would expect under the conditions of the uniform or the binomial model.

The use of results like those of Theorem 1 in cluster analysis together with the benefits and constraints, and the problems these models imply have been discussed in several papers (see for example Godehardt (1990, 1993, 1994), Ling (1973)). Quite often, these random graph models can be used very well in sociometry and psychometry where we have non-metric data and build a graph from a directed graph as described in Godehardt (1994).

2. A Probability Model for Intersection Graphs

In medical or biological research, we mostly deal with quantitative data from a t-dimensional space $\mathcal{R}^t$ with some metric which implies the similarity structure in the data set. Two vertices i and j of the graph generated from the data then are connected by an edge if and only if the mutual distance $d_{i,j}$ between i and j is not greater than a user-defined level d. Every such graph for a distance level d is called an *Intersection graph* $\mathcal{I}(d)$ (for an exact definition of intersection graphs and some applications see Roberts (1976)). Clusters, which are defined as special subgraphs of such a graph $\mathcal{I}(d)$ constructed from the data and a level d are called *clusters of level d*.

The triangle inequality for metric data is violated by the binomial and the uniform model. Moreover, for metric data, it is the data points that are considered as randomly drawn and not the distances—or edges—as in the $\mathcal{G}_{n,p}$-model. The presence of an edge in our classification model is determined by the distribution of the data points and the distance level d. Therefore, a probability model for *random intersection graphs* $\mathcal{I}(d)$ provides a model which is more appropriate for testing the hypothesis of homogeneity in a metric data set than the binomial or the uniform model.

2.1 The One-dimensional Case

Let us consider the one-dimensional case first. Let n points $x_1, \ldots, x_n$ be drawn at random and independently from the unit interval $[0;1]$ so that each point x_i is the realisation of a uniformly distributed random variable X_i. Let $d_{i,j} = |x_i - x_j|$ be the one-dimensional distances between two points, which are the realisations of the random variables $D_{i,j}$. Let d be a distance level $(0 < d < 1)$. A graph $\mathcal{I}_{n,d}$ defined by the vertex set $\mathcal{V} = \{1, \ldots, n\}$ (corresponding to the set $\mathcal{X} = \{x_1, \ldots, x_n\}$ of the n points), and by the edge set $\mathcal{E} = \{(i,j) : |x_i - x_j| =: d_{i,j} \leq d\}$ is called a *random intersection graph*. Let $U_{i,j}$ be the indicator for the presence or absence of an edge in $\mathcal{E}$. Then

$$\Pr(U_{i,j} = 1) = \Pr((i,j) \in \mathcal{E}) = \Pr(d_{i,j} \leq d) = 2\,d - d^2$$

follows directly for uniformly distributed points in $[0;1]$.

By $U_{i,\cdot} = \sum_{j=1}^{n} U_{i,j}$ we get the number of vertices connected to vertex i, that is the degree of this vertex, and $V = \frac{1}{2} U_{\cdot,\cdot} = \frac{1}{2} \sum_{i=1}^{n} \sum_{j=1}^{n} U_{i,j}$ is the number of edges in $\mathcal{I}_{n,d}$. While n is called the *order* of the graph $\mathcal{I}_{n,d}$, V is its *size*. As in the $\mathcal{G}_{n,p}$-model, the size is a random variable.

As with random graphs $\mathcal{G}_{n,p}$, the typical structure of random intersection graphs can be described by limit theorems. In the binomial and uniform models, we looked at graph properties for special threshold functions $N(n)$ or $p(n)$. In the random intersection graph model, we are interested in threshold functions for the distance level d. Since here, $p = 2\,d - d^2$ is the probabilty for the presence of an edge, we let $d = d(n) \to 0$ for $n \to \infty$. From our probability model for random intersection graphs, the following result for the limit distribution of the degree of any vertex i can be easily derived.

Theorem 2. *For sequences $(\mathcal{I}_{n,d})_{n\to\infty}$ of random intersection graphs with distance levels $d = d(n) = \frac{1}{2n}\,(c + o(1))$, the distribution of the degree of a* *vertex x_i in a random intersection graph $\mathcal{I}_{n,d}$ tends to a Poisson distribution with parameter $\lambda = c$.* •

Proof: Since the points $x_1, \ldots, x_n$ are drawn independently of each other, the distribution of the degree of any vertex i is given by:

$$\Pr(U_{i,\cdot} = k) = \binom{n-1}{k} (2d)^k (1-2d)^{n-k} + 2 \int_d^{2d} y^k (1-y)^{n-1-k} \, dy,$$

giving $\mathrm{Po}(\lambda)$ with $\lambda = c$ as the limit distribution for $n \to \infty$. ■

In Barbour (1992), a slightly different model is used: The n points are uniformly and independently thrown onto a circle of unit length with fixed origin. Let $x_{(1)} \le \cdots \le x_{(n)}$ be the ordered sequence of these points, and let $X_{(1)}, \ldots, X_{(n)}$ be the respective random variables. The arc-length distances $S_1 = d_{(1),(2)}, \ldots, S_{n-1} = d_{(n-1),(n)}, S_n = d_{(n),(1)}$ between successive points are called *spacings*. The *m-spacings* are defined as $S_{k,m} = \sum_{j=0}^{m-1} S_{k+j}$ for $k = 1, \ldots, n$ and $1 \le m \le n/2$, with $S_{n+j} = S_j$. Spacings larger than a given level d are called *big spacings*, spacings smaller than d are called *small spacings*. A.D. Barbour, L. Holst and S. Janson derive results on the asymptotic distribution of the numbers of small m-spacings and big m-spacings, respectively. These are defined as

$$W_{S,m} = \sum_{k=1}^{n} I(S_{k,m} < d), \quad W_{B,m} = \sum_{k=1}^{n} I(S_{k,m} > d) \quad (1 \le m \le n/2)$$

for any d with $0 < d < 1$ (here, I is the indicator variable of the respective property). Their results include the following two theorems.

Theorem 3. *Let $X_1, \ldots, X_n$ be independent and uniformly distributed on a cycle with circumference 1. Let $W_{S,1}$ and $W_{B,1}$ be the numbers of small and big spacings between two successive points. Then we have:*

1. *If $\mathrm{E}\, W_{S,1} \to \lambda \ge 0$ as $n \to \infty$, that is $nd = (\lambda + o(1))/n$, then the distribution of $W_{S,1}$ tends to a Poisson distribution with parameter λ.*
2. *If $\mathrm{E}\, W_{B,1} \to \lambda > 0$ as $n \to \infty$, that is $nd = \log(n) - \log(\lambda) + o(1)$, then the distribution of $W_{B,1}$ tends to a Poisson distribution with parameter λ.* •

Theorem 4. *Let $X_1, \ldots, X_n$ be independent and uniformly distributed on a cycle with circumference 1. Let $W_{S,m}$ and $W_{B,m}$ be the numbers of small and big m-spacings. Then for fixed $m > 1$ we have:*

1. *If $d \to 0$, $nd \to 0$ and $\mathrm{E}\, W_{S,m} \to \lambda \ge 0$ as $n \to \infty$, then the distribution of $W_{S,m}$ tends to a Poisson distribution with parameter λ.*
2. *If $d \to 0$, $nd \to \infty$ and $\mathrm{E}\, W_{B,m} \to \lambda \ge 0$ as $n \to \infty$, then the distribution of $W_{B,m}$ tends to a Poisson distribution with parameter λ.* •

For each of these cases also rates of convergence are given in Barbour (1992).

Now suppose that $W_{S,m}$ has a limit distribution F as $n \to \infty$. Let $\mathcal{A}_{n,m,k}$ denote a k-element random subset of all m-spacings and $\mathcal{B}_{n,m}$ the set of m-spacings covering the origin between $x_{(n)}$ and $x_{(1)}$. Then we have $\Pr(\mathcal{A}_{n,m,k} \cap \mathcal{B}_{n,m} = \emptyset) \geq 1 - k\,m/n \to 1$ as $n \to \infty$. From this it follows that the distribution of the number $\tilde{W}_{S,m}$ of small m-spacings not covering the origin also tends to F. Obviously, the number $\tilde{W}_{B,m}$ of big m-spacings not covering the origin can be dealt with in the same way. As the number V of edges in random intersection graphs $\mathcal{I}_{n,d}$ is identical to $\tilde{W}_{S,1}$, we get the following asymptotic result immediately from Theorem 3.

Theorem 5. *For sequences $(\mathcal{I}_{n,d})_{n\to\infty}$ of random intersection graphs with distance levels $d(n) = \frac{1}{2n^2}\,(c + o(1))$, the distribution of the number V of edges tends to a Poisson distribution with parameter $\lambda = c/2$.* •

For the limit distribution of the number of isolated vertices, we can prove the folllowing theorem.

Theorem 6. *For sequences $(\mathcal{I}_{n,d})_{n\to\infty}$ of random intersection graphs with distance levels $d(n) = \frac{1}{2n}\,(\log(n) + c + o(1))$, the distribution of the number X of isolated vertices tends to a Poisson distribution with parameter $\lambda = e^{-c}$.* •

Proof: First we note that for $d(n)$ as given above, the smallest and the largest point will asymptotically not be isolated:

$$\begin{aligned}\Pr(x_{(1)} \text{ or } x_{(n)} \text{ are isolated}) &\leq 2\Pr(x_{(1)} \text{ is isolated}) \leq (1 - d(n))^{n-1}\\ &\sim e^{-(n-1)d(n)} \to 0 \quad (n \to \infty).\end{aligned}$$

Furthermore, due to Lemma 7.3.1 in Barbour (1992), the distribution of the number of 3-spacings larger than $3\,d(n)$ converges to the degenerate distribution with total mass in 0 as $n \to \infty$, This gives for all $k \geq 0$:

$$\Pr(X = k) - \Pr(\{X = k\} \cap \mathcal{C}) \to 0$$

where $\mathcal{C}$ is the event that each two subsequent isolated points in a random intersection graph are seperated by at least one component with several points. Therefore, asymptotically each isolated point reduces the possible location of the remaining points by $2\,d$. From this the limit probability that a random intersection graph has no isolated points at all, can be found using a formula of inclusion and exclusion. We get

$$\begin{aligned}\Pr(X = 0) &= 1 + \sum_{k=1}^{n} \binom{n}{k} (-1)^k \Pr(x_1, \dots x_k \text{ are isolated})\\ &\sim \sum_{k=0} (-1)^k \binom{n-1-k}{k} (1 - 2kd)^n \sim e^{-e^{-c}}.\end{aligned}$$

The same shifting argument as for the $\mathcal{G}_{n,p}$-model (see Godehardt (1990)) gives the limit probability for getting k isolated vertices for $k \geq 1$. ■

From part (c) of Corollary 7.A.1 in Barbour (1992), it follows that for the distance level $d(n)$ given in Theorem 6, the number of components—which is essentially the number of big spacings larger than d—tends to infinity. The distance level $d(n)$, for which the limit distribution of the number of components in random intersection graphs tends to a Poisson distribution, is given in the following theorem.

Theorem 7. *Let Z be the number of components in random intersection graphs $\mathcal{I}_{n,d}$. For sequences $(\mathcal{I}_{n,d})_{n\to\infty}$ of random intersection graphs with distance levels $d(n) = (\log(n) + c + o(1))\,/n$, the distribution of $Z - 1$ tends to a Poisson distribution with parameter $\lambda = e^{-c}$.* •

Proof: We have $Z = \tilde{W}_{B,1} + 1$. For $d(n)$ as given above, $W_{B,1}$, and thus $\tilde{W}_{B,1}$, too, have $\mathrm{Po}(\lambda)$ with $\lambda = e^{-c}$ as the limit distribution. ■

From Theorems 6 and 7, we see that in the $\mathcal{I}_{n,d}$-model, the threshold functions for the number of isolated vertices and the number of components (diminished by 1) differ by a factor 2. For random graphs $\mathcal{G}_{n,p}$, however, we have the same threshold function for Poisson approximation for both random variables as can be seen from Theorem 1.

2.2 Multigraph Models for the Multidimensional Case

There are many results for the one-dimensional case, which have been derived and used especially in the theory of random clumping (see for example Roach (1968)), and which could be transferred to the more general model of random intersection graphs like Theorem 6 (and thus could be used in cluster analysis, for example, see the discussion later). For the 2- and 3-dimensional case, not so much is known in the theory of random clumps, if we use the Euclidean distance to define the structure in the data set; for t-dimensional data ($t > 3$), virtually no results for random clumpings are known (see Kennedy (1976)).

Often, especially for medical data, the scale levels vary considerably between the different items, that means between the dimensions of the data vectors. It then is questionable if not impossible to compute global distances $d_{i,j}$ over all t dimensions between the elements of a data set. The structure of a data set consisting of n t-dimensional vectors can be described better by a multigraph based on t local distances than by a graph. We therefore introduced a special type of multigraphs to describe this structure and to test the homogeneity within a data set (that means the hypothesis of "randomly generated" clusters).

Definition. *Let t graphs with identical vertex set $\mathcal{V}$ be superposed in t layers. This gives an undirected, completely labelled multigraph, where t edges, labelled by their two end points and by the number of their respective layers, can join the same pair of vertices. Let $\mathcal{G}_t = (\mathcal{V}, \mathcal{E})$ be an undirected,* *completely labelled multigraph with vertices* $i \in \mathcal{V}$, *and edges* $(i,j)_l \in \mathcal{E}$. *Every subset* $\mathcal{E}_{i,j} = \{(i,j)_1, \ldots, (i,j)_t\} \cap \mathcal{E} \neq \emptyset$ *is called a connection between* i *and* j. *Let* s *and* u *be two natural numbers with* $1 \leq s \leq u \leq t$. *If* $s \leq |\mathcal{E}_{i,j}| \leq u$ *holds (that means if at least* s *and at most* u *edges link the vertices* i *and* j *together) then* $\mathcal{E}_{i,j}$ *is called an* (s,u)*-connection.* •

Examples of such multigraphs are given in Godehardt (1990, 1993, 1994). With this definition, we define for natural numbers $1 \leq s \leq u \leq t$, the following mapping from multigraphs to graphs: The (s,u)*-projection* $\tilde{\mathcal{G}}$ of a multigraph $\mathcal{G}_t$ is the graph with the same vertex set and exactly those edges (i,j) for which $s \leq |\mathcal{E}_{i,j}| \leq u$ holds in the multigraph. This mapping $\mathcal{G}_t \rightarrow \tilde{\mathcal{G}}$ lets us generalize the definitions commonly used in graph theory to the multigraphs as defined above. For example an (s,u)*-component* of $\mathcal{G}_t$ is defined by the corresponding component in its (s,u)-projection. Since these generalizations are intuitively clear, we refer to Godehardt (1994) for details. For the case $u = t$, the (s,t)-projection is simply called s*-projection* and we also speak of s*-connections*, s*-isolated vertices*, ..., and so on.

There are several ways to generalize the uniform model and the binomial model of random graphs to probability models of random multigraphs, see Godehardt (1990, 1993, 1994). Their use in testing the randomness of clusters is limited if the cluster structure is based on a metric as in the papers cited before. However, we can define a model for random intersection multigraphs $\mathcal{I}_{t,n,(d_1,\ldots,d_t)}$ analogously: Let each layer l of a random intersection multigraph $\mathcal{I}_{t,n,(d_1,\ldots,d_t)}$ consist of a random intersection graph $\mathcal{I}_{n,d_l}$, $l = 1, \ldots, t$. Note that we have a vector $\vec{d}^T = (d_1, \ldots, d_t)$ of possibly different levels. Probability models for this type of random multigraphs are much more suitable for testing the hypothesis of homogeneity than models for random multigraphs $\mathcal{G}_{t,n,(p_1,\ldots,p_t)}$ with edge probabilities $p_1, \ldots, p_t$. Moreover, there is a geometrical interpretation of the t-projection of an intersection multigraph of level $\vec{d}^T = (d, \ldots, d)$: Two vertices i and j are connected by an edge if the maximum distance between the data points x_i and x_j is smaller than d. For intersection multigraphs of level $(d, \ldots, d)$, we write $\mathcal{I}_{t,n,d}$.

For some applications—especially in medical research—the case $s < t$ is useful, since then two points can be considered as similar if their differences are smaller than the level d_l in at least s of the dimensions $l = 1, \ldots, t$. In this case, the similarity of the points, and thus the cluster structure is not based on a metric as is discussed for example in Godehardt (1990). However, we quite often can argue that two objects should be in the same cluster if they are similar in many attributes, accepting that they may differ significantly in few attributes or dimensions of their data vectors. Thus, the components or the cliques of the s-projection of level $\vec{d}^T$ of the multigraph can define the single-linkage, and the complete-linkage clusters, respectively.

The (s,u)-projection makes little sense in cluster analysis. We do not know a case where two objects should be put into the same cluster if they are similar in at least s and at most u dimensions of the data vector. However, the definition of the (s,u)-projection is the most general case in our model; moreover, in the case of independent layers the asymptotic results for s-projections hold for all (s,u)-projections independent of u.

Let $T_{i,j,l}$ be the indicator that an edge connects i and j in the l-th layer of $\mathcal{I}_{t,n,d}$. Let $U_{(s,u),i,j}$ be the indicator for the edge (i,j) in the (s,u)-projection of $\mathcal{I}_{t,n,d}$, that means $U_{(s,u),i,j} = 1$ if $s \le \sum_{l=1}^{t} T_{i,j,l} \le u$, and $U_{(s,u),i,j} = 0$ else. Then by $U_{(s,u),i,\cdot}$ we get the (s,u)-degree of vertex i, and $V_{(s,u)} = \frac{1}{2} U_{(s,u),\cdot,\cdot}$ is the number of (s,u)-connections in $\mathcal{I}_{t,n,d}$, that is the number of edges in the (s,u)-projection. The following theorem shows the first result for the asymptotic behaviour of random intersection multigraphs.

Theorem 8. *For sequences $(\mathcal{I}_{t,n,d})_{n\to\infty}$ of random intersection multigraphs with t layers, n vertices and*

$$d(n) = \frac{1}{2\binom{t}{s}^{1/s}} \frac{1}{n^{2/s}} \left(c + o(1)\right)^{1/s}$$

as the distance level for each layer, the distribution of the number $V_{(s,u)}$ of edges in the (s,u)-projection of a random intersection multigraph $\mathcal{I}_{t,n,d}$ tends to a Poisson distribution with parameter $\lambda = c/2$. •

Proof: The probability p^* to get an edge (i,j) in the (s,u)-projection of a random intersection multigraph is the probability to get at least s and at most u of the t possible edges between i and j in the original multigraph. Since the random intersection graphs forming the t layers are mutually independent, this probability follows directly by a binomial argument as

$$p^* = \sum_{l=s}^{u} \binom{t}{l} p^l (1-p)^{t-l} \sim \binom{t}{s} p^s$$

with $p = 2d - d^2$. For $d = d(n)$ as given by the formula above, in the (s,u)-projection, the probability for an edge is $p^* = (c + o(1))/n^2$. Therefore the expected number of paths of length 2 is $\binom{n}{3} p^{*2} = O(1/n) \to 0$ (of course, for paths of length 3 or larger, the same holds). Thus, for large n we expect only isolated vertices and isolated edges (pairs of vertices). This gives

$$\begin{aligned} \Pr(V_{(s,u)} = k) &\sim \Pr(V_{(s,u)} = k \text{ and all edges are isolated}) \\ &= \binom{\binom{n}{2}}{k} p^{*k} (1-p^*)^{\binom{n}{2}-k} \end{aligned}$$

and $V_{(s,u)} \to \mathrm{Po}(c/2)$ for $\binom{n}{2} p \to c/2$. ■

3. Discussion of the Intersection Graph Model

Especially in medical research, where it is often tolerated that objects can differ in some variables if they are similar enough in the other variables, the concept of multigraphs is more appropriate than that of simple graphs. For $1 \leq s \leq t$, an s-projection defines the clusters in the sample (for example, as the components of that s-projection). For metric data, we consider the uniform distribution of the data points as the hypothesis of homogeneity. Instead of using the $\mathcal{G}_{n,p}$-model in each layer (and thus proposing a random drawing of edges as probability model), the edge set in each layer of a random intersection multigraph is defined by a level $\vec{d}^T$, which is exactly the way how we generate the multigraph from a data set. Therefore this will give us a probability model which is more appropriate for metric data.

In a first step, we propose a uniform distribution of the data points in each dimension and independence between the dimensions as the null hypothesis of homogeneity versus the alternative of clustered data. The proposition of a uniform distribution is no major disadvantage, since we can transform the data. We believe that the problem of dependent layers, which will prevent us from using our probability model for testing the hypothesis of homogeneity (that means of random clusters) can be solved in a similar way as has been described in Godehardt (1994). In a first step to solve this problem, we propose $d_1 = \cdots = d_t = d$ for the levels in each layer and identical pairwise correlation coefficients between the components of the data vectors (which is the analogon to the proposition of exchangeable indicators and the beta-binomial model used in Godehardt (1994) for random multigraphs $\mathcal{G}_{t,n,p}$.

There is a way to generalize the probability model: We consider the data points not as mathematical points but as centers of cubes with identical edge lengths $d_1, \ldots, d_t$. Two points now are connected by an edge if the respective cubes have a non-empty intersection. This produces the same intersection multigraph as the procedure described in Paragraph 2.2 using distance levels $d_1, \ldots, d_t$ (it additionally explains the name "intersection graph"). We now can consider the edge lengths as random variables $D_1, \ldots, D_t$ with expectations $d_1, \ldots, d_t$. Under additional conditions for the variances of the edge lengths, we expect the same asymptotic results for random edge lengths as for fixed edge lengths (or distance levels). This way of generating an intersection graph was used in Roach (1968). Using cubes with identical edge lengths $d_1 = \cdots = d_t = d$ is the same as generating an intersection graph using level d and the maximum distance; if we replace the cubes by balls with diameter d, then the Euclidean distance is used. For different edge lengths, the cubes or the balls can be replaced by cuboids or ellipsoids, respectively. The (s, u)-projection of an intersection graph can be considered as a generalization in that way that it contains the edges between those points, whose cuboids intersect in at least s and at most u of the t dimensions.

The assumption of a uniform distribution of the data points can be dispensed with. We can derive formulas for the moments of the number of edges in random intersection graphs for several distributions of the data points (this "work under progress" will be published by the authors in a joined paper with B. Harris from the Wisconsin State University in Madison).

Acknowledgements

The proofs of the results on random intersection graphs and multigraphs as presented in Chapter 2 are part of Armin Horsch's doctorial dissertation whose research is supported by DFG grant nos. Go 490/4-1 and Go 490/4-2. The authors wish to thank Jerzy Jaworski from the Adam-Mickiewicz-University and Krystyna T. Balińska from the Technical University (both in Poznań, Poland), and John W. Kennedy and L.V. Quintas from the Pace University in New York, for many helpful discussions during several research stays in Düsseldorf, Poznań and New York which were supported by DFG grant no. Go 490/4-1, too.

References:

BARBOUR, A.D., HOLST, L., JANSON, S. (1992): *Poisson approximations.* Clarendon Press, Oxford.

BOCK, H.H. (1980): Clusteranalyse – Überblick und neuere Entwicklungen. *OR Spektrum, 1, 211–232.*

BOLLOBÁS, B. (1985): *Random graphs.* Academic Press, London – New York – Tokyo.

ERDŐS, P., RÉNYI, A. (1960): On the evolution of random graphs. *Publications of the Mathematical Institute of the Hungarian Academy of Sciences, 5, 17–61.*

GILBERT, E.N. (1959): Random graphs. *Annals of Mathematical Statistics, 30, 1141–1144.*

GODEHARDT, E. (1990): *Graphs as structural models: The application of graphs and multigraphs in cluster analysis (Advances in systems analysis, Vol. 4).* Friedr. Vieweg & Sohn, Braunschweig – Wiesbaden.

GODEHARDT, E. (1993): Probability models for random multigraphs with applications in cluster analysis. *Annals of Discrete Mathematics, 55, 93–108.*

GODEHARDT, E., HORSCH, A. (1994): Testing of data structures with graph-theoretical models. in: Bock, H.H., Lenski, W., Richter, M.M. (eds.): *Information systems and data analysis (Proceedings 17th Annual Conference of the Gesellschaft für Klassifikation e.V., Kaiserslautern, March 3–5, 1993).* Springer, Berlin – Heidelberg – New York, 226–241

LING, R.F. (1973): A probability theory of cluster analysis. *Journal of the American Statistical Association, 68, 159–164.*

ŁUCZAK, T. (1990): On the equivalence of two basic models of random graphs. In: M. Karoński, J. Jaworski, A. Ruciński (eds.): *Random Graphs '87.* John Wiley & Sons, New York – Chichester – Brisbane, 151–157.

KENNEDY, J.W. (1976): Random clumps, graphs, and polymer solutions. In: Y. Alavi, D.R. Lick (eds.): *Theory and Applications of Graphs.* Springer, Berlin – Heidelberg – New York, 314–329.

ROACH, S.A. (1968): *The theory of random clumping.* Methuen & Co, London.

ROBERTS, F.S. (1976): *Discrete mathematical models.* Prentice-Hall, Englewood Cliffs.

Least Squares Multidimensional Scaling with Transformed Distances

Patrick J.F. Groenen[1], Jan de Leeuw[2] and Rudolf Mathar[3]

[1]Department of Data Theory, University of Leiden
P.O. Box 9555, 2300 RB Leiden, The Netherlands
[2]Interdivisional Program in Statistics, UCLA
[3]Institute of Statistics, Aachen University of Technology
Wüllnerstraße 3, D-5100 Aachen, Germany

Summary: We consider a general least squares loss function for multidimensional scaling. Special cases of this loss function are STRESS, S-STRESS, and MULTISCALE. Several analytic results are presented. In particular, we present the gradient and Hessian, and look at the differentiability at a local minimum. We also consider fulldimensional scaling and indicate when a global minimum can be obtained. Furthermore, we treat the problem of inverse multidimensional scaling, where the aim is to find those dissimilarity matrices for which a fixed configuration is a stationary point.

1. Introduction

Various loss functions exist for performing multidimensional scaling (MDS) that all aim at representing n objects in a p dimensional space such that the distances correspond in some optimal sense to fixed nonnegative dissimilarity measures δ_{ij} for every pair of objects i, j. Here, we consider the general least squares loss function

$$\sigma(\mathbf{X}, \Delta) = \sum_{i=1}^{n}\sum_{j=1}^{n} w_{ij}\left(f(d_{ij}(\mathbf{X})) - f(\delta_{ij})\right)^2. \tag{1}$$

It is convenient to express the squared distance between row i and row j of the $n \times p$ coordinate matrix $\mathbf{X}$ as $d_{ij}(\mathbf{X}) = \text{tr}\,(\mathbf{X}'\mathbf{A}_{ij}\mathbf{X})$, where $\mathbf{A}_{ij} = (\mathbf{e}_i - \mathbf{e}_j)(\mathbf{e}_i - \mathbf{e}_j)'$ with $\mathbf{e}_i$ equal to column i of the identity matrix. $w_{ij} = w_{ji}$ are fixed nonnegative weights with $w_{ii} = 0$. The function $f(z)$ could be any function from $\Re^1$ to $\Re^1$, although we shall assume that $f(z)$ is twice continuously differentiable over the domain $(0, \infty)$ and that the inverse function $f^{-1}(z)$ exists such that $f^{-1}(f(z)) = f(f^{-1}(z)) = z$. We focus on three particular examples of $f(z)$, i.e., $f(z) = z^{1/2}$ gives Kruskal's (1964) raw STRESS function, $f(z) = z$ gives S-STRESS (Takane, Young, and de Leeuw (1977)), and $f(z) = \log(z)$ gives Ramsay's (1977) MULTISCALE loss function. For these cases several algorithms for minimizing (1) over $\mathbf{X}$ exist, notably for S-STRESS the ALSCAL algorithm (Takane et al. (1977)), an algorithm of Glunt, Hayden, and Liu (1991), and the Newton–Raphson algorithm of Browne (1987). For the STRESS case the KYST algorithm (Kruskal, Young, and Seery (1977)) and SMACOF of de Leeuw and Heiser (1980) can be used. An algorithm based on a probabilistic version of (1) with replications was presented by Stoop, Heiser, and de Leeuw (1981).

In the next section we present the gradient and Hessian of the general least squares MDS loss function (1) and investigate several useful properties. Then we consider a special case, fulldimensional scaling, and indicate in what situations a global minimum can be obtained. One of the problems of the algorithms above is that they usually stop at a local minimum, which need not be the global minimum. In order to get a better understanding of the local minimum problem we also study its inverse problem; what dissimilarity matrices Δ have some given $\mathbf{X}$ as local minimum. This problem of *inverse scaling* has been discussed first in de Leeuw and Groenen (1993) using STRESS.

2. The Gradient and Hessian

For a local minimum $\mathbf{X}^*$ we need that, if it exists, the gradient equals zero and the Hessian is nonnegative definite. Explicit formulae for gradient and Hessian are given below.

A necessary condition for a stationary point $\mathbf{X}$ is that the gradient of $\sigma(\mathbf{X}, \Delta)$ is equal to zero. This gradient, if it exists, can be written as

$$\frac{\partial \sigma(\mathbf{X}, \Delta)}{\partial \mathbf{x}_s} = 4 \sum_{i=1}^{n} \sum_{j=1}^{n} w_{ij} \left(f(d_{ij}(\mathbf{X})) - f(\delta_{ij})\right) f'(d_{ij}(\mathbf{X})) \mathbf{A}_{ij} \mathbf{x}_s, \tag{2}$$

where $f'(z)$ denotes the first derivative of f at z and $\mathbf{x}_s$ is column s of $\mathbf{X}$. Since $\mathbf{A}_{ij}$ is double centered (has row and column sums equal to zero) we may assume that $\mathbf{X}$ also has column sum zero. It is not difficult to see that if $\mathbf{X}^*$ has zero gradient, $\mathbf{X}^*\mathbf{T}$ with $\mathbf{T}$ a rotation matrix ($\mathbf{TT}' = \mathbf{I}$) is also a stationary point since distances do not change under rotation of $\mathbf{X}$. Note that in general (2) may not be defined everywhere. Especially when zero distances occur $f(z)$ or $f'(z)$ may not be defined. For S-STRESS this problem does not occur which was an important reason for proposing this MDS loss function.

Of course, the set of configurations with zero gradient includes local minima, local maxima and saddle points. If the gradient of $\sigma(\mathbf{X}, \Delta)$ at $\mathbf{X}$ is zero and the Hessian $\mathbf{H}$ is positive definite, i.e., $\mathbf{y}'\mathbf{H}\mathbf{y} > 0$ for all $\mathbf{y} \neq 0$, then we have a strict local minimum at $\mathbf{X}$. Moreover, a necessary condition for a local minimum of σ is that the gradient vanishes and the Hessian $\mathbf{H}$ is positive semidefinite, i.e., $\mathbf{y}'\mathbf{H}\mathbf{y} \geq 0$ for all $\mathbf{y}$. The Hessian is a $p \times p$ partitioned block matrix with blocks

$$\begin{aligned}\mathbf{H}_{st} \;=\; & 4\beta^{st} \sum_{i=1}^{n} \sum_{j=1}^{n} w_{ij}(f(d_{ij}(\mathbf{X})) - f(\delta_{ij}))f'(d_{ij}(\mathbf{X}))\mathbf{A}_{ij} + \\ & 8 \sum_{i=1}^{n} \sum_{j=1}^{n} w_{ij}(f(d_{ij}(\mathbf{X})) - f(\delta_{ij}))f''(d_{ij}(\mathbf{X}))\mathbf{A}_{ij}\mathbf{x}_s\mathbf{x}_t'\mathbf{A}_{ij} +\end{aligned}$$

$$8\sum_{i=1}^{n}\sum_{j=1}^{n} w_{ij}(f'(d_{ij}(\mathbf{X})))^2\mathbf{A}_{ij}\mathbf{x}_s\mathbf{x}_t'\mathbf{A}_{ij} \tag{3}$$

of size $n \times n$, where $\beta^{st} = 1$ if $s = t$ and $\beta^{st} = 0$ otherwise. In Tab. 1 we give the particular gradients and in Tab. 2 the Hessians of STRESS, S-STRESS and MULTISCALE.

Some properties of the Hessian can be derived. If $\mathbf{H}$ is defined, then $\mathbf{H}$ has only real eigenvalues, since $\mathbf{H}$ is symmetric. Furthermore, $\mathbf{H}$ is rank deficient, which implies that $\mathbf{H}$ has zero eigenvalues. If the $np \times 1$ vector $\mathbf{y}$ is an eigenvector corresponding to a zero eigenvalue, then $\sum_{t=1}^{p} \mathbf{H}_{st}\mathbf{y}_t = \mathbf{0}$, where $\mathbf{y}$ is partitioned in p $n \times 1$ vectors $\mathbf{y}_t$. The Hessian $\mathbf{H}$ has p eigenvalues equal to zero corresponding to the p orthogonal eigenvectors for which $\mathbf{y}_s = \mathbf{1}$ if $s = t$ and $\mathbf{y}_s = \mathbf{0}$ if $s \neq t$. In addition, if $\mathbf{X}$ is a stationary point, then $\mathbf{H}$ has at least $p(p-1)/2$ additional eigenvalues equal to zero. Let $\mathbf{Y} = (\mathbf{y}_1|\ldots|\mathbf{y}_p) = \mathbf{XS}$ with $\mathbf{S}$ skewsymmetric, i.e., $\mathbf{S} = -\mathbf{S}'$. Without loss of generality we may assume that $\mathbf{X}$ is centered and of rank p. For any s consider $\sum_{t=1}^{p} \mathbf{H}_{st}\mathbf{y}_t$, which is equal to zero if $\mathbf{y}$ is the eigenvector corresponding to a zero eigenvalue. The first term of (3) becomes zero, because it is multiplied with linear combinations of columns of a stationary point $\mathbf{X}$ and the gradient is zero at stationary points. Furthermore, multiplying the last two terms of (3) by $\mathbf{y}_t$ gives

$$\begin{aligned}
&8\sum_{t=1}^{p}\sum_{i=1}^{n}\sum_{j=1}^{n} w_{ij}\left((f(d_{ij}(\mathbf{X}))-f(\delta_{ij}))f''(d_{ij}(\mathbf{X}))+(f'(d_{ij}(\mathbf{X})))^2\right)\\
&\mathbf{A}_{ij}\mathbf{x}_s\mathbf{x}_t'\mathbf{A}_{ij}\mathbf{y}_t =\\
&8\sum_{i=1}^{n}\sum_{j=1}^{n} w_{ij}\left((f(d_{ij}(\mathbf{X}))-f(\delta_{ij}))f''(d_{ij}(\mathbf{X}))+(f'(d_{ij}(\mathbf{X})))^2\right)\\
&\mathbf{A}_{ij}\mathbf{x}_s\sum_{t=1}^{p}\mathbf{x}_t'\mathbf{A}_{ij}\mathbf{y}_t.
\end{aligned} \tag{4}$$

The factor $\sum_{t=1}^{p} \mathbf{x}_t'\mathbf{A}_{ij}\mathbf{y}_t$ can be simplified into tr $(\mathbf{X}'\mathbf{A}_{ij}\mathbf{XS})$, which is zero, since it is the trace of the product of a symmetric and a skewsymmetric matrix. Thus all the terms that constitute $\sum_{t=1}^{p} \mathbf{H}_{st}\mathbf{y}_t$ are equal to zero, which proves that $\mathbf{y}$ is an eigenvector with zero eigenvalue. There are $p(p-1)/2$ linearly independent skewsymmetric matrices $\mathbf{S}$, which lead to linearly independent eigenvectors of the above type. This shows the assertion.

If at a stationary point $\mathbf{X}$ the Hessian $\mathbf{H}$ has exactly $p(p+1)/2$ zero eigenvalues and all other eigenvalues are positive, then we call $\mathbf{X}$ a *strict* local minimum.

3. Differentiability at a Local Minimum

To investigate differentiability of $\sigma(\mathbf{X}, \Delta)$ we set $f(z) = g(z^{1/2})$, where we assume the function $g : [0, \infty) \rightarrow \Re^1$ to be differentiable with right sided

$f(z)$	$f'(z)$	Name	Gradient
$z^{1/2}$	$\frac{1}{2}z^{-1/2}$	STRESS	$2\sum_{i=1}^{n}\sum_{j=1}^{n} w_{ij}(1-\sqrt{\delta_{ij}d_{ij}^{-1}(\mathbf{X})})\mathbf{A}_{ij}\mathbf{X}$
z	1	S-STRESS	$4\sum_{i=1}^{n}\sum_{j=1}^{n} w_{ij}(d_{ij}(\mathbf{X})-\delta_{ij})\mathbf{A}_{ij}\mathbf{X}$
$\log(z)$	z^{-1}	MULTISCALE	$4\sum_{i=1}^{n}\sum_{j=1}^{n} w_{ij}\log(d_{ij}(\mathbf{X})\delta_{ij}^{-1})d_{ij}^{-1}(\mathbf{X})\mathbf{A}_{ij}\mathbf{X}$

Tab. 1: The gradients of STRESS, S-STRESS, and MULTISCALE.

$f''(z)$	Name	$\mathbf{H}_{st}$
$-\frac{1}{4}z^{-3/2}$	STRESS	$2\beta^{st}\sum_{i=1}^{n}\sum_{j=1}^{n} w_{ij}(1-\delta_{ij}^{1/2}d_{ij}^{-1/2}(\mathbf{X}))\mathbf{A}_{ij}+$ $2\sum_{i=1}^{n}\sum_{j=1}^{n} w_{ij}\delta_{ij}^{1/2}d_{ij}^{-3/2}(\mathbf{X})\mathbf{A}_{ij}\mathbf{x}_s\mathbf{x}_t'\mathbf{A}_{ij}$
0	S-STRESS	$4\beta^{st}\sum_{i=1}^{n}\sum_{j=1}^{n} w_{ij}(d_{ij}(\mathbf{X})-\delta_{ij})\mathbf{A}_{ij}+$ $8\sum_{i=1}^{n}\sum_{j=1}^{n} w_{ij}\mathbf{A}_{ij}\mathbf{x}_s\mathbf{x}_t'\mathbf{A}_{ij}$
$-z^{-2}$	MULTI-SCALE	$4\beta^{st}\sum_{i=1}^{n}\sum_{j=1}^{n} w_{ij}\log(d_{ij}(\mathbf{X})\delta_{ij}^{-1}))d_{ij}^{-1}(\mathbf{X})\mathbf{A}_{ij}+$ $8\sum_{i=1}^{n}\sum_{j=1}^{n} w_{ij}(1-\log(d_{ij}(\mathbf{X})\delta_{ij}^{-1}))d_{ij}^{-2}(\mathbf{X})\mathbf{A}_{ij}\mathbf{x}_s\mathbf{x}_t'\mathbf{A}_{ij}$

Tab. 2: The Hessians of STRESS, S-STRESS, and MULTISCALE.

derivative $g'(0)$ at $z = 0$. Then $f(d_{ij}(\mathbf{X})) = g(\tilde{d}_{ij}(\mathbf{X}))$ with

$$\tilde{d}_{ij}(\mathbf{X}) = \left(\sum_{s=1}^{p}(x_{is} - x_{js})^2\right)^{1/2},$$

the distance between points i and j. It is quite natural to assume that $g(0) = 0$, i.e., zero distances and dissimilarities are not transformed to positive values, and $g'(z) \geq 0$ for all $z \geq 0$, which means that the transformation g of distances is monotone. In the limit, $g'(0) = \infty$ is also allowed.

Obviously, if $g'(0) = 0$ then $\sigma(\mathbf{X}, \Delta)$ is differentiable for all $\mathbf{X}$, no matter if zero distances occur. Thus we investigate the case that $g'(0) > 0$, such that nondifferentiable points may be encountered. Examples of corresponding transformations are the utility functions $g(z) = \ln(z+1)$ and $g(z) = 1-e^{-\lambda z}$, $\lambda > 0$, and furthermore the class of functions $g_\lambda(z) = z^\lambda$, $0 < \lambda \leq 1$.

$g_1(z) = z$, e.g., yields STRESS via $f(z) = z^{1/2}$. In this case de Leeuw (1984) has shown that STRESS is differentiable at a local minimum, provided $w_{ij}\delta_{ij} > 0$ for all $i \neq j$. He calls such data *usable*. This result has been extended to arbitrary Minkowski ℓ_p–distances by Groenen, Mathar, and Heiser (1992). We follow the basic idea to evaluate directional derivatives of $\sigma(\mathbf{X}, \Delta)$. The directional derivative of σ at $\mathbf{X}$ in direction $\mathbf{Y}$ is defined by

$$\nabla\sigma(\mathbf{X}; \mathbf{Y}) = \lim_{\varepsilon \downarrow 0} \frac{\sigma(\mathbf{X} + \varepsilon\mathbf{Y}, \Delta) - \sigma(\mathbf{X}, \Delta)}{\varepsilon},$$

and always exists if f is differentiable. The directional derivatives of the compositions $f \circ d_{ij} = g \circ \tilde{d}_{ij}$ and $f^2 \circ d_{ij} = g^2 \circ \tilde{d}_{ij}$ are given by

$$\nabla g \circ \tilde{d}_{ij}(\mathbf{X}; \mathbf{Y}) = \begin{cases} g'(\tilde{d}_{ij}(\mathbf{X}))\tilde{d}_{ij}(\mathbf{Y}), & \text{if } \tilde{d}_{ij}(\mathbf{X}) = 0 \\ \frac{g'(\tilde{d}_{ij}(\mathbf{X}))}{\tilde{d}_{ij}(\mathbf{X})} \sum_{s=1}^{p}(x_{is} - x_{js})(y_{is} - y_{js}), & \text{if } \tilde{d}_{ij}(\mathbf{X}) \neq 0, \end{cases}$$

and

$$\nabla g^2 \circ \tilde{d}_{ij}(\mathbf{X}; \mathbf{Y}) = \begin{cases} 2g(\tilde{d}_{ij}(\mathbf{X}))g'(\tilde{d}_{ij}(\mathbf{X}))\tilde{d}_{ij}(\mathbf{Y}), & \text{if } \tilde{d}_{ij}(\mathbf{X}) = 0 \\ \frac{2g(\tilde{d}_{ij}(\mathbf{X}))g'(\tilde{d}_{ij}(\mathbf{X}))\sum_{s=1}^{p}(x_{is}-x_{js})(y_{is}-y_{js})}{\tilde{d}_{ij}(\mathbf{X})}, & \text{if } \tilde{d}_{ij}(\mathbf{X}) \neq 0. \end{cases}$$

For $\mathbf{X} \in \Re^{n \times p}$ define $\mathcal{P} = \{(i,j) \mid i \neq j,\ \tilde{d}_{ij}(\mathbf{X}) \neq 0\}$ and correspondingly $\mathcal{Q} = \{(i,j) \mid i \neq j,\ \tilde{d}_{ij}(\mathbf{X}) = 0\}$. From the above representations we obtain the directional derivative of σ as

$$\begin{aligned} &\nabla\sigma(\mathbf{X}; \mathbf{Y}) \\ &= \sum_{i \neq j} w_{ij} \nabla g^2 \circ \tilde{d}_{ij}(\mathbf{X}; \mathbf{Y}) - 2\sum_{i \neq j} w_{ij} g(\delta_{ij}^{1/2}) \nabla g \circ \tilde{d}_{ij}(\mathbf{X}; \mathbf{Y}) \\ &= \sum_{(i,j) \in \mathcal{P}} \frac{2w_{ij}g'(\tilde{d}_{ij}(\mathbf{X}))\sum_{s=1}^{p}(x_{is} - x_{js})(y_{is} - y_{js})}{\tilde{d}_{ij}(\mathbf{X})}[g(\tilde{d}_{ij}(\mathbf{X})) - g(\delta_{ij}^{1/2})] \\ &\quad + \sum_{(i,j) \in \mathcal{Q}} 2w_{ij}g'(\tilde{d}_{ij}(\mathbf{X}))\tilde{d}_{ij}(\mathbf{Y})[g(\tilde{d}_{ij}(\mathbf{X})) - g(\delta_{ij}^{1/2})]. \end{aligned} \tag{5}$$

From this it easily follows that for all $\mathbf{X}, \mathbf{Y}$

$$\nabla\sigma(\mathbf{X};\mathbf{Y}) + \nabla\sigma(\mathbf{X};-\mathbf{Y}) = 4 \sum_{(i,j)\in\mathcal{Q}} w_{ij} g'(\tilde{d}_{ij}(\mathbf{X}))\tilde{d}_{ij}(\mathbf{Y})[g(\tilde{d}_{ij}(\mathbf{X})) - g(\delta_{ij}^{1/2})].$$

If $\mathbf{X}$ is a local minimum, the directional derivative in all directions is non-negative. This yields

$$\sum_{(i,j)\in\mathcal{Q}} w_{ij} g'(0)\tilde{d}_{ij}(\mathbf{Y})[g(0) - g(\delta_{ij}^{1/2})] \geq 0 \tag{6}$$

for all $\mathbf{Y}$. Now choose $\mathbf{Y}$ such that $\tilde{d}_{ij}(\mathbf{Y}) > 0$ for all $i \neq j$. Because of the assumptions $g'(0) > 0$ and $g(0) = 0$, (6) can happen for usable data only if $\mathcal{Q} = \emptyset$. Thus, at a local minimum $\mathbf{X}$ it holds that $\tilde{d}_{ij}(\mathbf{X}) > 0$ for all $i \neq j$.

In summary, we have shown the following result: for usable data ($w_{ij}\delta_{ij} > 0$ for all $i \neq j$), for any differentiable transformation g with $g(0) = 0$, $g'(0) > 0$, and $f(z) = g(z^{1/2})$, the general least squares loss function $\sigma(\mathbf{X}, \Delta)$ is differentiable at any local minimum $\mathbf{X}$. If $g'(0) = 0$ then $\sigma(\mathbf{X}, \Delta)$ is differentiable for all $\mathbf{X}$. Thus for usable data STRESS and S-STRESS are differentiable at a local minimum, but for MULTISCALE this need not be so.

4. Fulldimensional Scaling

For fulldimensional scaling, where $p = n-1$, every local minimum is a global minimum for some choices of f. This can be seen by using

$$d_{ij}(\mathbf{X}) = \text{tr } \mathbf{A}_{ij}\mathbf{X}\mathbf{X}' = \text{tr } \mathbf{A}_{ij}\mathbf{C} = c_{ii} + c_{jj} - 2c_{ij} \tag{7}$$

with the only requirement that $\mathbf{C}$ is in the cone of positive semi definite double centered symmetric (DCS) matrices. and rewriting (1) as

$$\sigma(\mathbf{C}, \Delta) = \sum_{i=1}^{n}\sum_{j=1}^{n}(f^2(\delta_{ij}) + f^2(\text{tr } \mathbf{A}_{ij}\mathbf{C}) - 2f(\text{tr } \mathbf{A}_{ij}\mathbf{C})f(\delta_{ij})). \tag{8}$$

Suppose that $f(z) \geq 0$ for $z \geq 0$ and that $f(z)$ is concave. This makes the third term of (8) a convex function in $\mathbf{C}$. If additionally $f^2(z)$ is convex, then $\sigma(\mathbf{C}, \Delta)$ is a convex function in $\mathbf{C}$. Thus, minimizing $\sigma(\mathbf{C}, \Delta)$ over $\mathbf{C}$ is minimizing a convex function over a convex set. Then any local minimum is a global minimum. It is easy to see that $f(z) = z$ and $f(z) = z^{1/2}$ satisfies these requirements, so that fulldimensional scaling for STRESS and S-STRESS results in a global minimum. In fact, $f(z) = z^{\lambda}$ with $\frac{1}{2} \leq \lambda \leq 1$ satisfies the requirement for a global minimum of fulldimensional scaling. Gaffke and Mathar (1989) proposed a special algorithm for S-STRESS with $p = n - 1$ based on cyclic projection.

Critchley (1986) and Bailey and Gower (1990) prove that the rank of the fulldimensional scaling solution of S-STRESS can never be larger than the number of positive eigenvalues of $-\frac{1}{2}\mathbf{J}\Delta\mathbf{J}$, where $\mathbf{J}$ is the centering operator $\mathbf{I} - n^{-1}\mathbf{1}\mathbf{1}'$. Numerical experiments with fulldimensional scaling of STRESS suggest that the same rank conditions also holds for STRESS, although no proof for this conjecture exists yet. It may even be the case that this assertion holds for all f for which $\sigma(\mathbf{C}, \Delta)$ is a convex function.

5. Inverse Scaling

Instead of finding the configurations which are optimal for given dissimilarities, we now look for dissimilarities for which a given configuration is optimal.

Let $f(\delta_{ij}) = f(d_{ij}(\mathbf{X})) - e_{ij}$. Inserting this in (2) gives

$$-\sum_{i=1}^{n} \sum_{j=1, j\neq i}^{n} w_{ij} e_{ij} f'(d_{ij}(\mathbf{X}))\mathbf{A}_{ij}\mathbf{X}. \tag{9}$$

By substituting $e_{ij} = t_{ij}/(w_{ij} f'(d_{ij}(\mathbf{X})))$ for $i \neq j$ and $e_{ii} = 0$ into (9) we have that $\mathbf{X}$ is a stationary point if the gradient (2) equals zero, or, equivalently, if

$$-\sum_{i=1}^{n} \sum_{j=1, j\neq i}^{n} t_{ij}\mathbf{A}_{ij}\mathbf{X} = \mathbf{0}. \tag{10}$$

Of course we assume that $f'(d_{ij}(\mathbf{X}))$ exists for all pairs i, j with $i \neq j$. Furthermore, we assume that $\mathbf{X}$ is centered, i.e., $\mathbf{X}'\mathbf{1} = \mathbf{0}$.

We should realize that the $\mathbf{A}_{ij}$ form a basis of the space of double centered symmetric (DCS) matrices. Condition (10) simply translates into $\mathbf{TX} = \mathbf{0}$ such that $\mathbf{T}$ is DCS. But any DCS matrix $\mathbf{T}$ satisfying $\mathbf{TX} = \mathbf{0}$ and $\mathbf{T1} = \mathbf{0}$ can be expressed as $\mathbf{KMK}'$ with $\mathbf{M}$ symmetric and $(\mathbf{K}|\frac{1}{\sqrt{n}}\mathbf{1})$ an orthonormal basis of the nullspace of $\mathbf{X}'$, i.e., $\mathbf{K}'\mathbf{X} = \mathbf{0}$ and $\mathbf{K}'\mathbf{1} = \mathbf{0}$. If r is the rank of $\mathbf{X}$ then the rank of $\mathbf{K}$ equals $n - r - 1$. Since $\mathbf{M}$ is symmetric there are $(n-r)(n-r-1)/2$ independent solutions. Note that the diagonal elements t_{ii} are left free, so that they can be chosen such that $\mathbf{T}$ becomes DCS.

Since the dissimilarities are required to be nonnegative and $f(\delta_{ij}) \in \Omega$ with $\Omega = \text{range}(f)$ must hold, certain restrictions on t_{ij} are necessary. In particular, it must be ensured that

$$f^{-1}\left(f(d_{ij}(\mathbf{X})) - \frac{t_{ij}}{w_{ij} f'(d_{ij}(\mathbf{X}))}\right) \geq 0 \tag{11}$$

and that

$$f(d_{ij}(\mathbf{X})) - \frac{t_{ij}}{w_{ij} f'(d_{ij}(\mathbf{X}))} \in \Omega. \tag{12}$$

For S-STRESS both requirements lead to $t_{ij} \leq w_{ij}d_{ij}(\mathbf{X})$. The second requirement imposes restrictions on $\mathbf{T}$ for STRESS. Since $\delta_{ij}^{1/2} \geq 0$ we have that $d_{ij}^{1/2}(\mathbf{X}) - t_{ij}/(2w_{ij}d_{ij}^{1/2}(\mathbf{X})) \geq 0$, or, equivalently, $t_{ij} \leq w_{ij}/2$. For MULTISCALE no restriction is needed on t_{ij}, because $\log(\delta_{ij})$ has $\Omega = \Re^1$ and

$$\log(\delta_{ij}) = \log\left(d_{ij}(\mathbf{X})e^{-t_{ij}w_{ij}^{-1}d_{ij}^{-1}(\mathrm{X})}\right) \tag{13}$$

so that the domain of right logarithm is positive for every t_{ij}.

De Leeuw and Groenen (1993) proved for STRESS by making use of the inequality constraints that inverse scaling defines a closed, convex polyhedron that contains the matrix of distances of $\mathbf{X}$. For S-STRESS a similar result can be proved, but not for MULTISCALE.

Thus we can find a dissimilarity matrix for which the gradient is zero, given a configuration $\mathbf{X}$. That only means that $\mathbf{X}$ is a stationary point for any of those dissimilarity matrices obtained by inverse scaling. But $\mathbf{X}$ can be a local minimum, a local maximum or a saddle point. If we wish to find only those dissimilarity matrices for which $\mathbf{X}$ is a strict local minimum, then we have to impose the additional constraint that the Hessian is positive semidefinite, where the only zero eigenvalues are those indicated in section 3. For more details of the STRESS case, we refer to de Leeuw and Groenen (1993).

References:

BAILEY, R., and GOWER, J. C. (1990): Approximating a symmetric matrix. *Psychometrika*, *55*, 665–675.

BROWNE, M. W. (1987): The Young-Householder algorithm and the least squares multidimensional scaling of squared distances. *Journal of Classification*, *4*, 175–190.

CRITCHLEY, F. (1986): Dimensionality theorems in multidimensional scaling and hierarchical cluster analysis. In: E. Diday, Y. Escoufier, L. Lebart, J. Lepage, Y. Schektman, and R. Tomassone (eds.), Informatics, IV, North-Holland, Amsterdam, 45–70.

DE LEEUW, J. (1984): Differentiability of Kruskal's Stress at a local minimum. *Psychometrika*, *49*, 111–113.

DE LEEUW, J., and GROENEN, P.J.F. (1993): Inverse scaling. Tech. rep. 144, UCLA Statistics Series, Interdivisonal Program in Statistics, UCLA, Los Angeles, California.

DE LEEUW, J., and HEISER, W. J. (1980): Multidimensional scaling with restrictions on the configuration. In: Krishnaiah, P. (ed.), *Multivariate Analysis, volume V*. North Holland, Amsterdam, 501–522.

GAFFKE, N., and MATHAR, R. (1989): A cyclic projection algorithm via duality. *Metrika*, *36*, 29–54.

GLUNT, W., HAYDEN, T., and LIU, W.-M. (1991): The embedding problem for predistance matrices. *Bulletin of Mathematical Biology*, *53*, 769–796.

GROENEN, P. J.F., MATHAR, R., and HEISER, W. J. (1992): The majorization approach to multidimensional scaling for Minkowski distances. Tech. rep. RR-92-11, Department of Data Theory, Leiden.

KRUSKAL, J. B. (1964): Multidimensional scaling by optimizing goodness of fit to a nonmetric hypothesis. *Psychometrika*, *29*, 1–27.

KRUSKAL, J. B., YOUNG, F. W., and SEERY, J. (1977): How to use KYST-2, a very flexible program to do multidimensional scaling. Tech. rep. AT&T Bell Laboratories, Murray Hill, New Jersey.

RAMSAY, J. O. (1977): Maximum likelihood estimation in MDS. *Psychometrika*, *42*, 241–266.

STOOP, I., HEISER, W.J., and DE LEEUW, J. (1981): How to use SMACOF-I A. Tech. rep. Department of Data Theory, Leiden.

TAKANE, Y., YOUNG, F. W., and DE LEEUW, J. (1977): Nonmetric individual differences in multidimensional scaling: An alternating least squares method with optimal scaling features. *Psychometrika*, *42*, 7–67.

Alternatives to Configural Frequency Analysis

Peter Ihm[1] and Ingeborg Küchler[2]

[1]Institut für Medizinische Biometrie, Philipps Universität, Bunsenstr. 3, D-35037 Marburg

[2]Institut für Biomathematik und Informatik, Charité, Humboldt Universität, Schumannstr. 20/21, D-10117 Berlin

Summary: The aim of Configural Frequency Analysis (CFA) is the search for outliers or 'types' (subdivided into 'types' and 'antitypes') in a sample of d-dimensional finite vectors, generally represented in a d-dimensional contingency table. Type search is done by analysis of residuals. It can be shown, however, that this technique may be misleading. The use of interpolated (deleted) residuals and/or other techniques will give better results. Deletion of entries results in incomplete tables. Expected values can be computed with the aid of Iterative Proportional Fitting (IPF). The analysis of logarithmic expectations leads to equation systems similar to those occurring in log-linear models. There is no restriction to the independence model assumed in CFA. The Markov chain as example of a more general but still simple model is treated in this paper.

1. Introduction

A sample of d-dimensional finite vectors can be represented in a d-dimensional contingency table. Let

$$\mathcal{N} = \{(i_1^{(1)}, \dots, i_1^{(d)}) \dots, (i_{m_1}^{(1)}, \dots, i_{m_d}^{(d)})\} \tag{1}$$

be the set of cells of the table. The model consists of a simple base model for entries in $\mathcal{B} \subset \mathcal{N}$ with the exception of (some) entries in a type set $\mathcal{T} = \mathcal{N} \backslash \mathcal{B}$. KRAUTH and LIENERT (1973), VON EYE (1990) and recently KRAUTH (1993) assume that the expectations $\mathrm{E}y_{i_1 \dots i_d}$ of entries $y_{i_1 \dots i_d}$ can be factorized as

$$\mathrm{E}y_{i_1 \dots i_d} = b_{i_1}^{(1)} \cdots b_{i_d}^{(d)} \tag{2}$$

in $\mathcal{B}$ but not in $\mathcal{T}$ where either

$$\mathrm{E}y_{i_1 \dots i_d} > b_{i_1}^{(1)} \cdots b_{i_d}^{(d)} \quad \text{(type)} \tag{3}$$

or

$$\mathrm{E}y_{i_1 \dots i_d} < b_{i_1}^{(1)} \cdots b_{i_d}^{(d)} \quad \text{(antitype).} \tag{4}$$

Since the factors $b_{i_j}^{(j)}$ in (2) are not uniquely determined,

$$b_1^{(j)} = 1, \qquad j = 2, \dots, d, \tag{5}$$

will be assumed. The parametrization (2) with (5) is convenient for the present purpose, instead of

$$\eta_{i_1 \ldots i_d} = \log(\mathrm{E}y_{i_1 \ldots i_d}) = \mu + \sum_j \alpha_{i_j}^{(j)},$$

with

$$\sum_{i_j} \alpha_{i_j}^{(j)} = 0, \quad j = 1, \ldots, d,$$

normally used in log-linear models.
Type search as done in CFA by analysis of residuals may be misleading. The use of interpolated (deleted) residuals and/or other techniques (cf. IHM (1986), VICTOR (1989) VICTOR & KIESER (1991)) will give better results. In practice, however, in order to avoid sparse tables, the cardinality $\mid \mathcal{N} \mid = N$ of $\mathcal{N}$ will only be moderately large. Then it will be possible to test all non-empty subsets $\mathcal{T}$ with cardinality $T = \mid \mathcal{T} \mid \leq T_{max}$ as possible type sets if T_{max} is small – contrarily to usual regression diagnostics where in common statistics packages $T_{max} = 1$.
Omission of entries in $\mathcal{T}$ results in incomplete tables. Estimates

$$\hat{y}_{i_1 \ldots i_d} = \hat{b}_{i_1}^{(1)} \cdots \hat{b}_{i_d}^{(d)} \tag{6}$$

of (2) with

$$\hat{b}_1^{(j)} = 1, \quad j = 2, \ldots, d \tag{7}$$

can be computed with the aid of the DEMING-STEPHAN algorithm (better known as Iterative Proportional Fitting). Unfortunately it does not give the factors $b_{i_j}^{(j)}$ in (6) directly. They can be obtained from the logarithms of the expectations leading to an equation system similar to that occurring in log-linear models. Since there is no restriction to the independence model (2) as base model the Markov chain as a more general but still simple one is treated.

2. Illustrative Examples

Before outlining the general theory some illustrative examples will be given. KIESER (1991) gives the following pairs of observed and expected tables:

Observed			Sum
10	1	1	12
1	1	1	3
1	1	1	3
12	3	3	18

Expected			Sum
8	2	2	12
2	.5	.5	3
2	.5	.5	3
12	3	3	18

with

z^2		
.5	.5	.5
.5	.5	.5
.5	.5	.5

and

Observed			Sum
1	10	10	21
10	10	10	30
10	10	370	390
21	30	390	441

Expected			Sum
1	1.4	18.6	21
1.4	2.1	26.5	30
18.6	26.5	344.9	390
21.0	30.0	390.0	441.0

with

z^2		
0	52.83	3.98
52.83	29.72	10.27
3.98	10.27	16.27

$z^2 = (observed-expected)^2/expected$. A significant z is used for the decision whether a cell entry is a type or antitype. Although in the first table (1,1) is obviously a type it is not discovered because of the small z^2-values. In the second table (1,1) is an antitype, (3,3) a type, but the z^2-values would have given an erroneous result.

The approach suggested in this paper will be illustrated with the aid of the following constructed example: Consider the 2×3-table

50	24	51	125
12	30	30	72
62	54	81	197

The indicator matrix of the table is

$$\begin{pmatrix} 1 & 0 & 1 & 0 & 0 \\ 1 & 0 & 0 & 1 & 0 \\ 1 & 0 & 0 & 0 & 1 \\ 0 & 1 & 1 & 0 & 0 \\ 0 & 1 & 0 & 1 & 0 \\ 0 & 1 & 0 & 0 & 1 \end{pmatrix} \tag{8}$$

Leaving out (1,1) as a potential type an incomplete table arises. Keeping the marginals constant IPF gives the following expected values:

.	30	45	75
12	24	36	72
12	54	81	147

(9)

The factorization is

.	15×2	15×3	75
12×1	12×2	12×3	72
12	54	81	147

but common IPF routines give only the expected values $\hat{y}_{i_1 \dots i_d}$ (6) and the factors $\hat{b}_{i_j}^{(j)}$ in (9) subject to (7) must be computed separately. Define

$$\hat{\eta} = (\log 30 \;\; \log 45 \;\; \log 12 \;\; \log 24 \;\; \log 36)'$$

and

$$\hat{\beta} = (\log b_1^{(1)} \log b_2^{(1)} \log b_2^{(2)} \log b_3^{(2)})'.$$

($\log b_1^{(2)}$ is missing because of (7)). The vector $\hat{\beta}$ of the factor logarithms is the solution of

$$X\hat{\beta} = \hat{\eta} \tag{10}$$

with a structure matrix X derived from the indicator matrix (8) by deletion of the first row and third column because of (7):

$$X = \begin{pmatrix} 1 & 0 & 1 & 0 \\ 1 & 0 & 0 & 1 \\ 0 & 1 & 0 & 0 \\ 0 & 1 & 1 & 0 \\ 0 & 1 & 0 & 1 \end{pmatrix} \tag{11}$$

The solution is

$$\hat{\beta} = (X'X)^{-1}X'\hat{\eta} = (\log 15 \log 12 \log 2 \log 3)'.$$

Having computed the factors the predicted value for cell (1,1) under H_0 is

$$\hat{y}_{(11)} = \exp\{(1\,0\,0\,0)\hat{\beta}\} = 15.$$

The subscript (11) in parentheses indicates that cell (1,1) was omitted in the computation of $\hat{\beta}$.

3. The General Type Model

3.1 Notation

It is convenient to label the cells of the table by $1, \ldots, N$ and to use $\mathcal{N} = \{1, \ldots, N\}$ instead of (1), likewise $\mathcal{T} = \{1, \ldots, T\}$ and $\mathcal{B} = \{T+1, \ldots, N\}$ for the sake of simplicity; $N \leq \prod_j m_{i_j}$ since there might be structural zeros. For a vector v the functions $\log v$ and $\exp(v)$ give the logarithm and the exponential function of every element of v.
The vectors of entries, expected values and their logarithms are $y = (y_1 \ldots y_N)'$, $\hat{y} = (\hat{y}_1 \ldots \hat{y}_N)'$ and $\hat{\eta} = (\hat{\eta}_1 \ldots \hat{\eta}_N)' = \log \hat{y}$.

3.2 The Model

The homogeneity (independence) model (2) as base model will be replaced by a more general log-normal one expressed by its structure matrix X. Suppose expected values were computed by IPF for all cells in $\mathcal{N}$. X, y and $\hat{\eta}$ can be split into two parts:

$$X = \begin{pmatrix} X_\mathcal{T} \\ X_\mathcal{B} \end{pmatrix}, \quad y = \begin{pmatrix} y_\mathcal{T} \\ y_\mathcal{B} \end{pmatrix} \quad \text{and} \quad \hat{\eta} = \begin{pmatrix} \hat{\eta}_\mathcal{T} \\ \hat{\eta}_\mathcal{B} \end{pmatrix}.$$

$X_\mathcal{T}$ is $T \times P$, $y_\mathcal{T}$ and $\eta_\mathcal{T}$ are $T \times 1$. The type model can be expressed as follows:

$$\mathrm{E}y = \exp(\eta)$$

with

$$\eta = \begin{pmatrix} I_T & X_T \\ O & X_\mathcal{B} \end{pmatrix} \beta \tag{12}$$

where β is a $(T+P) \times 1$ parameter vector and I_T the unit matrix of order T (VICTOR (1989), VICTOR & KIESER (1991).
Now suppose $\eta = \eta_{(\mathcal{T})}$ to be estimated from the incomplete table with cells in $\mathcal{T}$ omitted. Then

$$\hat{\beta}_{(\mathcal{T})} = (X'_\mathcal{B} X_\mathcal{B})^{-1} X'_\mathcal{B} \hat{\eta}_{(\mathcal{T})} = \log \hat{b}_{(\mathcal{T})}, \quad \text{say,}$$

is the solution of problem (6). Subscript $(\mathcal{T})$ indicates the omission of $\mathcal{T}$. Interpolated values in $\mathcal{T}$ under H_0 are

$$\hat{\eta}_{(\mathcal{T})} = X_\mathcal{T} \hat{\beta}_{(\mathcal{T})} = (\hat{\eta}_{k(\mathcal{T})}), \quad k \in \mathcal{T}.$$

Now

$$X'_\mathcal{B} X_\mathcal{B} = X'X - X'_\mathcal{T} X_\mathcal{T} = A - X'_\mathcal{T} X_\mathcal{T}, \quad \text{say,}$$

and

$$(A - X'_\mathcal{T} X_\mathcal{T})^{-1} = A^{-1} + A^{-1} X'_\mathcal{T} (I - X_\mathcal{T} A^{-1} X'_\mathcal{T})^{-1} X_\mathcal{T} A^{-1},$$

hence

$$\begin{aligned} \hat{\eta}_{(\mathcal{T})} &= X_\mathcal{T} \{A^{-1} + A^{-1} X'_\mathcal{T} (I - X_\mathcal{T} A^{-1} X'_\mathcal{T})^{-1} X_\mathcal{T} A^{-1}\} X_\mathcal{B} \hat{\eta}_\mathcal{B} \\ &= (I - H_{\mathcal{T}\mathcal{T}})^{-1} H_{\mathcal{T}\mathcal{B}} \hat{\eta}_\mathcal{B}. \end{aligned} \tag{13}$$

$H_{\mathcal{T}\mathcal{T}}$ and $H_{\mathcal{T}\mathcal{B}}$ follow from

$$H = X(X'X)^{-1}X' = \begin{pmatrix} H_{\mathcal{T}\mathcal{T}} & H_{\mathcal{T}\mathcal{B}} \\ H'_{\mathcal{T}\mathcal{B}} & H_{\mathcal{B}\mathcal{B}} \end{pmatrix}$$

where $H_{\mathcal{T}\mathcal{T}}$ is $T \times T$.

3.3 Markov Chains

In the notation of (2) the factorization of the expected values of a Markov chain is

$$\hat{y}_{i_1 \ldots i_d} = b^{(1,2)}_{i_1 i_2} b^{(2,3)}_{i_2 i_3} \cdots b^{(d-1,d)}_{i_{d-1} i_d} \tag{14}$$

with

$$b^{(j,j+1)}_{i_j 1} = 1, \qquad i_j = 1, \ldots, m_j, \qquad j = 2, \ldots, d. \tag{15}$$

Just like (2) with (5), (14) with (15) is an unusual parametrization but is sufficient for the present purpose.

4. Type Diagnostics

4.1 Iterative Proportional Fitting (IPF)

IPF was introduced by DEMING and STEPHAN (1940) to fit expected values to given marginal frequencies. Using notations (1) and (2) marginal sums are written additionally with a superscript list of the marginal components of the vector $(i_1 \dots i_d)'$, *i.e.* $\hat{y}_{i_j}^{(j)}, \dots, \hat{y}_{i_j i_k}^{(j,k)}, \dots$. The algorithm starts by defining an initial table $\hat{Y}_{(\mathcal{T})}$ where

$$\hat{y}_{i_1 \dots i_d} = \begin{cases} 0, & (i_1, \dots, i_d) \in \mathcal{T} \\ 1, & (i_1, \dots, i_d) \in \mathcal{B} \end{cases}$$

The expected values in (2) are obtained as follows: For $i_j = 1, \dots, m_j$ and $j = 1, \dots, d$ compute until convergence:

$$\hat{y}_{i_1, \dots, i_d} := \hat{y}_{i_1, \dots, i_d} \frac{y_{i_j}^{(j)}}{\hat{y}_{i_j}^{(j)}}.$$

The function `ipf(...)` in SAS/IML requires an indication of the marginal sums to be fitted and returns a table $\hat{Y}_{(\mathcal{T})}$ with the expected values for $\mathcal{N}$ and a period for missing values for $\mathcal{T}$. The function `shape(...)` allows the conversion of $\hat{Y}_{(\mathcal{T})}$ to $\hat{y}_{(\mathcal{T})}$.
IPF can be used for all log-linear models. In the case of the Markov model (14) the marginal sums $y_{i_j i_{j+1}}^{(j,j+1)}$, $j = 1, \dots, d-1$ must be kept constant.

4.2 Interpolated Residuals

Equation (13) is in close relationship to that of regression diagnostics except for variable $\hat{\eta}_{\mathcal{B}}$ as a function of $\mathcal{T}$. The interest of (13) lies in the fact that $A = X'X$ must be inverted only once and inversion of $(I - X_{\mathcal{T}} A^{-1} X'_{\mathcal{T}})$ is not too time-consuming if T_{max} is small. A suitable algorithm can easily be constructed:

1. Compute $A^{-1} = (X'X)^{-1}$ and $B = A^{-1}X'$.
2. For all non-empty subsets $\mathcal{T} \subset \mathcal{B}$ with cardinality $T \leq T_{max}$ do:
   - 2.1. Construct $X_{\mathcal{T}}$, the matrix of rows of X with row index $k \in \mathcal{T}$.
   - 2.2. Construct the incomplete table $Y_{(\mathcal{T})}$ with structural zeros for $k \in \mathcal{T}$.
   - 2.3. Compute $\hat{Y}_{(\mathcal{T})}$ by IPF.
   - 2.4. Express $\hat{Y}_{(\mathcal{T})}$ as vector $\hat{y}_{(\mathcal{T})}$.
   - 2.5. Define $\hat{\eta}$ by

$$\hat{\eta}_k = \begin{cases} 0, & k \in \mathcal{T}, \\ \log \hat{y}_k, & k \in \mathcal{B}. \end{cases}$$

   - 2.6. Compute $H_{\mathcal{T}\mathcal{T}} = X_{\mathcal{T}} A^{-1} X'_{\mathcal{T}}$.
   - 2.7. Compute $\hat{\eta}_{(\mathcal{T})} = (I - H_{\mathcal{T}\mathcal{T}})^{-1} X_{\mathcal{T}} B \hat{\eta}$ and $\hat{y}_{(\mathcal{T})} = \exp(\hat{\eta}_{(\mathcal{T})})$.

5. Statistical Tests

As suggested by VICTOR (1989) types will be tested by

$$z_{(k)} = \frac{y_k - \hat{y}_{(k)}}{\sqrt{\hat{y}_{(k)}}}, \tag{16}$$

assuming $z \sim \mathcal{N}(0,1)$. Although the test is anticonservative it was chosen in this paper because of the exploratory character of type search. KIESER and VICTOR (1991) derived an asymptotic test for the existence of one type (*i.e.* $T_{max} = 1$) using the deletion method described above. A generalization for $T_{max} > 1$ is suggested consisting in a partition of the total χ^2 into fractions corresponding to $\mathcal{T}$ and $\mathcal{B}$. The authors insist that not only types but also the base model should be tested.
An interesting approach for the identification of multiple outliers is due to HADI and SIMONOFF (1993). Applied to contingency table analysis the algorithm would be as follows:
Step 1: A type set $\mathcal{T}$ is constructed on the basis of all significant deleted residuals with $T_{max} = 1$, The result is a 'clean' set $\mathcal{B}$.
Step 2: Interpolated values for $k \in \mathcal{T}$ are computed on the basis of $\mathcal{B}$ and non-significant cells removed from $\mathcal{T}$ resulting in a new 'clean' set $\mathcal{B}$.
Step 3: Stop or continue with Step 2 until stability of the result.

6. Example: Genital Display of Sqirrel Monkeys

LIENERT, PLOOG and VON EYE (1993) analyzed frequencies of active and passive genital display in a colony of six squirrel monkeys given in the following table:

Active	Passive Display						Sum
Display	R (m)	S (m)	T (f)	U (f)	V (f)	W (im)	
R (m)	-	1	5	8	9	0	23
S (m)	29	-	14	46	4	0	93
T (f)	0	0	0	0	0	0	0
U (f)	2	3	1	-	38	2	46
V (f)	0	0	0	0	-	1	1
W (im)	9	25	4	6	13	-	57
Sum	40	29	24	60	64	3	220

(m = male, f = female, im = infant male, son of V). For the further analysis rows T, V and column W will be taken out; Three cells of the 4×5-subtable contain structural zeros. Each of the remaining 17 cells will be taken out in turn, i.e. $T = 1$. Interpolated values $\hat{y}_{(ij)}$ and $z_{(ij)} = (y_{ij} - \hat{y}_{(ij)})/\sqrt{\hat{y}_{(ij)}}$ (16) are computed. The critical BONFERRONI-adjusted 0.05-value is 2.976. The following table gives the values for significant z:

$i-j$	y_{ij}	$\hat{y}_{(ij)}$	$z_{(ij)}$
U - V	38	1.17	34.03
W - S	25	1.34	20.42
S - V	4	112.99	−10.25
S - V	46	12.11	9.74
S - R	29	7.86	7.54
U - S	3	30.72	−5.00
W - U	6	32.98	−4.70
U - R	2	20.11	−4.04

Among the quintuplets of deleted cells those corresponding to the first five rows of the above table the following expected/interpolated values were obtained:

Active	Passive Display						Sum
Display	R (m)	S (m)	T (f)	U (f)	V (f)	W (im)	
R (m)	-	1	5	8	9	-	23
	-	2.75	3.79	6.40	10.06	-	23.00
S (m)	29	-	14	46	4	-	93
	24.79	-	14.00	23.66	37.18	-	99.63
U (f)	2	3	1	-	38	-	44
	3.04	1.25	1.72	-	4.56	-	10.57
W (im)	9	25	4	6	13	-	57
	7.96	3.27	4.50	7.60	11.94	-	35.27
Sum	40	29	24	60	64	-	217
	35.79	7.27	24.01	37.66	63.74	-	168.47

$z_{(S-R)}$ is not significant. The expected/interpolated values correspond to random display, *i.e.* males and females show their penis or, respectively, their clitoris hapharzardly to other members of the colony. The table indicates, however, that male S prefers female U to female V. Female U prefers female V and infant male W prefers male S. Interesting enough, there is no symmetry in the behavior of the monkeys. The above result differs from that of LIENERT, PLOOG and VON EYE (1993) who did not discover type S-U.

7. Discussion

It is a reasonable principle to describe a sample by the parameters of the simplest model consistent with the data (*'Denkökonomie'* of ERNST MACH). For a sample of finite discrete vectors the simplest model is that of homogeneity (2)). Its parsimonious description is the one by marginal frequencies. A type model based on homogeneity plus a few outliers (types/antitypes) may still be parsimonious if the number of types is inferior to the supplementary parameters of a log-linear model with interactions of higher order. The number of parameters, however, does not play a fundamental role: The type model based on homogeneity assumes a logical structure of types – a syndrome – overlaid by some 'noise', *i.e.* a random component. This general idea may still be valid if 'noise' is caused by another simple random process. In the spirit outlined above CFA is based on a reasonable concept. The method of type search, however, does not fulfill the requirements of an efficient diagnostic procedure. Diagnostic methods based on systematic deletion of individuals of a sample have been known to be superior for some decades and it remains a secret why advocates of CFA ignore these methods. Deletion methods in contingency table methods are not new. The present paper should be understood as a contribution to computational statistics. Many aspects remain untreated, *e.g.* testing; the reader is referred to VICTOR and KIESER (1993). There also exist a number of heuristic methods useful for preliminary outlier search as robust estimation etc. (cf. IHM (1986)).

References:

DEMING, W. E., STEPHAN, F. F. (1940):
On a least squares adjustment of a sampled frequency table when the expected marginal totals are known. *Ann. Math. Statist., 11, 427–444.*

HADI, A. S., SIMONOFF, J. S. (1993):
Procedures for the identification of multiple outliers in linear models. *JASA, 88, 1254–1272.*

IHM, P. (1986):
Diagnostics and robust estimation in multivariate contingency tables. In: E. DIDAY *et al.* (eds.): *Data Analysis and Informatics. Proc. IVth Symp. on Data Analysis and Informatics, Versailles 1985.* North Holland, Amsterdam, *429–442.*

KIESER, M. (1991):
Identifikation von Syndromen und Typen mit Methoden der Kontingenztafelanalyse. Dissertation, Univ. Heidelberg.

KIESER, M., VICTOR, N. (1991):
A test procedure for an alternative approach to Configural Frequency Analysis. *Methodika 5, 87-97.*

KRAUTH, J. (1993):
Einführung in die Konfigurationsfrequenzanalyse (KFA). Ein multivariates nichtparametrisches Verfahren zum Nachweis und zur Interpretation von Typen und Syndromen. J. Beltz, Weinheim, Basel.

KRAUTH, J., LIENERT, G. A. (1973):
Die Konfigurationsfrequenzanalyse und ihre Anwendung in Psychologie und Medizin. K. Alber, Freiburg i. Br.

LIENERT, G. A., PLOOG, W. D., VON EYE, A. (1993):
Inverted configural types derived from incomplete contingency tables: Q-CFA. *Biometrical Journal, 35, 259–266.*

VON EYE, A. (1990):
Introduction to Configural Frequency Analysis. The Search For Types and Antitypes in Cross-Classifications. Cambridge Univ. Press, Cambridge.

VICTOR, N. (1989):
An alternative approach to Configural Frequency Analysis. *Methodika, 3, 61-73.*

VICTOR, N., KIESER, M. (1991):
Identification of types in contingency tables. *Bull. Int. Statist. Inst., 54.4, 691-692.*

Clustering Methods in Fuzzy Control

F. Klawonn, R. Kruse

Department of Computer Science, University of Braunschweig
D-38106 Braunschweig, Germany

Summary: Fuzzy controllers can be interpreted as an interpolation technique on the basis of fuzzy clusters of input/output pairs. It is therefore obvious that fuzzy clustering algorithms are a promising tool for supporting the design of a fuzzy controller when data of the process to be controlled are available.

This paper discusses the possibilities and limitations of fuzzy clustering for fuzzy control.

1. Introduction

Most of the classical control techniques are based on a mathematical or physical model of the process to be controlled, usually described by a set of difference or differential equations. These techniques are applicable when a suitable, not too complex mathematical description of the process can be provided. But there is a great number of processes with parameters for which no appropriate mathematical model is known. Nevertheless, many of these processes can be controlled by a human operator.

The idea behind fuzzy control is to model the behaviour of a human operator. Thus fuzzy controllers aim at determining a static control function on the basis of linguistic if–then–rules. The control function assigns to each tuple of measured input values (for example error and and change of error) a suitable control or output value that forces the process in the direction of the desired state. The control rules are of the form

$$\textit{If input}_1 \textit{ is } A_1 \textit{ and } \ldots \textit{ and input}_n \textit{ is } A_n \textit{ then output is } B \qquad (1)$$

where $A_1, \ldots, A_n, B$ are linguistic expressions of the form *approximately zero, (negative) small, (positive) big*, etc. that specify a vague description of the value of the corresponding variable. These linguistic expressions are represented by fuzzy sets.

In section 2 we will see that each fuzzy set can be interpreted as representing a crisp value in a vague environment admitting some small error. In this sense, the fuzzy sets are in some way fuzzy clusters. Section 3 gives a short review of basic fuzzy control techniques in the light of the above mentioned interpretation of fuzzy sets. Section 4 is devoted to the topic of generating fuzzy sets automatically by fuzzy clustering from a data set. It will be shown that fuzzy clustering can be useful for this task, but in many cases fuzzy clustering will not meet the requirements needed for fuzzy control.

2. Interpretation of the Fuzzy Sets

The idea behind fuzzy sets is to extend the concept of membership to a set to graded membership, i.e. not to restrict to the two membership degrees 1 ('belongs to the set') and 0 ('does not belong to the set'). Therefore, from a formal point of view, a fuzzy set μ on the domain X is a generalized characteristic function $\mu : X \to [0,1]$, assigning to each $x \in X$ its membership degree $\mu(x)$ to the fuzzy set μ, or, if μ is intended to represent a linguistic expression like *approximately zero*, the degree to which x satisfies the linguistic expression associated with μ.

Although it is very appealing to interpret the value $\mu(x)$ as the degree to which x belongs to the fuzzy set μ, there is a need for a concrete interpretation of membership degrees. Otherwise it is impossible to assign suitable membership degrees to elements or to compare or combine fuzzy sets specified by different people.

The approaches to the interpretation of membership degrees can be categorized into three types: uncertainty (for instance in the case when fuzzy sets are seen as possibility distributions like in Dubois and Prade (1988)), preference (as in Dubois and Prade (1993)), and similarity (see for example Trillas and Valverde (1984) or Kruse et al. (1993)).

In this paper where we only consider fuzzy control we restrict ourselves to the latter interpretation. In order to explain this interpretation of membership degrees in terms of similarity or indistinguishability, let us consider the following simple example.

Let δ be a (pseudo–)metric on X. Then

$$E_\delta : X \times X \to [0,1], \quad (x,y) \mapsto 1 - \min\{\delta(x,y), 1\}$$

is called an equality or similarity relation. The value $E_\delta(x,y)$ reflects the similarity of x and y. Note that it might be reasonable to apply a scaling to the metric δ. For example, in the case of a discrete metric δ which maps X into the natural numbers, one could use $c \cdot \delta$ instead of δ where $0 < c < 1$ is a suitable scaling factor. Otherwise the corresponding similarity relation would simply be the crisp equality.

Given a crisp element $x_0 \in X$, one should take the similarity relation into account and consider the 'equivalence class' of x_0 with respect to E_δ, i.e. the crisp set $\{x_0\}$ is extended to the fuzzy set

$$\mu_{x_0} : X \to [0,1], \quad x \mapsto E_\delta(x_0, x)$$

containing all elements that are similar to x_0. In this way, under consideration of the similarity relation E_δ or its dual concept, the metric δ, each crisp value x_0 induces a fuzzy set or fuzzy cluster μ_{x_0}. It is very important to note that in the case $X = \mathbb{R}$ and $\delta(x,y) = |x - y|$ the fuzzy set μ_{x_0} has a triangular shape like those fuzzy sets that are commonly used in fuzzy control.

Of course, the fuzzy sets appearing in fuzzy control are in general not so simple that they can be considered to be of the form μ_{x_0} with the standard metric on the real numbers as the underlying metric. Usually a transformation of the real line is assumed. The transformation is induced by a scaling function $c : \mathbb{R} \to [0, \infty[$ that assigns to each $x \in \mathbb{R}$ a scaling factor $c(x) \geq 0$. The greater this scaling factor is, the stronger is the distinguishability of values in the neighbourhood of x. Therefore, the underlying metric is given by the following formula.

$$\delta_c(x, y) \;=\; \left| \int_x^y c(s) ds \right|.$$

Note that generally in fuzzy control neither the scaling function c nor the the metric δ_c or its corresponding similarity relation E_{δ_c} is explictly considered. But it was shown by Klawonn and Kruse (1993) that in most cases an appropriate metric or even a scaling function can be found so that the fuzzy sets can be interpreted in the above mentioned way. The problem of finding such a metric or scaling function for a given fuzzy partition, i.e. a family of fuzzy sets is discussed by Höhle and Klawonn (1992) and Klawonn (1994).

Before we can relate these ideas to fuzzy control, we have to mention the problem of aggregating similarity relations on different domains. Let δ_i $(i = 1, 2)$ be a metric on X_i and let E_{δ_i} denote its corresponding similarity relation. When we consider the product space $X_1 \times X_2$ we have various possibilities to derive a similarity relation on $X_1 \times X_2$ from E_{δ_1} and E_{δ_2}. For reasons of simplicity, we restrict ourselves to

$$E : (X_1 \times X_2) \times (X_1 \times X_2) \to [0, 1],$$

$$((x_1, x_2), (y_1, y_2)) \;\mapsto\; \min\{E_{\delta_1}(x_1, y_1), E_{\delta_2}(x_2, y_2)\}.$$

Note that E is induced by the metric

$$\delta((x_1, x_2), (y_1, y_2)) \;=\; \max\{\delta_1(x_1, y_1), \delta_2(x_2, y_2)\}$$

and that δ is an ultrametric if δ_1 and δ_2 are ultrametrics.

3. Fuzzy Control

Fuzzy controllers are used to describe static control functions that assign to each tuple of measured input values of a process a suitable output value for the control variable that forces the system in the direction of the desired state. In order to achieve this, for each input variable and for the output variable a fuzzy partition of the corresponding domain is specified, i.e. a family of fuzzy sets for each domain has to be determined. A typical fuzzy partition is illustrated in Fig. 1.

Each of the fuzzy sets is associated with a linguistic expression like *approximately zero, (negative) small, (positive) big*, etc. These linguistic expressions are incorporated in the control rules of the form (1) mentioned in the

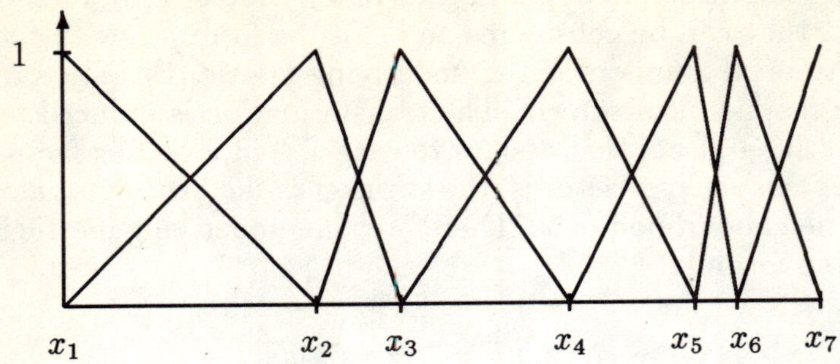

Fig. 1: A typical fuzzy partition

introduction.

For a given tuple $(\xi_1, \ldots, \xi_n)$ of measured inputs for each rule the 'matching degree' is computed, i.e. if the fuzzy set μ_i is associated with the linguistic expression A_i, then the matching degree of rule (1) for the input $(\xi_1, \ldots, \xi_n)$ is

$$\alpha_1 = \min\{\mu_1(\xi_1), \ldots, \mu_n(\xi_n)\}. \tag{2}$$

The matching degree determines how strong the rule is applicable and effects the output of the rule. Typically, the output of one rule is defined as the fuzzy set $\min\{\mu, \alpha_1\}$ where μ is the fuzzy set associated with the linguistic expression B and α_1 is the value defined in equation (2). The output fuzzy set of the system of rules is obtained by aggregating the output fuzzy sets of the single rules by the maximum. In order to get a crisp output value this fuzzy set has to be 'defuzzified' which is often done by the center–of gravity–method that takes the value under the center of gravity of the fuzzy set as output.

It would lead us to far to discuss here the details of the computations carried out in a fuzzy controller and we refer for an overview to Kruse et al. (1993). What is interesting from the viewpoint of cluster analysis is the following.

The fuzzy sets can be interpreted as fuzzy clusters, i.e. as representing crisp values or prototypes with respect to an underlying similarity relation. If the fuzzy set μ_i represents the value ξ_i, then the rule (1) specifies the output for the input tuple $(\xi_1, \ldots, \xi_n)$. Therefore, the rule base determines a partial function that assigns an output value to some input tuples. In this sense, fuzzy control can be interpreted as an interpolation method in the presence of indistinguishability characterized by similarity relations. For details see Klawonn and Kruse (1993).

Note that for each input domain and for the output domain we have a specific similarity relation. Taking equation (2) into account, that determines how the matching degree of an input tuple is calculated for one rule, one can show that the similarity relation on the different domains are aggregated to a similarity relation on their product space as proposed at the end of the previous section.

4. Fuzzy Clustering and Fuzzy Control

As we have explained in the previous section, the main ingredients for a fuzzy controller are fuzzy partitions of the output and input domains and a rule base. The fuzzy sets of the fuzzy partitions can be interpreted as crisp values in the presence of indistinguishability or as fuzzy clusters. The rule base corresponds to a partial function.

Although it is intuitively appealing to formulate the knowledge about the operator's behaviour in the form of if–then–rules involving linguistic expressions like *approximately zero* and to model these expressions with fuzzy sets, it is often very difficult to specify a suitable rule base and to determine appropriate fuzzy sets. Even if the principal shape of the fuzzy sets is known, it is not clear how to choose the exact values for the membership degrees. Thus an automatic generation of the fuzzy partitions and the rule base from data gained from observing the process and the operator's behaviour is desirable.

A fuzzy controller constructed in this way can be used to simulate the operator's behaviour for automatic control. But it can also be the basis of understanding the operator's control strategy in order to implement an improved strategy in a fuzzy controller with modified fuzzy partitions and a modified rule base.

It is near at hand to think of applying standard fuzzy clustering techniques like the fuzzy *c*–means algorithm. For a description of the fuzzy *c*–means algorithm see for example Bezdek (1973), Bezdek and Pal (1992), or Dunn (1974).

The problem is that in most cases the data are better suited for regression than for clustering as illustrated in Fig. 2, since the operator's actions are often more or less continuous distributed over the input domains.

There are, of course, exceptions, especially for chemical processes where the operator's action consists in adding half a ton of some liquid to the process. Such drastic actions are usually carried out, when the operator observes that a certain value exceeds or falls below some threshold. In this case, one may obtain data that have the characteristics of those in the left side of Fig. 2.

One might ask the question, why we insist on a fuzzy controller, when a regression technique might be applied easily. The reason is that although it might be very simple to obtain a suitable control function by such a technique, this function is difficult to interpret. This means, tuning and adjustments to changing parameters of the process are more or less impossible using the regression function. Since the behaviour of a fuzzy controller becomes clear from its fuzzy sets and its rule base, such changes can be carried out, when the control function is determined by a fuzzy controller.

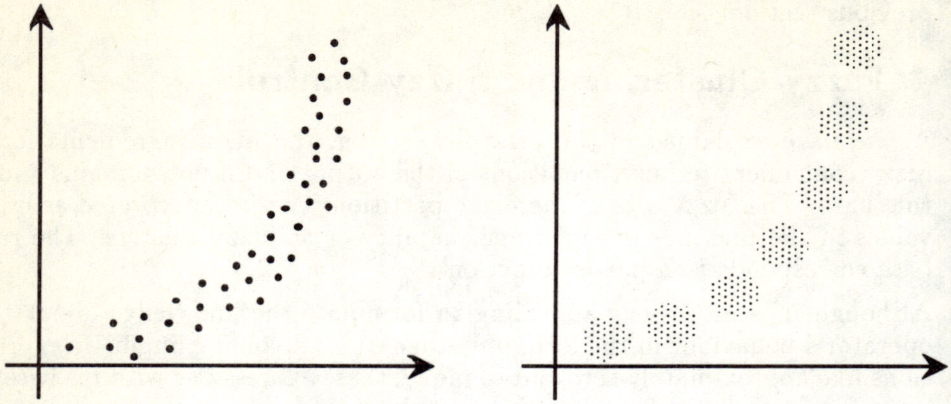

Fig. 2: Data that are better suited for regression (left side) and for clustering (right side).

Therefore, usually some heuristic filtering has to be applied to the data that extracts data in the form of those on the right side of Fig. 2 from data in the form of those on the left side.

Another problem is caused by the fact that for each input and output domain a separate fuzzy partition is needed in order to formulate the control rules in the form of (1). However, the data are in general observed in the product space of all input spaces and the output space. Projecting the data before applying fuzzy clustering will lead to severe problems, since (fuzzy) clusters that do not interfere in the product space might melt together when projected to some one–dimensional space so that the fuzzy clustering of the projected data comes up with clusters that cannot be associated with (fuzzy) clusters in the product space. On the other hand, applying a fuzzy clustering algorithm in the product space and then projecting the fuzzy clusters may yield projected clusters that strongly overlap. Nevertheless, although this overlapping (i.e. fuzzy partitions with strongly overlapping fuzzy sets) is not desirable, it is still possible to construct a fuzzy controller on the basis of these fuzzy partitions, whereas the other method – first projecting and then clustering – may lead to incoherent fuzzy partitions.

Again, some suitable preprocessing of the data that forces the (fuzzy) clusters to be near the grid points in the product space, can help avoiding this problem.

Another question is, whether the fuzzy c–means algorithm is really well–suited for generating fuzzy partitions for a fuzzy controller. One of the assumption of this algorithm is that for each element, the sum of its membership degrees to all (fuzzy) clusters is equal to one. This condition is very appealing, since in fuzzy control it is very common to assume that for any element, the sum of its membership degrees to all fuzzy sets of a fuzzy partition is also one. However, in fuzzy control there is in general the stronger assumption that usually the intersection of the supports of more than two fuzzy sets of a fuzzy partition should be empty – a condition which is normally not guaranteed by the fuzzy c–means algorithm.

An alternative to the fuzzy c–means algorithm is the possibilistic approach to fuzzy clustering proposed by Krishnapuram and Keller (1993). In opposition to the fuzzy c–means algorithm the requirement that the membership degrees to all fuzzy clusters sum up to one for each element is given up in possibilistic clustering. But possibilistic clustering can tend to non–overlapping fuzzy clusters. As a consequence for the corresponding fuzzy partitions, there can be areas which are covered by no fuzzy set at all. This means that for input values falling into such areas, no control action is specified which may lead to difficulties in controlling the process.

5. Conclusions

The discussion of fuzzy clustering and fuzzy control has shown that there are strong relations between these two fields. However, a lot of problems arise, when fuzzy clustering has to be applied to fuzzy control. Some of these problems can be overcome by some suitable filtering or preprocessing of the data, but there are still unsolved problems. Nevertheless, there are promising perspectives, even if one should not expect to generate an optimal fuzzy controller from data by some fuzzy clustering technique. In any case, it is reasonable to construct a rough design of a fuzzy controller by fuzzy clustering. Improvements and tuning of such a fuzzy controller can be carried out by hand or by other learning and optimization techniques in fuzzy control (for an overview see Nauck et al. (1994)), especially since most of these techniques do not learn from scratch but rely on some more or less well working first model.

References:

BEZDEK, J.C. (1973): Fuzzy Mathematics in Pattern Classification. Ph.D. Dissertation, Appl. Math., Cornell Univ., Ithaca, NY.

BEZDEK, J.C., and PAL, S.K. (1992): Fuzzy Models for Pattern Recognition. IEEE Press, New York.

DUBOIS, D., and PRADE, H. (1988): Possibility Theory. Plenum Press, New York.

DUBOIS, D., and PRADE, H. (1993): Possibility Theory, Belief Revision and Non–Monotonic Logic. Proc. EUFIT'93, Aachen, 714–719.

DUNN, J.C. (1974): A Fuzzy Relative of the ISODATA Process and its Use in Detecting Compact Well–Separated Clusters. *Journal of Cybernetics, 3, 32–57.*

HÖHLE, U., and KLAWONN, F. (1992): Fuzzy Control und Ununterscheidbarkeit. Proc. VDE–Fachtagung Technische Anwendungen von Fuzzy–Systemen, Dortmund, 3–9.

KLAWONN, F. (1994): Fuzzy Sets and Vague Environments. *Fuzzy Sets and Systems* (to appear).

KLAWONN, F., and KRUSE, R. (1993): Equality Relations as a Basis for Fuzzy Control. *Fuzzy Sets and Systems, 54, 147–156.*

KLAWONN, F., and KRUSE, R. (1993): Fuzzy Control as Interpolation on the Basis of Equality Relations. Proc. 2nd IEEE International Conference on Fuzzy Systems 1993, IEEE, San Francisco, 1125–1130.

KRISHNAPURAM, R., and KELLER, J.M. (1993): A Possibilistic Approach to Clustering. *IEEE Transactions on Fuzzy Systems, 1, 98–110.*

KRUSE, R., GEBHARDT, J., and KLAWONN, F. (1993): Fuzzy–Systeme. Teubner Stuttgart. (English translation: Foundations of Fuzzy Systems. Wiley, Chichester, 1994).

NAUCK, D., KLAWONN, F., and KRUSE, R. (1994): Neuronale Netze und Fuzzy–Systeme: Grundlagen des Konnektionismus, Neuronaler Netze und der Kopplung mit wissensbasierten Methoden. Vieweg, Braunschweig.

TRILLAS, E., and VALVERDE, L. (1984): An Inquiry into Indistinguishability Operators. In: H.J. Skala, S. Termini, E. Trillas (eds.): *Aspects of Vagueness.* Reidel, Dordrecht, 231–256.

The Excess Mass Approach and the Analysis of Multi-Modality

G. Sawitzki

StatLab Heidelberg, Im Neuenheimer Feld 294, D-69120 Heidelberg

Summary: The excess mass approach is a general approach to statistical analysis. It can be used to formulate a probabilistic model for clustering and can be applied to the analysis of multi-modality. Intuitively, a mode is present where an excess of probability mass is concentrated. This intuitive idea can be formalized directly by means of the excess mass functional. There is no need for intervening steps like initial density estimation. The excess mass measures the local difference of a given distribution to a reference model, usually the uniform distribution. The excess mass defines a functional which can be estimated efficiently from the data and can be used to test for multi-modality.

1. The Problem of Multi-Modality

We want to find the number of modes of a distribution in $\mathbf{R}^k$, based on a sample of n independent observations. There are many approaches to this problem. Any approach has to face an inherent difficulty of the modality-problem: the functional which associates the number of modes to a distribution is only semi-continuous. In any neighbourhood (with respect to the testing topology) of a given distribution, there are distributions with an arbitrarily large number of modes. As a consequence, any confidence interval for the number of modes with finite upper bound will have a confidence level zero (Donoho (1988), Theorem 2.1 and Theorem 2.2).

The impossibility of upper bounds is a combined effect of the semi-continuity, and the richness of the space of probability distributions. If we have restrictions on the family of distributions, upper bounds may be feasible. For example in finite-dimensional parametric families it may still be possible to give non-trivial upper bounds for the number of modes. Unfortunately the restrictions necessary to reduce the space of probability distributions are usually not empirically verifyable. In contrast to problems involving only continuous functionals, with only semi-continuity we cannot even derive approximate solutions for "nearly regular" distributions. Unless we resort to unverifyable assumptions of critical influence, the best we can do is to get **lower bounds** for the number of modes.

Getting lower bounds for the number of modes with guarantied confidence is the first task. Second, we can ask for the power of a procedure. When estimating the number of modes, the challenge is to avoid over-estimation.

2. The Excess Mass Functional

Any approach has to start with a proper definition of a mode. For a cluster analysis approach, a mode might be defined as a cluster center. For a density estimation based approach, a mode may be identified with a local maximum of the density. In a parametric mixture model, a mode might be related to a mixture component. We try to give here a truly nonparametric approach. Let F be our underlying distribution on $\mathbf{R}^k$. We assume that F has a (bounded, continuous) density $f, f > 0$.

Intuitively, a mode is present where probability mass is concentrated. A large value of the probability density is not enough to guarantee a high mass concentration: a distribution may have isolated spots with high densitiy values, but each with an arbitrarily small support. We may speak of modes of different strengths, depending on the probability mass contained in a mode.

A first step is to measure the mass concentration. Since 'high' mass concentration or 'low' mass concentration are relative properties, we have to take a reference measure. Using a λ-multiple of the Lebesgue measure $\mathbf{R}^k$ as a reference, we define the excess mass at level λ to be the integrated probability mass exceeding the Lebesgue density λ:

$$\mathbf{E}(\lambda) = \int (f(x) - \lambda)^+ dx \tag{1}$$

with $\mathbf{E}(\lambda) = 1$. At any level λ, the excess mass is the sum of contributions coming from the connectivity components $\mathrm{C_j}(\lambda) \subseteq \mathbf{R}^k$ of $\{f \geq \lambda\}$:

$$\mathbf{E}(\lambda) = \sum \int_{\mathrm{C_j}(\lambda)} (f(x) - \lambda)\, dx \tag{2}$$

For short, the connectivity components $\mathrm{C_j}(\lambda)$ of $\{f \geq \lambda\}$ are called λ-clusters. The λ-clusters can be characterized as sets maximizing (2).

$$\mathbf{E}(\lambda) = sup_{\mathrm{C} \in C_M} \int_{\mathrm{C}} (f(x) - \lambda)\, dx, \tag{3}$$

where $C_M = \{\mathrm{C} : \mathrm{C} = \mathrm{C_1} \cup \mathrm{C_2} \cup \ldots \cup \mathrm{C_M}; \mathrm{C_j} \subseteq \mathbf{R}^k, \mathrm{C_j} \neq \emptyset$, disjoint, connected$\}$ for some $\mathrm{M} \geq 1$. This leads to an immediate generalization. For any system of sets C, the excess mass at level λ with λ-clusters in C is defined as

$$E_{\mathrm{C}}(l) = sup_{\mathrm{C} \in C} \int_{\mathrm{C}} (f(x) - l)\, dx, \tag{4}$$

with $C = C_M$ as a special case. For a unimodal distribution, at any level λ we have exactly one λ-cluster. For an M-modal distribution, we will have at most M connected components, hence $\mathbf{E}_{C_M}(\lambda) = \mathbf{E}(\lambda)$ for any M-modal distribution F.

Equation (4) has an empirical version. With

$$H_\lambda := F - \lambda \cdot Leb, \tag{5}$$

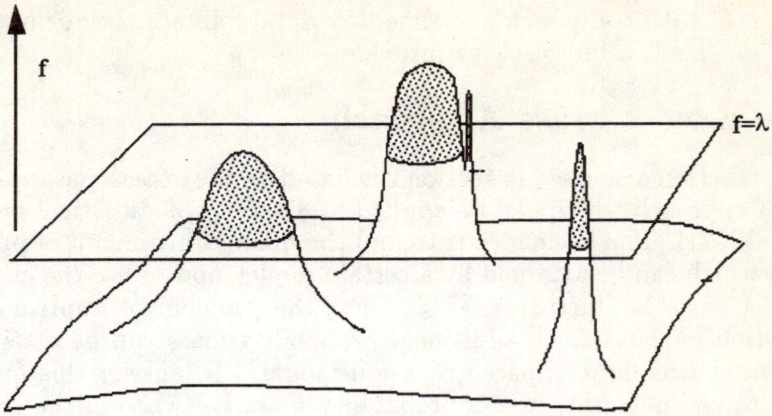

Fig. 1: Density and excess mass. The excess mass is the integrated probability mass exceeding a certain level λ

where Leb is the Lebesgue measure in $\mathbf{R}^k$, (4) can be written as $\mathbf{E}_C(\lambda) = sup_{C \in \mathcal{C}} H_\lambda(C)$. Using the empirical distribution function F_n in (5) yields an empirical version

$$H_{n,\lambda} := F_n - \lambda \cdot Leb, \tag{6}$$

leading to an empirical excess mass estimator

$$\mathbf{E}_{n,\mathcal{C}}(\lambda) = sup_{C \in \mathcal{C}} H_{n,\lambda}(C). \tag{7}$$

Various assumptions about the modality can be modeled using appropriate choices for $\mathcal{C}$, and tests for multi-modality can be based on the corresponding excess mass estimators. For example, a test for bi-modality can be based on the excess mass difference

$$D_n(\lambda) = E_{n,\mathcal{C}_2}(\lambda) - E_{n,\mathcal{C}_1}(\lambda), \tag{8}$$

using the maximal excess mass difference

$$sup_\lambda D_n(\lambda) = sup_\lambda (E_{n,\mathcal{C}_2}(\lambda) - E_{n,\mathcal{C}_1}(\lambda)). \tag{9}$$

as test statistics. Similar tests can be constructed for more general hypotheses and alternatives.

Since for any sets C, C′

$$F_n(C' \backslash C) = 0 \Rightarrow H_{n,\lambda}(C) \geq H_{n,\lambda}(C') \quad for \quad C \subset C'$$

and

$$H_{n,\lambda}(C \cup C') = H_{n,\lambda}(C) + H_{n,\lambda}(C') \quad for \quad C \cap C' = \emptyset$$

the calculation of the excess mass for usual choices of C amounts to a search for sets in C with components spanned by data points, maximizing (7). In most cases, this is a finite search problem.

3. The Excess Mass Approach

The construction discussed in section 2 is based on the excess mass approach, a general approach which can be applied to a variety of statistical problems (Müller (1992)). The basic idea is to find the maximum amount of probability mass which can be attained by a certain model, and to use the exceeding mass as a basis for further analysis. For the problem of multi-modality, the question is: how much additional probability mass can be attained by a multi-modal model compared to a uni-modal? To answer this question, we have to estimate this excess probability mass $\mathbf{E}_C(\lambda)$ from the data under specific assumptions about the number of modes, e. g. unimodality or bi-modality. To draw our conclusions, we have to study the stochastic behaviour of our excess mass estimator first. Then we can take the estimated excess probability mass as a decision basis. This approach yields diagnostic indices and statistics, which have an immediate empirical interpretation. The decision criterion is the amount of data not fitting a certain model.

While the excess mass approach can be used to find tests or estimators in the classical sense, for many of the practically interesting problems the classical framework is like a procrustean bed. For the multi-modality problem, almost any member of the nave null hypothesis described by the family of all uni-modal distributions, has most extreme alternatives in any neighbourhood. Defining a useful null hypothesis becomes a problem. The excess mass approach adds to the repertoire as discussed in Gordon (1994). The natural suggestion based on the excess mass approach is to start from the empirical distribution function, find best approximating unimodal models (i.e. distributions minimizing the total variation distance), and to compare the obtained test statistics with the distributions of the excess mass test statistics drawn from these models. As has been pointed out by Davies (1994), this kind of bootstrap fits well into a general framework of data-based inference which explicitly recognizes the approximate nature of probability models.

The excess mass approach has been first applied to the multi-modality problem in Müller and Sawitzki (1987) where the excess mass functional is introduced and first asymptotic results are given for the one-dimensional case. The resulting method is closely related to procedures suggested in Hartigan (1975), Hartigan and Hartigan (1985) and Hartigan (1987).

4. Analysis for Multi-Modality in One Dimension

In one dimension, the situation is simplified, as there is only one choice for the family of possible support sets C. If we have a continuous density, the λ-clusters for an M-modal distribution must be in C_M, the family of sets composed of at most M disjoint intervals. Given a data set, we can explicitely calculate the excess mass for any hypothetical number of modes M by searching for a set composed of at most M intervals with endpoints at data points, maximizing (7).

4.1 Excess Mass Algorithm in One Dimension

The excess mass $\mathbf{E}_{n,\mathrm{M}}(\lambda) = sup_{\mathrm{C}\in C_M} \mathrm{H}_{n,\lambda}(\mathrm{C})$ can be calculated stepwise using an iteration over the number of possible modes M. For M=1, this requires the search for an interval with endpoints at data points, i.e. C_1 = argmax $\mathrm{H}_{n,\lambda}(\mathrm{C})$. To pass from M to M+1, one of two cases may occur. Additional probability mass may be gained by splitting one of the intervals found in step M (by removing an open interval with endpoints at data points). Or additional probability mass may be gained by adding an interval in the complement of the intervals found at step M ("splitting lemma" in Müller and Sawitzki (1987)). Both possibilities must be explored, and the maximum contribution taken. The common computational problem resides in finding intervals with maximal ascent (or descent) of $\mathrm{H}_{\mathrm{n},\lambda}$. The complexity of this algorithm can be reduced by keeping a "hiker's record list": to find the maximum asscent on your trip, you must keep track of the lowest minimum you have seen so far, and compare the present relative height to the record obtained so far. This gives an algorithm of complexity O(n). More details and an explicit algorithm for the basic search algorithm is given in Müller and Sawitzki (1991).

As a by-product, the algorithm yields the empirical λ-clusters $\mathrm{C}_{\mathbf{n},\mathbf{j}}(\lambda)$, i.e. solutions of $\mathbf{E}_{\mathrm{n,M}}(\lambda) = \sum_{\mathrm{j=1...M}} \mathrm{H}_{\mathrm{n},\lambda}(\mathrm{C}_{\mathrm{n,j}}(\lambda))$, which can be plotted against λ to give a silhouette of the data set. In combination with the excess mass plot, the silhouette can be used for data analysis.

4.2 Asymptotic Stochastic Behaviour in One Dimension

A recommended test-statistics for bi-modality is

$$\mathbf{D}_n := sup\, \mathbf{D}_n(\lambda) = sup\, \mathbf{E}_{n,2}(\lambda) - \mathbf{E}_{n,1}(\lambda).$$

More generally we can allow for M modes as an alternative of interest. We have to investigate $\mathbf{E}_{n,\mathrm{M}}(\lambda)$ under a uni-modal F, but with M>1. Stochastic contributions to the estimation error $\mathbf{E}_{n,\mathrm{M}}(\lambda) - \mathbf{E}(\lambda)$ comes from two sources. There is the classical empirical fluctuation governing $\mathrm{H}_{n,\lambda} - \mathrm{H}_{\lambda}$. And there is an error since we use estimated λ-clusters instead of the true λ-clusters, hoping that $\{\mathrm{C}_{\mathbf{n},\mathbf{j}}(\lambda)\} \approx \{\mathrm{C}_{\mathrm{j}}(\lambda)\}$. The first asymptotic results can be summarized by:

Theorem (Müller and Sawitzki (1991), Theorem 1):
Let f be a smooth density on $\mathbf{R}$, and $x_O \in \mathbf{R}$ with derivative $f'(x) = 0$ only if $f(x) = 0$ or $x = x_0$. For all $\Lambda > 0, \mathrm{M} \geq 1$ the process $\lambda \rightarrow \sqrt{n(\mathbf{E}_{n,\mathrm{M}}(\lambda) - \mathbf{E}(\lambda))}$ converges weakly in $\mathcal{D}\,[/, *]$ to $\lambda \rightarrow \mathrm{B}(a_\lambda)$, B a standard-Brownian bridge, where $a_\lambda = \mathrm{P_F}\{x \mid f(x) \geq \lambda\}$.

This theorem guarantees a square root asymptotics for the excess mass es-

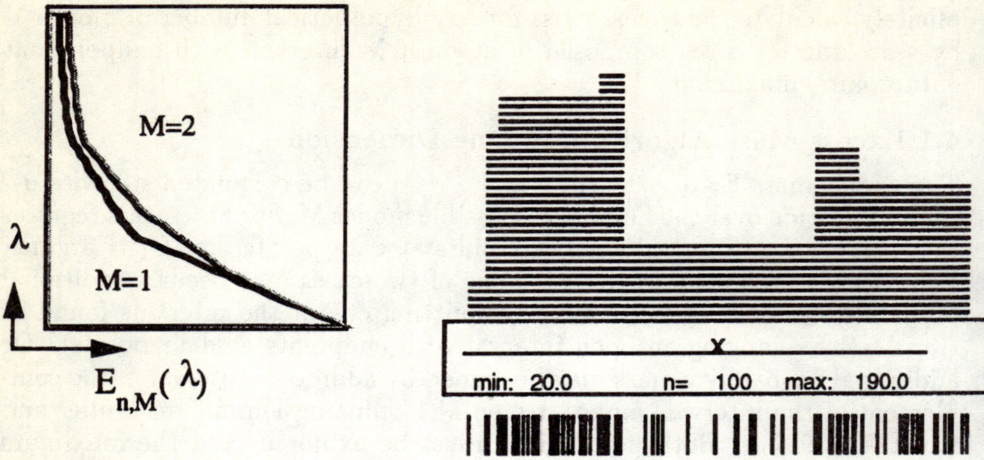

Fig. 2: Estimated excess mass under the assumption of uni-modality ($M = 1$) or bi-modality (M= 2) on the left. Silhouette and scatter plot of observed data on the right

timator under the unimodal hypothesis. This is a better rate than usually is achived. The key is that the excess mass functional contains information about mass concentration, but does not try to identify mass location. Separating the question of mass concentration from location allows a better error rate. Confidence bands can be constructed, using this theorem.

The behaviour of the suggested test statistics $\mathbf{D_n}$ is characterized by
Theorem (Müller and Sawitzki (1991), Theorem 2):
Let f be unimodal with $f'(x) = 0$ iff $f(x) = 0$ or $x = x_0$;
f' ultimately monotone in the tails;
f''' bounded in a neighbourhood of x_0, with $f''(x_0) < 0$.
Under these conditions:

(i) $\mathbf{D}_n(f(x_0)) = \mathbf{O}_P(n^{-3/5})$

(ii) $max_{\lambda \leq f(x_0)-\varepsilon}\mathbf{D}_n(\lambda) = \mathbf{O}_P(n^{-2/3}\log^{2/3} n) \quad (\varepsilon > 0)$

(iii) $max_{\lambda}\mathbf{D}_n(\lambda) = \mathbf{O}_P(n^{-3/5}\log^{3/5} n)$

This theorem tells that in the one-dimensional situation the essential stochastic contribution to the excess mass difference comes from the mode ($3/5 < 2/3$!). For the uniform distribution, we would have $max_{\lambda}\mathbf{D}_n(\lambda) = \mathbf{O}_P(n^{-1/2})$. The difference in order is sizeable: for a sample size of $n = 50$, the difference in order $n^{1/10}$ has a numeric value of 1.47.

5. Analysis for Multi-Modality in Higher Dimensions

In higher dimensions, additional difficulties occur. First, the family of possible λ-clusters is an open choice. While in one dimension any disjoint union of intervals are the obvious candidates, we have more freedom of choice in higher dimensions. Second, the tools at hand are restricted. In one dimension, the Komlós-Major-Tusnády machinery could be used to derive the asymptotic behaviour of the empirical excess mass differences. However this does not have an immediate extension to higher dimensions. Instead, empirical process theory must be used which requires a stricter control of the families of sets under discussion.

The choices of basic set families C_M in higher dimensions must be governed by two rationales. They must be sufficiently rich to allow at least for classical mixture models, like the mixture of normal distributions. On the other hand, they must be sufficiently sparse to allow empirical process theory, or allow for an adequate ad-hoc theory. Usual choices are sparse classes, like Vapnic-Cervonenkis classes, guaranteeing a small coverage dimension, or richer classes, like $conv^2$, the convex sets in the plane, as considered in Hartigan (1987).

For any choice of set systems $C_1 \subset C_2$ we can define empirical excess mass estimators $\mathbf{E}_{n,C_1}(\lambda), \mathbf{E}_{n,C_2}(\lambda)$ as above and use the excess mass difference $\mathbf{D}_n(\lambda) = \mathbf{E}_{n,C_2}(\lambda) - \mathbf{E}_{n,C_1}(\lambda)$ to define a test for the hypothesis $\{f \geq \lambda\} \in C_1$. To test against bimodaltity, C_1 will be chosen to have one connectivity component, and C_2 having two. But other choices modelling qualitative assumptions on the shape of the λ-clusters by appropriate choice of C_1 and C_2 are covered by the same framework (Polonik 1993a).

5.1 Asymptotic Stochastic Behaviour in Higher Dimensions

As in the one-dimensional case, a major step is to get hold of the estimation error involved in using an empirical λ-cluster $\mathrm{C_n}(\lambda)$ instead of the true set $\mathrm{C}(\lambda)$. A key tool is the inequality due to Polonik (1993):

$$Leb\{C(\lambda)\Delta C_n(\lambda)\} \leq$$
$$Leb\{x : |f(x) - \lambda| < \varepsilon\} + \varepsilon^{-1}\{(F_n - F)(C_n(\lambda)) - (F_n - F)(C(\lambda))\} \ \forall\varepsilon > 0.$$

This inequality separates analytical properties of the density f (first term) from oscillation of the process $\mathrm{F_n} - \mathrm{F}$ (second term).

The asymptotic behaviour of the excess mass difference is characterized by the following theorem (Polonik 1993):

Theorem Let f be regular unimodal density (i.e. elliptical at mode x_0+ regularity + rapidly decreasing tails). Then
(i) if C_2 is a VC-Class:
(dimension 1) $\quad max_\lambda \mathbf{D}_n(\lambda) = \mathbf{O}_P(n^{-3/5}\log^{3/5} n)$
(dimension > 1) $\quad max_\lambda \mathbf{D}_n(\lambda) = \mathbf{O}_P(n^{-2/3}\log^{2/3} n)$
(ii) if C_2 consists of finite unions of differences in $conv^2$:

$$max_\lambda \mathbf{D}_n(\lambda) = \mathbf{O}_P(n^{-4/7}).$$

In contrast to the one-dimensional situation, for any dimension > 1 there is no general dominating contribution from the modes since $Leb\{x :| f(x) - f(x_0) |< \varepsilon\} \approx \varepsilon^{1/2}$ for dimension one, but $Leb\{x :| f(x) - f(x_0) |< \varepsilon\} \approx \varepsilon^p$

with $p \geq 1$ in higher dimensions.

The excess mass difference for a uniform distribution on a bounded region has rates $\mathbf{O}_P(n^{-1/2})$, hence for VC-classes: the previous exponents differ at most by 1/6 (for illustration: $50^{1/6} = 1.919\ldots$).

5.2 Excess Mass Algorithms in Higher Algorithms

While the general algorithmic approach sketched above still holds in higher dimensions, general effective algorithms are not available in higher dimensions. The search space is defined by the choice of the model spaces C_M. For convex sets in two dimensions, the algorithm suggested by Hartigan (1987) can be applied. For ellipsoids, Nolan (1991) uses a variant of the Rousseuw and Leroy algorithm for minimal volume ellipsoids. Nason and Sibson (1992) suggest a combination of lower dimension search strategies with approaches from projection pursuit, like the grand tour method. But so far too little is known about appropriate search algorithms which can be applied here.

6. Tests for Multi-Modality

Despite the detailed asymptotics, the finite sample distribution of the excess mass difference is not yet sufficiently known. We can see three approaches to derive valid tests.

First, we can derive stochastic bounds. In one dimension, these bounds can be based on

$$sup_\lambda D_n(\lambda) \leq max_C |(F_n - F)(C)| \tag{10}$$

The right hand side is well-understood in one dimension (Müller and Sawitzki, (1991)). Unfortunately this bound appears to be very conservative. A similar bound is possible in higher dimensions (Polonik (1993a)).

Second, we can derive critical values from special model distributions. For one dimension, sample size $n = 50$ and a Gaussian, Cauchy and uniform model distribution, the resulting distribution of the test statistics is plotted in (Müller and Sawitzki, (1991)). For the uniform distribution, as an extremal case of unimodal distributions, the distribution is tabulated in (Müller and Sawitzki, (1991)).

Third, we can bootstrap the excess mass difference based on the estimator $f_n(x) = max\{\lambda \geq 0 : x \in \mathrm{C}_{n,1}(\lambda)\}$ as an estimator of the best-approximating unimodal distribution. Consistency and quality of this bootstrap approximation however still need further investigation.

References:

DAVIES, L. (1994): Data features. Manuscript. Essen 1994. To appear in *Statistica Nederlandica.*

DONOHO, D.L. (1988): One-sided inference about functionals of a density. *The Annals of Statistics, 16*, 1390-1420.

GORDON, A.D. (1994): Null models in cluster valiadation. In: W. Gaul, D. Pfeifer (eds.) From data to knowledge: Theoretical and practical aspects of classification, data analysis and knowledge organization. Proc. 18th Annual Conference of the GfKl, Univ. of Oldenburg, 1994. Springer Verlag, Heidelberg Berlin, 1994.

HARTIGAN, J.A. (1975): Clustering algorithms. Wiley, NewYork.

HARTIGAN, J.A., and HARTIGAN, P.M. (1985): The dip test of unimodality. *Annals of Statistics, 13*, 70-84.

HARTIGAN, J.A. (1987): Estimation of a convex density contour in two dimensions. *Journal of the American Statistical Association, 82*, 267-270.

MÜLLER, D.W., and SAWITZKI, G. (1987): Using excess mass estimates to investigate the modality of a distribution. Preprint Nr. 398, Januar 1987, Universität Heidelberg, Sonderforschungsbereich 123 Stochastische Mathematische Modelle.

MÜLLER, D.W., and SAWITZKI, G. (1991): Excess mass estimates and tests for multimodality. *Journal of the American Statistical Association, 86*, 738-746.

MÜLLER, D.W. (1992): The excess mass approach in statistics. *Beiträge zur Statistik 3. ftp: statlab.uni-heidelberg.de*

NASON, G.P., and SIBSON,R. (1992): Measuring multimodality. *Statistics and Computing 2*, 153-160.

NOLAN, D. (1991): The excess-mass ellipsoid. *Journal of Multivariate Analysis, 39*, 348-371.

POLONIK, W. (1993): Measuring mass concentration and estimating density contour clusters - an excess mass approach. *Beiträge zur Statistik 7. ftp: statlab.uni-heidelberg.de. Submitted to Annals of Statistics.*

POLONIK, W. (1993a): Density estimation under qualitative assumptions in higher dimensions. *Beiträge zur Statistik 15. ftp: statlab.uni-heidelberg.de*

Mode Extraction by Multivalue Morphology for Cluster Analysis

A. Sbihi[1], J. -G. Postaire[2]

[1] University of Kenitra, Morocco & "Centre d'Automatique" of U.S.T.L., France
[2] "Centre d'Automatique", University of Lille (U.S.T.L.), 59655, Villeneuve d'Ascq, France

Summary: The new statistical approach to unsupervised pattern classification, developed in this paper, consists to extending the multivalue morphological concepts to multidimensional functions in order to detect the modes of the underlying probability density function, particularly when no a priori information is available as to the number of clusters and their distribution.

1. Introduction

Many clustering procedures have been proposed, based on mode detection concepts. Regions of high local density, which might correspond to significant subclasses in the population, can be found from the peaks or modes of the estimated density [1]. The key problem is henceforth to detect the modes of the probability density function (p.d.f.) estimated from the input patterns [2].

Many clustering schemes are based on mode detection concepts, assuming implicitly that each mode of the p.d.f. can be considered as a good approximation of each cluster. When modes are regarded as local maxima of the p.d.f., they are generally detected by means of hill climbing procedures, using some kind of gradient search technique [3]. These methods are known to be sensitive to details in the data structure and, as a result, they tend to generate a greater number of modes than the true p.d.f. [4].

Another approach is based on the analysis of the convexity properties of the underlying p.d.f. [5]. Modes are then considered as concave domains of this function and are detected using integral operators instead of gradient differential operators. Although this approach yields more robust results than the previous one, it remains sensitive to local irregularities in the pattern distribution, espacially for small data sets [6].

Independently from these researches in pattern classification and cluster analysis, mathematical morphology has been developed as an algebra of set theoretic operations for quantitative description of geometrical structures in Euclidean spaces. As introduced by Matheron and Serra [7-8], this approach is mainly concerned with image analysis [9]. Binary morphology has recently found a development in the area of cluster analysis [10]

The intent of the present paper is to introduce a morphological transformation of the underlying density function, as a help for cluster analysis.

Since the language of mathematical morphology is that of set theory, it is first shown how the underlying p.d.f. can be considered as discrete set of integer values (section 2). Section 3 gives the fundamentals of grayscale morphology as applied to high dimensional discrete density function and section 4 presents a new algorithm for clustering based of combination of the basic morphological transformations and morphological segmentation tools applied to the underlying density function. The interest of this procedure for cluster analysis is demonstrated, in section 5, using an artificially generated data set.

2. Discretisation of the Underlying Density Function

Let $X_q = [x_{1,q}, x_{2,q}, \ldots, x_{N,q}]^T, q = 1, 2, \ldots, Q$, the Q available observations, defined as points in R^N. The range of each component is normalised to the interval $[0, R]$, with an integer $R \geq 2$(resolution), by the transformation defined as:

$$y_{n,q} = R.(X_{n,q} - min_{q'}x_{n,q'})/(max_{q'}x_{n,q'} - min_{q'}x_{n,q'})$$

Let $Y_q = [y_{1,q}, \ldots, y_{n,q}, \ldots, y_{N,q}]^T$ the new observation obtained from X_q by this transformation. Dividing each axis of the new space of representation of the data into R adjacent intervals of unit length, we define an hypercubic lattice. Each observation Y_q falls into a non-empty hypercube with R integer coordinates defined as:

$$H = [\mathrm{INT}(y_{1,q}), \ldots, \mathrm{INT}(y_{n,q}), \ldots, \mathrm{INT}(y_{N,q})]^T$$

when $\mathrm{INT}(y_{n,q})$ is the integer part of $y_{n,q}$.

This simple procedure yields directly the list of the non-empty hypercubes. If several observations fall into the same hypercube, this one appears several times in the list of non empty hypercubes. It is easy to find the number of observations falling into each hypercube by counting the number of times it appears in that list.

Let $H = [h_1, \ldots, h_n, \ldots, h_N]^T$ be the integer coordinates of the non-empty hypercube H to which we associate a point X such as: $X = [x_1, \ldots, x_n, \ldots, x_N]^T$ with $x_n = h_n, n = 1, 2, \ldots, N$. Let $p(X)$ denote the number of observations falling into H. It is easy to show that $p(X)$ is nothing else than a discrete estimate of the underlying p.d.f. [11].

3. Morphological Transformations on Density Function

Before the presentation of these transformations, we introduce some of the notational conventions used in the sequel. Underlined block capital letters "$\underline{\mathbf{X}}, \underline{\mathbf{Y}}, \ldots$" denote discrete sets in $(Z^+)^N, Z^+$ is the set of the positive integers. The elements of these sets are denoted by the corresponding capital letters "$X, Y, \ldots$", such as $X = [x_1, \ldots, x_n, \ldots, x_N]^T$ with $X \in \underline{\mathbf{X}}$, $Y = [y_1, \ldots, y_n, \ldots, y_N]^T$ with $Y \in \underline{\mathbf{Y}}$.

3.1 Elementary Morphological Operations on Binary Sets

Binary morphological transformations consist in combining a discrete binary set $\underline{\mathbf{X}}$ with a "structuring element" $\underline{\mathbf{S}}$ [7-8]. In all what follows, it is assumed that this structuring element is symmetrical with respect to its centre.

Under these conditions, the dilatation of $\underline{\mathbf{X}}$ by $\underline{\mathbf{S}}$ is the Minkowski addition of $\underline{\mathbf{X}}$ and $\underline{\mathbf{S}}$:

$$\underline{\mathbf{X}} \oplus \underline{\mathbf{S}} = \{Y \in (Z^+)^N : Y = X + S, X \in \underline{\mathbf{X}}, S \in \underline{\mathbf{S}}\}$$

The erosion, which is the Minkowski set subtraction of $\underline{\mathbf{S}}$ from $\underline{\mathbf{X}}$ is defined by:

$$\underline{\mathbf{X}} \ominus \underline{\mathbf{S}} = \{Y' \in (Z^+)^N : Y' + S = X, X \in \underline{\mathbf{X}}, S \in \underline{\mathbf{S}}\}$$

3.2 Basic Morphological Operations on the Density Function

The function $p(X)$ which takes value in Z^+ and is defined onto subset E of $(Z^+)^N$ is represented by a finite set, called its "umbra", which is defined on $(Z^+)^{N+1}$. To be more specific, the umbra $\underline{\mathbf{O(p)}}$ of $p(X)$ is the subset of $(Z^+)^{N+1}$ defined as: $\underline{\mathbf{O(p)}} = \{(X,t) \in (Z^+)^{N+1} : p(X) \geq t\}$

The function $p(X)$ can be considered as the "top" of the umbra. For each value of X, the value of the top is the greatest value of t i.e.: $\mathrm{SUP}(\underline{\mathbf{X}}) = sup\{t \in Z^+ : (X,t) \in \underline{\mathbf{X}}\}$

Dilatation and erosion can be applied to the umbra of the function $p(X)$. Let $\underline{\mathbf{O(s)}}$ be the umbra of the structuring element $\underline{\mathbf{S}}$. The dilatation of $p(X)$ by $\underline{\mathbf{S}}$ is given by:

$$p \oplus \underline{\mathbf{S}} = \mathrm{SUP}[\underline{\mathbf{O(p)}} \oplus \underline{\mathbf{O(s)}}]$$

Let δ denote the value of the dilation of $p(X)$ by a "structuring element" $\underline{\mathbf{S}}$. The transformation is often implemented as [9]:

$$\delta = (p \oplus \underline{\mathbf{S}})(X) = max_{S \in \underline{\mathbf{S}}}\{p(X - S)\}$$

Similarly, let ε denote the value of the erosion of $p(X)$ by $\underline{\mathbf{S}}$. The erosion is implemented as:

$$\varepsilon = (p \ominus \underline{\mathbf{S}})(X) = min_{S \in \underline{\mathbf{S}}}\{p(X - S)\}$$

The opening operator, $\gamma(X)$, and closing operator, $\varphi(X)$, of $p(X)$ are given respectively by [9]:

$$\gamma(X) = p \circ \underline{\mathbf{S}} = [(p \ominus \underline{\mathbf{S}}) \oplus \underline{\mathbf{S}}]$$
$$\varphi(X) = p \odot \underline{\mathbf{S}} = [(p \oplus \underline{\mathbf{S}}) \ominus \underline{\mathbf{S}}]$$

3.3 Morphological Thinning and Thickening on Density Function

Let $\underline{\mathbf{S}} = (\underline{\mathbf{S^0}}, \underline{\mathbf{S^1}})$ be a flat structuring element [8], where $\underline{\mathbf{S^0}}$ and $\underline{\mathbf{S^1}}$ are two given sets. The multivalue morphological thinning of a local function $p(X)$ by $\underline{\mathbf{S}}$, written as $(p \circ \underline{\mathbf{S}})(X)$, is a mapping which provides a new function [8] defined as follows:

if: $\mathrm{Sup}\{[p(Y)]/Y \in \underline{\mathbf{S}_{\mathbf{x}}^{0}}\} < p(X) \leq \mathrm{Inf}\{[p(Y)]/Y \in \underline{\mathbf{S}_{\mathbf{x}}^{1}}\}$

then: $(p \circ \underline{\mathbf{S}})(X) = \mathrm{Sup}\{[p(Y)]/Y \in \underline{\mathbf{S}_{\mathbf{x}}^{0}}\} = (p \oplus \underline{\mathbf{S}_{\mathbf{x}}^{0}})(X)$

otherwise: $(p \circ \underline{\mathbf{S}})(X) = p(X)$

$\underline{\mathbf{S}_{\mathbf{x}}^{0}}$ and $\underline{\mathbf{S}_{\mathbf{x}}^{1}}$ means that the origin of $\underline{\mathbf{S}}$ is set on X.

The multivalue morphological thickening of a local function $p(X)$ by $\underline{\mathbf{S}} = (\underline{\mathbf{S}^{0}}, \underline{\mathbf{S}^{1}})$, written $(p \bullet \underline{\mathbf{S}})(X)$, is the dual of morphological thinning. It provides a new function defined as [8]:

if: $\mathrm{Sup}\{[p(Y)]/Y \in \underline{\mathbf{S}_{\mathbf{x}}^{1}}\} < p(X) \leq \mathrm{Inf}\{[p(Y)]/Y \in \underline{\mathbf{S}_{\mathbf{x}}^{0}}\}$

then: $(p \bullet \underline{\mathbf{S}})(X) = \mathrm{Inf}\{[p(Y)]/Y \in \underline{\mathbf{S}_{\mathbf{x}}^{0}}\} = (p \oplus \underline{\mathbf{S}_{\mathbf{x}}^{0}})(X)$

otherwise: $(p \bullet \underline{\mathbf{S}})(X) = p(X)$

4. Clustering Algorithm

A clustering algorithm, which takes advantage of the morphological mapping, is presented using a two dimensional data set in order to make the results easy to display and interpret. The observations for this model example are shown in Fig. 1 and drawn from three artificially generated data sets.

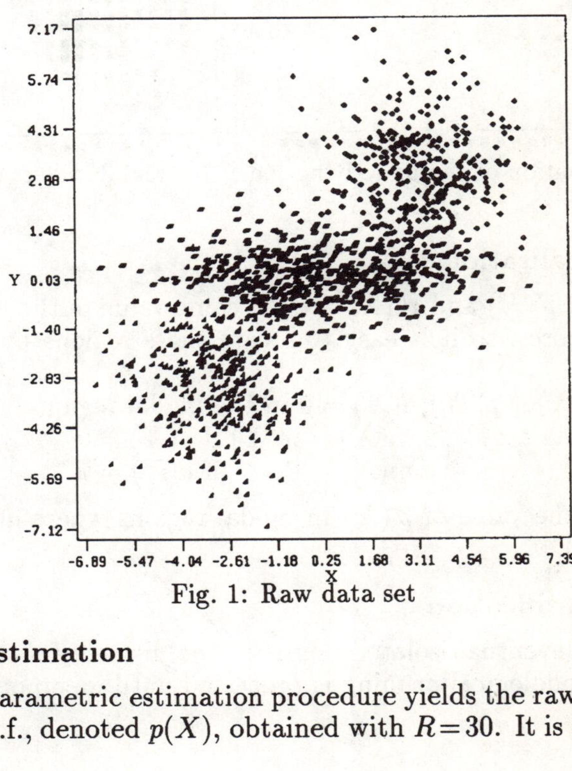

Fig. 1: Raw data set

4.1 P.D.F. Estimation

The fast non-parametric estimation procedure yields the raw estimate of the underlying p.d.f., denoted $p(X)$, obtained with $R=30$. It is shown in Fig. 2.

4.2 Smoothing

The discrete function is systematically followed by a filtering process which smoothes the shape of the modes (see Fig. 3). This filtered function, denoted $p'(X)$, is obtained as: $p'(X) = \gamma[\varphi(\gamma)](X)$. Where γ and φ are the opening and closing operators, respectively.

4.3 Dilation

The filtered function $p'(X)$ is dilated in order to enlarge the modes like is shown in Fig. 4. Let us denote $p''(X)$ the result of this dilation: $p''(X) = (p' \oplus \underline{\mathbf{S}})(X)$.

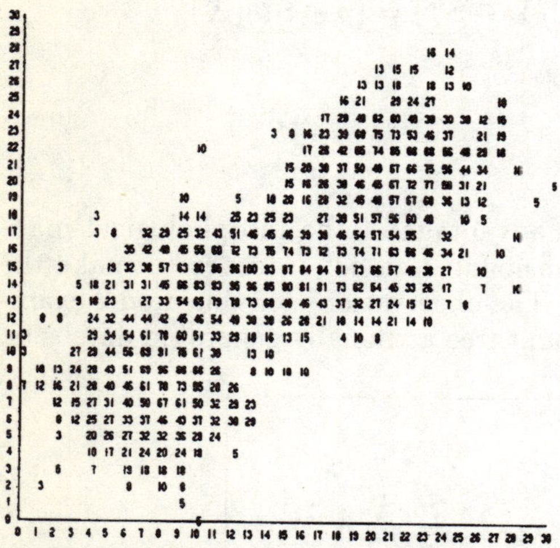

Fig. 2: Estimation of the underlying p.d.f.

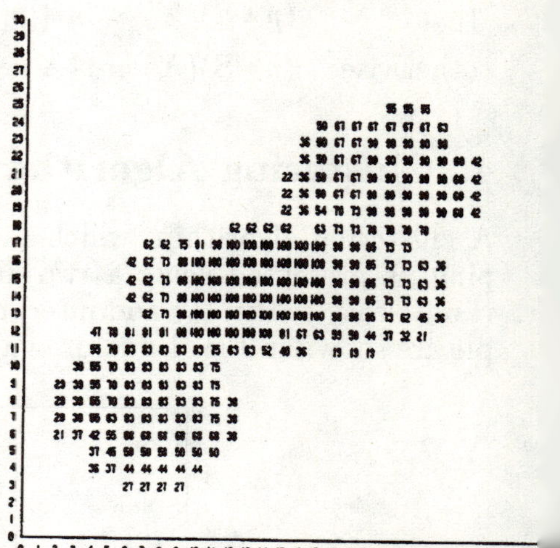

Fig. 3: Effect of smoothing

4.4 Mode Localisation

We remark that $p'(X)$ and $p''(X)$ are the same value in the regions situated in the modes' cores, so it is easy to locate these regions by subtraction of $p'(X)$ by $p''(X)$.
Let $d(X) = p''(X) - p'(X), d(X) = 0$ in the modal regions.
Let $m(X)$ defined as:

$$m(X) = p'(X) \text{ if } d(X) = 0$$
$$m(X) = 0 \quad \text{ if } d(X) \neq 0.$$

$m(X)$ restores the value of $p'(X)$ in modal regions where $d(X)$ is equal to zero (see Fig. 5).

4.5 Mode Identification

For eliminating eventual isolated points in the discrete function $m(X)$, the multivalue morphological pruning is processed until idempotence.

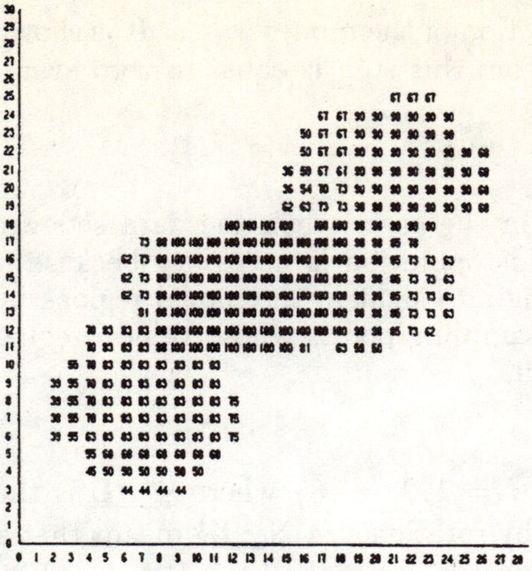

Fig. 4: Effect of dilation

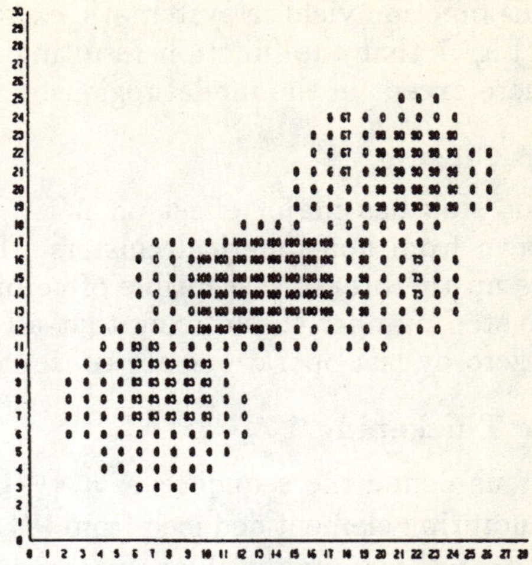

Fig. 5: Localisation of the modes' cores

To be more specific, let us, at first, see in Fig. 6.1 the non-homotopic configuration, denoted $\underline{^{(1)}\mathbf{E}}$, ($\underline{\mathbf{S}} = \underline{^{(1)}\mathbf{E}}$), with the 8-connectivity neighbourhood of point (up to rotation $\pi/4$). $\underline{^{(1)}\mathbf{E}}$ is considered as two-phases structuring element, $\underline{^{(1)}\mathbf{E}} = (\underline{^{(1)}\mathbf{E_0}}, \underline{^{(1)}\mathbf{E_1}})$. In the configuration shown in Fig. 6.1, "1" is used to specify elements belonging to $\underline{^{(1)}\mathbf{E_1}}$ and "0" to $\underline{^{(1)}\mathbf{E_0}}$. Points "*" means the elements belonging neither to $\underline{^{(1)}\mathbf{E_1}}$ nor to $\underline{^{(1)}\mathbf{E_0}}$. Let $\xi = \{\underline{^{(i)}\mathbf{E}}, j = 1, 2, \ldots, 8\}$ be the family of configurations $\underline{^{(j)}\mathbf{E}}$ such as $\underline{^{(j+1)}\mathbf{E}}$ is obtained from $\underline{^{(j)}\mathbf{E}}$ by rotation $\pi/4$.

0	0	0
0	1	*
0	0	*

Fig. 6.1: $\underline{^{(1)}\mathbf{E}} = (\underline{^{(1)}\mathbf{E_0}}, \underline{^{(1)}\mathbf{E_1}})$

0	0	0
*	1	*
1	1	1

Fig. 6.2: $\underline{^{(1)}\mathbf{L}} = \underline{^{(1)}\mathbf{L_0}}, \underline{^{(1)}\mathbf{L_1}}$

Noticing that pruning a function is nothing but thinning it with such configurations, we understand that sequential prunings lead to eliminate the isolated points.

In the multidimensional data space case, pruning is made plane by plane by taking the axes two by two. So, in N-dimensional data space cases, we consider the $(N \times (N-1))/2$ planes perpendicular to the space axes.

Let us now perform the succesive pruning (thinning with ξ) until idempotence on the function $m(X)$ by using the sequence $\{\underline{^{(j)}\mathbf{E}}\}$. This operation leads to the function $M(X)$.

$$M(X) = [m(X) \circ \{\underline{^{(j)}\mathbf{E}}\}]_\infty = [(((m(X) \circ \underline{^{(1)}\mathbf{E}}) \circ \underline{^{(2)}\mathbf{E}}) \circ \ldots) \circ \underline{^{(8)}\mathbf{E}}]_\infty$$

This function yields a systematic extraction of the modes' cores. It is shown in Fig. 7 that the function resulting from this step is equal to zero everywhere except in the modal regions.

4.6 Closing

This step has major effect on $M(X)$ in the cases when the data sets are drawn from non-spherical clusters. This operation is necessary because it fills up the breaks susceptible of coming into sight in the modal regions by the step number 4. Noticing that all sampling points whose value is equal to zero by last operations are conserved.

4.7 Thickening

Let us define the sequence $\Gamma = \{{}^{(j)}\underline{\mathbf{L}}, j = 1, 2, \ldots, 8\}$ where ${}^{(j+1)}\underline{\mathbf{L}}$ is the structuring element deduced from ${}^{(j)}\underline{\mathbf{L}}$ by rotation $\pi/4$. $\{{}^{(j)}\underline{\mathbf{L}}\}$ means that a sequential mapping with the successive rotations of the basic ${}^{(1)}\underline{\mathbf{L}}$ (see Fig. 6.2) must be accomplished. These structuring elements have a homotopic configurations [8].

The modes' cores, identified by the previous steps, then go through an iterative numerical homotopic thickening, processed until idempotence on each plane by using the sequence Γ, so as to optimise the unsupervised classification process according to the minimisation of the error rate, i.e. the decision theory (see Fig. 8).

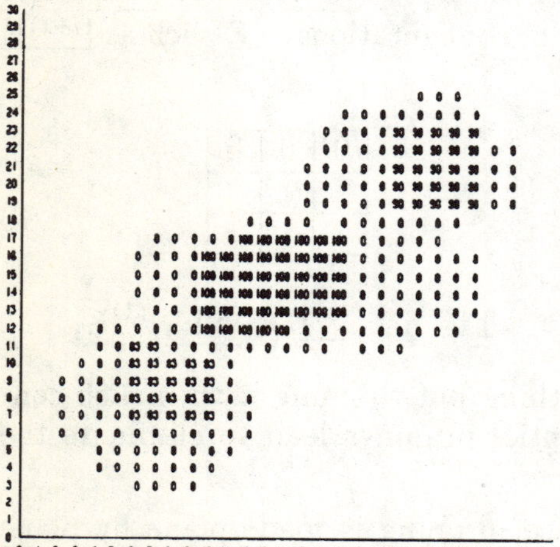

Fig. 7: Cores modes filtered

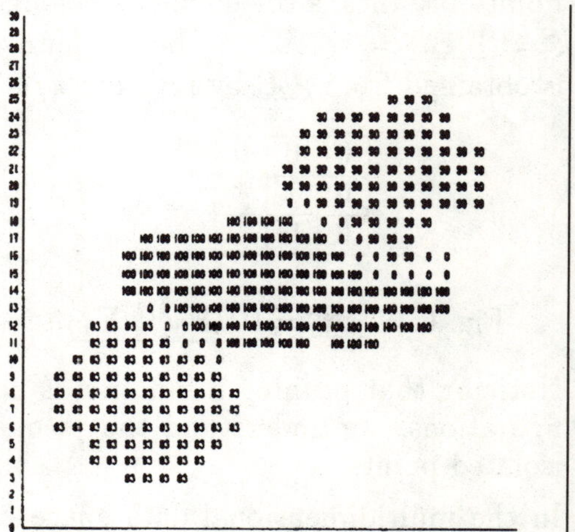

Fig. 8: Modal subsets

This operation is made plane by plane by taken the axes of the discrete space, $\underline{\mathbf{X}}$, two by two. The result of this process shows that the density function can be easily segmented into connected modal regions by eliminating the hypercubes where the value of density function is equal to zero.

4.8 Classification

Once the different modes of the p.d.f. are extracted, the data points falling into each of them are considered to be prototypes. The remaining observations can be assigned to their respective clusters by means of the nearest neighbour classification rule [10]. The result achieved by this procedure, shown in Fig. 9, can be compared to raw data set.

The error rate, estimated as the ratio of misclassified data points to the total number of valuable observations is equal to 8.15%, while the theoretical minimum error rate of the classification is 7.42%.

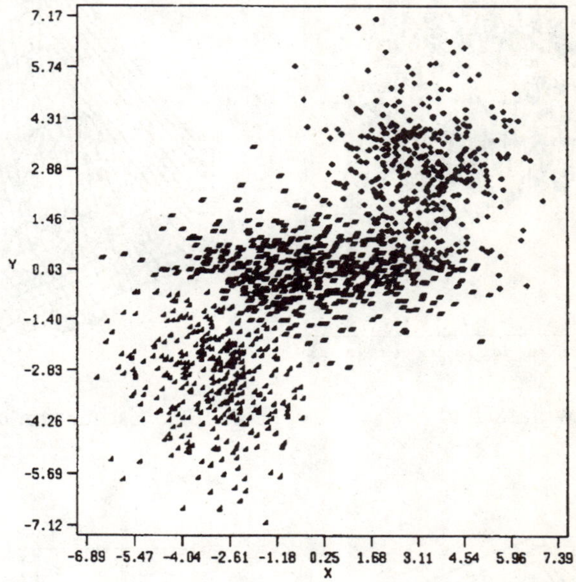

Fig. 9: Result of clustering

4.9 Tuning the Algorithm

The behaviour of the procedure depends on the adjustment of the resolution R of the discretization process. If R is too small, the procedure will suffer from too little resolution and small clusters will not be detected. If R is too large, the procedure will tend to find out a great number of non significant clusters. However, it can be expected that when true clusters exist, stable connected subsets corresponding to these clusters appear for the wide range of values of R. Based on this assumption, the adjustment of R is governed by the concept of cluster stability [11]. Choosing such a parameter in the middle of the largest range when the number of detected clusters remains constant has proved to be a good procedure to optimise a number of classification algorithm.

For above example, the longest stable string occurs between $R = 24$ and $R = 36$. It is the reason why R has been taken equal to 30, which is the middle of that string.

5. Example

The bivariate data set of Fig. 10.1, which is composed of three non-gaussian and non-linearly separable clusters, is presented to demonstrate the ability of the procedure to detect modes of various shapes. Fig. 10.2 shows the raw estimate of the underlying p.d.f. and Fig. 10.3 is a display of the three detected modes obtained with $R = 25$.

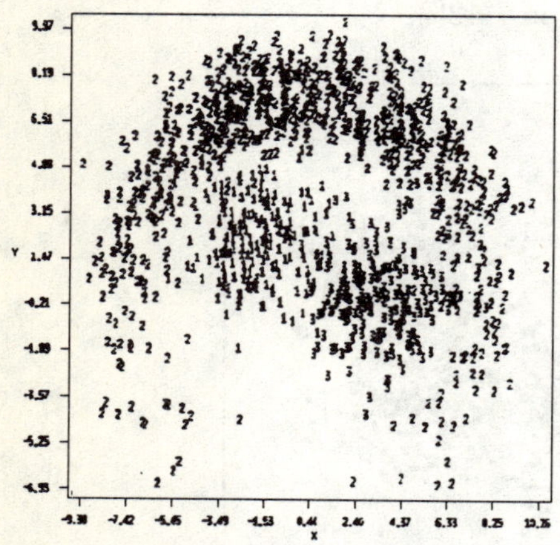

Fig. 10.1: Raw data set

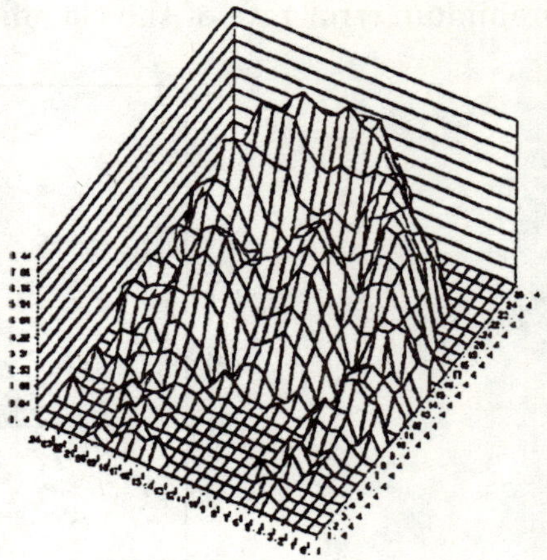

Fig. 10.2: Estimation of the underlying p.d.f.

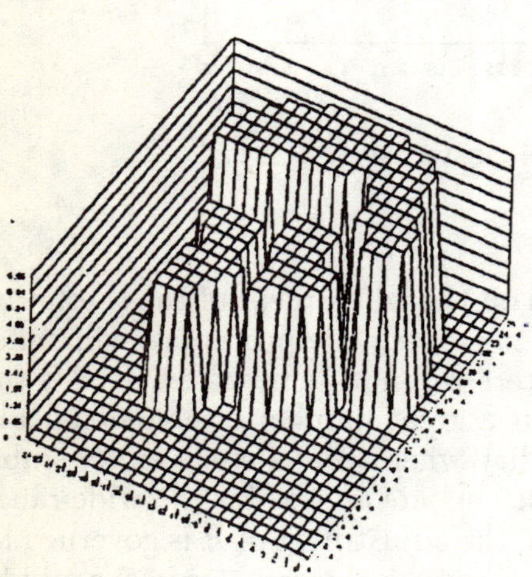

Fig. 10.3: Modal subsets

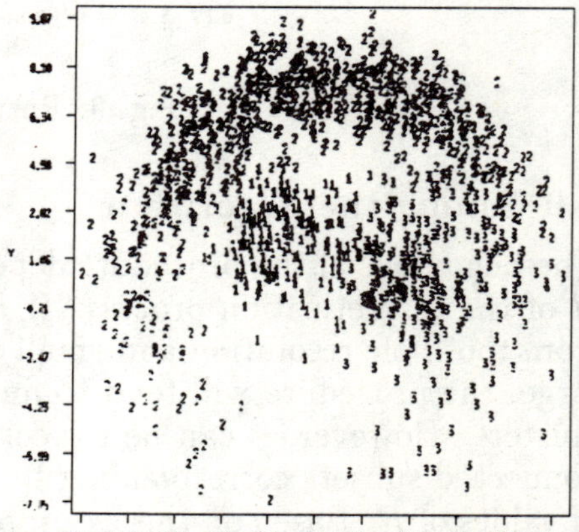

Fig. 10.4: Result of clustering

The error rate associated with the final classification of Fig. 10.4 is 7.82%, while it reaches 26.37% when the data are classified with the well known ISODATA algorithm [12]. This example shows that when the clusters cannot be represented globally by central points, the proposed technique, which takes into account the local structure of the input data, performs much better than well established procedures like the ISODATA algorithm.

Conclusion

An analysis of the perspective described in this paper proves that the multivalue morphological mapping is much more efficient in difficult clustering situations such as non-spherical clusters, linearly non-separable clusters, clusters of unequal weights, and clusters with bridges between them, particularly when no a priori information is available. Thus it appears that multivalue morphology has found a new development in the area of cluster analysis.

References:

[1] DEVIJVER, P.A. et KITTLER, J. (1982) Pattern recognition: A statistical approach. Prentice-Hall, Englewood Cliffs, 448

[2] GITMAN, I. (1973) An algorithm for nonsupervised pattern classification. *IEEE Trans. Syst. Man. Cybern.*, SCM-3, 66-74

[3] FUKUNAGA, K. et HOSTLER, L.D. (1975) The estimation of the gradient of a density function with applications in pattern recognition. *IEEE Trans. Inf. Theory*, IT-21, 1, 32-40

[4] KITTLER J. (1976) A local sensitive method for cluster analysis. *Pattern Recognition*, 8, 23-33

[5] VASSEUR, C.P.A. et POSTAIRE, J.G. (1980) A convexity testing method for cluster analysis. *IEEE Trans. Syst. Man. Cyber.*, SMC-10, 3, 145-179

[6] VASSEUR, C.P.A. et POSTAIRE, J.G. (1981) An approximate solution to normal mixture identification with application to unsupervised pattern classification. *IEEE Trans. Pattern Anal. Machine Intell.*, PAMI-3, 2, 163-179

[7] MATHERON, G. (1975) Random sets and integral geometry. Wiley, New York

[8] SERRA, J. (1982) Image analysis and mathematical morphology. Academic Press, New York

[9] HARALICK, R.M., STENBERG, S.R. et ZHUNG, X. (1987) Image analysis using mathematical morphology. *IEEE Trans. Pattern Anal. Machine Intell.*, 9, 532-550

[10] POSTAIRE, J.G., ZHANG, R.D. et BOTTE-LECOCQ, C. (1993) Cluster analysis by binary morphology. *IEEE Trans. Pattern Anal. Machine Intell.*, 15, 2, 170-180

[11] POSTAIRE, J.G. et VASSEUR, C.P.A. (1982) A fast algorithm for non parametric probability density estimation. *IEEE Trans. Pattern Anal. Machine Intell.*, PAMI-4, 6, 663-666

[12] BALL, G.H. et HALL, J.D. (1967) A clustering technique for summarizing multivariate data. *Behavioral Sci.*, 12, 135-155

On the Longest Edge of the Minimal Spanning Tree

Evangelos Tabakis

Mathematisches Institut, LS Mathematik VII
Universität Bayreuth, D-95440 Bayreuth
Federal Republic of Germany

Summary: As part of the effort to (a) robustify the single link clustering method and (b) develop new inference approaches in clustering in the spirit of Bock (1985), we introduce upper and lower bounds for the length M_n of the longest edge of the minimal spanning tree on n iid random variables drawn from a probability measure P, having Lebesgue density f. We prove that with probability converging to 1 as $n \to \infty$:

$$k_L \cdot \frac{1}{\Delta} \cdot \frac{\log n}{n} \leq M_n^d \leq k_U \cdot \frac{1}{\delta} \cdot \frac{\log n}{n},$$

where $\delta = \inf\{f(x), x \in \mathrm{supp}(P)\}$, $\Delta = \sup\{f(x), x \in \mathrm{supp}(P)\}$ and k_L, k_U are constants depending only on $\partial\mathrm{supp}(P)$.

1. Minimal Spanning Trees and Clustering

Minimal spanning trees (MSTs) are combinatorial structures widely used in combinatorial optimization, operations research and multivariate (nonparametric) statistics. They can be generally defined as follows:

Definition 1.1: Given a graph $G = (V, E)$ with vertices v in V and edges e in E:

(1) a tree $T = (V_T, E_T)$ is a subgraph of G, (i.e. $V_T \subseteq V$ and $E_T \subseteq E$) which is connected and contains no cycles,
(2) a spanning tree is a tree for which $V_T = V$,
(3) a weight function is a function $w : E \longrightarrow \mathbf{R}^+$,
(4) the weight of a tree is $w(T) = \sum_{e \in E_T} w(e)$ and
(5) a minimal spanning tree of (G, w) is a tree T_0 such that $w(T_0) = \min_T w(T)$ where the minimum is taken over all spanning trees T of G.

An important property of the MST is given by the following:

Lemma 1.1: Consider a graph $G = (V, E)$ and a weight function w. For every pair $x, y \in V$ and any path p from x to y on G define $\mathrm{size}(p) = \max\{w(e), e \in p\}$. Let T be a minimal spanning tree on a (G, w) and let p_T be the unique path from x to y on T. Then, $\mathrm{size}(p_T) = \min \mathrm{size}(p)$ where the minimum is taken over all possible paths p from x to y.
Proof: Lebart et al. (1984), pages 143-145. □

In what follows, the topology is the usual topology in $\mathbf{R}^d$ and when considering MSTs for points in $\mathbf{R}^d$, we will always use the euclidean metric $\rho(x,y) = \| x - y \|$.

Definition 1.2: Let *CFC* denote the class of all subsets of $\mathbf{R}^d$ that are compact and have a finite number of connected components. If $K \in CFC$ and $\mathcal{C} = \{C_1, C_2, \ldots, C_k\}$ is the set of connected components of K, $\mathcal{C}$ will be called the *clustering* of K.

We now introduce an (intensionally oversimplified) probabilistic clustering structure by defining the following probability measures. Let supp(P) denote the support of a Borel probability measure P, i.e. the smallest closed set having P-probability 1.

Definition 1.3: Let $\mathcal{P}_{CFC}$ denote all Borel probability measures on $\mathbf{R}^d$ for which supp$(P) \in CFC$. In this case, let $\mathcal{C}(P)$ be the clustering of supp(P). These measures will be called clustered measures. Note that when $X_1, \ldots X_n$ are iid $\sim P$ and $P_n = \frac{1}{n}\sum_{i=1}^{n} \delta_{X_i}$ is the empirical measure of the X_i then $P_n \in \mathcal{P}_{CFC}$, even if $P \notin \mathcal{P}_{CFC}$.

Having such a clustered measure P, we can define an MST on $\mathcal{C}(P)$.

Definition 1.4: Given a measure $P \in \mathcal{P}_{CFC}$, consider the complete graph $G = (\mathcal{C}(P), \mathcal{C}(P) \times \mathcal{C}(P))$. Define further the weight function w on the edge set $\mathcal{C}(P) \times \mathcal{C}(P)$ by $w(A,B) = \min_{a \in A, b \in B} \rho(a,b)$ for any $A, B \in \mathcal{C}(P)$. Finally, let $M(P)$ be the length of the longest edge of any MST on (G, w). In particular: $M_n := M(P_n)$.

Theorem 1.1: (Consistency of M_n). Let P be a clustered measure and let $X_1, X_2, \ldots, X_n \sim P$. Let P_n be the empirical measure. Then: $\lim_{n\to\infty} M_n = M(P)$ a.s.

Proof: This is one of those "obvious" facts whose detailed proof requires a couple of pages. A proof of a more general result (involving consistency of all single link distances) was given in Tabakis (1992), page 42. □

2. Chaining and Breakdown Points

The MST can be turned into a divisive clustering technique in which one removes the longest edge of the tree on the n observations to obtain a first level partition, then the next longest edge to obtain the next and continues this way (if desired) up to the trivial partition into n one-element "clusters". The consistency result of the last section can be considered as the reason that this method works in situations where the "clusters" C_i belong to the class *CFC* and are, therefore, nicely separated: Edges of the MST within the "clusters" converge eventually to 0 leaving the positive distances between "clusters" standing out. The main drawback of the method is the *chaining effect.* A few observations between 'true" clusters can create a *chain*, i.e. a path of short edges joining the clusters and thus postponing their recovery to a much later stage of the method, making the resulting partitions nonsensical or misleading.

We can describe this effect more precisely using the terminology of robust statistics. Let $\mathcal{P}_C$ denote the family of clustered probability measures on $\mathbf{R}^d$ that have at least two clusters. Then $M(P) > 0$ for any $P \in \mathcal{P}_C$, where, by definition 1.4, $M(P)$ denotes the length of the longest edge of any minimal spanning tree on $\mathcal{C}(P)$ (the clustering of P). By theorem 1.1, we have $M_n = M(P_n) \longrightarrow M(P)$ a.s. as $n \to \infty$.

Recall now the definition of the gross-error breakdown point of a sequence of estimators (see also Huber (1981), page 13 or Hampel et al. (1986), page 97):

Definition 2.1: Let P be a Borel probability measure on $\mathbf{R}^d$. Suppose that a parameter $\theta \in \Theta \subset \mathbf{R}^k$ is estimated by a sequence of estimators of the form $T_n(X_1 \ldots X_n)$ where $T_n : \mathbf{R}^{d \times n} \mapsto \mathbf{R}^k$. The gross-error breakdown point of T_n at P is defined to be:

$$b(P, \Theta, T_n) := \sup \left\{ \epsilon \leq 1 : \begin{array}{l} \exists \text{ a compact set } K_\epsilon \subset \Theta \text{ such that:} \\ \forall \text{probability measure } R \text{ and} \\ X_1, X_2, \ldots, X_n \text{ iid } \sim Q, Q = (1-\epsilon)P + \epsilon R : \\ \lim_n \Pr_Q(T_n(X_1, X_2, \ldots, X_n) \in K_\epsilon) = 1 \end{array} \right\}$$

When $P \in \mathcal{P}_C$ then $M(P) > 0$. Therefore, the natural parameter space Θ for $\theta = M(P)$ is $\Theta = (0, \infty)$. We now have the following:

Theorem 2.1: For every $P \in \mathcal{P}_C$, $b(P, (0, \infty), M_n) = 0$.
Proof: For every $0 < \epsilon \leq 1$, we can find a probability measure R such that the measure $Q_\epsilon = (1-\epsilon)P + \epsilon R$ has compact and connected support. If $X_1, X_2, \ldots, X_n iid \sim Q_\epsilon$ and $Q_{\epsilon,n} = \frac{1}{n} \sum_{i=1}^n \delta_{X_i}$ is the empirical measure, then $\lim_n M(Q_{\epsilon,n}) = 0$ a.s. by theorem 1.1. Any compact set K_ϵ in $\Theta = (0, \infty)$ must have a positive distance from 0 and so $\Pr(M(Q_{\epsilon,n}) \in K_\epsilon) \longrightarrow 0$ as $n \to \infty$. So, the breakdown point is 0. □

When dealing with real data, one is more likely to come across a *contaminated* version of a $P \in \mathcal{P}_C$ than P itself. Ideally, one would like to "strip" the contamination ϵR_ϵ from the measure Q_ϵ and *then* estimate the functional $M(P)$. The idea is similar to the use of *trimmed means* as location estimators (instead of simply using the non-robust sample mean). In the case of location estimators the type of contamination we mostly want to avoid is that of long-tailed distributions. Therefore, in the sample version we choose to remove the extreme observations and compute the mean of the rest. In our clustering case, the contamination to be avoided is that of *low density areas.* It would then seem appropriate to remove low-density observations and compute the minimal spanning tree on those observations X_i that have density $f(X_i)$ larger than or equal to (say) δ.

Let us fix $\delta > 0$ and suppose that we have succeeded in isolating the measure P from any *low-density* contamination so that we are able to assume we are dealing exclusively with observations from a central model (to further pursue the robustness analogy in terminology) having the following two properties: (i) The model is a subset of $\mathcal{P}_{CFC}$ and (ii) all P in the model have a (Lebesgue) density f where $f \geq \delta$ on supp(P). A test of the existence of cluster structure on this model would test the null hypothesis $k = 1$ (compact and connected support) against the alternative $k > 1$ (compact and disconnected support). We would like to investigate the behavior of our test statistic $M(P_n)$ under the null hypothesis. This leads us to the formulation of our main question: When n iid observations are drawn from a distribution with compact and connected support and having a density larger than or equal to $\delta > 0$ on that support, how large is the longest edge of the minimal spanning tree having these observations as vertices? We will try to answer this question in the remaining of this paper.

3. Relations to Combinatorial Optimization

We will now summarize some results (mainly referring to problems in combinatorial optimization) that relate to our MST problem:

1. The case $d = 1$ would involve a distribution whose support is an interval and has density $f \geq \delta > 0$ on that interval. Then the longest edge of the minimal spanning tree is the quantity $Z_n = \max_{1 \leq i \leq n-1} X_{(i+1)} - X_{(i)}$ (i.e. the largest spacing). For the case of the uniform distribution on $[0, 1]$, it is known that $\lim_{n \to \infty} \Pr(nZ_n - \log n < t) = \exp(-e^{-t})$, $\forall t \in \mathbf{R}$ (see Lévy (1939)). It is not difficult to generalize to other densities and to prove that $Z_n = O_p(\log n/n)$ (see Tabakis (1992), page 51). This would hint towards an M_n of the size $O_p((\log n/n)^{1/d})$ in the d-dimensional case.

2. The total length of the MST on n random points is known to be asymptotically given by the quantity $k_d n^{\frac{d-1}{d}} \int f(x)^{\frac{d-1}{d}} dx$, where k_d is a constant depending only on the dimension (Steele (1988)). Therefore, the average length of the MST edges is asymptotically a quantity of the order $O_p(n^{-1/d})$.

3. If x_2 is the nearest neighbor of x_1 in the set $\{x_1, x_2, \ldots, x_n\} \subset \mathbf{R}^d$ then (lemma 1.1) the edge (x_1, x_2) must be an edge of the MST. So M_n must be at least as large as the maximal nearest neighbor, a quantity known to be of the order $O_p((\log n/n)^{1/d})$ (see eg. Henze (1983), Steele and Tierney (1988), Dette and Henze (1989)). This argument will be used later to give a precise lower bound for M_n.

4. In Shor and Yukich (1991) the following remarkable result is proved: Let $n = k^d$, for some positive integer k and let $X_1, X_2, \ldots, X_n$ be drawn from the uniform distribution on the unit cube in $d > 2$ dimensions. Let $x_1, x_2, \ldots, x_n$ be the n points in the unit cube with coordinates $(\frac{i_1}{k+1}, \frac{i_2}{k+1}, \ldots, \frac{i_d}{k+1})$, where $1 \leq i_1, i_2, \ldots, i_d \leq k$. Then, there exists a matching $(X_1, x_{\sigma_1}), \ldots, (X_n, x_{\sigma_n})$ so that: $\max_{1 \leq i \leq n} \| X_i - x_{\sigma_i} \| = O_p((\log n/n)^{1/d})$. An MST can now be constructed on $\{x_1, x_2, \ldots, x_n\}$ having edges of size $O_p(n^{-1/d})$ by simply joining suitable neighboring points on the grid. Then, a path on $\{X_1, X_2, \ldots, X_n\}$ can be constructed for every pair (X_i, X_j) by using the MST path on the x's and transferring it to corresponding points in $\{X_1, X_2, \ldots, X_n\}$ according to the optimal matching. The triangle inequality then shows that no edge on the latter path (and therefore not even M_n by lemma 1.1) can exceed $O_p((\log/n)^{1/d})$.

5. If the MST on $S_n = \{X_1, X_2, \ldots, X_n\}$ includes an edge of length s then there must be an open set of d-volume $v_d s^d$ containing none of the points in S_n (v_d a constant). A precise formulation and proof of this fact will be given in the next section (lemma 4.1). This, however, connects our problem to the theory of multivariate spacings (see Deheuvels et al. (1988)). In brief, multivariate spacings are the largest possible members of a certain class of sets not containing any point of an iid sample. Under suitable assumptions, the d-volume of such sets is asymptotic to $\log n/n$, a result which bounds M_n by a quantity of the order $O_p((\log n/n)^{1/d})$.

4. Upper and Lower Bounds

Definition 4.1: Let $B(a, r)$ denote the open ball $\{x \in \mathbf{R}^d : \| x - a \| < r\}$ for $a \in \mathbf{R}^d$, $r > 0$. Then let $D_{a,b} = B(a, \| a - b \|) \cap B(b, \| a - b \|)$, $a, b \in \mathbf{R}^d$.

Definition 4.2: A class $\mathcal{G}$ of subsets of $\mathbf{R}^d$ is going to be called *ball-like* if there is a functional $r : \mathcal{G} \longrightarrow \mathbf{R}^+$ such that: (1) $r(G_1) = sr(G_2) \Rightarrow \lambda(G_1) = s^d \lambda(G_2)$ for $G_1, G_2 \in \mathcal{G}$ and $s > 0$, and (2) $\exists G_0 \in \mathcal{G}$ such that $r(G_0) = 1$. For such an r, and for $G \in \mathcal{G}$, $r(G)$ will be called the radius of G. The class $\mathcal{G} = \{D_{a,b},\ a, b \in \mathbf{R}^d\}$ is *ball-like* with $r(D_{a,b}) = \| a - b \|$.

The following is a crucial property of the (euclidean) MST:

Lemma 4.1: Let $V = \{x_1, x_2, \ldots, x_n\} \subset \mathbf{R}^d$ and let $T = (V, E)$ be an MST of V. Then, $\forall i, j$ such that $(x_i, x_j) \in E$: $D_{x_i, x_j} \cap \{x_1, x_2, \ldots, x_n\} = \emptyset$.
Proof: Since $(x_i, x_j) \in E$, (x_i, x_j) is the path of minimal size from x_i to x_j (lemma 1.1). If $x_k \in D_{x_i, x_j}$ then the path $p = (x_i, x_k, x_j)$ has size equal to $\max\{\| x_i - x_k \|, \| x_j - x_k \|\} < \| x_i - x_j \|$, a contradiction. □

Now we can give the following upper bound:

Theorem 4.1: Let P satisfy the following assumptions:
(1) supp(P) is compact and connected $\subset \mathbf{R}^d$,
(2) P has a density f (w.r.t. Lebesgue measure λ) such that:
$\inf\{f(x),\ x \in \text{supp}(P)\} = \delta > 0$, and
(3) there is a class $\mathcal{G}$ of *ball-like* sets with radius $r(\cdot)$, a $G_0 \in \mathcal{G}$ with $r(G_0) = 1$ and an $\epsilon > 0$ such that $\forall x, y \in \text{supp}(P)$, with $\| x - y \| < \epsilon$: $\exists G \in \mathcal{G}$, $G \subset D_{x,y} \cap \text{supp}(P)$ and $r(G) = \| x - y \|$.

Let $X_1, X_2, \ldots, X_n$ *iid* $\sim P$, and let M_n be the length of the longest edge of any minimal spanning tree of $\{X_1, X_2, \ldots, X_n\}$. Then for all $k > 2$:

$$\lim_n \Pr\left(M_n^d \leq \frac{k}{\delta \lambda(G_0)} \cdot \frac{\log n}{n} \right) = 1.$$

Proof: Theorem 1.1 guarantees that, outside a set of probability 0, there exists a certain n_ϵ so that for any $n \geq n_\epsilon$ all edges of the MST on $\{X_1, \ldots, X_n\}$

are smaller than ϵ. Outside this null set and given any edge (X_i, X_j) of the MST, let $G_{i,j} \in \mathcal{G}$ be the set for which $r(G_{i,j}) = \parallel X_i - X_j \parallel$ and $G_{i,j} \subset D_{X_i,X_j} \cap \mathrm{supp}(P)$ as guaranteed by the third assumption of the theorem. Let $r_n := \frac{k}{\delta\lambda(G_0)} \cdot \frac{\log n}{n}$. If $M_n^d > r_n$ then there must exist an edge (X_i, X_j) such that $\parallel X_i - X_j \parallel^d > r_n$ and (because of lemma 4.1) $D_{X_i,X_j} \cap \{X_1, X_2, \ldots, X_n\} = \emptyset$. So, for $n \geq n_\epsilon$, $\Pr(M_n^d > r_n)$ is bounded by:

$$
\begin{aligned}
& \textstyle\sum_{1\leq i<j\leq n} \Pr(\parallel X_i - X_j \parallel^d > r_n \text{ and } D_{X_i,X_j} \cap \{X_1, X_2, \ldots, X_n\} = \emptyset) \\
\leq\ & \textstyle\sum_{1\leq i<j\leq n} \Pr(D_{X_i,X_j} \cap \{X_1, X_2, \ldots, X_n\} = \emptyset \,\Big|\, \parallel X_i - X_j \parallel^d > r_n) \\
\leq\ & \textstyle\sum_{1\leq i<j\leq n} \Pr(G_{i,j} \cap \{X_1, X_2, \ldots, X_n\} = \emptyset \,\Big|\, \parallel X_i - X_j \parallel^d > r_n) \\
=\ & \binom{n}{2} \Pr(G_{1,2} \cap \{X_3, \ldots, X_n\} = \emptyset \,\Big|\, \parallel X_1 - X_2 \parallel^d > r_n) \\
& \text{(by symmetry and the fact that, by definition, } X_1, X_2 \notin G_{1,2}) \\
=\ & \binom{n}{2} \left[\Pr(X_3 \notin G_{1,2} \,\Big|\, \parallel X_1 - X_2 \parallel^d > r_n)\right]^{n-2} \quad (X_3, \ldots, X_n \text{ are iid}) \\
=\ & \binom{n}{2} \left[E(1 - 1_{G_{1,2}}(X_3) \,\Big|\, \parallel X_1 - X_2 \parallel^d > r_n)\right]^{n-2} \\
\leq\ & \binom{n}{2} \left[E(1 - \delta\lambda(G_{1,2})) \,\Big|\, \parallel X_1 - X_2 \parallel^d > r_n)\right]^{n-2} \\
=\ & \binom{n}{2} \left[E(1 - \delta \parallel X_1 - X_2 \parallel^d \lambda(G_0) \,\Big|\, \parallel X_1 - X_2 \parallel^d > r_n)\right]^{n-2} \\
\leq\ & \binom{n}{2} (1 - \delta r_n \lambda(G_0))^{n-2} \\
=\ & \binom{n}{2} (1 - \tfrac{k \log n}{n})^{n-2} \sim \binom{n}{2} \tfrac{1}{n^k} \stackrel{n\to\infty}{\longrightarrow} 0 \quad \text{since} \quad k > 2. \ \square
\end{aligned}
$$

We will now sketch the proof of a lower bound and refer the reader to section 5.3 of Tabakis (1992) for the details. In the previous section we mentioned that the longest edge of the MST must be at least as large as the largest nearest neighbor distance. A related quantity has been studied in Henze (1983):

Definition 4.3: Let P have a density f with respect to Lebesgue measure. Define $D_n := \max_{1\leq i\leq n} f(X_i)^{1/d} \min\{\min_{j\neq i} \rho(X_i, X_j), \rho(X_i, \partial\mathrm{supp}(P))\}$.

Let M_n^* be the length of the longest edge of the MST on $\{X_1, X_2, \ldots, X_n\}$ and $\partial\mathrm{supp}(P)$. By the above mentioned argument $\Delta M_n^{*d} \geq D_n^d$ where $\Delta = \sup\{f(x), x \in \mathrm{supp}(f)\}$. However, the size of the optimal path from X_i to X_j can only be decreased when $\partial\mathrm{supp}(P)$ is used as an additional "point". This means that $M_n^* \leq \max\{M_n, B_n\}$ where B_n is defined and bounded by the following:

Lemma 4.2: Let P have compact support and density f (with respect to Lebesgue measure) such that $\inf\{f(x), x \in \mathrm{supp}(P)\} = \delta > 0$. Let $X_1, X_2, \ldots, X_n$ *iid* $\sim P$ and $B_n := \min_{1\leq i\leq n} \rho(X_i, \partial\mathrm{supp}(P))$. Then $B_n = O_p(1/n)$ as $n \to \infty$.
Proof: See Tabakis (1992) pages 66-67. $\square$

We can now make use of the following result available for D_n:

Proposition 4.1: Let P be concentrated on an open bounded subset U of $\mathbf{R}^d$ whose closure is supp(P) and assume it has a continuous density f satisfying:
(1) $\inf\{f(x),\ x \in \text{supp}(P)\} > 0$ and
(2) $\sup\{|f(x) - f(y)|\ :\ x, y \in U,\ |x - y| < s\} = o((-\log s)^{-1})$ as $s \to 0$.
Let ω_d denote the volume of the unit ball B_d in $\mathbf{R}^d$ and let $X_1, X_2, \ldots, X_n$ $iid \sim P$. Then: $\lim_n \Pr(n\omega_d D_n^d - \log n < \xi) = \exp(-e^{-\xi})$, $\forall \xi \in \mathbf{R}$.
Proof: This is the main result of Henze (1983). □

Piecing everything together we arrive at the following lower bound:

Theorem 4.2: Let P satisfy the same assumptions as in proposition 4.1. Let $X_1, X_2, \ldots, X_n$ be $iid \sim P$, $\Delta := \max\{f(x),\ x \in \text{supp}(P)\}$ and ω_d the volume of the unit ball in $\mathbf{R}^d$. Then, for every $k < 1$:

$$\lim_n \Pr\left(M_n^d \geq \frac{k}{\Delta\omega_d} \cdot \frac{\log n}{n}\right) = 1.$$

5. Simulations, Conclusions, Conjectures

As a concrete example we will use the uniform distribution on the unit ball B_d in $\mathbf{R}^d$. For the upper bound we will need an appropriate class of sets $\mathcal{G}$. The following remark suggests such a class.

Remark 5.1: If $x, y \in B_d$ and $\| x - y \|$ is small enough, then at least 1/2 of the set $D_{x,y}$ is contained in B_d, for $d > 1$. This worst case appear when $\overline{xy}$ is close to the boundary of B_d. Furthermore, a routine integral calculation shows that the volume of the set $D_{a,b} \subset \mathbf{R}^d$ defined in definition 4.1 is $|a-b|$, for $d = 1$ and $2\omega_{d-1} \cdot \| a - b \|^d \cdot c_d$ for $d > 1$ where $c_d := \int_{\pi/6}^{\pi/2} \cos^d(\theta) d\theta$.

Theorem 5.1: If $X_1, X_2, \ldots, X_n$ $iid \sim U(B_d)$ and $c_L^B < 1, c_U^B > 1$ then:

$$\lim_n \Pr\left(c_L^B \left(\frac{\log n}{n}\right)^{1/d} \leq M_n \leq c_U^B \left(\frac{2\omega_d}{\omega_{d-1} c_d}\right)^{1/d} \left(\frac{\log n}{n}\right)^{1/d}\right) = 1.$$

Proof: We apply theorem 4.1: For $P = U(B_d)$ we choose the class $\mathcal{G}$ of the theorem to be the *halves* in which the sets $\{D_{a,b}, a, b \in \mathbf{R}^d\}$ are divided by the $(d-1)$-dimensional ball with diameter $\overline{ab}$ and use radius $r = \| a - b \|$. Then $\lambda(G_0) = \omega_{d-1} c_d$ (remark 5.1) while $\delta = \min\{f(x), x \in \text{supp}(P)\} = \omega_d^{-1}$. The upper bound follows. We now apply theorem 4.2: For $P = U(B_d)$ we have $\Delta = \omega_d^{-1}$ and the lower bound follows. □

To assess the validity of these bounds for sample sizes of $n = 1000$ we engaged in a small simulation study. For $2 \leq d \leq 6$ we drew 100 iid samples of size $n = 1000$ out of $U(B_d)$ and computed the length M_n of the longest edge of the minimal spanning tree. The results are summarized in Tab. 1. The

dim	Lower bound	Minimum	Median	Maximum	Upper Bound
2	0.0831	0.0804	0.0967	0.1377	0.2658
3	0.1904	0.1956	0.2202	0.2835	0.4455
4	0.2883	0.3071	0.3317	0.3783	0.5748
5	0.3697	0.3869	0.4341	0.4788	0.6685
6	0.4364	0.4753	0.5197	0.6717	0.7393

Tab. 1: Simulation results for $U(B_d)$ with $n = 1000$.

first column shows the dimension d of the space. The second and sixth (last) contain the lower and upper bound for M_n as predicted by theorem 5.1. The third, fourth and fifth contain the minimum, median and maximum of the 100 observed M_n for each dimension d.

From the table we conclude that the lower bound is quite sharp, while the upper bound seems to be rather conservative. This suggests that, at least for the case of the uniform distribution, there is little difference between the longest edge of the MST and the largest nearest neighbor distance on which the lower bound was based (something to be expected from a unimodal distribution). There is certainly room for improvement for both bounds. In the case of the lower bound, it seems plausible to substitute the supremum Δ of the density by the infimum δ since that is where the largest nearest neighbor distance is more likely to appear. The case of the upper bound is certainly more difficult and will require a different approach. In fact, at this point one could conjecture the following:
Conjecture: Let $X_1, X_2, \ldots, X_n iid \sim f$ and let $M_n(\delta)$ be the length of the longest edge of the MST on $S_n(\delta) = \{X_i : f(X_i) \geq \delta\}$. Then, under the same assumptions as those of theorem 4.1, $M_n(\delta)^d \sim (k/\delta) \cdot (\log n/n)$ as $n \to \infty$ for some constant k.

The most promising approach to the proof of this conjecture seems to be that of Aldous and Steele (1992), who study the asymptotic behavior of the whole tree structure of the MST. Research in this direction is continuing.

Let us reflect a little on the possible uses of the conjecture. Suppose P is a measure with unimodal Lebesgue density f. For our purposes this could be defined to mean that the conditional measure P_δ of P given $f \geq \delta$ has a compact and connected support for every $\delta > 0$. Let us further assume that the measures P_δ all satisfy assumption (3) of Thm. 4.1 with the same family of sets $\mathcal{G}$. This could be interpreted to mean that the contours of f all have the same general shape.

We can use the stochastic process $M_n(\delta)$ to test whether a sample $X_1, \ldots, X_n$ comes from a P with the above mentioned properties. If it does, then the aforementioned conjecture predicts that $M_n(\delta)^d$ goes to 0 like the hyperbola $1/\delta$ as a function of δ. Given a density estimator f_n to approximate $S_n(\delta)$ by $\hat{S}_n(\delta) = \{X_i : f_n(X_i) \geq \delta\}$, we can assess the "unimodality" of f by fitting a curve of the form $c \cdot 1/\delta$ to the stochastic process $\hat{M}_n(\delta)^d$ corresponding to $\hat{S}_n(\delta)$. Often, a graph of $\hat{M}_n(\delta)^d$ against δ will be enough for a visualization of the cluster structure. Note that such a graphical test would be able to detect both multimodal densities ($\hat{M}_n(\delta)^d$ would start increasing instead of decreasing when δ reaches the first "valley" between modes) and "needles in the haystack" (dense subclouds within a point cloud manifesting themselves as change points on the $\hat{M}_n(\delta)^d$ curve). Empirical evaluations of the method based on both simulated and real data will be published elsewhere.

References:

ALDOUS, D., and STEELE, J.M. (1992): Asymptotics for euclidean minimal spanning trees on random points. *Probab. Theory Relat. Fields, 92, 247-258.*

BOCK, H.H. (1985): On some significance tests in cluster analysis. *Journal of Classification, 2, 77–108.*

DEHEUVELS, P., EINMAHL, J., MASON, D., and RUYMGAART, F.H. (1988): The almost sure behavior of maximal and minimal multivariate k_n spacings. *Journal of Multivariate Analysis, 24, 155–176.*

DETTE, H., and HENZE, N. (1989): The limit distribution of the largest nearest-neighbour link in the unit d-cube. *Journal of Applied Probability, 26, 67–80.*

HENZE, N. (1983): Ein asymptotischer Satz über den maximalen Minimalabstand von unabhängigen Zufallsvektoren mit Anwendung auf einen Anpassungstest im R^d und auf der Kugel. *Metrika, 30, 245–259.*

HAMPEL, F.R., RONCHETTI, E.M., ROUSSEEUW, P.J., and STAHEL, W.A. (1986): Robust Statistics: The approach based on Influence Functions. Wiley, New York.

HUBER, P.J. (1981): Robust Statistics. Wiley, New York.

LEBART, L., MORINEAU, A. and WARWICK, K.M. (1984): Multivariate Descriptive Statistical Analysis: Correspondence Analysis and Related Techniques for Large Matrices. Wiley, New York.

LEVY, P. (1939): Sur la division d'un segment par des points choisis au hasard. *C.R.Acad.Sci. Paris, 208, 147–149*

SHOR, P.W., and YUKICH, J.E. (1991): Minimax grid matching and empirical measures. *Annals of Probability, 19, 1338–1348.*

STEELE, J.M. (1988): Growth rates of euclidean minimal spanning trees with power weighted edges. *Annals of Probability, 16, 1767–1787.*

STEELE, J.M., and TIERNEY, L. (1988): Boundary domination and the distribution of the largest nearest neighbor link in higher dimensions. *Journal of Applied Probability, 23, 524–528.*

TABAKIS, E. (1992): Asymptotic and Computational Problems in Single-Link Clustering. PhD thesis, Massachusetts Institute of Technology.

Detection of Blocks in a Binary Matrix – A Bayesian Approach

W. Vach[1], K.W. Alt[2]

[1] Institute of Medical Biometry and Medical Informatics, University of Freiburg, Stefan Meier Str. 26, D-79104 Freiburg

[2] Institute of Forensic Medicine, Düsseldorf University, Moorenstr. 5, D-40255 Düsseldorf

Summary: A binary matrix representing the presence of traits and objects is given. We consider the task to detect a subblock of traits and objects such that for each trait its frequency of occurrence within these objects is highly increased. A Bayesian framework is specified and the Gibbs sampler is used to approximate the posterior distribution. The necessary extensions to use this procedure for kinship analysis in prehistoric anthropology are outlined.

1. Introduction

In prehistoric anthropology it is of great interest to reconstruct genetic relationships among individuals based on their skeletal remains. For this task one approach is to look for coincidences in rare, discrete traits which are known to be inherited to some degree. A typical example for such a trait is the aplasia of the wisdom teeth, which is known to show increased frequencies within some families. Alt (1991) provided a catalog of mainly dental (odontological) traits which may be useful for this task. As many of these traits can be observed at several teeth and some of them have several variants we have finally to investigate more than 1000 binary traits with respect to coincidences among individuals. Statistical techniques for this type of kinship analysis depend on the size of the population and the available preinformation; different approaches are discussed in Alt & Vach (1991, 1992) and Vach & Alt (1993).

In this paper we consider an approach to detect a family, if this can be characterized by some traits typical for its members. Hence the basic data analytical task is to find a group C of individuals and a set $\widetilde{C}$ of traits such that for each trait the frequency of occurrence within the group is high and the frequency outside of the group is small. If we find such a group and such a set then we interpret C as a family and $\widetilde{C}$ as the set of typical traits (here "family" means genetically related individuals, which may be part of a larger "family" determined by social rules). This interpretation is of course only possible if we know that the traits are genetically independent. The

The authors are indebted to Walter Gilks, Cambridge, for his valuable introduction into this type of Bayesian cluster analysis data analytical task is a special case of the general problem to find suspicious blocks in a binary matrix. Methods for binary matrices like correspondence analysis (Greenacre 1984) are not specific for this task as they aim to find a global structure of the matrix. In this paper we present a Bayesian approach for the general problem. Sections 2 to 4 present the basic steps and Section 5 considers a simple example. To use this approach in prehistoric kinship analysis some extensions are necessary which are outlined in Section 6.

2. Stochastic Model and Bayesian Framework

Let I be a set of N objects and J a set of M traits. $Y := (Y_{ij})_{i \in I, j \in J}$ is a binary matrix such that $Y_{ij} = 1$ indicates the presence of trait j at object i. We assume that there exist K blocks $B_k = (C_k, \widetilde{C}_k)$ with an object cluster $C_k \subseteq I$ and a trait cluster $\widetilde{C}_k \subseteq J$. The object clusters $C_1, \ldots, C_K$ and the trait clusters $\widetilde{C}_1, \ldots, \widetilde{C}_K$ should be disjoint but need not to build a partition of I or J, respectively. With respect to the law generating Y one may assume that all components are independent and that within a block the probability of occurrence is highly increased, i.e.,

$$P(Y_{ij} = 1) = \begin{cases} \pi_j & \text{if } i \in C_k \text{ and } j \in \widetilde{C}_k \text{ for some } k \in \{1,..,K\} \\ \tau_j & \text{otherwise} \end{cases}$$

with $\pi_j > \tau_j$. However, we want to be sure that each trait of a trait cluster $\widetilde{C}_k$ should occur for at least two objects of the corresponding object cluster C_k. Hence we assume that if $j \in \widetilde{C}_k$ the distribution of $(Y_{ij})_{i \in C_k}$ is equal to the conditional distribution of $(Y'_{ij})_{i \in C_k}$ given $\sum_{i \in C_k} Y'_{ij} \geq 2$, where $(Y'_{ij})_{i \in C_k}$ are independent and identically distributed with $P(Y'_{ij} = 1) = \pi_j$. Finally by conditioning on the trait margins $K_j := \sum_{i \in I} Y_{ij}$ we can reduce the $2 \times M$ parameters $(\pi_j, \tau_j)_{j \in J}$ to the M parameters $\rho_j := \frac{\pi_j}{1-\pi_j} / \frac{\tau_j}{1-\tau_j}$. Summarizing the considered stochastic model assumes independence between traits and the joint distribution of trait j within all objects is expressed by

$$P((Y_{ij})_{i \in I} | K_j) = \begin{cases} h(\rho_j, k_j, S_k, K_j, N) & \text{if } j \in \widetilde{C}_k \text{ for some } k \in \{1,..., K\} \\ \binom{N}{K_j}^{-1} & \text{otherwise} \end{cases} \tag{1}$$

$$\text{with } h(\rho, k', n, K, N) := \begin{cases} 0 & \text{if } k' < 2 \text{ or } k' > \min(n, K) \\ \frac{\rho^{k'}}{\sum_{l=2}^{\min(n,K)} \binom{n}{l} \binom{N-n}{K-l} \rho^l} & \text{otherwise} \end{cases}$$

where $S_k := |C_k|$ denotes the size of the cluster C_k and $k_j := \sum_{i \in C_k} Y_{ij}$ is the number of occurrences of trait j within cluster C_k. This stochastic model depends on the parameter $\theta := (\mathcal{B}, \rho)$ with $\mathcal{B} := (B_k)_{k=1,\ldots,K}$ and $\rho := (\rho_j)_{j \in J}$.

For a Bayesian analysis we have to specify a prior distribution for θ. K is fixed and we assume that the three random variables $(C_k)_{k=1,\ldots,K}$, $(\widetilde{C}_k)_{k=1,\ldots,K}$ and ρ are independent. To describe the joint distribution of $(C_k)_{k=1,\ldots,K}$ we introduce the membership indicators $m_i^k := \begin{cases} 1 & \text{if } i \in C_k \\ 0 & \text{otherwise} \end{cases}$ such that $S_k = \sum_{i \in I} m_i^k$. We assume that $S_1, \ldots, S_K$ are independent identically distributed and that $(m_i^k)_{i \in I, k=1,\ldots,K}$ given $S_1, \ldots, S_K$ is uniformly distributed among all possible values satisfying $\sum_{i \in I} m_i^k = S_k \; \forall k = 1, \ldots, K$ and $m_i^k = 1$ for at most one $k \; \forall i \in I$. Hence it suffices to specify the prior distribution of a single S_k. We suggest trapezoid distributions described by five parameters (s_1, s_2, s_3, s_4, f) with $\gamma_s := P(S_k = s)$ given by

$$\gamma_s := \begin{cases} p + \frac{s - s_1}{s_2 - s_1}(f-1)p & \text{if } s_1 \leq s \leq s_2 \\ f\,p & \text{if } s_2 < s < s_3 \\ p + \frac{s_4 - s}{s_4 - s_3}(f-1)p & \text{if } s_3 \leq s \leq s_4 \\ 0 & \text{otherwise} \end{cases}$$

where p is implicitly defined by $\sum_{s=s_1}^{s_4} \gamma(s) = 1$. s_1 and s_4 are the minimal and maximal object cluster sizes we allow, and f indicates that object sizes between s_2 and s_3 are f times more likely than the extreme sizes s_1 and s_4. Analogously we introduce membership indicators $\widetilde{m}_j^k := \begin{cases} 1 & \text{if } j \in \widetilde{C}_k \\ 0 & \text{otherwise} \end{cases}$ and the parameters $(\widetilde{s}_1, \widetilde{s}_2, \widetilde{s}_3, \widetilde{s}_4, \widetilde{f})$ to describe the prior distribution of the trait cluster size $\widetilde{S}_k$. We further assume that $Ks_4 \leq N$ and $K\widetilde{s}_4 \leq M$, because otherwise the assumption of independence of $S_1, ..., S_K$ and of $\widetilde{S}_1, ..., \widetilde{S}_K$, respectively, cannot be maintained.

With respect to the prior distribution of ρ we assume that all $\rho_j, j \in J$, are independent and identically distributed. We are only interested in blocks such that ρ_j is distinctly larger than 1, hence the prior distribution should give no mass to small values of ρ_j. For computational reasons we consider a discrete distribution on the values $r_1, \ldots, r_{20}$ with $r_l := \exp(2 + \frac{l}{4})$ with probabilities $p_l := P(\rho_j = r_l)$ increasing linear in l such that $p_{20} = g\,p_1$. The parameter g describes to which degree we prefer large values of ρ.

3. Computation of Posterior Distribution Using Gibbs Sampling

For a Bayesian analysis we have to compute the posterior distribution

$$p(\theta|Y) := \frac{l(Y|\theta)p(\theta)}{\int l(Y|\theta)p(\theta)\,d\theta}$$

where $p(\theta)$ is the prior distribution and $l(Y|\theta)$ is the likelihood based on (1). In the sequel we identify θ with the membership indicators $(m_i^k), (\widetilde{m}_j^k)$ and with ρ, such that θ has $(N+M) \times K + M$ components. The Gibbs Sampler (Geman & Geman 1984, Gelfand & Smith 1990) allows us to approximate the posterior distribution by the empirical distribution of a sequence $(\theta^{(n)})_{n \in \mathbb{N}}$. This sequence is a realisation of a Markov Chain for which the stationary distribution coincides with the posterior distribution. In the n-th step we cycle through the components of θ and we draw $\theta_r^{(n)}$ from the conditional distribution of θ_r given Y, $(\theta_s^{(n-1)})_{s>r}$ and $(\theta_s^{(n)})_{s<r}$. Hence in our application implementation of the Gibbs sampler requires to compute conditional probabilities of the following type:

a) $P(m_i^k = 1|Y, \theta \setminus \{m_i^k\}) =$

$$\begin{cases} 0 & \text{if } m_i^{k'}=1 \text{ for some } k'\neq k \\ 1 & \text{if } k_j'=1 \text{ for some } j\in\widetilde{C}_k \\ \frac{\prod_{j\in\widetilde{C}_k} h(\rho_j,k_j'+1,s+1,K_j,N)q}{\prod_{j\in\widetilde{C}_k} h(\rho_j,k_j',s,K_j,N)(1-q)+\prod_{j\in\widetilde{C}_k} h(\rho_j,k_j'+1,s+1,K_j,N)q} & \text{otherwise} \end{cases}$$

where $k_j' := \sum_{i'\neq i} Y_{i'j} m_{i'}^k$, $s := \sum_{i'\neq i} m_{i'}^k$ and with $N' := N - \sum_{k'\neq k} S_k$

$$q := P(m_i^k = 1|(m_{i'}^{k'})_{i'\in I,k'=1,\dots,K} \setminus \{m_i^k\}) = \frac{\gamma(s+1)/\binom{N'}{s+1}}{\gamma(s)/\binom{N'}{s}+\gamma(s+1)/\binom{N'}{s+1}}$$

b) $P(\widetilde{m}_j^k = 1|Y, \theta \setminus \{m_j^k\}) =$

$$\begin{cases} 0 & \text{if } m_j^{k'} = 1 \text{ for some } k' \neq k \\ \frac{h(\rho_j,k_j,S_k,K_j,N)\widetilde{q}}{\binom{N}{k_j}^{-1}(1-\widetilde{q})+h(\rho_j,k_j,S_k,K_j,N)\widetilde{q}} & \text{otherwise} \end{cases}$$

where with $M' := M - \sum_{k'\neq k} \widetilde{S}_k$ and $\widetilde{s} := \sum_{j'\neq j} \widetilde{m}_{j'}^k$

$$\widetilde{q} := P(\widetilde{m}_j^k = 1|(\widetilde{m}_{j'}^{k'})_{j'\in J,k'=1,\dots,K} \setminus \{\widetilde{m}_j^k\}) = \frac{\widetilde{\gamma}(\widetilde{s}+1)/\binom{\widetilde{M'}}{\widetilde{s}+1}}{\widetilde{\gamma}(\widetilde{s})/\binom{\widetilde{M'}}{\widetilde{s}}+\widetilde{\gamma}(\widetilde{s}+1)/\binom{\widetilde{M'}}{\widetilde{s}+1}}$$

c) $P(\rho_j = r_l|Y, \theta \setminus \{\rho_j\}) \propto \begin{cases} h(r_l, k_j, S_k, K_j, N)p_l & \text{if } j \in \widetilde{C}_k \\ p_l & \text{otherwise} \end{cases}$

In a) the second condition is implied by the requirement that within a block each trait should occur for each element of the object cluster at least twice.

Starting the Gibbs sampler with an arbitrary initial value (which must be of course within the support of the posterior distribution) in its first repetitions the Markov chain may be far away from its stationary distribution. Hence we can use the empirical distribution only after removing the first iteration values. The length of this burn-in phase has to be chosen in an appropriate way. The size of the Gibbs sample is the second choice to be made. We must be aware that in our specific setting the Gibbs sampler may be rather slowly mixing; and hence we may require rather long runs to achieve a sufficient approximation of the posterior distribution. Especially if we consider an analysis based on the assumption $K = 1$, but $K > 1$ is true, the posterior distribution may have several distinct modes (corresponding to the blocks $B_1, \dots, B_K$). For the Gibbs sampler it may be hard to leave one of these modes. This problem is aggravated by our requirement that each trait of a block is shown by at least two elements of its object cluster. As a consequence a careful examination of the generated sequence is necessary to ensure that the approximation is sufficient; cf. the examples in Section 5.

4. Analysis of Posterior Distribution

The posterior distribution tells us where to find the blocks for which the expected object and trait cluster sizes have been described by the prior distribution. As the stochastic model and the prior distribution is invariant under permutations of the blocks $(B_1, \ldots, B_K)$ we can look at posterior distributions like $P(m_i^k = 1 \text{ for one } k \in \{1, \ldots, K\}|Y)$ indicating whether object i is a member of some cluster, $P(\exists k \text{ with } \widetilde{m}_j^k = 1 \wedge \widetilde{m}_{j'}^k = 1|Y)$ indicating whether traits j and j' belong to the same cluster, $P(\exists k \text{ with } m_i^k = 1 \wedge \widetilde{m}_j^k = 1|Y)$ indicating whether object i and trait j are members of the same block, etc. Approximations for these probabilities can be easily computed from the Gibbs sample. Regarding posterior distributions (or Bayes factors) for pairs of objects and/or traits as a similarity measure clustering procedures can be used to identify blocks.

Alternatively we suggest to compute approximate Bayes estimates for single blocks B^* minimizing the Bayes risk $E_{\mathcal{B}|Y}\ L(B^*, \mathcal{B})$ with respect to the loss function $L(B^*, (B_1, \ldots, B_K)) := \min_k \Delta(B^*, B_k)$ with

$$\begin{aligned}\Delta((D,\widetilde{D}),(C,\widetilde{C})) := \max\ \Big(& h(\tfrac{|D-C|}{|D|}, 0.5) + h(\tfrac{|C-D|}{|C|}, 0.33), \\ & h(\tfrac{|\widetilde{D}-\widetilde{C}|}{|\widetilde{D}|}, 0.5) + h(\tfrac{|\widetilde{C}-\widetilde{D}|}{|\widetilde{C}|}, 0.33)\Big) \qquad (1)\end{aligned}$$

$$\text{and } h(a,b) := \begin{cases} 0 & \text{if } a \le b \\ (\frac{a-b}{1-b})^2 & \text{if } a > b \end{cases} .$$

The loss function is large, if for each block of $\mathcal{B}$ the object cluster or the trait cluster differs from that of B. It moreover reflects that overestimating a block (with respect to the object/trait cluster size) is less harmful than underestimating, because for any estimated block we can look at the distribution of Y within the block to identify objects or traits whose membership is questionable.

The exact computation of a Bayes estimate is cumbersome as it involves the computation of the Bayes risk (with respect to the empirical distribution of the Gibbs sample) for a lot of blocks. The computation of the Bayes risk for a single block B is already a computer time intensive task as we have to compute $L(B^*, \mathcal{B}^{(n)})$ for all n of our sample. Hence we suggest a heuristic procedure to compute Bayes estimates approximatively. The basic step of the procedure is the following completion rule: For a given block B we can look for the completion by a single object or trait with minimal Bayes risk. To avoid the computation of the Bayes risk for any possible completion we compute for any object/trait outside of B and any object/trait inside of

B the conditional posteriori probability that the outside element is in the same cluster as the inside element. Then we compute for any outside element the average over the inside elements. For the three objects and three traits with maximal average we compute the Bayes risk as and choose that with minimal Bayes risk for completion. Starting with the pair (i, j) with maximal posterior probability $P(\exists(C, \widetilde{C}) \in \mathcal{B} \text{ with } i \in C \wedge j \in \widetilde{C}|Y)$ we build a sequence of blocks with increasing size by repeated application of the completion rule. We stop this process, if in three consecutive steps the minimal Bayes risks within the sequence is not improved and store the block with minimal Bayes risk. Then we repeat the procedure with the pair with the second largest posterior probability, and so on. We omit pairs, which are included in more than two of the previously stored blocks, and we stop the process if the posterior probability becomes too small (e.g. smaller than the prior probability.) The stored blocks are ordered by increasing Bayes risk, and any block in this sequence which is disjoint (or nearly disjoint) to any of its predecessors is an (approximate) Bayes estimate.

It is a crucial question how many of these estimates we want to accept at the end. We emphasize that this question has little to do with the original choice of K. From a theoretical point of view we can always choose $K = 1$ and we can identify any members of (true) clusters as modes in the posterior distribution. Choosing $K > 1$ is similar to running K sequences of the Gibbs sampler with $K = 1$ adding the restriction that the sequences are in different modes. Choosing K too small or too large (compared with the true number of clusters) we have only to pay the price that the modes are less distinct in the posterior distribution and that hence we need longer sequences of the Gibbs sampler to achieve a sufficient approximation to detect them.

5. A Simple Example

In Fig. 1 a binary matrix with 40 objects and 40 traits is shown. For our analysis we choose $K = 2$. The prior distributions of the object cluster sizes and the trait cluster sizes are described by the parameters $(s_1, s_2, s_3, s_4, f) = (\widetilde{s}_1, \widetilde{s}_2, \widetilde{s}_3, \widetilde{s}_4, \widetilde{f}) = (2, 3, 6, 8, 2.0)$, and we choose $g = 3.0$. After a burn-in phase of 1000 repetitions and additional 10000 repetitions of the Gibbs sampler we found two Bayes estimates for blocks. These are shown in Tab. 1. One may argue that some of the traits and objects with a small posterior probability are superfluous, but this results from the intended choice of the loss function. Note that in Fig. 1 traits and objects are already sorted by the scores from correspondence analysis, but that the blocks found are not recognizable directly in this representation.

As mentioned above it is necessary to examine the Gibbs sampling sequence in order to exclude artificial results due to insufficient approximation of the posterior distribution. In our investigation we observe that within the 10000 pairs of blocks generated there was none without at least one coincidence of a trait or object with the first block of Tab. 1 and only two small subsequences with no coincidences with the second block. This does not allow us

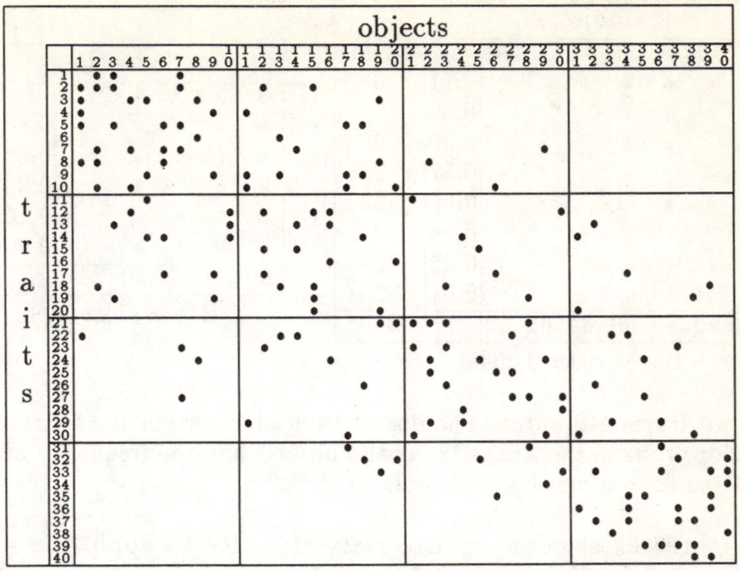

Fig. 1: A binary data matrix

to exclude that we may have overlooked a third block. However, an analysis with $K = 3$ reveals no additional blocks. With respect to the estimation of posterior probabilities within the blocks the mixing within the sequence seems to be sufficient. For Example trait 38 with highest posterior probability shows a maximal run length of 546 and is removed and reselected 123 times.

We further investigated the sensitivity of our results to the specific choice of the prior distributions. Enlarging the maximal cluster size we find additional members with a small posterior probability. Similar alternative choices of g do not change the core of the estimated blocks, but for g near 1.0 the second block shows a smaller Bayes risk than the first.

6. Extensions for Kinship Analysis in Prehistoric Anthropology

The Bayesian framework is especially useful for kinship analysis in skeletal remains from burial grounds, because here some preinformation on cluster sizes are available. As we cannot assume that family-typical traits are inherited for a lot of generations the possible number of members of a group of genetically related individuals is limited. Additionally demographic features of a prehistoric population give us some information about the average number of siblings in one generation and the archaeological context can provide information about migration rates. With respect to the trait cluster sizes first experiences with small populations suggest to expect between 3 and 10 typical traits for a family within the catalogue of Alt (1991).

traits	38	39	34	37	31	32	36		outside
36		•	•	•	•			0.71	0
40	•	•						0.69	0
37	•		•	•		•		0.69	0
39	•			•			•	0.57	1
35		•	•					0.56	2
30	•				•			0.06	3
33		•				•	•	0.05	3
31							•	0.04	2
	.91	.90	.90	.90	.64	.61	.56		

(objects: columns 38–36)

Bayes risk=0.054

traits	7	3	2	1	15	6		outside
1	•	•	•				0.86	0
2	•	•	•	•	•		0.81	1
5	•	•		•		•	0.39	2
7	•		•			•	0.16	3
8			•	•		•	0.06	2
	.90	.85	.63	.40	.26			

(objects: columns 7–6)

Bayes risk=0.082

Tab. 1: Two Bayes estimates: The distribution of Y within the estimated block, the posterior probabilities of the traits and objects and the frequency of the traits outside of the estimated object cluster

However, three extensions are necessary in order to apply the above approach. First, the traits are usually not binary but consist of several binary components reflecting one genetic information. Typical examples are symmetric traits on the left and right side of the jaw. Additionally the single components are affected by missing values. Second, there exist groups of odontological traits which are (by definition) not independent. For example aplasia of a single tooth is related to microsymptoms like size reduction and it is necessary to consider aplasia as one trait as well as aplasia or microsymptoms. Such pairs of concurring traits may never become members of a block simultaneously because otherwise a block does not reflect a family but dependencies between traits. Third, we have often some preknowledge or hypotheses about special subgroups of individuals based on archaeological information. For example chronological information may allow to identify individuals of the founding generation of a settlement and one may ask whether these individuals constitute a family. Also we may identify a few individuals with a special jewel or a special ornament on their arms and we may ask whether these individuals are part of family. The first two points require an extension of the stochastic model, whereas the third can be covered by an appropriate choice of the prior distribution.

7. Conclusions

The suggested Bayesian framework seems to be a useful tool for the detection of blocks. It can be used in situations where others approaches to detect blocks in a binary matrix may fail. E.g. the approach of Hartigan (1972) and its refinement by Duffy & Quiroz (1991) assume that blocks can be detected by analysing the marginal distribution of the rows or columns, which need not to be true. Moreover, this approach and most others described by Arabie, Boorman & Levitt (1978) or by Opitz & Schader (1984) have in common that the complete matrix is decomposed into blocks, whereas we only assume a block structure for a part of the matrix. The same difference holds with respect to block seriation methods (e.g. Marcotorchino 1987). As a further basic difference we allow different degrees of membership expressed as posterior probabilities. This way we can reflect the fuzzy nature of blocks in many applications.

Contrary to other procedures to analyse binary matrices like correspondence analysis our approach lacks symmetry with respect to the handling of rows and columns. Especially in the stochastic model we only condition on the trait frequencies but we do not adjust for differences in the object totals. This is justified in kinship analysis, because each individual has the same chance to show a trait, but it might be not justified in other applications.

References:

ALT, K.W. (1991): Verwandschaftsanalyse an Skelettmaterial – Methodenentwicklung auf der Basis odontologischer Merkmale. Medizinische Habilitationsschrift Freiburg. Gustav Fischer Verlag (in press).

ALT, K.W. & VACH, W. (1991): The Reconstruction of "Genetic Kinship" in Prehistoric Burial Complexes - Problems and Statistics. In: Bock, H.H., Ihm, P.(eds): *Classification, Data Analysis, and Knowledge Organization.* Springer, 299-310.

ALT, K.W. & VACH, W. (1992): Non-Spatial Analysis of "Genetic Kinship" in Skeletal Remains. In: Schader, M. (ed): *Analysing and Modeling Data and Knowledge.* Springer, 247-256.

ARABIE, P., BOORMAN, S.A. & LEVITT, P.R. (1978): Constructing block models: How and why. *Journal of Mathematical Psychology, 17, 21-63.*

DUFFY, D.E. & QUIROZ, A.F. (1991): A permutation-based algorithm for block clustering. *Journal of Classification, 8, 65-91.*

GELFAND, A.E. & SMITH, A.F.M. (1990): Sampling-based approaches to calculating marginal densities. *Journal of the Amer. Statist. Assoc., 85, 398-409.*

GEMAN, S. & GEMAN, D. (1984): Stochastic relaxation, Gibbs distributions and the Bayesian restoration of images. *IEEE Transactions on Pattern Analysis and Machine Intelligence, 6, 721-741.*

GREENACRE, M. (1984): Theory and application of correspondence analysis. Academic Press.

HARTIGAN, J.A. (1972): Direct Clustering of a Data Matrix. *Journal of the Amer. Statist. Assoc., 67, 123-129.*

MARCOTORCHINO, F. (1987): Block Seriation Problems: A Unified Approach. *Applied Stochastic Models and Data Analysis, 3, 73-91.*

OPITZ, O. & SCHADER, M. (1984): Analyse qualitativer Daten: Einführung und Übersicht, Teil 1. *OR Spektrum, 6, 67-83.*

VACH, W. & ALT, K.W. (1993): Detection of kinship structures in prehistoric burial sites. In: Andresen, J., Madsen. T., Scollar, I. (eds): *Computing the Past.* Aarhus University Press, 287-292.

Spatial Data Analysis

Detection of Spatial Discontinuities in Vegetation Data by a Moving Window Algorithm

Heiko Balzter, Paul Braun, Wolfgang Köhler

Institut für Pflanzenbau und Pflanzenzüchtung II, Biometrie und Populationsgenetik, Justus-Liebig-Universität Gießen, 35390 Gießen

Summary: In order to detect boundaries in vegetation the moving window algorithm (MWA) was applied to point-quadrat data obtained from a meadow classified as *Lolio-Cynosuretum.* Squared Euclidean distance was used as a dissimilarity measure. The phytosociological classification into a mown and an unmown area by character species was confirmed by the results of MWA. MWA is a suitable multivariate method for the detection of spatial discontinuities in vegetation data.

1. Boundaries in Vegetation

Vegetation is one main object of ecological research. The plant cover of our planet's surface plays probably the most important role in ecosystems because its primary production of biomass is a prerequisite for the evolution of higher taxa. In the past the distinct approaches to vegetation analysis by American Scientists and by the European Zrich-Montpellier school caused considerable discussion. In American research various methods of ordination by resemblance matrices combined with numerical classifications have been used, while in Europe the phytosociological system of Braun-Blanquet (1964) achieved great importance. Great care must be exercised when attempting to compare the vegetation of Europe with that of America. Because of the cultivation of nearly all of the land surface in Europe, boundaries between different plant communities are probably clearer than in natural vegetation in America. However, if the intention is to classify vegetation into zones or different plant communities, objective methods should be used to detect boundaries, because any visual classification may be subjective. One major problem in classification of vegetation is the spatial distribution of plants. Different species follow different spatial distributions that depend on their biology and morphology. Theoretical distributions are discussed by Knapp (1984). Greig-Smith (1983) pointed out that in the field the detection of plant clusters may depend on the underlying spatial scale. Despite the complicated nature of vegetation structure, 'vegetation is the best measure of environment' (Greig-Smith (1979)), because its composition is determined by the habitat. Methods commonly used for vegetation surveys are the phytosociological relevés of Braun-Blanquet (1964), the use of small quadrats (Pakarinen (1984)) and the point-quadrat methods (Goodall (1952), Grant (1981)). We used these methods to study a meadow of the nursery of the agricultural school in Giessen, Germany, and in addition a moving window algorithm was applied to the pointquadrat data in order to examine its suitability to detect boundaries in vegetation (Balzter (1993)).

2. Methods

2.1 Point-Quadrat Method

Point-quadrats can be explained as sample quadrats being as small as possible. General reviews of common point-quadrat methods are given by Goodall (1952) and Grant (1981). The shapes of point-quadrat frames described in the literature vary widely. Although sometimes various kinds of optical apparatus are used (Woolhouse (1976)), point-quadrat frames are the most common instruments. In general they are constructed as follows: a number of pins is held by the frame above the vegetation. During sampling each pin passes through a guide channel down to the ground and the number of contacts of the pins to each plant species is counted. This procedure is repeated several times to achieve appropriate accuracy by surveying sufficient points. For field sampling a point-quadrat frame with three vertical pins in a row was constructed (Fig. 1), similar to the frame described by Kreeb (1983).

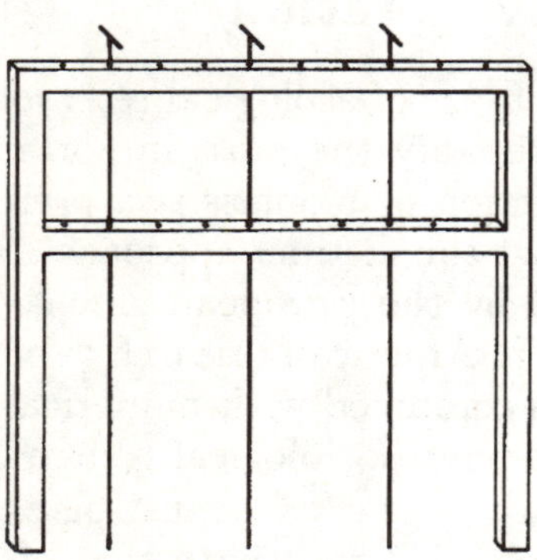

Fig. 1: Typical point-quadrat frame described by Kreeb (1983). The three metal pins are passed through the vegetation counting the contacts of the pins to each species. A similar frame was used in this study.

Because species with small percentage cover have small probabilities of being hit by a pin, in the second replicate (in June) only 28 species were discovered by this method, in contrast to a comparative study using phytosociological relevés which yielded 50 species within the same area. Another problem of the point-quadrat method is the bias caused by the diameter of the pins. Thick pins tend to overestimate the true percentage cover of the species, as discussed by Goodall (1952). In our study a pin diameter of 1 mm was used. The frame was moved along 10 parallel transects over the study area letting down the three pins 26 times on every transect, which resulted in 78 single points per transect and a total of 780 points over the entire study area. In Fig. 2 the positions of the 10 transects, the unmown area in the centre of the meadow and the two first pairwise comparisons of the MWA are marked. These points are ordered in the two dimensions x and y. In direction x the transects are numbered from 1 to 10 and in direction y the data points are numbered from 1 to 78 along each transect. Every single point is described by a vector of 28 variates, whose values are the frequencies of contact to the pins of the 28 species encountered. This procedure was done in May and June, and because the results obtained were very similar and the detected boundaries were identical, only the data of the second replicate are presented here.

		mown ?:	y	y	y	y	y	n	n	n
		relevé no.:	1	2	3	4	5	6	7	8
	species:	size:	24	24	24	24	24	13	13	13
		total cover %:	90	90	90	80	80	100	100	100
AC	Lolium perenne		3	2	2	3	2			
VC	Crepis capillaris		+	+	1	+	+			
VC	Trifolium repens		3	3	3	3	2			
VC	Phleum pratense		2	1	1	1	1	1	1	1
OC	Taraxacum officinale		+	+	+	+	+			
KC	Festuca pratensis		+	+	+			+		
KC	Plantago lanceolata				+	+	+			
KC	Poa pratensis		2	2	2	2	3	+	+	+
KC	Poa trivialis		1	2	1	1	2	1	1	+
Dm	Festuca rubra		+	+	1	1	1			
Dm	Poa annua		1	2	2	2	2			
Dm	Rorippa sylvestris		+	+	+	1	+			
Dm	Veronica arvensis			+	+	+	+			
Du	Medicago x varia							5	5	5
Du	Urtica dioica							1	1	2
Du	Cirsium arvense			+	1	+	1	2	2	2
Du	Rumex obtusifolius		+					2	3	2

Tab. 1: Phytosociological classification of the eight relevés in June of 1993. AC: character species of the association *Lolio-Cynosuretum*; VC: character species of the alliance *Cynosurion*; OC: character species of the order *Arrhenatheretalia*; KC: character species of the class *Molinio-Arrhenatheretea*; (Oberdorfer (1990)).
Dm: differential species of the mown area; Du: differential species of the unmown area. The 30 accompanying species are not listed here. Numbers in the table above refer to cover-abundance of the species present in the relevé and have the following meaning:

5	species is covering more than 75% of the area;
4	any number of individuals covering 50% - 75% of the area;
3	any number of individuals covering 25% - 50% of the area;
2	individuals are very numerous or covering at least 5% of the area;
1	species is plentiful, but of small cover value (less than 5%);
+	the species is present sparsely or very sparsely, cover very small.

The examined meadow in Giessen is regularly mown apart from an area in its centre. The mown plant community was classified as *Lolio-Cynosuretum* following from the phytosociological system of Braun-Blanquet (1964), whereas the unmown plant community had a quite different composition. Tab. 1 shows a comparison of the species composition of both plant communities in June 1993, presented as percentage cover estimated using the scale of Braun-Blanquet (1964); differential species are marked with 'Dm' for the mown and 'Du' for the unmown area.

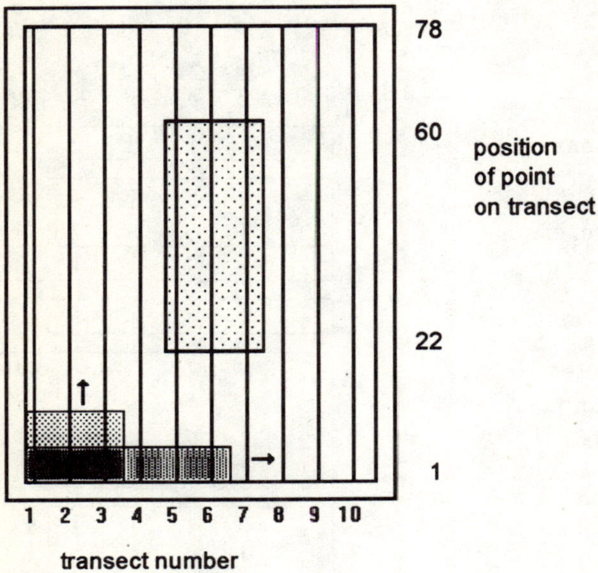

Fig. 2: Shape of the study area with the unmown area in its centre, the positions of the 10 transects and the 3 central pixels of the 2 first pairwise comparisons of the moving window algorithm (MWA) in direction x and y (arrows). Note that only the transects 5, 6 and 7 cross the unmown area, and that the discontinuities detected by MWA match the boundaries between the mown and the unmown areas.

The main advantage of the point-quadrat method is the objective positioning of the points, at which plants are recorded, without preconceptions of the observer (Goodall & Feoli (1991)). Furthermore the point-quadrat method yields quantitative data that can be used for many numerical methods of data analysis as demonstrated below. As Goodall (1952) pointed out, the results of the point-quadrat method can be expressed in terms of the proportion of the ground covered by each species (percentage cover), the average number of layers of foliage covering each point of the ground (cover repetition) and the proportion of each species in the vegetation as a whole, which is sometimes used as an estimate for the dry weight of the species (percentage of sward).

2.2 Moving Window Algorithm

A simple and common method for locating discontinuities in a data set is the moving window algorithm (MWA) for two-dimensional data, which is sometimes called 'moving split-window method' for one-dimensional data. MWA is used to analyse data obtained from field sampling, aerial photography and satellite remote sensing (Johnston et al. (1992)). Its basic procedure is: (1) obtain multivariate ordered data; (2) bracket a two-dimensional set of sampling points into a data window of preassigned width; (3) compute the vector of arithmetic means for each variate and assign it to a central pixel in the window midpoint; (4) move the window one position forward and bracket the next set of sampling points into a data window of equal width, continue until the end of the data set is reached; (5) compute a distance or dissimilarity measure pairwisely between all adjacent central pixels (that must not overlap); (6) repeat (2)-(5) for the second dimension; and (7) after computing comparisons for the whole data set plot the distances against their positions in the data space. High peaks identify the location of discontinuities. The distance measure commonly used is the squared Euclidean distance (SED), as it tends to agree with field observations (Wierenga et al. (1987)). Brunt & Conley (1990) examine the behaviour of the SED by applying it to simulated data with known properties. If window-width is increased peaks become more distinct from background noise. MWA itself is a descriptive method. It is possible to determine the statistical significance of the detected discontinuities by specific test procedures (Cornelius & Reynolds (1991)), but this program was not available to us. The MWA can be applied to data ordered in one or two dimensions, for example spatially in two directions or in time and in space (Ludwig & Cornelius (1987)). Kerr et al. (1992) improved an MWA in order to achieve accurate estimates for land surface temperature from satellite data. Wierenga et al. (1987) used an MWA to classify soil and vegetation zones along a 3km transect in the Chihuahuan desert in New Mexico. They computed the Euclidean distance for the vegetation data and the Hotelling-Lawley trace following a principal component analysis for the soil data. Absolute ground cover of 19 perennial species was estimated and evaluated (Wierenga et al. (1987)). Considering the limitations of their study it seems to be necessary to focus attention on more detailed applications of MWA to spatially ordered vegetation data.

In our study for MWA a window width of $3 * 6 = 18$ points was chosen. In Fig. 2 the location of the two first pairwise comparisons in direction x and y are presented.

3. Results

The unmown area was covered mainly by *Medicago x varia* (90%), *Cirsium arvense* (8%) and *Rumex obtusifolius* (8%), while the mown area was dominated by *Poa pratensis* (47%), *Lolium perenne* (45%) and *Trifolium repens* (31%); the percentage cover was estimated from the point-quadrat data. The computed squared Euclidean distances (SED) between adjacent central

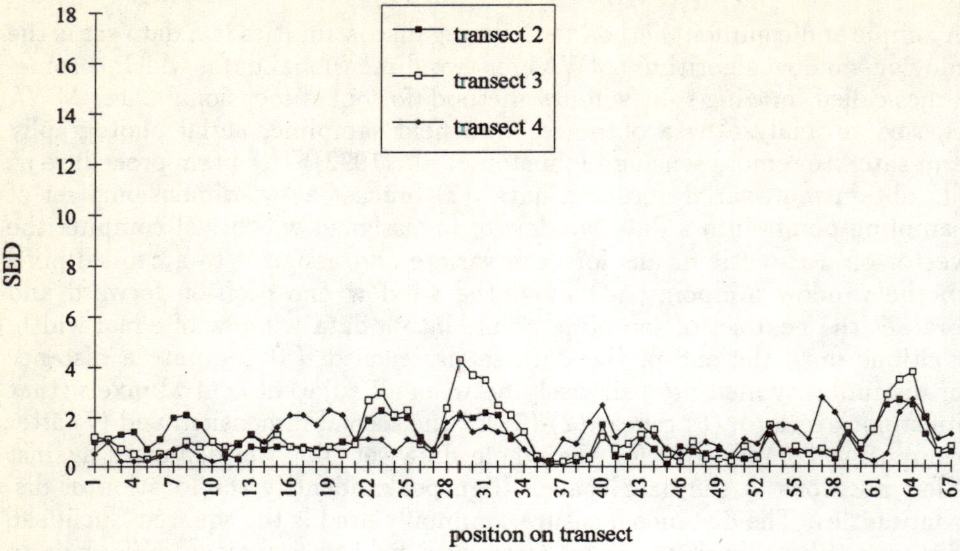

Fig. 3: Squared Euclidean distance (SED) plotted against position of the compared central pixels in transect direction. The data series are located on transects 2, 3 and 4 and do not show clear peaks; all data points are located on the mown area.

pixels in direction y are spatially located on transects 2 - 9 and presented graphically in Fig's. 3, 4 and 5. In Fig. 3 the SED series located on transects 2, 3 and 4 are plotted against their positions on the transects. These data series contain points that are located on the mown area. Fig. 4 graphically presents the SED series located on transects 5, 6 and 7, which are the transects that cross the unmown area in the centre of the meadow. Two peaks tower above the background noise. Fig. 5 seperately presents the SED series located on transects 8 and 9, which sampled the mown area. The three data series in Fig. 4 show high peaks at positions 16 and 55, while in the graphical presentations of the data series of only the mown area (Fig's. 3 and 5) no clear peaks are recognizable. From the sampling design and the systematically distributed points of the point-quadrat method it is possible to draw some conclusions about the peaks in Fig. 4. The first peak (at position 16) represents the distance between the two central pixels that are calculated from data points 16 - 21 in comparison with points 22 - 27 in the y-direction. As known from the sampling design the border between these central pixels is located exactly at the boundary between the mown and the unmown plant community. The second peak detected (at position 55) stands for the distance between the central pixels calculated from data points 55 - 60 and 61 - 66 respectively, which again exactly matches the boundary between the unmown and the mown plant communities (see Fig. 2). Note that the SED series located on transect 6 towers above the other series (Fig. 4), because for the calculation of the arithmetic means exclusively point-quadrat data

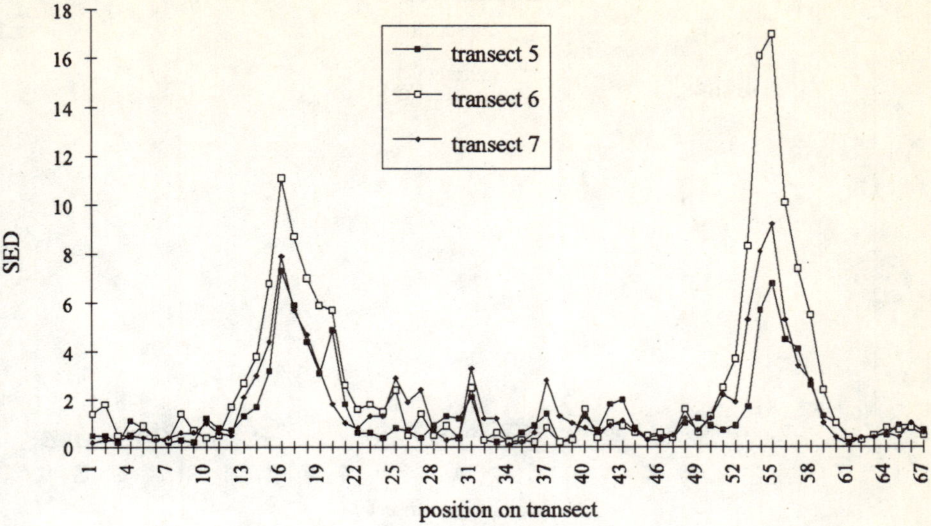

Fig. 4: Squared Euclidean distance (SED) plotted against position of the compared central pixels in transect direction. The data series are located on transects 5, 6 and 7. Two peaks tower above the data noise in all three data series at the boundaries mown/unmown and unmown/mown, because these transects all cross the unmown centre of the meadow.

from the unmown area have been used, in contrast to the data series located on transects 5 and 7, which include some points from the mown area located on transects 4 and 8 respectively (see Fig. 2).
In Fig. 6 the results of the MWA in direction x are presented. Two data series tower above the others in the interval between distances 22 and 55: first the series of distances bracketing and comparing transects 2 - 4 with 5 - 7 and second those bracketing and comparing transects 5 - 7 with 8 - 10. These two series describe the dissimilarity between central pixels computed from data points located exclusively on the mown area with central pixels from data points located exclusively on the unmown area. The interval between distances 22 and 55 can be explained from the distribution of sampling points as the set of comparisons between points from the unmown and the mown area (see Fig. 2). Distance 22 was calculated from points 22 - 25 and distance 55 from points 55 - 60, which are all located on the unmown area.

The detected discontinuities can be summarized as the line between transects 4 and 5 on the left and between transects 7 and 8 on the right, both boundaries limited in direction y by points 22 and 60. Applied to both data sets sampled in May and June these discontinuities exactly match the observed boundaries between the unmown central area and the mown area and therefore confirm the phytosociological classification by differential species shown in Tab. 1.

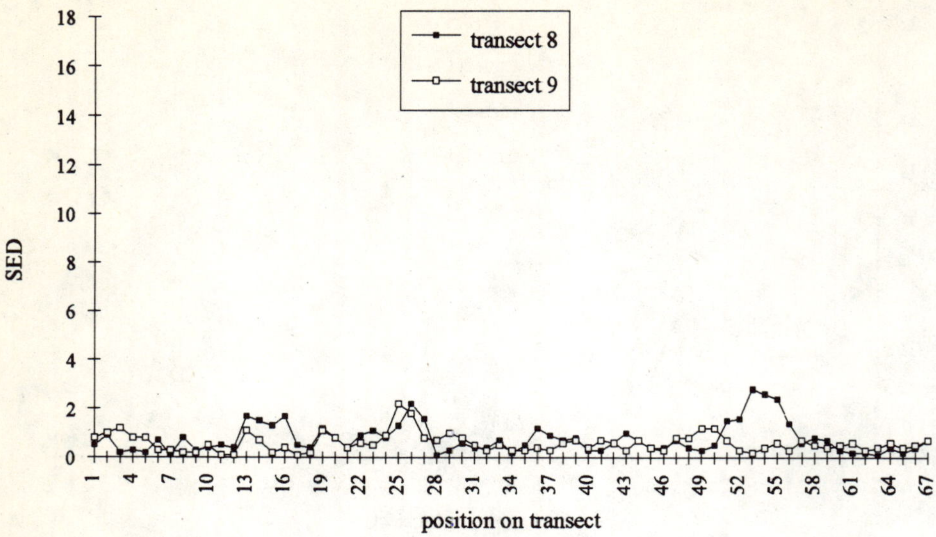

Fig. 5: Squared Euclidean distance (SED) plotted against position of the compared central pixels on the transect in transect direction. The data series are located on transects 8 and 9 and do not show clear peaks, because all the data points are located on the mown area.

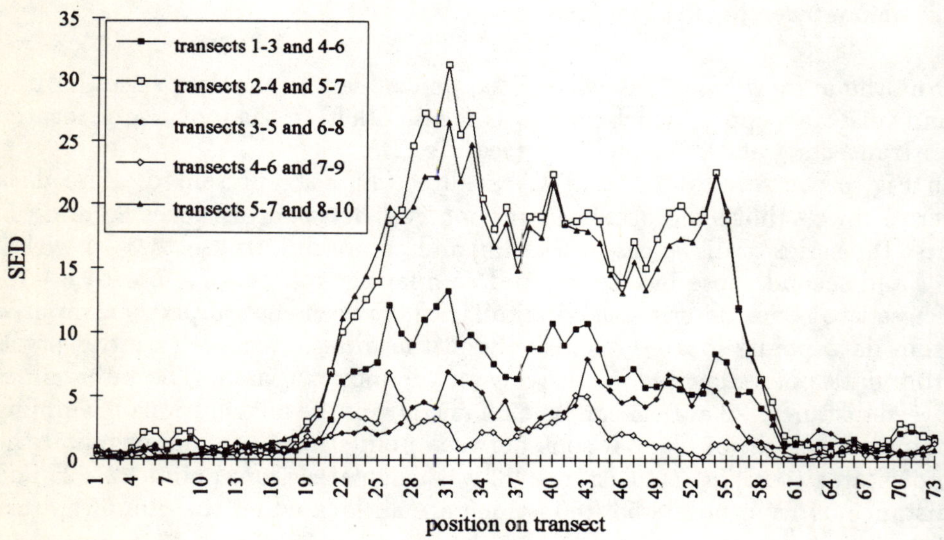

Fig. 6: Squared Euclidean distance (SED) plotted against position of the compared central pixels perpendicular on the transects. The two data series comparing data points from transects 2-4 with 5-7 and 5-7 with 8-10 respectively tower above the other data series in an interval between positions 22 and 55. Note that this is the interval, where the central pixels of the mown and the unmown area are compared by the SED.

4. Discussion

Point-quadrat methods are probably the most accurate methods for vegetation surveys because they give very detailed data. This study confirms the usefulness of the moving window algorithm (MWA) applied to vegetation data obtained by a point-quadrat method. The detected discontinuities exactly match the observed boundary between the mown and the unmown areas of the meadow. The previous phytosociological classification of relevés by differential species can be confirmed by these results. The squared Euclidean distance turned out to be a dissimilarity measure that matches observations in the field, as is often the case (Brunt & Conley (1990)). Using MWA Wierenga et al. (1987) found boundaries between zones based on soil morphological characteristics, that closely coincided with boundaries based on soil water content and between vegetation zones. In our study the suitability of MWA for the detection of discontinuities in vegetation data was demonstrated using more detailed vegetation data for calculation than Wierenga et al. (1987). Because of the known boundary between the mown and the unmown areas it was possible to confirm that MWA is indeed a useful method to detect discontinuities in vegetation.

References:

BALZTER, H. (1993, unpubl.): Anwendung und Vergleich verschiedener Verfahren zur Vegetationsaufnahme. Diplomarbeit, Institut für Pflanzenbau und Pflanzenzüchtung II, Biometrie und Populationsgenetik, Universität Gießen.

BRAUN-BLANQUET, J. (1964): Pflanzensoziologie. Springer, Wien & New York.

BRUNT, J.W. & CONLEY, W. (1990): Behaviour of a Multivariate Algorithm for Ecological Edge Detection. *Ecological Modelling 49, 179-203.*

CORNELIUS, J.M. & REYNOLDS, J.F. (1991): On Determining the Statistical Significance of Discontinuities within Ordered Ecological Data. *Ecology 72 (6), 2057-2070.*

GOODALL, D.W. & FEOLI, E. (1991): Application of Probabilistic Methods in the Analysis of Phytosociological Data. In: E. Feoli and L. Orloci (eds.): *Handbook of Vegetation Science 11: Computer Assisted Vegetation Analysis.* Kluwer, Dordrecht, 137-146.

GOODALL, D.W. (1952): Some Considerations in the Use of Point Quadrats for the Analysis of Vegetation. *Australian Journal of Scientific Research, B. 5, 1-41.*

GRANT, S.A. (1981): Sward Components. In: J. Hodgson, R.D. Baker, A. Davies, A.S. Laidlaw & J.D. Leaver (eds.): *Sward Measurement Handbook.* The British Grassland Society, Hurley, Maidenhead, Berkshire, 71-92.

GREIG-SMITH, P. (1983): Quantitative Plant Ecology. Blackwell, Oxord.

GREIG-SMITH, P. (1979): Pattern in Vegetation. *Journal of Ecology 67 (3), 755-779.*

JOHNSTON, C.A., PASTOR, J. & PINAY, G. (1992): Quantitative Methods for Studying Landscape Boundaries. In: A.J. Hansen & F. Di Castri (eds.): *Landscape Boundaries. Consequences for Biotic Diversity and Ecological Flows.* Springer, New York, 107-125.

KERR, Y.H., LAGOUARDE, J.P. & IMBERNON, J. (1992): Accurate Land Surface Temperature Retrieval from AVHRR Data with Use of an Improved Split Window Algorithm. *Remote Sensing of Environment 41 (2/3), 197-209.*

KNAPP, R. (1984): Considerations on Quantitative Parameters and Qualitative Attributes in Vegetation Analysis and in Phytosociological Relevés. In: R. Knapp (ed.): *Handbook of Vegetation Science 4: Sampling Methods and Taxon Analysis in Vegetation Science.* Junk, The Hague, 77-100.

KREEB, K.H. (1983): Vegetationskunde. Ulmer, Stuttgart.

LUDWIG, J.A. & CORNELIUS, J.M. (1987): Locating Discontinuities along Ecological Gradients. *Ecology 68 (2), 448-450.*

OBERDORFER, E. (1990): Pflanzensoziologische Exkursionsflora. Ulmer, Stuttgart.

PAKARINEN, P. (1984): Cover Estimation and Sampling of Boreal Vegetation in Northern Europe. In: R. Knapp (ed.): *Handbook of Vegetation Science 4: Sampling Methods and Taxon Analysis in Vegetation Science.* Junk, The Hague, 35-44.

WIERENGA, P.J., HENDRICKX, J.M.H., NASH, M.H., LUDWIG, J. & DAUGHERTY, L.A. (1987): Variation of Soil and Vegetation with Distance along a Transect in the Chihuahuan Desert. *Journal of Arid Environment 13, 53-63.*

WOOLHOUSE, A.R. (1976): Estimation of Ground Cover in Turfgrass Communities. *Journal of the Sports Turf Research Institute 52, 64-76.*

Spatial Clustering of Neurons by Hypergeometric Disjoint Statistics

J. Krauth

Department of Psychology, University of Düsseldorf
Universitätsstr. 1, D–40225 Düsseldorf, F.R.G.

Summary: Grimson and Rose (1991) suggested the use of a join–count statistic for detecting spatial clusters of neurons. We observe certain practical and theoretical difficulties in following this approach and propose instead the use of a maximum statistic. For this statistic, we derive in a similar way as for the disjoint statistic in Krauth (1991) exact upper and lower bounds for the upper tail probabilities. The procedure is illustrated by real data examples.

1. Join-Count Statistics and Neuron Clustering

In the neurosciences it is often tried to identify subdivisions of parts of the brain in order to locate neuron populations with specific functions. A particular problem is the clustering of cell bodies in sensory ganglia for cells projecting through the same peripheral nerve. Grimson and Rose (1991) mentioned that many such reports are based on subjective analyses. Therefore, the authors offered a statistical approach for the analysis of clustering of labeled cell bodies.

This approach is based on the so–called join–count statistics. These were introduced by Moran (1947, 1948) though according to this author the fundamental ideas seem to be much older. The properties and use of these statistics are described in many books on spatial analysis (Cliff and Haggett (1988); Cliff, Haggett, Ord, Bassett and Davies (1975); Cliff, Haggett, Ord and Versey (1981); Cliff and Ord (1973, 1981); Haining (1990); Upton and Fingleton (1988)).

The definition and some properties of the join-count statistics are given as follows: We assume that n objects $o_1, \ldots, o_n$ exist. For each pair (o_i, o_j) of objects we set $\delta_{ij} = 1$ if o_i and o_j are neighbours and $\delta_{ij} = 0$ otherwise. In addition, we set $\delta_{ii} = 0$. We define random variables $Z_1, \ldots, Z_n$ with $Z_i = 1$ if object o_i is labeled and $Z_i = 0$ otherwise, for $1 \leq i \leq n$. In the so–called free sampling or binomial sampling model we assume that $Z_1, \ldots, Z_n$ are independent identically distributed random variables. In the nonfree sampling or hypergeometric sampling model we assume that the number of labeled objects, i.e. the sum $m = Z_1 + \ldots + Z_n$, is fixed. The following join–count statistics are considered:

$$X_1 = \frac{1}{2}\sum_{i=1}^{n}\sum_{j=1}^{n}\delta_{ij}Z_iZ_j,$$

i.e. the number of pairs of neighbour objects where both objects are labeled,

$$X_2 = \frac{1}{2}\sum_{i=1}^{n}\sum_{j=1}^{n}\delta_{ij}(Z_i - Z_j)^2,$$

i.e. the number of pairs of neighbour objects where one object is labeled but the other not, and

$$X_3 = \frac{1}{2}\sum_{i=1}^{n}\sum_{j=1}^{n}\delta_{ij}(1 - Z_i)(1 - Z_j),$$

i.e. the number of pairs of neighbour objects where both objects are not labeled.

Since the number of pairs of neighbour objects is fixed and given by

$$X_1 + X_2 + X_3 = \frac{1}{2}\sum_{i=1}^{n}\sum_{j=1}^{n}\delta_{ij},$$

only X_1 and X_2 must be considered.

For both the free and the nonfree sampling model, the first four moments were derived. Cliff and Ord (1981) discussed conditions on the neighbourhood structure, i.e. on the matrix for the δ_{ij}, for which asymptotic normality or the Poisson limit can be derived for X_1 and X_2. However, the authors also gave examples for neighbourhood structures where these asymptotic results did not hold and their empirical investigations showed that the asymptotic results may not provide a reasonable approximation for small numbers of objects.

With respect to efficiency when testing for clusters, Cliff and Ord (1981) showed that the test based on X_2 is superior to any linear combination of X_1 and X_2 including the test based on X_1.

It is obvious that the objects $o_1, \ldots, o_n$ may be identified with the vertices of a graph, and for $\delta_{ij} = 1$ for a pair (o_i, o_j) we may assume that there exists an edge between o_i and o_j. By a random process (free or nonfree sampling) a label is assigned to a vertice or not. This kind of interpretation has been used for descriptive purposes in Cliff and Haggett (1988).

Grimson and Rose (1991) considered the join–count statistic X_1 under the nonfree sampling model for testing for clusters of labeled neurons. They assumed $\delta_{ij} = 1$ for two neurons o_i and o_j if they were not separated by other neurons. A neuron was considered labeled if dye was transported by the peripheral projections of this neuron to its cell body where it then became visible.

The number (δ_{ij}) of neighbour cells of cell o_i is given by

$$\delta_i = \sum_{j=1}^{n} \delta_{ij}, \qquad 1 \le i \le n.$$

The mean of these numbers is given by

$$\mu_\delta = \frac{1}{n}\sum_{i=1}^{n} \delta_i = \frac{1}{n}\sum_{i=1}^{n}\sum_{j=1}^{n} \delta_{ij}$$

and its variance by

$$\sigma_\delta^2 = \frac{1}{n}\sum_{i=1}^{n}(\delta_i - \mu_\delta)^2 = \frac{1}{n}\sum_{i=1}^{n}\left(\sum_{j=1}^{n}\delta_{ij}\right)^2 - \frac{1}{n^2}\left(\sum_{i=1}^{n}\sum_{j=1}^{n}\delta_{ij}\right)^2.$$

For the nonfree sampling model we have a fixed number $m = Z_1 + \ldots + Z_n$ of labeled neurons. With the notation

$$x^{(k)} = x(x-1)\ldots(x-k+1)$$

the mean and the variance of X_1 are derived as functions of μ_δ and σ_δ^2:

$$E[X_1] = \mu_\delta \frac{m^{(2)}}{n-1},$$

$$Var[X_1] = E[X_1]\left\{1 + \frac{2(\mu_\delta - 1)(m-2)}{n-2} + \frac{(n\mu_\delta - 4\mu_\delta + 2)(m-2)(m-3)}{2(n-2)(n-3)} - E[X_1]\right\} + \sigma_\delta^2\left(\frac{m^{(3)}}{(n-1)^{(2)}} - \frac{m^{(4)}}{(n-1)^{(3)}}\right).$$

Grimson and Rose (1991) argued that the mean (μ_δ) and the variance (σ_δ^2) of the numbers of neighbour cells could be estimated via random samples of cells from the considered cell population. From this, estimates for $E[X_1]$ and $Var[X_1]$ can be obtained. For values of the estimates which are approximately of the same size, Poisson approximation was recommended, otherwise a normal approximation.

In contrast to the authors' optimism we doubt that in this way reliable tests for neuron clusters can be derived. First, considering the usual quality of photomicrographs we do not believe that in practical problems reliable estimates of μ_δ and σ_δ^2 can be derived. As a consequence the results of the statistical tests cannot be trusted. Second, typically it is not known if for the neighbourhood structure of real neuron populations normal or Poisson approximations are justified. Thus again the results of the corresponding tests may be quite misleading.

2. Neuron Clustering with Maximum Statistics

While we agree that it might be difficult to base tests for clusters of labeled neurons on photomicrographs, this seems to be much easier for camera lucida reconstructions. Because these give only the position of the cell somata, it is difficult, however, if not impossible to identify the neighbourhood structure. Therefore, we propose a procedure which does not presume the knowledge of this structure.

In Krauth (1991) bounds for the upper tail probabilities of the multivariate disjoint test were derived. The idea was to consider n independent identically distributed k–dimensional random vectors $X_1, \ldots, X_n$ with a distribution that is concentrated on a bounded Borel set B in the k–dimensional Euclidean space. In our case, B might be a set bounded by a two–dimensional contour encircling a camera lucida reconstruction. A vector X_i might then denote the coordinates of the point corresponding to the location of the soma of the neuron o_i. Under the null hypothesis (H_0) of randomness we assume a uniform distribution over B, under the alternative hypothesis we assume clusters. The set B is dissected into $k \geq 2$ disjoint Borel subsets $C_1, \ldots, C_k$. For each subset we observe the total number (n_i) of neurons with $n = n_1 + \ldots + n_k$ and the number (M_i) of labeled neurons, for $1 \leq i \leq k$. We consider the maximum statistic

$$S = \max_{1 \leq i \leq k} M_i.$$

In contrast to Krauth (1991) we do not assume that the probability of the occurrence of the event that an X falls into C_i is known under H_0. Rather we assume a fixed number $m = M_1 + \ldots + M_k$ of labeled neurons resulting in a multivariate hypergeometric distribution of $M_1, \ldots, M_k$ under H_0 (Jogdeo and Patil (1975)):

$$P(M_1 = m_1, \ldots, M_k = m_k) = \frac{\binom{n_1}{m_1} \cdots \binom{n_k}{m_k}}{\binom{n}{m}}.$$

With the definition $A_i = \{M_i \geq s\}, 1 \leq i \leq k$, we derive

$$P(S \geq s) = P\left(\bigcup_{i=1}^{k} A_i\right), \qquad s \in \{0, 1, \ldots, n\}.$$

From Jogdeo and Patil (1975) we conclude that all marginals are again hypergeometric and that

$$P(A_i \cap A_j) \leq P(A_i)P(A_j), \qquad 1 \leq i, j \leq k.$$

We use the notation

$$q_i = P(A_i) = 0, \text{ for } s > n_i,$$

$$q_i = P(A_i) = \sum_{u=s}^{n} \frac{\binom{n_i}{u}\binom{n-n_i}{m-u}}{\binom{n}{m}} \text{ otherwise, } 1 \le i \le k,$$

$$q_{ij} = P(A_i \cap A_j) = 0, \text{ for } s > \min\{n_i, n_j\},$$

$$q_{ij} = P(A_i \cap A_j) = 0, \text{ for } 2s > m,$$

$$q_{ij} = P(A_i \cap A_j) = \sum_{u=s}^{\min\{n_i,m\}} \sum_{v=s}^{\min\{n_j,m-n\}} \frac{\binom{n_i}{u}\binom{n_i}{v}\binom{n-n_i-n_j}{m-u-v}}{\binom{n}{m}}$$

$$\text{otherwise, } 1 \le i,j \le n \qquad i \ne j.$$

In particular, for $2s > m$ we have the exact result

$$P(S \ge s) = \sum_{i=1}^{k} q_i.$$

Just as in Krauth (1991) we derive upper and lower bounds for $P(S \ge s)$ in the general case. For this we define

$$T_1 = \sum_{i=1}^{k} q_i, \qquad T_2 = \sum_{i=2}^{k}\sum_{j=1}^{i-1} q_{ij} \qquad T_2' = \frac{1}{2}\,(T_1^2 - \sum_{i=1}^{k} q_i^2).$$

The best linear upper bound of degree one for $P(S \ge s)$ is given by

$$U_1 = \min\{1, T_1\}$$

(Kounias and Marin (1974)). If only the values of T_1 and T_2 are known, the best linear upper bound of degree two is given by

$$\min\left\{1, T_1 - \frac{2}{k}T_2\right\}$$

(Kwerel (1975)). Because the probabilities q_{ij} are known in our case, we prefer the upper bound

$$U_2 = \min\left\{1, T_1 - \max_{1\le j\le k} \sum_{\substack{i=1\\ i\ne j}}^{k} q_{ij}\right\}$$

of degree two (Kounias (1968)) which proved to be empirically better than the Kwerel bound.

In Kwerel (1975) it was proved that the bound

$$L_2 = \frac{2}{w(w+1)}(wT_1 - T_2), \qquad w = 1 + [2T_2/T_1]$$

is the best linear lower bound of degree two for $P(S \geq s)$ if only the values of T_1 and T_2 are known. Here, $[x]$ denotes the integer part of x.

The expression for L_2 gives lower bounds for $P(S \geq s)$ for arbitrary positive integer values of w (Galambos (1977)). Thus, by considering that $q_{ij} \leq q_i q_j$ holds as mentioned above, we derive the lower bound

$$L_1 = \frac{2}{w(w+1)}(wT_1 - T_2'), \qquad w = 1 + [2T_2'/T_1]$$

of degree one.

3. Testing for Clusters

In applications we may dissect the given camera lucida or other reconstruction by means of rectangles or other contours and count the total numbers of neurons (n_i) and the numbers of labeled neurons (m_i) for each contour. For the maximum number of labeled neurons (s) we compute the upper bound U_2. If this bound is smaller than the given value of the significance level (α), we can conclude that the labeled neurons cluster. The lower bound L_2 may be used to evaluate the goodness of the approximation of the true P–value $P(S \geq s)$ by the upper bound. In general, this P–value will be nearer to the upper bound than to the lower one. This follows from the fact that for large values of s the upper bound equals the true P–value.

The bounds U_1 and L_1 of degree one are just given for comparison. They may be used if only a pocket calculator is at hand. Otherwise, the better bounds U_2 und L_2 should always be used.

A difficult problem is the appropriate choice of a dissection. Obviously, the power of this test for clusters cannot be high if the total numbers $(n_1, \ldots, n_k)$ of neurons differ much. On the other side, the dissection must be based on objective criteria, i.e. without any knowledge of the numbers $(m_1, \ldots, m_k)$ of labeled neurons, because otherwise a subjective bias of $P(S \geq s)$ results.

4. Applications

As a first example we consider Figure 2 in Pilar, Landmesser and Burstein (1980). Here, a camera lucida reconstruction shows the distribution of labeled ciliary cells (dots) and unlabeled choroid cells (crosses) in a ciliary ganglion from a white leghorn chick embryo. We choose our dissection by circumscribing the given contour by a rectangle parallel to the sides of the page and being tangent of the contour. We divide each side of the rectangle into four equal parts. By linking corresponding points by a line we thus gain a dissection into 16 equalsized rectangles. The counts are given in Tab. 1. Thus we observe $n = 651$ neurons altogether, $m = 360$ labeled neurons and a maximum of $s = 50$ labeled neurons. The bounds are given in this example by

$$L_1 = 0.00270548, U_1 = 0.00270564, L_2 = 0.00270561, U_2 = 0.00270561.$$

14,14	51,50	38,34	8, 7
51,29	49,43	54,47	48,25
33,20	52,37	64,25	70,7
11,5	30,11	49,6	29,0

Tab. 1: Counts for the chick embryo example with the total number of cells (first) and the number of labeled cells (second) for each rectangle

Assuming $\alpha = 0.05$ we observe $U_2 \leq \alpha$, i.e. the hypothesis of randomness can be rejected. From $L_2 = U_2$ we conclude that in this case U_2 equals the exact P–value $P(S \geq 50)$. A comparison with L_1 and U_1 shows that in this example with a very small P–value the simpler bounds L_1 and U_1 would have been close enough to the true P–value.

As a second example we consider Figure 12C in Sun and Cassell (1993). Here, gamma–aminobutyric acid–immunoreactive (GABA–IR) neurons (open circles) and horseradish peroxidose (HRP)–labeled neurons (filled circles) in the central amygdaloid nucleus of the rat are shown in a camera–lucida drawing. We proceed in the same way as above. The results are given in Tab. 2.

6, 1	14,8	12,9	1, 1
8, 4	38,20	16,12	3, 3
1, 1	13,6	5, 5	1, 1
0, 0	1, 1	0, 0	0, 0

Tab. 2: Counts for the rat brain example with the total number of neurons (first) and the number of labeled neurons (second) for each rectangle

The bounds for this example are given by

$$L_1 = U_1 = L_2 = U_2 = 0.89787693.$$

In this case, all four bounds are identical but no significant result is derived. A reason for this might be the low power of the test due to the variation of $n_1, \ldots, n_{16}$. By combining adjacent rectangles to get approximately equal values of the n_i we find

$$n_1 = 40, m_1 = 22; n_2 = 38, m_2 = 20; n_3 = 41, m_3 = 29.$$

Here, the first subset corresponds to the combination of the four rectangles (1,1), (1,2), (1,3) and (2,1), the second subset to the rectangle (2,2) and the third subset to the combination of all other rectangles. We have $s = 29$ and

$$L_1 = 0.09371850, U_1 = 0.09627124, L_2 = 0.09623749, U_2 = 0.09623920.$$

In this case, the approximate matching of the numbers of neurons yields much smaller bounds though the result is still not significant.

As a third example we consider data which we got by personal communication from Professor Jrgen Mai (Department of Neuroanatomy, University of Dsseldorf). This concerns a frontal section of the human hypothalamus, in particular a part of the nucleus paraventricularis. In this example, all cells are labeled by using antibodies against vasopressin while a subpopulation of cells is in addition CD15–positive. For our dissection we consider four sections in one direction and eight sections in the other direction, yielding 32 rectangles with the counts given in Tab. 3.

7,2	21,7	19,10	13,7	9,5	2,1	10,7	0,0
1,1	62,29	55,30	30,12	5,5	4,2	9,8	5,3
0,0	16,12	54,24	43,17	11,4	9,6	3,3	4,4
0,0	0,0	1,1	1,1	1,0	2,1	1,1	1,1

Tab. 3: Counts for the human brain example with the total number of neurons (first) and the number of labeled neurons (second) for each rectangle

The bounds for this example are given by

$$L_1 = 0.79822460, U_1 = 1.00000000, L_2 = 0.86251746, U_2 = 0.94552227.$$

Obviously, the hypothesis of randomness cannot be rejected. By combining again adjacent rectangles we consider $k = 7$ subsets with

$$n_1 = 64, m_1 = 32; n_2 = 62, m_2 = 29; n_3 = 55, m_3 = 30; n_4 = 54, m_4 = 24;$$

$$n_5 = 57, m_5 = 23; n_6 = 54, m_6 = 25; n_7 = 53, m_7 = 41.$$

From this results $s = 41$ and

$$L_1 = 0.02429276, U_1 = 0.02442929, L_2 = 0.02440157, U_2 = 0.02440225.$$

Here, the hypothesis of randomness can be rejected and a clustering of CD 15–positive neurons be assumed.

References:

CLIFF, A.D. and HAGGETT, P. (1988): *Atlas of Disease Distributions. Analytic Approaches to Epidemiological Data.* Blackwell, Oxford.

CLIFF, A.D., HAGGETT, P., ORD, J.K., BASSETT, K.A. and DAVIES,R.B. (1975): *Elements of Spatial Structure. A Quantitative Approach.* Cambridge University Press, Cambridge etc.

CLIFF, A.D., HAGGETT, P., ORD, J.K. and VERSEY, G.R. (1981): *Spatial Diffusion. A Historical Geography of Epidemics in an Island Community.* Cambridge University Press, Cambridge etc.

CLIFF, A.D. and Ord, J.K. (1973): *Spatial Autocorrelation.* Pion, London.

CLIFF, A.D. and ORD, J.K. (1981): *Spatial Processes. Models & Applications.*Pion, London.

GALAMBOS, J. (1977): Bonferroni inequalities. *Annals of Probability, 5, 577–581.*

GRIMSON, R.C. and ROSE, R.D. (1991): A versatile test for clustering and a proximity analysis of neurons. *Methods of Information in Medicine, 30, 299–303.*

HAINING, R. (1990): *Spatial data analysis in the social and environmental sciences.* Cambridge University Press, Cambridge etc.

JOGDEO, K. and PATIL, G.P. (1975): Probability inequalities for certain multivariate discrete distribution. *Sankyā, Series B, 37, 158–164.*

KOUNIAS, E.G. (1968): Bounds for the probability of a union with applications. *Annals of Mathematical Statistics, 39, 2154–2158.*

KOUNIAS, E. and MARIN, D. (1974): Best linear Bonferroni bounds, in: *Proceedings of the Prague Symposium on Asymptotic Statistics,* Vol. II, Charles University, Prague, 179–213.

KRAUTH, J. (1991): Bounds for the upper tail probabilities of the multivariate disjoint test. *Biometrie und Informatik in Medizin und Biologie, 22, 147–155.*

KWEREL, S. (1975): Most stringent bounds on aggregated probabilities of partially specified dependent probability systems. *Journal of the American Statistical Association, 70, 472–479.*

MORAN, P.A.P. (1947): Random associations on a lattice. *Proceedings of the Cambridge Philosophical Society, 43, 321–328.*

MORAN, P.A.P. (1948): The interpretation of statistical maps. *Journal of the Royal Statistical Society, Series B, 10, 243–251.*

PILAR, G., LANDMESSER, L. and BURSTEIN, L. (1980): Competition for surviving among developing ciliary ganglion cells. *Journal of Neurophysiology, 43, 233–254.*

SUN, N. and CASSELL, M.D. (1993): Intrinsic GABAergic neurons in the rat central extended amygdala. *Journal of Comparative Neurology, 330, 381–404.*

UPTON, G.J.G. and FINGLETON, B. (1988): *Spatial Data Analysis by Example. Vol. I. Point Pattern and Quantitative Data.* John Wiley & Sons, Chichester etc.

A New Approach of Regionalisation by Classifying Hydrological Quantities

K.-P. Nieschulz, O. Richter, B. Diekkrüger, A. Lücke

Institut für Geographie und Geoökologie, TU Braunschweig,
Langer Kamp 19c, D-38106 Braunschweig, FRG

Summary: To calculate the water balance of an agricultural used landscape based on a deterministic site model one is faced with many soil, weather and plant parameters used within the nonlinear process models describing the water transport in the unsaturated zoneof the soil. Investigating evapotranspiration and ground water recharge our mathematical concept consists of classifying model soil columns (ms) with prescribed texture by regionalisation indicators ($ri \in \Re^n$). The grouping tool is hierarchical cluster analysis by several (dis-)similarity measures.

1. Regionalisation?

Regionalisation in *hydrology* means to investigate **spatial** processes affecting the water fluxes below and at the surface. E.g. there are both processes relevant only at the **micro-scale** (scale: $\leq 100\text{m}, \leq 10^{-1}\text{km}^2$) like macro pore flow and processes having an effect only on the **macro-scale** (scale: $\geq 30\text{km}, \geq 10^3\text{km}^2$) like advective rainfalls beside processes important on all scales.

Classical tasks of regionalisation (often synonymous "scaling") are
- describing mathematically spatial **properties**,
- up and down scaling (changing the scale) of **properties** and
- transferring **properties** from one geographical site to another.

Possible **properties** are precipitation, evaporation, transpiration, soil water run off, ground water recharge and model parameters of stochastic or deterministic models.

2. Processes

Our research interest is the unsaturated zone of the soil in agricultural used regions. The soil is naturally divided into horizons. Each horizon is described by a set of soil properties (P). These input variables are represented in a matrix

$$(P_{ij})_{j=1,k}^{i=1,m} := \begin{pmatrix} p_{11} & p_{12} & \cdots & p_{1k} \\ p_{21} & p_{22} & & p_{2k} \\ \vdots & & \ddots & \vdots \\ p_{m1} & p_{m2} & \cdots & p_{mk} \end{pmatrix}, \tag{1}$$

with p_{ij} the variable j in horizon i.

We are interested in examining the quantities evaporation (E), transpiration (T) and ground water recharge (Gw), our output variables. The simulations are carried out by the Deterministic Site Model (DESIM), cf. Diekkrger (1992). They are performed for 1989 from April, 16th till October, 15th using real weather data. The numerical calculations are based on the coupling of two deterministic processes:

2.1 The (Vertical) Water Flux in the Soil

Out of given soil texture the parameters of the water retention curve $\theta(\Psi)$ and the hydraulic conductivity curve $K(\theta)$ are evaluated by a so-called pedo transfer function (PTF), a non-linear regression function (cf. Rawls and Brakensiek (1985)). The water retention curve describes the functional relationship between the two unknown quantities water content θ and water suction Ψ of the Fokker-Planck-equation

$$\frac{\partial \theta}{\partial t} = \frac{\partial}{\partial z}\left(K(\Psi)\frac{\partial \Psi}{\partial z} - K(\Psi)\right) + S(\Psi) \tag{2}$$

with - θ, the soil water content [% by volume],
- Ψ, the water suction [hPa],
- t, z, the time and space variables[d,cm],
- $K = K(\Psi, z, t)$, the unsaturated hydraulic conductivity [cm/d] and
- S, the sink-source term [% by volume/d].

The hydraulic conductivity curve is a parameter of this partial differential equation (2). This equation is solved for appropriate initial and boundary conditions.

2.2 The Evapotranspiration

Evapotranspiration is calculated by the Penman-Monteith-equation considering temporal variable climate and plant properties (cf. Braden (1990)).

2.3 Mathematical Formulation

Mathematically we define the water balance by the mapping $\Omega := (\Omega_1, \Omega_2, \ldots, \Omega_n)$ from soil properties into hydrological quantities for n soil columns with m horizons and k variables by

$$\Omega_j : \begin{cases} \mathbf{A} := [a_1, b_1]^m \times [a_2, b_2]^m \times \ldots \times [a_k, b_k]^m \quad \rightarrow \quad \Re \times \Re \times \Re \\ \Omega_j((S_{1j}, To_{1j}, Pv_{1j}, B_{1j}^1, \ldots, B_{1j}^{k-3}), (S_{sj}, To_{2j}, Pv_{2j}, B_{2j}^1, \ldots, B_{2j}^{k-3}), \ldots, \\ \qquad (S_{mj}, To_{mj}, Pv_{mj}, B_{mj}^1, \ldots, B_{mj}^{k-3})) := (E_j, T_j, Gw_j), \\ a_i, b_i \in \Re, \quad i = 1, 2, \ldots, k, \quad j = 1, 2, \ldots, n, \end{cases} \tag{3}$$

with - S_{ij}, the sand content of horizon i in column j,
- To_{ij}, the clay content of horizon i in column j,
- Pv_{ij}, the porosity of horizon i in column j,
- $B_{ij}^1, \ldots, B_{ij}^{k-3}$, $k-3$ variables of horizon i in column j, $i = 1, 2, \ldots, m$, e.g. organic carbon content, thickness of horizons, etc.,
- E_j, the evaporation of column j,

- T_j, the transpiration of column j and
- Gw_j, the ground water recharge for column $j, j = 1, 2, \ldots, n$.

3. The Regionalisation Approach

3.1 Problem

Calculate the water balance evaporation (E), transpiration (T) and ground water recharge (Gw) in a given region as values of Ω at points of given soil texture.

3.2 Situation

In an investigated region with known soil physical properties (e.g. sand-, silt content, porosity) and agricultural landuse we define an "ecotope" as a homogeneous area generated by intersection of pedological units with landuse. Each "ecotope" is represented by a soil column. We call this column a "real" one. In our research catchment (16 km^2) in the quaternary region of Northern Germany thus about 1700 ecotopes are generated from which 500 different ecotopes are identified. In this study it is assumed that only winter wheat is grown.

The calculation of the water balance for this catchment requires an effort of about 500×95 CPU-sec ≈ 13,2 CPU-h on an IBM 3090.

If one intends to compute water balances of larger catchments (thousands of km^2) it is necessary to combine deterministic with statistical tools to group the ecotopes with respect to "equal" hydrological properties. "Representative ecotopes" will be simulated to obtain the water balance.

3.3 Regionalisation Steps

The prescribed regionalisation problem is solved by the following steps (cf. Nieschulz et al. 1994):

R1) Define "**Model soil columns**" (in contrast to real soil columns) on the domain of the simulations, the set $A = [a_1, b_1]^m \times [a_2, b_2]^m \times \ldots \times [a_k, b_k]^m$ (cf. Tab. 2).

R2) Calculate the **water balances** Ω_j :=(evaporation (e_J), transpiration (t_J), ground water recharge (Gw_j)) of the model soil columns using the simulation model (cf. Tab. 2).

R3) Carry out a **cluster analysis** of the water balances Ω_j dependent on a prescribed accuracy $||\Omega_j|| = ||(E_j, T_j, Gw_j)||$ which results in q clusters $C_1, C_2, \ldots, C_q$, $q \in \mathbf{N}$ (cf. chapter 4).

R4) Define **regionalisation indicators** $(RI)\, ri := (ri_1, ri_2, \ldots, ri_p) \in \Re^p, p \in \mathbf{N}$ (cf. chapter 3.4). Each soil column, represented by a matrix P_{ij}, is assigned to an ri.

R5) **Calculate** the regionalisation indicators for

a) the model soil columns. The result is a set $RI_M \subset \Re^p$ (cf. Tab. 2, Fig. 2).

b) the real soil columns in the examined region. The result is a set $RI_R \subset \Re^p$ (cf. Fig. 2). The regionalisation indicators are defined that holds $RI_R \subseteq RI_M$.

R6) **Identify** in RI_M the clusters $C_1, C_2, \ldots, C_q, q \in \mathbf{N}$, and transfer these structure to RI_R.

R7) Define depending on the structure in RI_R "representative elements", the "**representative ecotopes**".

R8) Calculate the regional **water balance** Ω by simulating the representative ecotopes weighted by area and frequency of the real soil columns.

3.4 Definition of Regionalisation Indicators

We are defining two regionalisation indicators:
3.4.1 The Hilbert-Norm (ri_1): The idea: To compare two objects we have to measure their length. Mathematically we have to define a norm on the $m \times k$-matrices P. Define

$$\begin{aligned} ri_1 &:= Hi(P) \\ &:= \kappa_{max}(P) \\ &:= \text{largest singular value of P} \\ &:= +\sqrt{(\lambda_{max}(P^*P))} \\ &= \text{positive square root of the largest eigen value } (EW) \text{ of } P^*P (4) \end{aligned}$$

3.4.2 Water available for plants (WAP; ri_2): The first regionalisation indicator is a pure mathematical one. To take soil scientific knowledge into account, we are using the water available for plants as a second regionalisation indicator. It indicates how much water is at the plants disposal. This quantity is evaluable at the base of the water retention curve $\theta(\Psi)$ as a difference of two values (field capacity (fc $\approx$ 60 hPa) and the permanent wilting point (pwp $\approx$ 15800 hPa)).

$$\text{WAP} := \theta(60\text{hPa}) - \theta(15800\text{hPa}) \quad (5)$$

4. Clustering the Model Soil Columns

To execute step (R3) the hydrological quantities $(E_j, T_j, Gw_j) = \Omega_j$ have to be grouped. The grouping tool is hierarchical cluster analysis. The cluster analysis is carried out with the **MultiVariate Statistical Package Plus-Version 2.1** (MVSP) (KOVACH (1993)). This software is often used in mathematical biology and related topics (cf. the references in KOVACH (1993)). In this package common distances and cluster procedures are available like:

distances:

- squared Euclidean distance (SED),
- standardised Euclidean distance (STE),
- cosine theta distance (normalised Euclidean distance(CTD)),
- Manhattan metric distance (l_1-norm) (MMD) and
- chord distance (CD+).

cluster procedures:

- unweighted pair group (U),
- unweighted centroid (C),
- weighted pair group (W),
- weighted centroid/median (M),
- minimum variance (V),
- nearest neighbour (N) and
- farthest neighbour (F).

The objective of this chapter is to group the model soil columns. In Tab. 1 the procedures and distances taken into account are listed.

Procedures → Distances ↓	U	C	W	M	V	N	F
SED	X		X	X	X	X	X
STE	X					X	
CTD		XI		XI			
MMD		X					X
CD+		X	X				

X: Chosen method, I: Inversion

Tab. 1: Combination of procedures and distances

The following subdivision of the model soil columns throughout the methods has been calculated: The soils with 0-10%, 15-20%, 25-30%, 35-40%, and 45% silt, respectively, are forming five clusters, except for the combinations CTD+C, CTD+M (cf. Fig. 1d) with inversion and the algorithms STE+U, MMD+F. This result is shown exemplary for two methods in Fig. 1a and Fig. 1b. Even the procedures STE+U (cf. Fig. 1c)) and CTD+M (inversion!; not shown) result in a subdivision nearby the previous mentioned.
We conclude that the arrangement of the defined model soil columns is

(1) almost independent of distances and procedures - in spite of their individual characteristics - and

(2) comprehensible for soil scientists and hydrologists.

This **stable grouping of the model soil columns** is supported by regarding the **clustering of the real soil columns** on the same distance level as the model soil columns. The second conclusion is supported by the fact that the spatial distribution of the resulting clusters corresponds with the spatial distribution of the silt fraction in the soil.

The subsequent regionalisation steps are not presented here in detail. This paper is mainly concerned with cluster aspects within our regionalisation approach. Especially Fig. 2 gives an idea how (R6) works. For defining the representative ecotopes (R7) and the calculation of the water balance (R8) see detailed description in Nieschulz et al. (1994).

ident-ification no.	soil properties p_{i1} sand [% by weight]	p_{i2} clay [% by weight]	p_{i3} porosit y [cm³/cm³]	hydrological quantity Ω [cm] evaporation E	transpiration T	ground water recharge Gw	regionalisation indicator WAP ri_2 [cm]	Hi(P) ri_1 [-]
2000	90.0	10.0	0.48	-20.040	-16.444	9.163	8.85	3.7860
2001	80.0	10.0	0.48	-20.038	-17.706	10.821	10.91	3.1097
2002	70.0	10.0	0.51	-20.782	-18.682	13.184	13.29	2.0955
2003	60.0	10.0	0.51	-21.255	-19.751	16.238	15.98	2.0013
2004	50.0	10.0	0.51	-21.917	-20.627	19.638	18.92	1.5705
2005	80.0	20.0	0.48	-19.420	-17.046	9.085	8.51	3.2289
2006	70.0	20.0	0.48	-19.997	-18.175	10.106	10.41	2.6334
2007	60.0	20.0	0.51	-20.463	-19.276	12.155	12.65	2.1194
2008	50.0	20.0	0.51	-20.767	-20.407	14.828	15.18	1.6875
2009	90.0	5.0	0.48	-20.296	-17.163	11.147	10.68	3.7562
2010	80.0	5.0	0.48	-20.938	-18.153	13.423	13.03	3.0799
2011	70.0	5.0	0.51	-21.321	-19.168	16.372	15.66	2.4849
2012	60.0	5.0	0.51	-21.553	-20.234	19.677	18.53	1.9718
2013	50.0	5.0	0.51	-21.813	-21.324	23.182	21.59	1.5414
2014	70.0	30.0	0.56	-19.341	-17.976	9.344	8.77	2.8317
2015	60.0	30.0	0.56	-19.626	-19.261	10.333	10.61	2.3168
2016	50.0	30.0	0.56	-20.029	-20.328	11.658	12.73	1.8833

Tab. 2: Model soil columns with soil properties, computed hydrological quantity Ω and regionalisation indicator (ri_1, ri_2)

5. Discussion

In problems concerning regionalisation of hydrology it is necessary to summarise information. It is required to group the information with little loss of basic physical information.

The method presented here calculates the water balances of a defined region by a deterministic model combined with statistical elements.

The classification of the model soil columns by regionalisation indicators is performed with hierarchical cluster analysis. The first results are satisfying both soil scientists and hydrologists. To generalise the method the whole domain of possible soil properties (0-100% sand, etc., called "soil texture triangle") has to be covered with model soil columns to define "zones of similar hydrological characteristics" in this domain. A statistical analysis of the resulting clusters carefully has to be carried out.

Acknowledgements

This research is supported by the Deutsche Forschungsgemeinschaft.

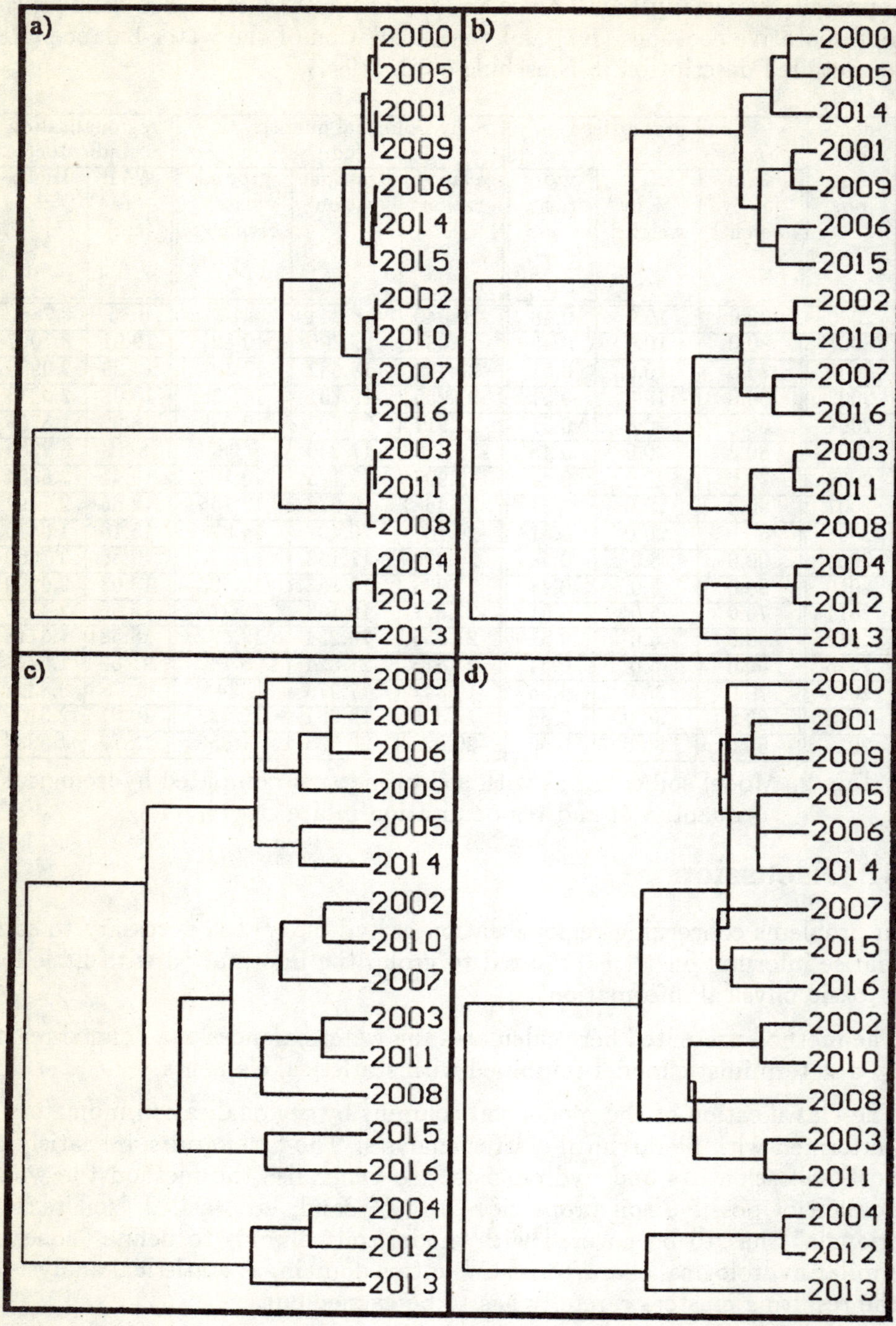

Fig. 1: Four cluster results of the water balance vector of the model soil columns a) SED+F, b) CD++W, c) STE+U, and d) CTD+C

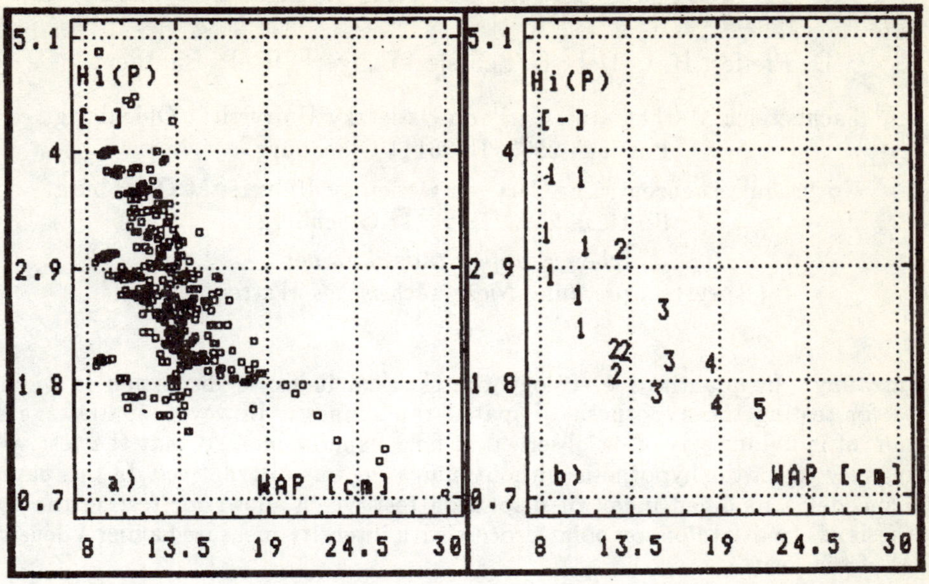

Fig. 2: Regionalisation indicators (ri_2, ri_1) of a) the real soil columns (RI_R) and b) the model soil columns including five clusters (RI_M)

The data used in this study are provided by the Soil survey of Lower Saxony, Germany (NLfB) and by the Collaborative Research Program (CRP) 179 "Water and Matter Dynamics in Agroecosystems" of the TU Braunschweig. They are stored in Geographical Information System (GIS) ARC/INFO in connection with the ORACLE data base management system.

References:

BRADEN, H. (1990) Modellierung der Wasser- und Stoffdynamik in Agrarökosystemen mit Hilfe des Agrarmeteorologischen Modells AMBETI. Bayer. Landwirtsch. Jb., 67, Sonderheft, 145-156.

DIEKKRÜGER, B. (1992): Standort- und Gebietsmodelle zur Simulation der Wasserbewegung in Agrarökosystemen. *Landschaftsökologie und Umweltforschung* 19, 169.

KOVACH, W.L. (1993): MultiVariate Statistical Package, PLUS - Version 2.1 User's Manual, Kovach Computing Services, Pentraeth, Wales, UK.

NIESCHULZ, K.-P., RICHTER, O.,DIEKKRÜGER, B., and LÜCKE, A. (1994): A New Method for Upscaling Hydrological Processes, Annales Geophysicae, Supplement II to vol. 12, C 396.

RAWLS, W. J., BRAKENSIEK, D. L. (1985) Prediction of soil water properties for hydrological modeling. In: Proc. of the Symp. Watershed management in the eighties. 293-299.

The Index–of–Dispersion Test Revisited

D. Pfeifer, H. Ortleb, U. Schleier–Langer[1], H.–P. Bäumer[2]

[1] Fachbereich Mathematik, Carl von Ossietzky Universität Oldenburg, Postfach 2503, D–26111 Oldenburg

[2] Hochschulrechenzentrum, Carl von Ossietzky Universität Oldenburg, Postfach 2503, D–26111 Oldenburg

Arbeitsgruppe Statistik in der Ökosystemforschung Niedersächsisches Wattenmeer

Summary: In quantitative ecology the classical index–of–dispersion is widely used for testing the hypothesis of spatial randomness. However, spatial aggregation of individuals is often observed in field experiments, so that the test will frequently reject the hypothesis without indicating any alternatives. In this paper we consider a modified index–of–dispersion test which allows for testing the hypothesis of a spatial Poisson point process with intensity measure having a density $\lambda(\mathbf{x})$ of the form

$$\lambda(\mathbf{x}) = \sum_{j=1}^{n} a_j f_j(\mathbf{x}), \quad \mathbf{x} \in \mathbb{R}^2$$

with known non–negative regression functions $f_j(\mathbf{x})$ and unknown non–negative parameters a_j which are to be estimated by the observed data. This model includes the classical case for $n = 1$ and $f_1(\mathbf{x}) \equiv 1$. Further applications to testing local geographical influences on health data are also pointed out.

1. Introduction

In quantitative ecology the classical index–of–dispersion (normalized variance–to–mean ratio of quadrat counts) is widely used for testing the hypothesis of spatial randomness (see e.g. Greig–Smith (1983), p. 61ff, or Diggle (1983), p. 23ff). One reason for this is the fact that the experimental estimation of spatial distribution and abundance of animals living on small scales is usually faced with taking samples by means of physical devices (quadrat sampling; cf. eg. Krebs (1985), p. 160ff). Due to individual aggregation effects, patchiness of the spatial distribution can be observed even on those scales: the following figure represents two examples of 5×5 multicorer samples of benthic fauna *(Harpacticus obscurus)* in the Wadden Sea, taken from Pfeifer, Schleier–Langer, and Bäumer (1994).

95	1	3	0	42
1	1	1	4	2
0	5	8	81	24
11	1	6	71	116
1	5	116	2	10

165	22	1	94	68
11	82	111	97	153
0	0	24	13	15
31	1	46	22	11
2	0	5	8	6

Fig. 1

The aggregation effect observed in these samples can e.g. be explained by oxidation and fertilization of the surrounding sediment by lugworms *(Arenicola marina)* or the polychaete *Lanice conchilega*, attracting meiofauna in high numbers around their burrows and funnels (cf. Reise (1985), p.126ff; Ekschmitt (1993) contains a more general discussion of aggregation effects). It is therefore clear that the hypothesis of a random spatial distribution of meio– and microfauna must in general be rejected, although the spatial dispersion of lugworms and polychaetes – as the "parent" point pattern – is in good coincidence with the Poisson process assumption (Pfeifer, Bäumer and Albrecht (1993)). However, it might be that the spatial distributional pattern of individuals within each cluster (represented by a parent polychaete) may possibly stem from a Poisson process itself and is hence "locally random". In this paper we therefore suggest a modification of the classical index–of–dispersion test which allows for testing the hypothesis of a spatial Poisson point process with intensity measure possessing a density $\lambda(\mathbf{x})$ of the form

$$\lambda(\mathbf{x}) = \sum_{j=1}^{n} a_j f_j(\mathbf{x}), \quad \mathbf{x} \in \mathbb{R}^2 \tag{1}$$

with known non–negative regression functions $f_j(\mathbf{x})$ and unknown non–negative parameters a_j. As usual in linear statistical models, these parameters are to be estimated by the observed data via suitable least–squares–methods. This model includes the classical case for $n = 1$ and $f_1(\mathbf{x}) \equiv 1$. (Note that this procedure is not restricted to the two–dimensional Euclidean space, but is generally applicable in any dimension.)

2. The Model

Consider a spatial (in general non–homogeneous) Poisson point process ξ with an intensity (= Lebesgue–density for its intensity measure) of type (1). Suppose further that observations $Z_1, \ldots, Z_m$ are taken with $m > n$, representing aggregate point counts within pairwise disjoint subsets $B_1, \ldots, B_m$ of $\mathbb{R}^2$, i.e.

$$Z_i = \xi(B_i), \quad i = 1, \ldots, m. \tag{2}$$

Then

$$\mu_i := E(Z_i) = \int_{B_i} \lambda(\mathbf{x})\, d\mathbf{x} = \sum_{j=1}^{n} a_j \int_{B_i} f_j(\mathbf{x})\, d\mathbf{x} = (\mathbf{Wa})_i, \; i = 1, \ldots, m \tag{3}$$

where $\mathbf{W}$ is the weight matrix $\mathbf{W} = (w_{ij})$ with

$$w_{ij} = \int_{B_i} f_j(\mathbf{x})\, d\mathbf{x}, \quad i = 1, \ldots, m, \; j = 1, \ldots, n \tag{4}$$

and $\mathbf{a}$ is the column vector of parameters a_j. By the central limit theorem, if the μ_i are sufficiently large, the column vector $\mathbf{Z}$ consisting of the components Z_i is approximately normally distributed as $\mathcal{N}(\mathbf{Wa}, \Sigma)$ with a variance–covariance matrix $\Sigma = \Delta(\mathbf{Wa})$ of diagonal form where

$$\Delta(\mathbf{b})_{ij} = \begin{cases} b_i, & \text{if } i = j \\ 0, & \text{otherwise,} \end{cases} \quad 1 \leq i, j \leq m \tag{5}$$

for any column vector $\mathbf{b} = (b_1, \ldots, b_m)^{\text{tr}} \in \mathbb{R}^m$. Hence

$$\hat{\mathbf{a}} = (\mathbf{W}^{\text{tr}}\mathbf{W})^{-1}\mathbf{W}^{\text{tr}}\mathbf{Z}, \quad \hat{\mu} = \mathbf{W}\hat{\mathbf{a}} \tag{6}$$

are appropriate least–squares–estimates for the parameter vectors $\mathbf{a}$ and μ, resp., and the *generalized index–of–dispersion* D is just the sum of squares of residuals:

$$D = \sum_{i=1}^{m} \frac{(Z_i - \hat{\mu}_i)^2}{\hat{\mu}_i} = (\mathbf{Z} - \mathbf{W}\hat{\mathbf{a}})^{\text{tr}}\hat{\Sigma}^{-1}(\mathbf{Z} - \mathbf{W}\hat{\mathbf{a}}) \tag{7}$$

where $\hat{\Sigma} = \Delta(\hat{\mu}) = \Delta(\mathbf{W}\hat{\mathbf{a}})$ is the estimated diagonal matrix with entries $\hat{\mu}_i$, $i = 1, \ldots, m$. Hence D is approximately χ^2–distributed with $m - n$ degrees of freedom if $\mathbf{W}$ has full rank n and can thus be used as a test statistic for the null–hypothesis that the underlying point process is Poisson with an intensity of the form (1) (for technical details, see e.g. Fahrmeir and Hamerle (1984), p.84ff).

However, the estimate $\hat{\mathbf{a}}$ given in (6) is not optimal; a better estimate would be

$$\mathbf{a}^* = (\mathbf{W}^{\text{tr}}\Sigma^{-1}\mathbf{W})^{-1}\mathbf{W}^{\text{tr}}\Sigma^{-1}\mathbf{Z} \tag{8}$$

if Σ was known. The problem here is that Σ is *unknown*, so that we should use an iterative procedure in order to get a better estimate of $\mathbf{a}$, inserting consecutively the updated estimate $\hat{\Sigma}$ in (8):

$$\mathbf{a}_1^* := \hat{\mathbf{a}}, \quad \mathbf{a}_{k+1}^* = (\mathbf{W}^{\text{tr}}\Delta(\mathbf{W}\mathbf{a}_k^*)^{-1}\mathbf{W})^{-1}\mathbf{W}^{\text{tr}}\Delta(\mathbf{W}\mathbf{a}_k^*)^{-1}\mathbf{Z}, \; k \in \mathbb{N}. \tag{9}$$

It is not always ensured that this iterative procedure will give meaningful results. However, as long as the estimates of the parameters a_j remain non-negative, it seems that in most cases the corresponding dispersion indices according to (7) will decrease and thus by boundedness converge to a lower limit which would then also imply convergence of the estimator sequence $\{\mathbf{a}_k^*\}$. [If by chance some of the a_{kj}^* become negative, they should be set to zero, omitting thus the corresponding regression functions in the basic setup.]

3. Examples

For simplicity we shall here assume $m = 4$, $n = 2$ and regression functions f_j of the following type:

$$f_j(\mathbf{x}) = \sum_{k=1}^{4} c_{jk} I_{B_k}(\mathbf{x}), \quad \mathbf{x} \in \mathbb{R}^2, \; j = 1, 2 \tag{10}$$

with disjoint sets B_k of unit area each (for example, a 2×2 square subdivided into 4 equally large subsquares), indicator functions I_B defined by

$$I_B(\mathbf{x}) = \begin{cases} 1, & \text{if } \mathbf{x} \in B \\ 0, & \text{otherwise} \end{cases} \tag{11}$$

for any Borel subset $B \subseteq \mathbb{R}^2$, and coefficient matrix $\mathbf{C} = (c_{jk})$ given by

$$\mathbf{C} = \begin{pmatrix} 7 & 4 & 3 & 1 \\ 1 & 2 & 5 & 9 \end{pmatrix}. \tag{12}$$

If we think of the B_k forming a 2×2 gridded square, these coefficients could also be thought of being arranged in matrix form:

$$\mathbf{C}_1 = \begin{bmatrix} 7 & 4 \\ 3 & 1 \end{bmatrix} \qquad \mathbf{C}_2 = \begin{bmatrix} 1 & 2 \\ 5 & 9 \end{bmatrix} \tag{13}$$

Poisson point processes with intensities f_1 and f_2, resp. would hence place the majority of points in the north–west and the south–east corner of the square, resp. By our assumptions, it follows that the weight matrix $\mathbf{W}$ coincides with the coefficient matrix $\mathbf{C}^{tr}$ since all B_j have equal area of unit measure. By (6), we obtain

$$\mathbf{W}^{tr}\mathbf{W} = \begin{pmatrix} 75 & 39 \\ 39 & 111 \end{pmatrix}$$

and hence

$$(\mathbf{W}^{tr}\mathbf{W})^{-1}\mathbf{W}^{tr} = \frac{1}{1134}\begin{pmatrix} 123 & 61 & 23 & -40 \\ -33 & -1 & 43 & 106 \end{pmatrix} \tag{14}$$

Suppose we have obtained the sample $\mathbf{Z} = (41\ 36\ 57\ 95)^{tr}$ from quadrat counts, or, in matrix form (corresponding to the gridded square above):

$$\mathbf{Z} \simeq \begin{bmatrix} 41 & 36 \\ 57 & 95 \end{bmatrix}. \tag{15}$$

Then (14) yields the estimate

$$\hat{\mathbf{a}} = (2375/567\ \ 5566/567) = (4.188\ \ 9.816) \tag{16}$$

with an index–of–dispersion of

$$D = 0.508. \tag{17}$$

Iteration of the procedure gives the following table for $\mathbf{a}_k^*$:

k	1	2	3	4
$\mathbf{a}_k^*$	(4.188 9.816)	(4.290 9.684)	(4.287 9.687)	(4.287 9.687)
D	0.508	0.488	0.488	0.488

Fig. 2

A stabilisation of the recursive estimates of **a** is visible here already for values of k smaller than 5; the corresponding χ^2–test would further not reject the null–hypothesis at any reasonable level of significance while the ordinary index–of–dispersion gives a value of 233.706 which is far beyond the 0.999–quantile of the χ^2_3–distribution.

For the sample $\mathbf{Z} = (75\ 71\ 79\ 68)$ or, in matrix form,

$$\mathbf{Z} \simeq \begin{bmatrix} 75 & 71 \\ 79 & 68 \end{bmatrix} \tag{18}$$

the situation is different: here (14) gives the initial estimate

$$\hat{\mathbf{a}} = (12653/1134 \quad 8059/1134) = (11.157 \quad 7.106) \tag{19}$$

with $D = 5.856$ and iterations

k	1	2	3	4
$\mathbf{a}^*_k$	(11.175 7.106)	(11.496 7.091)	(11.510 7.079)	(11.511 7.078)
D	5.856	5.6363	5.6346	5.6344

Fig. 3

such that the χ^2–test would reject the null–hypothesis at a significance level of 0.10 with a critical value of 4.61 (from the χ^2_2–distribution). Note that the ordinary index–of–dispersion is 5.866 here which is below the 0.90–quantile of the χ^2_3–distribution, hence the hypothesis of a purely random spatial distribution would not be rejected at this level.

An application of this method for instance to benthos data seems reasonable if clustering of meio– and microfauna is caused by several superposed parent macrofauna patterns with different intensities of species attraction. For such a kind of analysis, it is therefore necessary to register macrofauna data from the same samples as well.

Distributional patterns as in Fig. 1 are in good coincidence with simulated patterns from Thomas–processes with radially symmetric bivariate normal distributions for the daughter points (see e.g. Diggle (1983), p. 54ff.). Due to the possibly quadrat–overlapping variability of such a normal distribution, quadrats should, however, be grouped again before further investigation. For example, the left hand sample in Fig. 1 could be grouped as follows:

95	1	3	0	42
1	1	1	4	2
0	5	8	81	24
11	1	6	71	116
1	5	116	2	10

or, in matrix form :

$$\begin{bmatrix} 115 & 153 \\ 140 & 199 \end{bmatrix} \tag{20}$$

A closer look at several of the obtained benthos data sets suggests that clustering of meio– and microfauna occurs typically around the high peaks within the quadrats, hence the following assumptions could be justified:

$$\mathbf{C}_1 = \begin{bmatrix} 4 & 6 \\ 5 & 7 \end{bmatrix} \qquad \mathbf{C}_2 = \begin{bmatrix} 9 & 6 \\ 6 & 4 \end{bmatrix} \tag{21}$$

Here $\mathbf{C}_1$ corresponds to the local intensities for the parent point pattern of polychaetes attracting meiofauna [not actually counted in this sample], while $\mathbf{C}_2$ corresponds to a pure background noise component of homogeneous Poisson type per quadrat. The method described above gives here

k	1	2	3	4
$\mathbf{a}_k^*$	(27.316 0.223)	(27.024 0.498)	(27.033 0.491)	(27.022 0.488)
D	1.309	1.290	1.290	1.290

Fig. 4

which means that the average cluster size (within a quadrat) is about 27 individuals with a superimposed Poisson noise of about one individual per two quadrats on average. The modified index–of–dispersion test would not reject the null–hypothesis at any reasonable level here, while the ordinary index–of–dispersion for these four data points gives a value of 24.53.

A similar procedure for the right hand sample in Fig. 1 yields:

165	22	1	94	68
11	82	111	97	153
0	0	24	13	15
31	1	46	22	11
2	0	5	8	6

or, in matrix form :

$$\begin{bmatrix} 392 & 412 \\ 109 & 75 \end{bmatrix} \tag{22}$$

Choosing

$$\mathbf{C}_1 = \begin{bmatrix} 12 & 13 \\ 3 & 2 \end{bmatrix} \qquad \mathbf{C}_2 = \begin{bmatrix} 6 & 4 \\ 9 & 6 \end{bmatrix} \tag{23}$$

we obtain

k	1	2	3	4
$\mathbf{a}_k^*$	(31.393 1.857)	(31.420 1.815)	(31.420 1.815)	(31.420 1.815)
D	0.122	0.120	0.120	0.120

Fig. 5

The average cluster size (within a quadrat) is slightly larger as before, while the Poisson background noise here is about 2 individuals per quadrat. The modified index–of–dispersion test would again not reject the null hypothesis, while the simple index–of–dispersion gives a value of 392.21.

4. Further Applications

A particular case of interest is $n = 1$, i.e. there is only one parent macrofauna type causing patchiness in the spatial distribution. In this case, the calculations simplify to a great extent: let f denote the regression density and a the parameter to be estimated. The design matrix $\mathbf{W}$ then reduces to a column vector $\mathbf{W} = (w_1, \ldots, w_m)^{\text{tr}}$, with $w_i = \int_{B_i} f(\mathbf{x})\, d\mathbf{x}$, hence by (6),

$$\hat{a} = \frac{\sum_{i=1}^{m} w_i Z_i}{\sum_{i=1}^{m} w_i^2} \tag{24}$$

and the index–of–dispersion is given by

$$D = \sum_{i=1}^{m} \frac{(Z_i - \hat{a} w_i)^2}{\hat{a} w_i}. \tag{25}$$

Note that since here $\mathbf{a}$ is one–dimensional only, there is no difference between the estimate $\hat{a}$ and the estimate a^* according to (8).

Another more natural application of the latter model to geographic health data was incorporated into the CARLOS–project recently (**Ca**ncer **R**egistry **Lo**wer **S**axony; see Appelrath et al. (1993)). The establishment of a country cancer registry, based on a homogeneous, area–wide, population–based registration of cancer cases, is pushed by the social ministry of Lower Saxony (FRG) since 1992. CARLOS is the respective project for the Weser–Ems region, providing also statistical tests besides regional information and the calculation of epidemiological indices. Questions whether for instance the presence of nuclear power plants rises the risk of childhood Leukaemia are usually discussed controversely (see Gardner (1993) for a recent U.K. study). The simple index–of–dispersion test cannot directly be applied here due to the different population structures in the various geographic regions so that adjustments have to be made in advance (see Kafadar and Tukey (1993) for an alternative approach). For instance, the weights w_i could be chosen proportional to the local population densities n_i, $i = 1, \ldots, m$, for the m local districts in the country. Another problem is that not all cancer cases are registered with the same registration probability p_i, $i = 1, \ldots, m$. This means that in the medical data actually a *thinned* (Poisson) process is observed, so that the weights w_i should be chosen proportional to $p_i n_i$. The problem of a reliable estimation of the p_i is, however, an unsolved problem until now, since it requires further considerations other than mere statistical investigations.

The multi–parameter regression approach as in (1) might likewise be fruitful for the analysis of cancer data, for instance, if it is reasonable to consider age classes or other personal or social factors.

On the other hand, local influences on health data could be tested by this approach as well, by increasing the values of the regression functions over the corresponding areas. Experience with such kind of statistical procedures is presently in progress.

Acknowledgements:
Financial support was given by the Ministry of Research and Technology, FRG, within the research project "Ökosystemforschung Niedersächsisches Wattenmeer, Hauptphase B (ELAWAT)". The contribution of H. Ortleb was, in particular, financed by Terramare e.V., Wilhelmshaven.

References:

APPELRATH, H.-J., BEHRENDS, H., JASPER, H., ORTLEB, H. ET AL. (1993): Endbericht der Projektgruppe "Aktive Informationssysteme", Bericht IS 15, Teil A. Fachbereich Informatik, Universität Oldenburg.

DIGGLE, P.J. (1983): *Statistical Analysis of Spatial Point Patterns.* Mathematics in Biology. Ac. Press, N.Y.

EKSCHMITT, K. (1993): Über die räumliche Verteilung von Bodentieren. Zur ökologischen Interpretation der Aggregation und zur Probenstatistik. Dissertation, Universität Bremen.

FAHRMEIR, L., AND HAMERLE, A. (Eds.) (1984): *Multivariate statistische Verfahren.* W. de Gruyter, Berlin.

GARDNER, M.J. (1993): Investigating childhood leukaemia rates around the Sellafield nuclear plant. *Int. Stat. Rev.* 61, 231 – 244.

GREIG–SMITH, P. (1983): *Quantitative Plant Ecology.* Studies in Ecology, Vol. 9. 3rd edition, Blackwell Scientific Publ., Oxford.

KAFADAR, K., AND TUKEY, J.W. (1993): U.S. cancer death rates: a simple adjustment for urbanization. *Int. Stat. Rev.* 61, 257 – 281.

KREBS, C.J. (1985): *Ecology.* The Experimental Analysis of Distribution and Abundance. 3rd edition, Harper & Row, N.Y.

PFEIFER, D., BÄUMER, H.–P., AND ALBRECHT, M. (1992): Spatial point processes and their applications to biology and ecology. *Modeling of Geo–Biosphere Processes* 1, 145 – 161.

PFEIFER, D., SCHLEIER–LANGER, U., AND BÄUMER, H.–P. (1994): The analysis of spatial data from marine ecosystems. To appear in: H.–H. Bock, W. Lenski, and M.M. Richter (eds.): *Information Systems and Data Analysis.* Prospects – Foundations – Applications. Studies in Classification, Data Analysis, and Knowledge Organization. Springer, N.Y.

REISE, K. (1985): *Tidal Flat Ecology.* An experimental approach to species interactions. Ecological Studies 54, Springer, N.Y.

Comparing Spatio–Temporal Patterns from Defaunization Experiments in Marine Ecology

R. Wilhelm, A. Tecklenborg[1]

[1] Institut für Zoophysiologie, Westfälische Wilhelms–Universität Münster, Hindenburgplatz 55, 48143 Münster

Summary: The time–space–relationship of the fauna of the zoobenthos is studied with regard to the possibility of its regeneration after disturbance. Recolonization of defaunated areas can easily be investigated in terms of single species as well as of the whole community. In contrast to this the forecast of the end of this development is very difficult: It is a problem of the comparison of complicated patterns.

1. Introduction

In the present paper we shall describe some problems that have arisen in thecontext of the research project "Ökosystemforschung Niedersächsisches Wattenmeer" while investigating the composition of zoobenthos communities on the Gröninger Plate, comparing spatial patterns before and after certain defaunization experiments.

The term "macrozoobenthos" covers animals of a defined size, living in the sediment or on the sediment of the Wadden Sea. Benthic fauna in general is characterized by a huge amount of biomass. Within the ecosystem "Wadden Sea" it serves mainly as a nutrient source for higher living beings (e.g., birds and fish). The species composition and population density of the zoobenthos varies in general over time, as well as its temporal settlement and spatial distributional pattern. The causal reasons for this are only vaguely known. Due to seasonal variations it can happen that abiotic factors (like storm tides and ice winters) and biotic factors (like mats of green algae) lead to a drastic reduction or even elimination of the whole benthic population. In this case we speak of a "natural defaunization" with a subsequent recolonization.

One aim of the forementioned research project is a precise description of those parameters which govern the emergence of typical structures of macrozoobenthos communities, on the basis of "artificial defaunization" experiments. These experiments are carried out in the eulitoral part of the Wadden Sea, a part that falls periodically dry during low tides. At first sight this region appears to be rather homogeneous; however, it is shaped by several gradients that play an important role for the communities ("resource patchiness"). Among them one finds general environment parameters like sediment characteristics (grain size, grain size distribution, water content), content of organic material, salinity, nutrient content and nutrient dispersal, and – most important – the oxygen availability. Further, climatic factors are important (wind, temperature, precipitation). Statements about resource patchiness in habitats are, however, both dependent on the observer (i.e. on the variables considered and the scale chosen) as well as on the organisms themselves (see Jax, Vareschi, Zauke (1991)).

Besides general environmental parameters and their seasonal variability, also biological interactions play an important role for the emergence of spatio–temporal distributional patterns, such as states of succession, occupation of niches, dominance structures, to mention some. Beukema (1989) has investigated the Netherlands' part of the Wadden Sea during 20 years with respect to the number of species as well as the total biomass. Due to climatic factors, he found different spatial densities of 29 particular species during the years. Stecher and Dörjes (1993) provided species lists for the summer / winter season at the sea–side of the island of Spiekeroog which allow for a classification of types according to life–strategic communities. In the eulitoral part of the Wadden Sea essentially three types of flats can be distinguished, with respect to the main sediment components silt (mud), sand and a mixture of both. In each of these parts different community structures and also different patterns of settlement and spatial dispersial are to be found. Artificial disturbances of these areas thus allow for the observation and investigation of factors which are essential for the emergence of zoobenthic communities. In this paper we briefly report some of the results obtained here, together with some mathematical discussions.

2. The Defaunization Experiments

In order to obtain suitable defaunated areas we covered rectangular areas of size 20 m^2 (4 m $\times$ 5 m) with a black foil of 0.8 mm thickness during the months of March and April of the year 1992. In the sand and mixed flats each two experimental fields were chosen. After 4 weeks the experimental fields were uncovered again; this time period was sufficient for an almost complete defaunization, even for the species of the benthic fauna that are euryoxic (i.e. can live for a longer time period with a reduced or no oxygen supply). Probes were taken subsequently within the fields according to a sampling plan radially from the outside to the interior, and also outside the fields for reference. (Due to lack of space, a more detailed biological discussion of these experiments will be published elsewhere.)

Some key questions to be answered by the experiments were, among others:
- Which species recolonize the defaunated areas, at what time scale?
- What are the spatial patterns of recolonization?
- Which role play the "life states" of the species (larvae, juvenile, adults)?
- How long do differences between experimental and reference sites occur?

The following two figures show the development of the number of species and the total number of individuals found during April (= month 4) to October (= month 10) in 1992. The data are split according to juvenile and adult individuals, for the defaunated (=internal) and non–defaunated (=external) sampling sites.

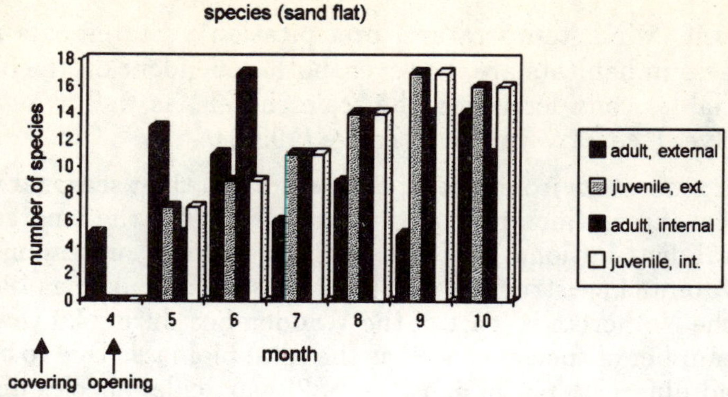

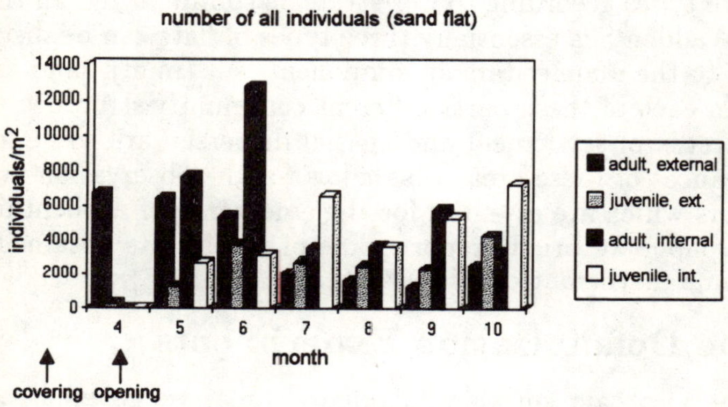

Although these figures suggest that the recolonization of the defaunated areas look similar, a more detailed investigation shows that there are great differences with respect to particular species. The following four figures show the development of numbers of individuals for the species *Corophium volutator*, *Pygospio elegans*, *Nephtys hombergii* and *Hydrobia ulvae*, for the sand flat areas:

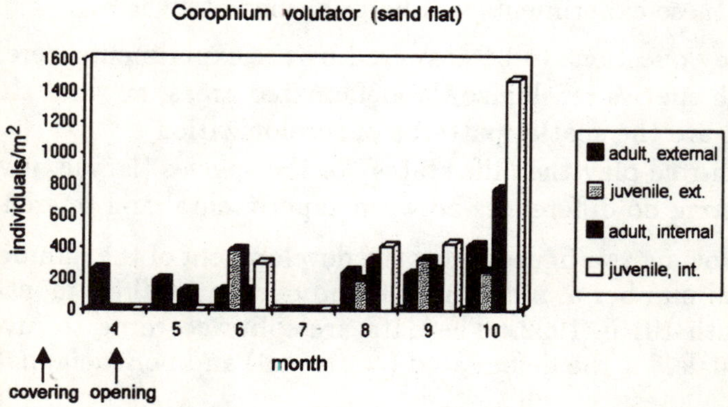

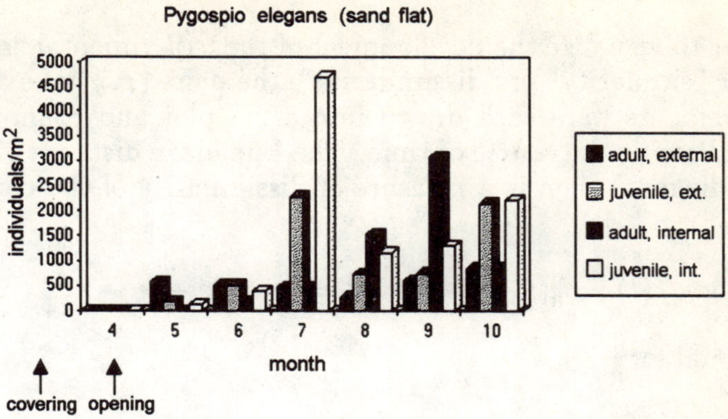

Pygospio elegans (sand flat)
individuals/m2
5000
4500
4000
3500
3000
2500
2000
1500
1000
500
0
4
5
6
7
8
9
10
month
adult, external
juvenile, ext.
adult, internal
juvenile, int.
covering
opening

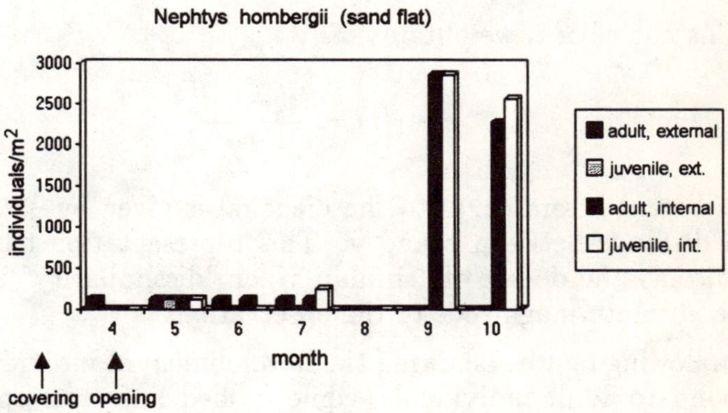

Nephtys hombergii (sand flat)
individuals/m2
3000
2500
2000
1500
1000
500
0
4
5
6
7
8
9
10
month
adult, external
juvenile, ext.
adult, internal
juvenile, int.
covering
opening

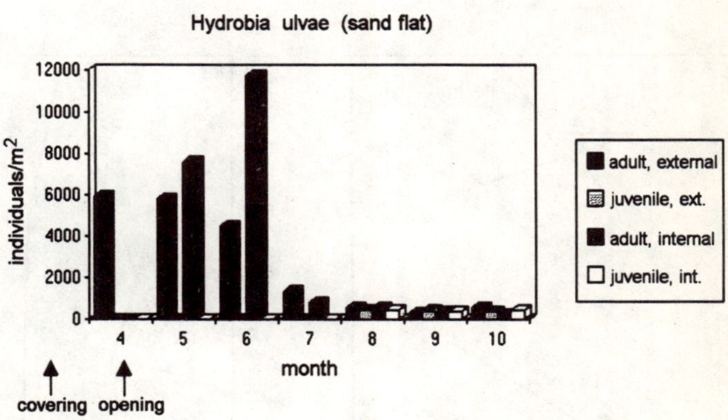

Hydrobia ulvae (sand flat)
individuals/m2
12000
10000
8000
6000
4000
2000
0
4
5
6
7
8
9
10
month
adult, external
juvenile, ext.
adult, internal
juvenile, int.
covering
opening

3. A Graphical Analysis

In order to visualize the development of the experimental fields with respect to their "similarity" or "dissimilarity", the pairs (x, y) of external / internal measurements were each drawn into a 2D–plot and connected by straight lines, following the course of time. The Euclidean distance of the point (x, y) to the diagonal then is a measure of dissimilarity of the components x and y since

$$f(t) := \|(x,y) - (t,t)\|^2 = (x-t)^2 + (y-t)^2, \quad t \in \mathbf{R}$$

is minimal for

$$f'(t) = 0 \quad \Longleftrightarrow \quad 2(x - t + y - t) = 0, \text{ i.e. } t = \frac{x+y}{2}.$$

With this choice of t, we obtain

$$f(t) = \frac{(x-y)^2}{2},$$

i.e. the distance from (x, y) to the diagonal is given by $\frac{1}{2}\sqrt{2}$ times the Euclidean distance between x and y. This representation has the advantage that not only the degree of "similarity" or "dissimilarity" is visualized but also the absolute magnitude of the observations.

In the following figures (showing the development of individuals), solid lines correspond to adult individuals, while dashed lines correspond to juvenile individuals.

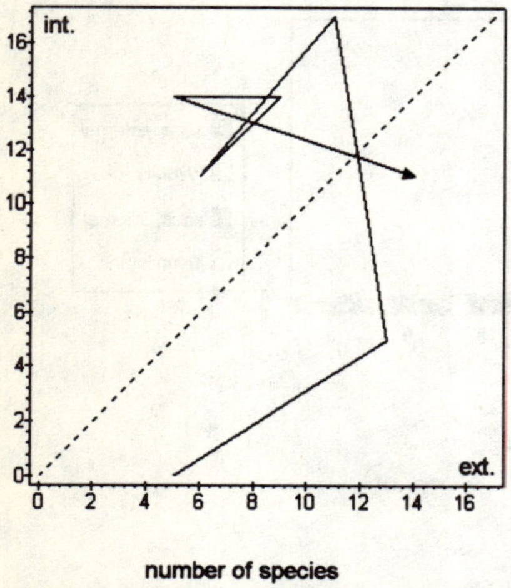

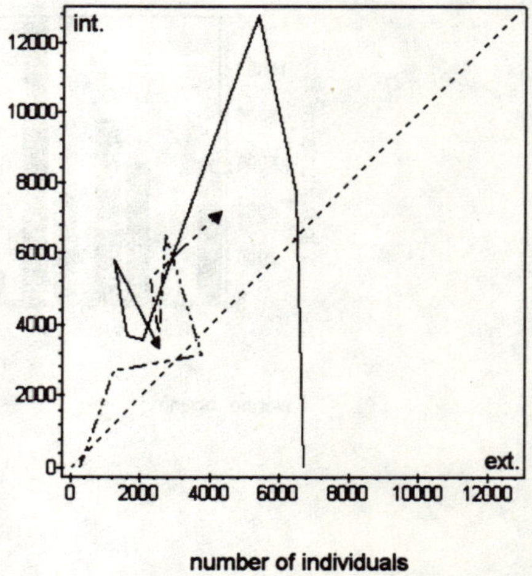

While these two figures show clearly that the defaunated and non–defaunated areas become more "similar" with respect to the number of species and the total number of individuals over time (cf. the corresponding remark above), this is no longer true for particular species. The following figures represent the development over time for the species *Corophium volutator*, *Pygospio elegans*, *Nephtys hombergii* and *Hydrobia ulvae*.

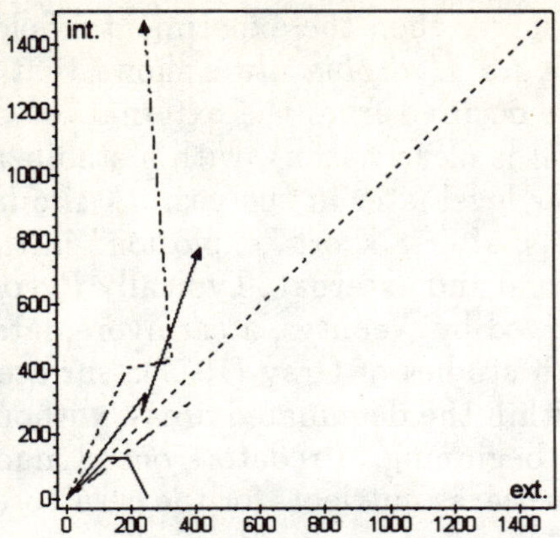

Corophium volutator

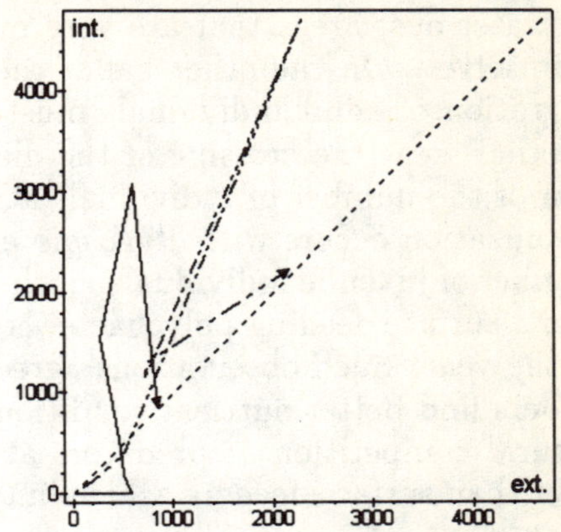

Pygospio elegans

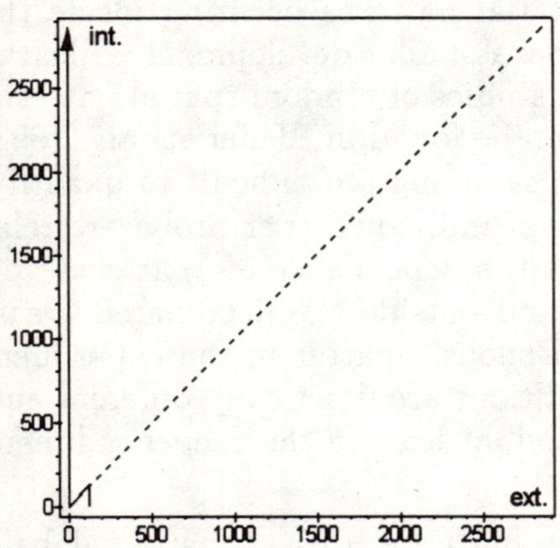

Nephtys hombergii

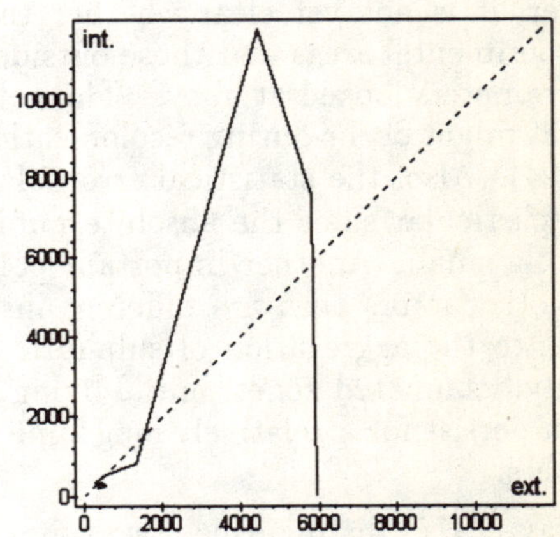

Hydrobia ulvae

4. Comments

At a first glance the figures above show clearly that the development over time w.r.t. adult and juvenile individuals is completely different. (Note that in the numbers of juvenile individuals of *Nephtys hombergii* and *Hydrobia ulvae* there is so little variation that the plot cannot give a reasonable picture for them.) The greatest differences occur with *Corophium volutator* where the defaunated area has eventually much more juvenile individuals than the reference area, and with *Nephtys hombergii* where the defaunated area has eventually much more adults (and also juveniles) than the reference area. Note also that the plots for the total number of individuals and the development of *Hydrobia ulvae* look quite similar, due to the fact that early in the year, *Hydrobia* is the most abundant species.

In general it can be said that the lack of a "real" (i.e. not merely statistical) difference between the defaunated and non–defaunated areas w.r.t. the juvenile individuals comes from the fact that the larvae of the species are driven by water over areas that are very much larger then the experimental fields themselves. On the other hand, the plot for *Hydrobia ulvae* shows that a migration of adult individuals must have occured from the external to the internal area (the crossing of the diagonal is clearly seen), with a stabilization of the number of individuals at a low level later in the year. A similar stabilization occurs with *Pygospio elegans*, after a short "explosion" in the number of juvenile individuals, both internal and external. Typically *Pygospio*, a surface–feeding polychaete, is followed by *Nephtys*, a carnivore, later in the year. Such observations agree with studies of Gray (1977): surface–feeders find better nutrient conditions within the defaunated areas, without natural competition or predators at the beginning. Predators occur much later than surface–feeders, which in turn serve as nutrients for the carnivores themselves.

The experiments that have been carried out so far during the "Pilotphase" of the ÖSF can thus answer some of the key questions posed above; however, it is not yet clear whether the spatial patterns occuring inside the experimental areas and those outside show a similar development – like the parameters looked at above. Theoretical studies of random spatial patterns that might occur during recolonization can be found in Pfeifer et. al. (1993, 1994). Also, the statistical errors due to sampling are difficult to quantify, in particular since the absolute numbers of individuals per probe are relatively small. Another important point is the dependence of "patterns" on abiotic factors that are different inside and outside the defaunated areas: due to the aggregation of sulphuric "poisonous" matter in the experimentally defaunated zones, initial living conditions are drastically different and can persist for a relatively long time period, at least in the deeper sediment layers.

Gray (1977) distinguishes two types of stability for ecosystems: a "global" and a "neighbourhood" stability; the latter is perhaps the most appropriate notion for the Wadden Sea. He considers a "species set", i.e. a certain dominance structure in the community. After a disturbance of local character all vectors tend to reach an equilibrium; the disturbed area hence represents a central attractor in this concept. However, Gray only considers a simple predator–prey system and a homogeneous area of a relatively large size of several hundred square meters. The situation is quite different for much smaller areas, say of one square meter only (cf. Gödde (1990)). The relative closeness of the reference area here prohibits the developments of spatial patterns in general. For the size of the experimental fields described in this paper (20 m^2), a particular development of specific spatial patterns, however, can be expected. It is a task for future experiments to find appropriate sampling designs that will be able to describe and analyze such spatial patterns. A further development of Gray's notion of a "species set" might be a solution.

Acknowledgements

Financial support was given by the Ministry of Research and Technology, FRG, within the research project "Ökosystemforschung Niedersächsisches Wattenmeer" (FZ 03F0023G). Special thanks are further given to the Statistics Group in the Mathematics Department of the Carl von Ossietzky Universität Oldenburg (D. Pfeifer, H.-P. Bäumer, H. Ortleb and U. Schleier) for statistical consulting and advice.

References:

BEUKEMA, J.J. (1989): Long-term changes in macrozoobenthic abundance of the western part of the Dutch Wadden Sea. *Helgol. wiss. Meeresunters.* 43, 405-415.

STECHER, J. AND DÖRJES,J. (1993): Das Makrozoobenthos im Vorfeld der Insel Spiekeroog. *Senckenbergiana* 23, 49-66.

GRAY, J.S. (1977): The stability of benthic ecosystems. *Helgol. wiss. Meeresunters.* 30, 427-444.

GÖDDE, F. (1990): Untersuchungen zur Wiederbesiedlung künstlich verödeteter Wattflächen. Schriftl. Hausarbeit im Rahmen der Ersten Staatsprüfung f.d. Lehramt Sek.II, Inst. f. Zoophysiologie der Univ. Münster.

JAX, K. , VARESCHI, E. AND ZAUKE, G. (1991): Entwicklung eines theoretischen Konzepts zur Ökosystemforschung Wattenmeer. Ökosystemforschung Wattenmeer, Teilvorhaben "Niedersächsisches Wattenmeer", Forschungsbericht: 108 02 085/02, Umweltforschungsplan des Bundesministers für Umwelt, Naturschutz und Reaktorsicherheit, Universität Oldenburg.

PFEIFER, D., BÄUMER, H.-P., AND ALBRECHT, M. (1993): Moving point patterns: the Poisson case. In: O. Opitz and B. Lausen (eds.): *Information and Classification: Concepts, Methods and Applications.* Studies in Classification, Data Analysis, and Knowledge Organization, 248 - 257. Springer, N.Y.

PFEIFER, D., BÄUMER, H.-P., AND SCHLEIER-LANGER, U. (1994): The analysis of spatial data in marine ecosystems. In: H.-H. Bock, W. Lenski, and M.M Richter (eds.): *Information Systems and Data Analysis.* Prospects - Foundations - Applications. Studies in Classification, Data Analysis, and Knowledge Organization, 340 - 349. Springer, N.Y.

Applications in Economics

A Unifying Approach to Benefit Segmentation and Product Line Design Based on Rank Order Conjoint Data

E. Aust, W. Gaul

Institut für Entscheidungstheorie und Unternehmensforschung,
Universität Karlsruhe, Postfach 6980, D-76128 Karlsruhe

Summary: Simultaneous part-worths estimation, benefit segmentation, repositioning of established products, and product line design can be achieved by estimating the parameters of a constrained latent class model. In an application concerning swimming pool design questions the output of the new approach is contrasted both with the results of traditional benefit segmentation and product line design modeling.

1. Introduction

Benefit segmentation (Gaul, Lutz, Aust (1994)) and product line design (Aust and Gaul (1994)) are major but so far quite distinct directions of research in conjoint analysis. A unifying approach with respect to both fields is proposed based on a constrained latent class model (Formann (1985)) with rank order input data. The basic idea is, that a benefit segment is (partly) characterized by its ideal product profile, obtained by segment-specific selection of the levels with maximum part-worth for every attribute in the conjoint design. This ideal product profile constitutes the first choice alternative for members of the specific segment independent of competitive offerings. It is an obvious assumption that every product in the market is an ideal product for at least a certain number of individuals. This knowledge can be put to use in a segmentation process by fixing ex-ante a number of segments each assigned to a specific product of the market. The part-worths of these segments can be constrained to ensure that the assigned product is the segment specific ideal product. In addition, a number of segments with unconstrained part-worths can be presupposed in which individuals can be grouped whose stated preferences cannot be reconciled with the preferences of the established product segments. The ideal products of these segments – if not yet available in the market – are proposed as the new products to be introduced, being both an optimal solution for the buyers' and for the share of choices' problem (Kohli and Sukumar (1990)). Individuals are assigned to segments according to the a-posteriori membership probability derived from their stated rank order. Thus, this unifying approach has the often desired property of probabilistic choice on an individual level. Part-worths estimation, segmentation, and (implicitly) product line design are performed simultaneously via an EM-Algorithm.

2. Methodology

The following notation is used:

i	=	$1,\ldots,I$ individuals,
k	=	$1,\ldots,K$ attributes,
j	=	$1,\ldots,J_k$ levels of attribute k,
l	=	$1,\ldots,F+R$ segments with F given segments (a-priori reserved for buyers of established products) and R additional segments (for the remaining individuals),
j'_{kl}	=	specific level j' of attribute k for segment l,
j^*_{kl}	=	most preferred level j^* of attribute k for segment l,
s	=	$1,\ldots,S$ (conjoint) profiles,
t	=	$(t_1,t_2,\ldots,t_S)$ rank order vector, where t_1 denotes the profile ranked at the first postion, t_2 the profile ranked at the second position, etc.,
x_{jks}	=	binary (conjoint) variable (equal to 1 if and only if profile s features attribute k at level j),
X	=	$((x_{jks}))$ (conjoint) design matrix,
y_{it}	=	binary individual rank order variable (equal to 1 if and only if individual i reports rank order vector t),
y_i	=	vector of y_{it} for individual i,
Y	=	$((y_{it}))$ matrix of y_{it},
u_{jkl}	=	part-worth with respect to level j of attribute k for segment l,
U_l	=	$((u_{jkl}))$ matrix of part-worths for segment l,
U	=	$(U_1,\ldots,U_{F+R})$ matrix of part-worths over all segments,
U_{t_sl}	=	deterministic utility with respect to profile t_s for segment l,
λ_l	=	a-priori membership probability for segment l,
λ	=	vector of λ_l.

The deterministic utility component with respect to profile t_s for segment l is modeled as additive main effects model

$$U_{t_sl} = \sum_{k=1}^{K}\sum_{j=1}^{J_k} u_{jkl}x_{jkt_s}. \tag{1}$$

Adding an *iid* Weibull error term to the U_{t_sl} and assuming that the ranking process can be decomposed into $S-1$ independent choices employing the rank explosion rule (Chapman and Staelin (1982), Ben-Akiva et al. (1992)) the probability that a member of segment l reports rank order vector t can be modeled by (Ogawa (1987))

$$P_{t|l} = \prod_{s=1}^{S-1} \frac{\exp(U_{t_sl})}{\sum\limits_{s'=s}^{S}\exp(U_{t_{s'}l})}. \tag{2}$$

Thus, the likelihood of observing y_i is

$$L_{i|l}(y_i \mid X, U) = \prod_{t}\prod_{s=1}^{S-1} \frac{(\exp(U_{t_sl}))^{y_{it}}}{(\sum\limits_{s'=s}^{S}\exp(U_{t_{s'}l}))^{y_{it}}} \tag{3}$$

conditional on individual i belonging to segment l. The unconditional likelihood of observing y_i is given by

$$L_i(y_i \mid X, U, \lambda) = \sum_{l=1}^{F+R} \lambda_l L_{i|l}(y_i \mid X, U). \tag{4}$$

Within this framework the aim of traditional benefit segmentation is to obtain estimates $\hat{U}$ and $\hat{\lambda}$ of U and λ respectively s.t.

$$L(\hat{U}, \hat{\lambda} \mid X, Y) = \max_{U,\lambda} \prod_{i=1}^{I} \sum_{l=1}^{F+R} \lambda_l L_{i|l}(y_i \mid X, U) \tag{5}$$

$$\sum_{l=1}^{F+R} \lambda_l = 1 \tag{6}$$

$$\lambda_l \geq 0 \quad \forall l \tag{7}$$

$$u_{j'_{kl}kl} = 0 \quad \forall k, l \tag{8}$$

where constraints (6) and (7) ensure that the a-priori probabilities for segment membership are non-negative and sum to 1 while constraints (8) provide a standardization for the part-worths. Given estimates $\hat{U}$ and $\hat{\lambda}$ for U and λ one obtains a fuzzy classification of the individuals into $F + R$ segments by calculating the a-posteriori probability

$$\hat{p}_{il} = \frac{\hat{\lambda}_l L_{i|l}(y_i \mid X, \hat{U})}{\sum_{l=1}^{F+R} \hat{\lambda}_l L_{i|l}(y_i \mid X, \hat{U})} \tag{9}$$

for segment membership. If the approach is to be used for simultaneous benefit segmentation and product line design it is assumed that $j'_{kl}=j^*_{kl}$, $\forall k$ and $l \leq F$. For given $\varepsilon > 0$ the part-worths of non ideal levels are constrained to be negative for all attributes

$$u_{jkl} \leq -\varepsilon \quad j \neq j'_{kl}, \ \forall\, k \text{ and } l \leq F \tag{10}$$

where $K\varepsilon$ is the minimum utility difference necessary for a member of an established product segment to change to a new product. Simultaneous repositioning of a number of established products $l \leq F$ can be integrated by enforcing (10) only for a subset of the attributes

$$u_{jkl} \leq -\varepsilon \quad j \neq j'_{kl}, \ k \in \Omega_l \text{ and } l \leq F \tag{10'}$$

where $\Omega_l \subseteq \{1, \ldots, K\}$ is the subset of attributes that may not be changed during the repositioning of product l.

3. Estimation Within an EM-Framework

To allow for a solution via an EM-Algorithm unobserved data have to be introduced for the unknown segment membership of the individuals. Let be:

- z_{il} = unobserved binary segment membership variable (equal to 1 if and only if individual i belongs to segment l),
- z_i = $(z_{i1}, \ldots, z_{i(F+R)})'$ unobserved individual segment membership vector,
- Z = $(z_i, \ldots, z_I)$ unobserved segment membership matrix.

Assuming that segment membership is *iid* multinomial according to λ, the likelihood of individual i to be in the segment determined by z_i is

$$L_i(z_i \mid \lambda) = \prod_{l=1}^{F+R} \lambda_l^{z_{il}}. \tag{11}$$

Thus, the likelihood that y_i and z_i jointly occur is

$$L_i(y_i, z_i \mid X, U, \lambda) = \prod_{l=1}^{F+R} \lambda_l^{z_{il}} L_{i|l}(y_i \mid X, U)^{z_{il}} \tag{12}$$

which is consistent with (4). Given the likelihood for all individuals of the set I

$$L_I(Y, Z \mid X, U, \lambda) = \prod_{i}^{I} L_i(y_i, z_i \mid X, U, \lambda) \tag{13}$$

the ln likelihood

$$\ln(L_I(Y, Z \mid X, U, \lambda)) = \sum_{i=1}^{I} \sum_{l=1}^{F+R} z_{il} \ln(L_{i|l}(y_i \mid X, U)) + \sum_{i=1}^{I} \sum_{l=1}^{F+R} z_{il} \ln(\lambda_l) \tag{14}$$

can easily be derived. Additionally, from formula (12) the conditional distribution of z_{il}

$$f(z_{il} \mid X, y_i, U, \lambda) = \frac{\prod_{l=1}^{F+R} (\lambda_l L_{i|l}(y_i \mid X, U))^{z_{il}}}{\sum_{l=1}^{F+R} \lambda_l L_{i|l}(y_i \mid X, U)} \tag{15}$$

can be obtained by employing Bayes' rule. The conditional expectancy of z_{il}

$$p_{il}(U, \lambda) := E(z_{il} \mid X, y_i, U, \lambda) = \frac{\lambda_l L_{i|l}(y_i \mid X, U)}{\sum_{l=1}^{F+R} \lambda_l L_{i|l}(y_i \mid X, U)} \tag{16}$$

is of crucial importance for the E-Step of the EM-Algorithm.

The **E-Step** is used for the evaluation of the expectation of $\ln(L_I(Y, Z \mid X, U, \lambda))$ over the conditional distribution of the unobserved data Z, based upon the observed rank order data Y, the design matrix X and provisional estimates $\hat{\lambda}$ and $\hat{U}$. As this expectation is calculated with respect to the conditional distribution of Z we denote it by E_Z. Using (16) the following expression can be obtained

$$E_Z(\ln(L_I(Y, Z \mid X, U, \lambda)) \mid \hat{U}, \hat{\lambda}) =$$
$$\sum_{i=1}^{I} \sum_{l=1}^{F+R} p_{il}(\hat{U}, \hat{\lambda}) \ln(L_{i|l}(y_i \mid X, U)) + \sum_{i=1}^{I} \sum_{l=1}^{F+R} p_{il}(\hat{U}, \hat{\lambda}) \ln(\lambda_l) \,. \tag{17}$$

Thus, the E-Step basically amounts to replacing z_{il} by its conditional expectancy.

The **M-Step** is dedicated to the maximization of $E_Z(\ln(L_I(Y, Z \mid X, U, \lambda)) \mid \hat{U}, \hat{\lambda})$ with respect to λ and U subject to the constraints (6), (7), (8), and (10). For abbreviation, the resulting nonlinear constrained optimization problem is denoted as MSP(M-Step Problem) in the following. We introduce the augmented transformed ln likelihood function

$$\begin{aligned}\tilde{\Phi} := \sum_{i=1}^{I}\Bigg[&\sum_{l=1}^{F} p_{il}(\hat{U},\hat{\lambda}) \sum_t y_{it} \sum_{s=1}^{S-1} \Bigg\{ \Bigg(\sum_{k=1}^{K} \sum_{j \neq j^*_{kl}} (-q^2_{jkl} - \varepsilon) x_{jkt_s} \Bigg) - \\ &\ln \Bigg(\sum_{s'=s}^{S} \exp(\sum_{k=1}^{K} \sum_{j \neq j^*_{kl}} (-q^2_{jkl} - \varepsilon) x_{jkt_{s'}}) \Bigg) \Bigg\} + \\ &\sum_{l=F+1}^{F+R} p_{il}(\hat{U},\hat{\lambda}) \sum_t y_{it} \sum_{s=1}^{S-1} \Bigg\{ (\sum_{k=1}^{K} \sum_{j \neq j'_{kl}} u_{jkl} x_{jkt_s}) - \\ &\ln(\sum_{s'=s}^{S} \exp(\sum_{k=1}^{K} \sum_{j \neq j'_{kl}} u_{jkl} x_{jkt_{s'}})) \Bigg\} + \sum_{l=1}^{F+R} p_{il}(\hat{U},\hat{\lambda}) \ln(\lambda_l) \Bigg] - v(\sum_{l=1}^{F+R} \lambda_l - 1)\end{aligned} \tag{18}$$

that features a Lagrange Multiplier for the sum constraint (6) on λ and in which the constrained independent variables u_{jkl}, $j \neq j^*_{kl}$, $\forall k$ and $l \leq F$, are substituted by $-q^2_{jkl} - \varepsilon$ to account for constraints of type (10). This kind of problem transformation allows for a solution of MSP without resorting to constrained nonlinear programming techniques and has been shown to be numerically efficient (Box (1966)). Summing the stationary equation of $\tilde{\Phi}$ with respect to v over l using also the stationary equation of $\tilde{\Phi}$ with respect to λ yields $v = I$. Thus, new estimates

$$\hat{\lambda}'_l = \sum_{i=1}^{I} p_{il}(\hat{U},\hat{\lambda})/I \qquad \forall l \tag{19}$$

can be obtained for λ_l, $\forall l$. Using the remaining stationary equations of $\tilde{\Phi}$ combined with nonlinear programming techniques one may obtain a set of parameter estimates $\hat{q}^*_{jkl}$, $j \neq j^*_{kl}, \forall k$ and $l \leq F$, and $\hat{u}^*_{jkl}$, $j \neq j'_{kl}, \forall k$ and $l > F$, that can readily be transformed back to yield new parameter estimates

$$\hat{u}'_{jkl} = -(\hat{q}^*_{jkl})^2 - \varepsilon \qquad j \neq j^*_{kl}, \forall k \text{ and } l \leq F \tag{20}$$

$$\hat{u}'_{jkl} = \hat{u}^*_{jkl} \qquad j \neq j'_{kl}, \forall k \text{ and } l > F \tag{21}$$

$$\hat{u}'_{j'_{kl}kl} = 0 \qquad \forall k, l \tag{22}$$

for the original MSP. At least locally optimal parameter estimates can be obtained within a finite number of iterations by alternating E- and M-Steps starting, e.g., from a random solution (Dempster et al. (1977)).

One of the least satisfactorily solved problems in finite mixture estimation problems is the determination of the number of components, i.e., benefit segments in the present case. Very common is a heuristic approach (Ramaswamy et al. (1993)) based on an information criterion, e.g., the selection of the model with minimum CAIC-value, that may be calculated as

$$\text{CAIC}=2L(\hat{\lambda},\hat{U} \mid X,Y) + ((F+R)(\sum_{k=1}^{K}(J_k - 1) + 1) - 1)(\ln((S-1)I) + 1) \tag{23}$$

for the basic model. Alternative procedures for model selection are discussed by McLachlan and Basford (1988).

4. Application

Data are taken from a recent survey with respect to consumers' preferences on public baths. 4 attributes for the description of swimming pools were considered: shape (square, round, curved), depth (no swimmer, sloping, swimmer), waves (yes, no), and water (normal, thermal). The stimuli consisted of 12 peeling stickers with full-profile descriptions illustrated in a way as partly displayed in Fig. 1. A subset of 9 profiles constituted a reduced orthogonal main effects design, whereas the 3 remaining ones served as holdout profiles. By asking for successive first choices, finally, a rank order consistent with the assumptions of the rank explosion rule was obtained. In total, 120 individuals responded to the survey.

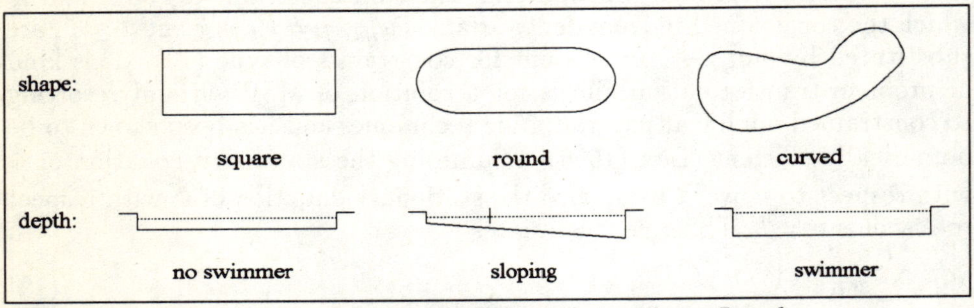

Fig. 1: Illustrations of Selected Attribute Levels

Pool	Profile (shape, depth, water, waves)
Rheinstrandbad 1	(square, sloping, normal, yes)
Rheinstrandbad 2 and 3, Vierordtbad 1, Ehrmannbad 1	(square, sloping, normal, no)
Fächerbad 1	(round, no swimmer, normal, no)
Fächerbad 2, Tullabad 1	(square, swimmer, normal, no)

Tab. 1: Status Quo Market of Public Baths in Karlsruhe

Tab. 1 gives the status quo market of public baths in Karlsruhe. Currently, there are 4 different types of pools available. The CAIC-values indicate an 8-segments solution for the unifying approach as given in Fig. 2. 63% of the considered individuals are adequately supplied for with the current line of pools. For the 4 remaining segments a substantial increase in welfare could be obtained by introducing their ideal products marked with black (standardized) part-worths bars. Especially striking is segment 8 for which the best status quo market product (round, no swimmer, normal, no) has only 27% of the utility of its ideal product. It is interesting to observe that the largest established product segment (segment 2) is assigned to the product most frequently featured in the status quo market which underscores the strong interdependence between status quo market products and consumers' preferences. Although building on quite different modeling assumptions the PROLIN (Aust and Gaul (1994) with $\alpha = 2$, $R = 2$) approach on the basis of MONANOVA individual level part-worths estimates proposes the same product line for introduction in the case of welfare maximization. The solution has high face validity as in Wörth, a community boardering on Karlsruhe, a public bath successfully operates which features all of the proposed pools but for the one with thermal water.

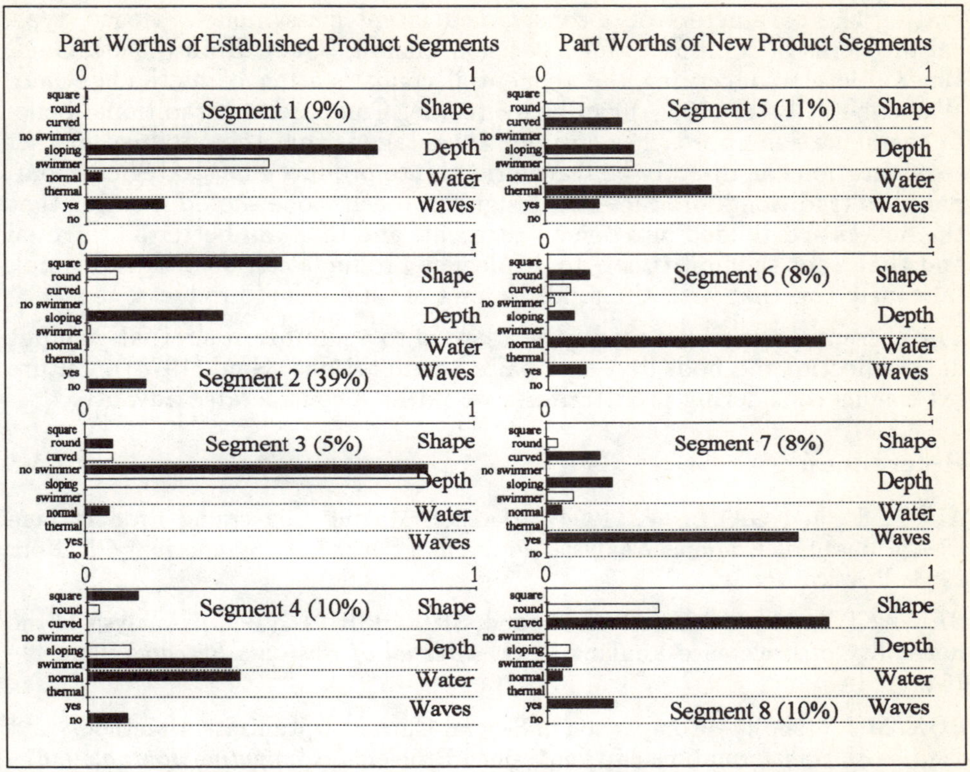

Fig. 2: 8-Segments Solution of the Unifying Approach

An important criterion to assess any benefit segmentation should be its predictive validity (Tab. 2). In the present case the 8-segments solution of the unifying approach achieves a higher first choice hit ratio than likewise the 7-segments solution of a traditional WARD post hoc benefit segmentation (Green and Krieger (1991)) as indicated by the elbow criterion (not displayed explicitly) and the 7-segments solution of the unconstrained latent class model as indicated by the CAIC-values.

Segmentation	Measure of Fit	$F+R$=6	$F+R$=7	$F+R$=8	$F+R$=9
Unifying Approach Solution	CAIC	3021.91	3008.14	2995.00	3006.67
	First Choice Hit Ratio	57.60%	74.75%	93.30%	94.56%
Ward Solution	CAIC	3404.10	3521.81	3577.14	3703.15
	First Choice Hit Ratio	59.27%	73.22%	80.20%	80.20%
Unconstrained L.C. Solution	CAIC	3006.03	2990.55	2993.08	3005.06
	First Choice Hit Ratio	73.22%	90.65%	94.28%	89.26%

Tab. 2: Measures of Fit for Varying Numbers of Segments

5. Conclusions and Outlook

Simultaneous part-worths estimation, benefit segmentation, repositioning of established products and product line design can be achieved by estimating the parameters of a constrained latent class model. Hence, typical problems of computer aided decision making (Gaul, Both (1990), Gaul, Baier (1994)) concerning the sequential concatenation of methods (Gaul, Wartenberg, Baier (1994)) can be overcome. Compared to traditional benefit segmentation approaches a major advantage is that the obtained market segmentation can directly be transformed into product line strategies. Compared to traditional product line design approaches one should mention that the buyers are divided into benefit segments and thus can better be targeted and that – for the first time – multiple new products can be designed simultaneously together with the repositioning of established products.

The present research can readily be adapted to a number of alternative input data collection methods like, e.g., paired comparisons (Gaul (1978)). Future extensions considering profit criteria would be a considerable advance.

References:

AUST, E., and GAUL, W. (1994): Decision Making Concerning Product Line Design Based on Conjoint Analysis. *Proceedings of the 18. Symposium OR*, Köln 1993. Physica Verlag.

BEN-AKIVA, M., MORIKAWA, T., and, SHIROISHI, F. (1992): Analysis of the Reliability of Preference Ranking Data. *Journal of Business Research,* 24, 149–164.

BOX, M.J. (1966): A Comparison of Several Current Optimization Methods, and the Use of Transformations in Constrained Problems. *Computing Journal,* 9, 67–77.

CHAPMAN, R.G., and STAELIN, R. (1982): Exploiting Rank Order Choice Set Data Within the Stochastic Utility Model. *Journal of Marketing Research,* 19, 288–301.

DEMPSTER, A.P., LAIRD, N.M., and RUBIN, D.B. (1977): Maximum Likelihood from Incomplete Data via the EM Algorithm. *Journal of the Royal Statistical Society,* B39, 1–38.

FORMANN, A.K. (1985): Constrained Latent Class Models: Theory and Applications. *British Journal of Mathematical and Statistical Psychology,* 38, 87–111.

GAUL, W. (1978): Zur Methode der paarweisen Vergleiche und ihrer Anwendung im Marketingbereich. *Methods of Operations Research,* 35, 123–139.

GAUL, W., and BAIER, D. (1994): Marktforschung und Marketing Management. Oldenbourg Verlag. München, 2. Auflage.

GAUL, W., and BOTH, M. (1990): Computergestütztes Marketing. Springer, Berlin, Heidelberg, New York, etc.

GAUL, W., LUTZ, U., and AUST, E. (1994): Goodwill Towards Domestic Products as Segmentation Criterion: An Empirical Study Within the Scope of Research on Country-of-Origin Effects. *Studies in Classification, Data Analysis, and*

Knowledge Organization, 4, 421–430.

GAUL, W., WARTENBERG, F., and BAIER, D. (1994): Comparing Proposals for the Solution of Data Analysis Problems in a Knowledge-Based-System. *Annals of OR,* 52, 131–150.

GREEN, P.E., and KRIEGER, A.M. (1991): Segmenting Markets with Conjoint Analysis. *Journal of Marketing,* 55, 20–31.

KOHLI, R., and SUKUMAR, R. (1990): Heuristics for Product-Line Design Using Conjoint Analysis. *Management Science,* 36, 12, 1464–1478.

MCLACHLAN, G.J., and BASFORD, K.E. (1988): Mixture Models: Inference and Applications to Clustering. Marcel Dekker, New York.

OGAWA, K. (1987): An Approach to Simultaneous Estimation and Segmentation in Conjoint Analysis. *Marketing Science,* 6, 1, 66–81.

RAMASWAMY, V., DESARBO, W., REIBSTEIN, D.J., and ROBINSON, W.T. (1993): An Empirical Pooling Approach for Estimating Marketing Mix Elasticities With PIMS Data. *Marketing Science,* 12, 1, 103–124.

Classification and Representation Using Conjoint Data

D. Baier, W. Gaul

Institut für Entscheidungstheorie und Unternehmensforschung, Universität Karlsruhe (TH), Postfach 6980, D-76128 Karlsruhe

Summary: We present new approaches to the analysis of conjoint data. One part of this paper deals with classification, another with representation issues. Both parts start with an overview of available approaches and then introduce new approaches. A real-world application concerning the introduction of a new product in the European air freight market shows advantages of the presented approaches.

1. Introduction

Conjoint analysis is the label attached to a research tool for measuring subjects' tradeoffs among competing objects via rank order or rating scale responses to constructed multiattribute stimuli (see, e.g., Green, Srinivasan (1990)). Surveys on the commercial use of conjoint analysis in the United States (Wittink, Cattin, (1989)) or in Europe (Wittink, Vriens, Burhenne (1994)) indicate that, since the first papers on the applicability of this methodology to marketing problems (see, e.g., Green, Rao, (1971)), conjoint analysis has become a popular research tool within many applications.

Contrary to early definitions of conjoint analysis and contrary to commercial usage reported in earlier surveys where nonmetric procedures like LINMAP or MONANOVA were preferred, more recent surveys show that metric procedures like OLS (Ordinary Least Squares) applied to rank order or rating scale responses are most frequently named. As OLS provides similar results to nonmetric procedures when applied to rank order responses (see, e.g., Green, Srinivasan (1978) for respective references) and the part-worth approach is commonly used for preference modeling (Green, Srinivasan (1990)), the following presentation will be based on part-worth estimation using OLS:

Let be i an index for N respondents, j an index for n stimuli, v an index for V attributes, and w an index for W_v levels of the v-th attribute. With this notation, typical conjoint data are (binary) profile data $B_{111}, \ldots, B_{nVW_V}$ (where B_{jvw} indicates whether stimulus j has level w for attribute v (=1) or not (=0)) and response data $y_{11}, \ldots, y_{nN}$ (where y_{ji} describes the observed preference value for stimulus j obtained from respondent i). Model parameters are the respondents' part-worths $u_{111}, \ldots, u_{NVW_V}$, which are estimated in such a way that the least squares loss function

$$Z = \sum_{i=1}^{N} \sum_{j=1}^{n} (y_{ji} - \hat{y}_{ji})^2 \quad \text{with} \quad \hat{y}_{ji} = \sum_{v=1}^{V} \sum_{w=1}^{W_v} B_{jvw} u_{ivw} \quad \forall j, i \tag{1}$$

is minimized. Because of

$$\sum_{w=1}^{W_v} B_{jvw} = 1 \quad \forall j, v \tag{2}$$

dummy-variable coding can be used for the design matrix $\dot{\mathbf{B}}$ with elements

$$\dot{B}_{jl} = \begin{cases} 1 & \text{if } l = 1 \\ B_{jv_0w_0} & \text{else, } \begin{array}{l} v_0 = \max\{v|l>1 + (W_1\text{-}1) + \cdots + (W_{v\text{-}1}\text{-}1)\}, \\ w_0 = l - (1 + (W_1\text{-}1) + \cdots + (W_{v_0\text{-}1}\text{-}1)) \end{array} \end{cases} \forall j, \tag{3}$$

which we obtain from the profile data by introducing one intercept column and omitting one superfluous level column for each attribute.

Since

$$\hat{y}_{ji} = \underbrace{\sum_{v=1}^{V} u_{ivW_v}}_{=:c_{1i}} + \sum_{v=1}^{V} \sum_{w=1}^{W_v-1} B_{jvw} \underbrace{(u_{ivw} - u_{ivW_v})}_{=:c_{(1+(W_1-1)+\cdots+(W_{v-1}-1)+w)i}} = \sum_{l=1}^{1+(W_1-1)+\cdots+(W_V-1)} \dot{B}_{jl}c_{li}, \quad \forall j, i \tag{4}$$

we get OLS estimates (Note that the existence of $(\dot{\mathbf{B}}'\dot{\mathbf{B}})^{-1}$ is assumed.)

$$\hat{\mathbf{Y}} = ((\hat{y}_{ji})) = \dot{\mathbf{B}}\mathbf{C} \quad \text{with} \quad \mathbf{C} = ((c_{li})) = (\dot{\mathbf{B}}'\dot{\mathbf{B}})^{-1}\dot{\mathbf{B}}'\mathbf{Y} \quad \text{and} \quad \mathbf{Y} = ((y_{ji})) \tag{5}$$

at the disaggregate level from which respondents' part-worths can easily be calculated according to

$$u_{ivw} = \begin{cases} \frac{c_{1i}}{V} + c_{(1+(W_1-1)+\cdots+(W_{v-1}-1)+w)i} & \text{if } w \neq W_v \\ \frac{c_{1i}}{V} & \text{else} \end{cases} \quad \forall i, v, w. \tag{6}$$

Some problems with conjoint analysis applications – even within a so-called average commercial study (n=16 stimuli, V=8 attributes, $W_1 = \ldots = W_V = 3$ levels (Wittink, Cattin (1989), Green, Srinivasan (1990))) – are as follows: Firstly, response data is observed and model parameters are estimated, both at the disaggregate level. Due to the usage of reduced designs for stimuli construction, the few degrees of freedom cause a problem. In the average commercial study mentioned, there are 16 observations (one for each stimulus) per respondent and 17 model parameters (the intercept and two coefficients for each attribute) per respondent resulting in overparametrization. Secondly, due to the few degrees of freedom, the response prediction for attribute-level-combinations not included in the data collection step may be insufficient and cause a so-called predictive accuracy problem. Thirdly, whereas tabular and graphical displays of the resulting part-worths may be appropriate for applications with few respondents, information overload problems occur when data from hundreds of respondents has to be analyzed. In these cases, modified forms of analyses or other graphical display forms – as discussed in the following parts – may be helpful.

2. Classification Using Conjoint Data

2.1 Overview

Various classification approaches have been proposed in order to prevent the above mentioned overparametrization and predictive accuracy problems with conjoint models at the disaggregate level by combining information across respondents. Moreover, these approaches have been used to derive so-called benefit segments (see, e.g., Green, Krieger (1991) for an overview).

At the moment, according to the already mentioned commercial application surveys, so-called sequential approaches seem to be most popular: Segments are formed either with or without usage of cluster analysis based on, e.g., respondents' background characteristics or part-worths estimated at the disaggregate level. Afterwards, segment-specific model parameters are estimated by aggregation or by group level procedures (see, e.g., Moore (1980) for an overview).

In the newer so-called simultaneous approaches, segmentation parameters and segment-specific model parameters are simultaneously estimated. Some of these procedures (e.g., Hagerty (1985), Kamakura (1988), DeSarbo, Oliver, Rangaswamy (1989), Wedel, Kistemaker (1989)) generalize known clusterwise regression procedures (Bock (1969), Späth (1983), DeSarbo, Cron (1988)) to conjoint analysis applications. The new clusterwise regression procedure, which will be presented in the next section, differs only with respect to response data description and parameter estimation from known approaches. Contrary to, e.g., Späth's model for one-mode one-way response data, two-mode two-way response data can be analyzed and the well-known iterative minimum-distance algorithm – instead of some exchange algorithm – is applied for parameter estimation.

2.2 Iterative Minimum-Distance Clusterwise Regression

2.2.1 The Model: We use the same notation as in the introduction, but add an index t for T segments or homogeneous groups of respondents. Input data are again the (binary) profile data $B_{111}, \ldots, B_{nVW_V}$ and the individual response data $y_{11}, \ldots, y_{nN}$.

Model parameters are now the segment membership indicators $h_{11}, \ldots, h_{TN}$, where h_{ti} denotes whether respondent i belongs to segment t (=1) or not (=0), and segment-specific part-worths $u_{111}, \ldots, u_{TVW_V}$. Again, we use a loss function as given in formula (1), but now the individual response estimates are replaced by respective segment-specific response estimates:

$$Z = \sum_{i=1}^{N} \sum_{j=1}^{n} (y_{ji} - \hat{y}_{ji})^2 = \sum_{i=1}^{N} \sum_{t=1}^{T} h_{ti} \sum_{j=1}^{n} (y_{ji} - u_{jt})^2 \rightarrow \min! \qquad (7)$$

with

$$\hat{y}_{ji} = \sum_{t=1}^{T} h_{ti} u_{jt} \quad \forall j, i, \qquad u_{jt} = \sum_{v=1}^{V} \sum_{w=1}^{W_v} B_{jvw} u_{tvw} \quad \forall j, t, \qquad (8)$$

$$h_{ti} \in \{0,1\} \quad \forall t,i, \quad \sum_{t=1}^{T} h_{ti} = 1 \quad \forall i, \quad \sum_{i=1}^{N} h_{ti} > 0 \quad \forall t, \tag{9}$$

where the segmentation schemes are restricted to nonoverlapping.

2.2.2 Parameter Estimation with Given Segmentation Matrix: For the proposed algorithm we use some computational simplifications concerning parameter estimation when response data $\mathbf{Y}$, the already mentioned design matrix $\dot{\mathbf{B}}$, and, additionally, a segmentation matrix $\mathbf{H}=((h_{ti}))$ are given (see also Hagerty (1985), Kamakura (1988)):

We get individual response estimates $\hat{\mathbf{Y}}=\mathbf{UH}$ and segment-specific response estimates $\mathbf{U}=\dot{\mathbf{B}}\mathbf{C}$ by weighting the OLS results obtained from (5):

$$\mathbf{C} = (\dot{\mathbf{B}}'\dot{\mathbf{B}})^{-1}\dot{\mathbf{B}}'\mathbf{Y}\,\underbrace{\mathbf{H}'(\mathbf{HH}')^{-1}}_{=:\,\mathbf{G}} \tag{10}$$

The elements of matrix $\mathbf{G}$ (the weights) can be easily computed via

$$g_{it} = \begin{cases} \frac{1}{N_t} & \text{if } h_{ti} = 1 \\ 0 & \text{else} \end{cases} \qquad \forall t,i \ \text{ with } \ N_t = \sum_{i=1}^{N} h_{ti} \quad \forall t. \tag{11}$$

2.2.3 The Algorithm: Our proposed iterative minimum-distance algorithm is given in Tab. 1: In the initialization phase we start with design matrix $\dot{\mathbf{B}}$ and an arbitrary segmentation matrix $\mathbf{H}$. Next, we estimate segment-specific response data $\mathbf{U}$ using OLS estimates based on our dummy-variable coding at the disaggregate level and the corresponding matrix of weights. Additionally, the initial loss function value is computed. In the two-step iteration phase we repeatedly reallocate respondents to segments and estimate segment-specific response data $\mathbf{U}$ in order to minimize the loss function until some stopping criterion is fulfilled. In the final phase, segment-specific part-worths are computed.

Empirical results obtained so far show that typical problems with the iterative minimum-distance algorithm – like, e.g., the reduction of class numbers (see, e.g., Späth (1983)) – are not relevant to this setting.

3. Representation Using Conjoint Data

3.1 Overview

As already mentioned in the introduction, problems sometimes occur with the presentation of conjoint analysis results. Here, the incorporation of MDS (MultiDimensional Scaling), or specifically multidimensional unfolding, to derive joint spaces – with acknowledged display power – may be useful.

Several approaches have already been proposed, incorporating, e.g., constrained multidimensional unfolding (see, e.g., DeSarbo, Rao (1986)) or multiple correspondence analysis (see, e.g., Green, Krieger, Carroll (1987)). The latter uses Green-Carroll-Schaffer-scaling to display results obtained by usage of conjoint analysis and choice simulation. Within other approaches (see, e.g., Carroll, Green, Kim (1989)), first MDS and then conjoint analysis is applied to conjoint data. The methodology discussed in the next section differs from the Carroll, Green, Kim (1989) approach with respect to the concrete models applied and the additional simulation/optimization phase.

{Initialization phase:}

Set $\dot{\mathbf{B}} := \begin{pmatrix} 1 & B_{111} & \cdots & B_{11(W_1-1)} & \cdots & B_{1V1} & \cdots & B_{1V(W_V-1)} \\ \vdots & \vdots & & \vdots & & \vdots & & \vdots \\ 1 & B_{n11} & \cdots & B_{n1(W_1-1)} & \cdots & B_{nV1} & \cdots & B_{nV(W_V-1)} \end{pmatrix}.$

Set $s := 0$. Choose an arbitrary segmentation matrix $\mathbf{H}^{(0)}$ and $\epsilon > 0$.

Set $\mathbf{U}^{(0)} := \dot{\mathbf{B}}(\dot{\mathbf{B}}'\dot{\mathbf{B}})^{-1}\dot{\mathbf{B}}'\mathbf{Y}\mathbf{H}^{(0)\prime}(\mathbf{H}^{(0)}\mathbf{H}^{(0)\prime})^{-1}.$

Set $Z^{(0)} := \sum_{i=1}^{N}\sum_{t=1}^{T} h_{ti}^{(0)} \sum_{j=1}^{n}(y_{ji} - u_{jt}^{(0)})^2.$

{Iteration phase:}

Repeat {Step 1 (Reallocation):}

Set $h_{ti}^{(s+1)} = \begin{cases} 1 & \text{if } \sum_{j=1}^{n}(y_{ji}\text{-}u_{jt}^{(s)})^2 = \min_{t'=1,\ldots,T}\{\sum_{j=1}^{n}(y_{ji}\text{-}u_{jt'}^{(s)})^2\} \\ 0 & \text{else} \end{cases} \quad \forall t, i.$

{Step 2 (Estimation):}

Set $\mathbf{U}^{(s+1)} := \dot{\mathbf{B}}(\dot{\mathbf{B}}'\dot{\mathbf{B}})^{-1}\dot{\mathbf{B}}'\mathbf{Y}\mathbf{H}^{(s+1)\prime}(\mathbf{H}^{(s+1)}\mathbf{H}^{(s+1)\prime})^{-1}.$

Set $Z^{(s+1)} := \sum_{i=1}^{N}\sum_{t=1}^{T} h_{ti}^{(s+1)} \sum_{j=1}^{n}(y_{ji} - u_{jt}^{(s+1)})^2$ and $s := s + 1$.

Until $Z^{(s)} - Z^{(s+1)} < \epsilon.$

{Final phase:}

Set $\mathbf{C} := (\dot{\mathbf{B}}'\dot{\mathbf{B}})^{-1}\dot{\mathbf{B}}'\mathbf{Y}\mathbf{H}^{(s)\prime}(\mathbf{H}^{(s)}\mathbf{H}^{(s)\prime})^{-1}.$

Set $u_{tvw} := \begin{cases} \frac{c_{1t}}{V} + c_{(1+(W_1-1)+\cdots+(W_{v'-1}-1)+w)t} & \text{if } w \neq W_v \\ \frac{c_{1t}}{V} & \text{else} \end{cases} \quad \forall t, v, w.$

Tab. 1: Iterative minimum-distance clusterwise regression: The algorithm

3.2 A Combined MDS/Conjoint Analysis Methodology

3.2.1 The Model: Again, the same notation with N respondents, n stimuli, V attributes and $W_1,\ldots,W_V$ levels is used. Additionally, we employ $\tilde{j}$ as an index for $\tilde{n}$ competing objects (e.g., brands, products) and p as an index for r space dimensions. Input data are the already mentioned profile data $B_{111}, \ldots, B_{nVW_V}$ and the response data $y_{11}, \ldots, y_{nN}$, but now we use, additionally, profile data $\tilde{B}_{111}, \ldots, \tilde{B}_{\tilde{n}VW_V}$ for the $\tilde{n}$ competing objects.

This time, model parameters are stimulus point coordinates $x_{11}, \ldots, x_{nr}$, respondents' ideal point coordinates $v_{11}, \ldots, v_{Nr}$, dimension-specific regression coefficients $\boldsymbol{\beta}_1, \ldots, \boldsymbol{\beta}_r$, and object point coordinates $\tilde{x}_{11}, \ldots, \tilde{x}_{\tilde{n}r}$.

3.2.2 The Algorithm: The algorithm distinguishes three phases, a MDS, a conjoint analysis, and a simulation/optimization phase as shown in Tab. 2.

{MDS phase:}
Set $d_{ijk} := \|y_{ji} - y_{ki}\| \quad \forall i,j,k$.
Estimate stimulus point coordinates $\mathbf{X}=((x_{jp}))$
    using (weighted) MDS based on $d_{111}, \ldots, d_{Nnn}$.
Estimate ideal point coordinates $\mathbf{V}=((v_{ip}))$
    using external multidimensional unfolding based on $\mathbf{X}$ and $\mathbf{Y}$.

{Conjoint analysis phase:}
Estimate regression coefficients $\boldsymbol{\beta}_1, \ldots, \boldsymbol{\beta}_r$ (with $\boldsymbol{\beta}_p = ((\beta_{pl}))$)
    using OLS via $\mathbf{X} = (\dot{\mathbf{B}}\boldsymbol{\beta}_1 \cdots \dot{\mathbf{B}}\boldsymbol{\beta}_r)$ based on $\mathbf{X}$ and $\dot{\mathbf{B}}$.
Estimate object point coordinates $\tilde{\mathbf{X}}=((\tilde{x}_{\tilde{j}p}))$
    using $\tilde{\mathbf{X}} := (\dot{\tilde{\mathbf{B}}}\boldsymbol{\beta}_1 \cdots \dot{\tilde{\mathbf{B}}}\boldsymbol{\beta}_r)$ based on $\boldsymbol{\beta}_1, \ldots, \boldsymbol{\beta}_r$, and $\dot{\tilde{\mathbf{B}}}=((\dot{\tilde{b}}_{\tilde{j}l}))$.

{Simulation/optimization phase:}
Estimate shares of choices or other aggregate response measures
    using choice simulators.
Find attribute-level-combinations
    maximizing share of choices or other aggregate response measures
    using optimal positioning methods.

Tab. 2: A combined MDS/conjoint analysis methodology: The algorithm

In the MDS phase, we estimate a joint space representation of the stimuli and the respondents applying the INDSCAL-model for MDS and the GENFOLD-model for external multidimensional unfolding based on the individual response data $\mathbf{Y}$. In the conjoint analysis phase, we regress the stimulus coordinates $\mathbf{X}$ on the dummy-variables of the design matrix $\dot{\mathbf{B}}$ in a first step. A second step is used to estimate object point coordinates $\tilde{\mathbf{X}}$ based on the estimated regression coefficients $\boldsymbol{\beta}_1, \ldots, \boldsymbol{\beta}_r$ and the objects' design matrix $\dot{\tilde{\mathbf{B}}}$. As a result of this second phase we have a joint space representation of respondents, stimuli, and competing objects. The last phase – the simulation and optimization phase – can now be used to predict shares of choices for the competing objects applying conventional choice simulators or to find suitable attribute-level-combinations for new or modified objects in the competitive context applying optimal positioning methods.

4. Application to the European Air Freight Market

A major European airline company planned the introduction of a new overnight parcel service concerning house-to-airport delivery in the European air freight market. A conjoint analysis application was used in order to analyze the preference structure of potential customers, to derive benefit segments, and to find suitable attribute-level-combinations for the new service. Pretests showed that the attributes 'collection time', 'agency type', 'price' (for a 10 kg parcel with European destination), 'transport control', and 'delivery time' should be considered (see also Baier (1994), Mengen (1993)).

In total, 150 people responsible for parcel delivery within German companies sending more than 25 air freight parcels per month within Europe were personally interviewed. Typical conjoint data was collected with respect to a reduced design with 18 stimuli as given in Tab. 3. Abbreviations of the attribute levels are used as stimulus short names, e.g., the short name '16C160A10' of the first stimulus indicates that attribute 'collection time' has level '16:30', attribute 'agency type' has level 'airline company', attribute 'price' has level '160 DM', attribute 'transport control' has level active, and attribute 'delivery time' has level '10:30'. Additionally, data on company characteristics, on return from attribute-level-combinations and attribute-levels from six competing services ('product A' to 'product F') was collected.

	attributes				
stimulus short name	collection time	agency type	price	transport control	delivery time
16C160A10	16:30	airline company	160DM	active	10:30
16C200P10	16:30	airline company	200DM	passive	10:30
16I200A13	16:30	integrator	200DM	active	13:30
16I240P13	16:30	integrator	240DM	passive	13:30
16S160A12	16:30	forwarding agency	160DM	active	12:00
16S240A12	16:30	forwarding agency	240DM	active	12:00
17C160A13	17:30	airline company	160DM	active	13:30
17C240A13	17:30	airline company	240DM	active	13:30
17I160P12	17:30	integrator	160DM	passive	12:00
17I200A12	17:30	integrator	200DM	active	12:00
17S200A10	17:30	forwarding agency	200DM	active	10:30
17S240P10	17:30	forwarding agency	240DM	passive	10:30
18C200P12	18:30	airline company	200DM	passive	12:00
18C240A12	18:30	airline company	240DM	active	12:00
18I160A10	18:30	integrator	160DM	active	10:30
18I240A10	18:30	integrator	240DM	active	10:30
18S160P13	18:30	forwarding agency	160DM	passive	13:30
18S200A13	18:30	forwarding agency	200DM	active	13:30

Tab. 3: Reduced design with 18 stimuli in the European air freight market

Application of the presented iterative minimum-distance clusterwise regression procedure for simultaneous segmentation and estimation led to the results shown in Tab. 4: For nearly half of the sample, i.e. segments 3 and 4 ('segm.3 (10.0%)', 'segm.4 (38.7%)') the attribute 'collection time' contributes mostly to overall response. Two segments focus on the attributes 'transport control' ('segm.2 (17.3%)') and 'price' ('segm.5 (14.7%)'). The five-segment solution was selected on basis of an elbow criterion with respect to the R^2-measure with values 0.2413, 0.4645, 0.5305, 0.5868, 0.6336, 0.6579, 0.6832 for the one-, two-, ..., seven-segment solutions (see Wedel, Kistemaker (1989) for a similar decision).

Next, the available conjoint data was analyzed by the presented combined MDS/conjoint analysis methodology. For space restrictions, we only discuss the results from the MDS phase and refer to Baier (1994) for results from the remaining two phases: A four-dimensional joint space representation with stimuli's points and respondents' ideal points was derived by application of

INDSCAL for MDS (correlation coefficient R=0.665) and GENFOLD for external multidimensional unfolding (correlation coefficient R=0.760). For comparisons the affiliation of respondents to the above derived five segments ('segm.1' to 'segm.5') is indicated by the corresponding segment numbers ('1' to '5') in Fig. 1. From the stimulus short names we can see that 'dimension1' can be interpreted as collection time dimension, 'dimension2' as price dimension, and 'dimension3' as transport control dimension, whereas 'dimension4' does not allow such an obvious interpretation. From the ideal point positions we recognize two groups of respondents, where the respondents of the larger group are in favour of earlier collection times and the respondents of the smaller group prefer lower prices.

attribute	level	segm.1 (19.3%)	segm.2 (17.3%)	segm.3 (10.0%)	segm.4 (38.7%)	segm.5 (14.7%)
collection time	16:30	0.210	0.000	0.000	0.901	0.100
	17:30	0.316	0.004	0.696	0.237	0.026
	18:30	0.000	0.022	0.717	0.000	0.000
agency type	airline company	0.000	0.118	0.026	0.000	0.128
	integrator	0.002	0.191	0.000	0.000	0.035
	forwarding agency	0.145	0.000	0.006	0.023	0.000
price	160DM	0.148	0.100	0.083	0.033	0.645
	200DM	0.062	0.038	0.049	0.015	0.166
	240DM	0.000	0.000	0.000	0.000	0.000
transport control	active	0.105	0.673	0.053	0.020	0.053
	passive	0.000	0.000	0.000	0.000	0.000
delivery time	10:30	0.286	0.000	0.099	0.022	0.074
	12:00	0.111	0.004	0.121	0.009	0.028
	13:30	0.000	0.022	0.000	0.000	0.000

Tab. 4: (Standardized) part-worths in the European air freight market

As we can see, both approaches try to overcome the three problems with respect to conjoint analysis applications as pointed out in the introduction: Compared to the traditional part-worth estimation at the disaggregate level, the number of model parameters was substantially reduced which leads to more degrees of freedom and – hopefully – to better predictive accuracy. (Note that, e.g., the predictive power of methods at the group level may be lower than that of methods at the disaggregate level, as shown by, e.g., Moore (1980).) Moreover, both approaches provide results in a form that can be easily communicated.

5. Conclusions

New ways of classification and representation using conjoint data offer advantages over traditional approaches with respect to various aspects like overparametrization, predictive power, and communication of results. In this paper, we have only been able to demonstrate some of these advantages within one application example (see Baier (1994) for a more detailed description). Further research on comparisons concerning simultaneous vs. sequential approaches is in preparation.

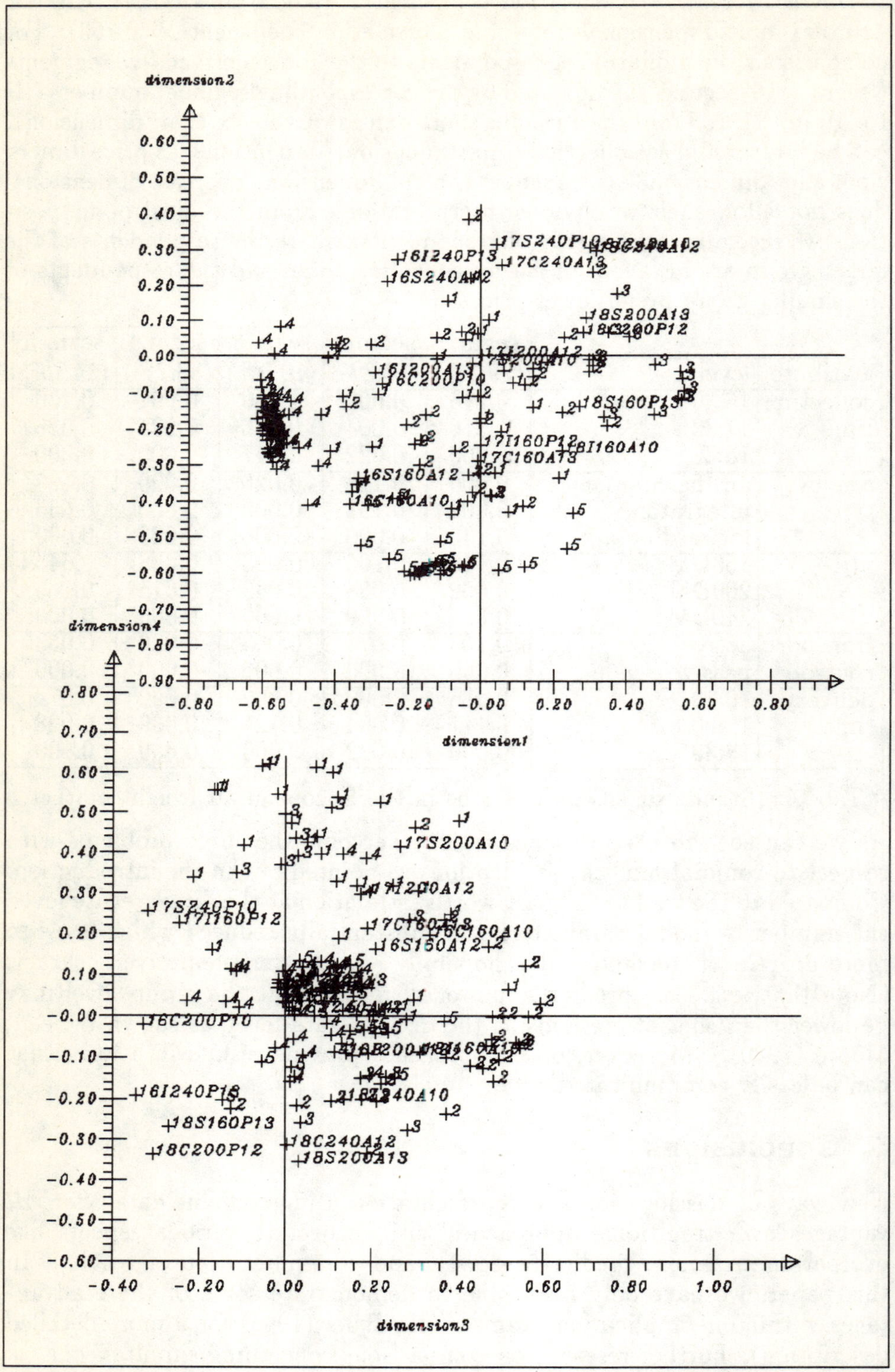

Fig. 1: Joint space representation in the European air freight market

References:

BAIER, D. (1994): Konzipierung und Realisierung einer Unterstützung des kombinierten Einsatzes von Methoden bei der Positionierungsanalyse. Lang, Frankfurt.

BOCK, H.H. (1969): The Equivalence of Two Extremal Problems and its Application to the Iterative Classification of Multivariate Data. In Report on the Conference "Medizinische Statistik", Forschungsinstitut Oberwolfach.

CARROLL, J.D., GREEN, P.E., and KIM, J. (1989): Preference Mapping of Conjoint-Based Profiles: An INDSCAL Approach. *Journal of the Academy of Marketing Science*, 17, 273-281.

DESARBO, W.S., and CRON, W.L. (1988): A Maximum Likelihood Methodology for Clusterwise Regression. *Journal of Classification*, 5, 249-282.

DESARBO, W.S., OLIVER, R., and RANGASWAMY, A. (1989): A Simulated Annealing Methodology for Clusterwise Linear Regression. *Psychometrika*, 54, 707-736.

DESARBO, W.S., and RAO, V.R. (1986): A Constrained Unfolding Methodology for Product Positioning. *Marketing Science*, 5, 1-19.

GREEN, P.E., and KRIEGER, A.M. (1991): Segmenting Markets With Conjoint Analysis. *Journal of Marketing*, 55, 20-31.

GREEN, P.E., KRIEGER, A.M., and CARROLL, J.D. (1987): Conjoint Analysis and Multidimensional Scaling: A Complementary Approach. *Journal of Advertising Research*, 21-26.

GREEN, P.E., and RAO, V.R. (1971): Conjoint Measurement for Quantifying Judgemental Data. *Journal of Marketing Research*, 8, 355-363.

GREEN, P.E., and SRINIVASAN, V. (1978): Conjoint Analysis in Consumer Research: Issues and Outlook. *Journal of Consumer Research*, 5, 103-123.

GREEN, P.E., and SRINIVASAN, V. (1990): Conjoint Analysis in Marketing: New Developments With Implications for Research and Practice. *Journal of Marketing*, 54, October, 3-19.

HAGERTY, M.R. (1985): Improving the Predictive Power of Conjoint Analysis: The Use of Factor Analysis and Cluster Analysis. *Journal of Marketing Research*, 22, 168-184.

KAMAKURA, W.A. (1988): A Least Squares Procedure for Benefit Segmentation with Conjoint Experiments. *Journal of Marketing Research*, 25, 157-167.

MENGEN, A. (1993): Konzeptgestaltung von Dienstleistungsprodukten. Schäffer-Poeschel, Stuttgart.

MOORE, W.L. (1980): Levels of Aggregation in Conjoint Analysis: An Experimental Comparison. *Journal of Marketing Research*, 17, 516-523.

SPÄTH, H. (1983): Cluster-Formation und -Analyse. Oldenbourg, München.

WEDEL, M., and KISTEMAKER, C. (1989): Consumer Benefit Segmentation Using Clusterwise Linear Regression. *International Journal of Research in Marketing*, 6, 45-49.

WITTINK, D.R., and CATTIN, P. (1989): Commercial Use of Conjoint Analysis: An Update. *Journal of Marketing*, 53, Summer, 91-96.

WITTINK, D.R., VRIENS, M., and BURHENNE, W. (1994): Commercial Use of Conjoint Analysis in Europe: Results and Critical Reflections. *International Journal of Research in Marketing*, 11, 41-52.

Overlapping Clustering of Statistical Software Packages for PC

Rainer Lasch

Institut für Statistik und Mathematische Wirtschaftstheorie,
Universität Augsburg, D-86135 Augsburg

Summary: Cluster analysis is specially concerned with algorithms for computing non- overlapping classifications, for example partitions or hierarchies, on given object sets. For several economic problems the determination of non-overlapping classifications representing the structure of data is too specific and narrow. In opposition to that given natural overlappings should not be suppressed because the construction of overlapping clusters gives a better insight into the structure of data. In this paper several principles of constructing overlapping clusters e.g. maximal cliques, fuzzy clustering, quasi-hierarchies and pyramidal classification are presented. The advantages and disadvantages of these clumping techniques are discussed in an overlapping clustering of selected software packages.

1. Clumping Techniques

Subsequently let $N = \{1, ..., n\}$ be a finite **set of objects** and $A = (a_{ik})_{n,m}$ a **data matrix**, with rows containing the values of all variables for the objects. In spite of different data levels a complete ordering $\preceq$ is often justified on pairs of objects presented by a **distance index** compatible to a given preordering

$$d : N \times N \longrightarrow \mathbb{R}_+ \text{ with } \begin{array}{ll} d_{ii} = 0 & \text{for all } i \in N \\ d_{ij} = d_{ji} & \text{for all } i,j \in N. \end{array} \tag{1}$$

1.1 Maximal c-Cliques

Several non-partitioning classification algorithms without fixing the number of classes are based on the principle of maximal c-cliques.

Definition 1.1

A subset $A \subset N$ is called a **maximal c-clique** *with $c \geq 0$, if*

$$d_{ij} \quad \leq \quad c \qquad \text{for all } i,j \in A, \tag{2}$$

$$\max_{i \in A} d_{ij} \quad > \quad c \text{ for all } j \in N - A. \tag{3}$$

This definition of a maximal c-clique uses only the complete preordering property of the distance index d. Therefore the diameter c of a class gives a reasonable measure of the within class dissimilarity for mixed data too. For an increasing c one gets more sizeable classes or cliques with more overlappings, for a decreasing c one gets more homogeneous classes with less overlappings.

The notation of a maximal c-clique allows also a graph-theoretic interpretation. Considering a complete graph G, the n objects $O_1, ..., O_n$ are defined as vertices and the dissimilarities d_{ij} are the weights of the edges of this graph. For a given threshold level c all edges with weights $d_{ij} > c$ are deleted from G. In this resulting partial graph $G(c)$ a maximal c-clique can be represented by a maximal complete subgraph G^S. A classification based on maximal c-cliques corresponds to the problem of finding all maximal complete subgraphs G^S of the partial graph $G(c)$ of a given graph G.

Example 1.2

Considering the following dissimilarity matrix for a given object set $N = \{1, ..., 5\}$

$$D = \begin{bmatrix} 0 & & & & \\ 4 & 0 & & & \\ 2 & 9 & 0 & & \\ 2 & 9 & 4 & 0 & \\ 4 & 2 & 7 & 7 & 0 \end{bmatrix}.$$

There exist two maximal 4-cliques $\{1,2,5\}, \{1,3,4\}$ and two maximal 7-cliques $\{1,2,5\}, \{1,3,4,5\}$.

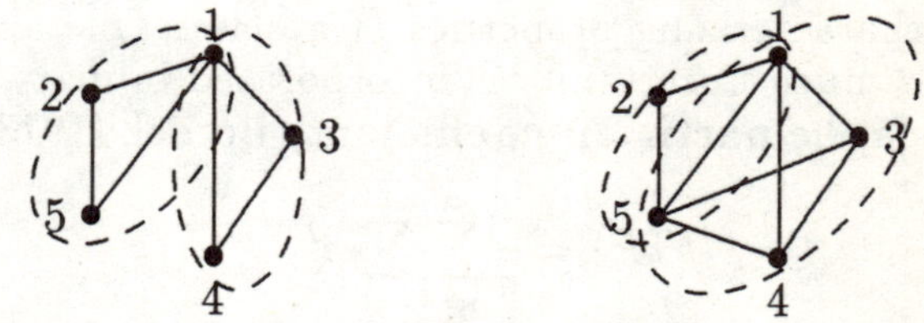

Fig. 1: The partial graphs $G(4)$ and $G(7)$ with its maximal cliques

1.2 Fuzzy Sets

All clustering techniques suffer from a common central problem: lack of a unique classification criteria whereby optimal clusters can be identified. The inherent non-uniqueness of clustering argue for adopting fuzziness, since fuzzy membership values incorporate a reflection of non-statistical uncertainty about structure in the data.

Definition 1.3

Let $\mathcal{K} = \{A_1, ..., A_s\}$ be a classification of the object set $N = \{1, ..., n\}$ in s non-empty classes.

a) *A class A_l characterized by the vector $(p_{1l}, ..., p_{nl})$, which consists of grades of membership p_{il}, is called* **fuzzy cluster**, *if*

$$p_{il} \geq 0, \tag{4}$$

$$\sum_{l=1}^{s} p_{il} = 1 \quad (i = 1, ..., n). \tag{5}$$

b) *Summarizing the vectors $A_l = (p_{1l}, ..., p_{nl})$ $(l = 1, ..., s)$ of all fuzzy clusters in a $n \times s$- matrix P, then P is called a* **fuzzy s-classification** *of N:*

$$P = \begin{array}{c} \\ \\ \end{array}\begin{array}{ccc} A_1 & \cdots & A_s \end{array}$$

$$P = \left[\begin{array}{ccc} p_{11} & \cdots & p_{1s} \\ \vdots & & \vdots \\ p_{n1} & \cdots & p_{ns} \end{array}\right]\begin{array}{c} 1 \\ \vdots \\ n \end{array}$$

Fuzziness allows an object in N the option of partial membership in two or more fuzzy clusters. An object i with a grade of membership $0.7 \leq p_{il} \leq 1$ is a typical representative (fuzzy center) of the cluster A_l. Objects with grades of membership $0.3 < p_{il} < 0.7$ belong to the intersection of two or more classes. An object i lying at the border of a class A_l has a grade of membership $0 \leq p_{il} \leq 0.3$.
It exists a philosophical distinction between probabilistic and fuzzy uncertainty. A grade of membership $p_{il} = 0.25$ should not indicate a belief that the likelihood of i belonging to A_l is 0.25; rather, it asserts that k shares about 0.25 of the qualities requisite for unequivocal membership in class A_l. Fuzzy uncertainty reflects how distinctive an object is, comparing to others closely related, in the characterizing properties of a class. One approach towards determining how much uncertainty is incorporated in fuzzy s- classification of N has been via the **partition coefficient** (Bezdek (1973)):

$$F(P) = \frac{1}{n}\sum_{i=1}^{n}\sum_{l=1}^{s} p_{il}^2 \tag{6}$$

The use of $F(P)$ in connection with clustering is exemplified in Bezdek (1974a,b), where it is shown that $F(P)$ satisfies

$$1/s \leq F(P) \leq 1 \tag{7}$$

$$F(P) = 1/s \iff p_{il} = 1/s \quad \text{for all} \quad i, l \tag{8}$$

$$F(P) = 1 \iff P \quad \text{is a hard partition.} \tag{9}$$

Values of $F(P)$ are inversely proportional to the average content of fuzzy intersections of P. $F(P)$ maximizes as unshared membership increases and pairwise fuzzy intersections reduce to minimal membership content. $F(P)$ thus affords a measure by which competing s-classifications of N can be compared as regard the amount of given uncertainty.

1.3 Quasi-Hierarchies Using Maximal c-Cliques

The construction of a classification based on maximal c-cliques or a fuzzy s-classification requires the a priori determination of a proper within class dissimilarity c or the number of classes s. Analyzing empirical data there exist more easy indicators for the number of classes that a nested sequence of overlapping classifications provides a better insight into the possible structure of data.

Definition 1.4

Let N be a finite set of objects and Q a set consisting of non-empty parts of N. Q is called a **quasi-hierarchy**, *if*

(1) $N \in Q$

(2) for all $i \in N : \{i\} \in \mathcal{Q}$

(3) For any element $A \in \mathcal{Q}$ $(|A| > 1)$ *there exist elements* $A_i \in \mathcal{Q}$, $A_i \subsetneqq A$ *with*

$$A := \bigcup_{A_i \subsetneqq A} .$$

Any class $A \in \mathcal{Q}$ of a quasi-hierarchy based on maximal c-cliques fulfills condition (2) and (3) of definition 1.1.

Example 1.5

A quasi-hierarchy allows a graphical representation as can be seen in Fig. 2.

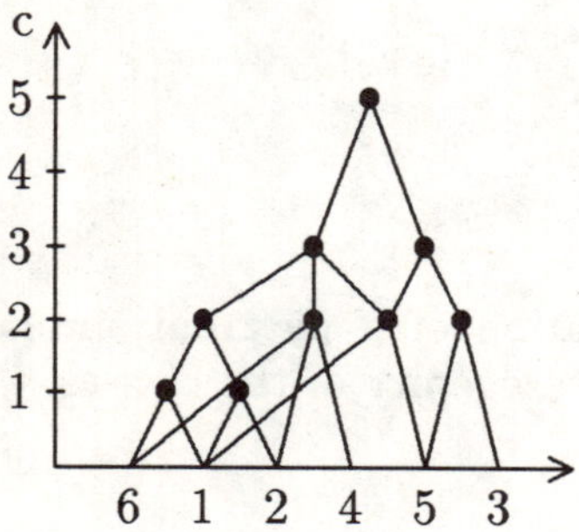

Q = { {1},{2},{3},{4},{5},{6},

{1,2},{1,6},{1,2,6},

{2,4,6},{1,5},{3,5},

{1,2,4,5,6},{1,3,5},N }

Fig. 2: Quasi-hierarchy of 6 objects

Each level $c > 0$ of the quasi-hierarchy represents a non-disjoint classification with many overlappings or nearly identical classes. Jardine and Sibson (1971) claim that the number of overlapping objects in each of two classes are restricted. The technique which they call $\mathcal{B}_k$ allows the overlap between clusters at the same threshold level c to contain not more than $k-1$ objects. At each threshold level $c > 0$ Jardine/Sibson construct a system $\mathcal{B}_k(c)$ called **k-partition** by fusing each of two classes which overlap in k or more objects. The sequence of systems $\mathcal{B}_k(c)$ can be given a graph-theoretic description. A graph is drawn whose edges join just those pairs of vertices which represent objects with dissimilarity $\leq c$. The maximal complete subgraphs are marked, and wherever the vertex sets of two such subgraphs intersect in at least k vertices further edges are drawn in to make the union of the two vertex sets into a complete subgraph. The process is repeated until there is no further alteration.

1.4 Pyramidal Classification

An other classification method generating only moderate overlappings is the pyramidal classification.

Definition 1.6

a) *Given a set* $A \subset N$ *let* i_{min}, $i_{max} \in A$ *be the minimal and maximal elements of* A *according to an order* Θ. *The set* A *is* **connex** *according to* Θ, *if*

$$(i \in N \textit{ is according to } \Theta \textit{ between } i_{min} \text{ and } i_{max}) \Longleftrightarrow (i \in A).$$

b) Let $\mathcal{S}$ be a set of non-empty parts of N. An order Θ is **compatible** *with $\mathcal{S}$, if all subsets $A \subset N$ ($A \in \mathcal{S}$) are connex according to Θ.*

Definition 1.7

Denote by $\mathcal{P}$ a set of non-empty parts of N which will be called a **pyramid,** *if*

(1) $N \in \mathcal{P}$

(2) $\forall\, i \in N : \{i\} \in \mathcal{P}$

(3) $\forall\, A, B \in \mathcal{P}\, (A \neq B): \quad A \cap B = \emptyset \quad \text{or} \quad A \cap B \in \mathcal{P}$

(4) There exists an order Θ compatible with $\mathcal{P}$.

Example 1.8

There exists a useful graphical representation of a pyramid similar to a dendrogram. The index f indicates the heterogenity of the classes.

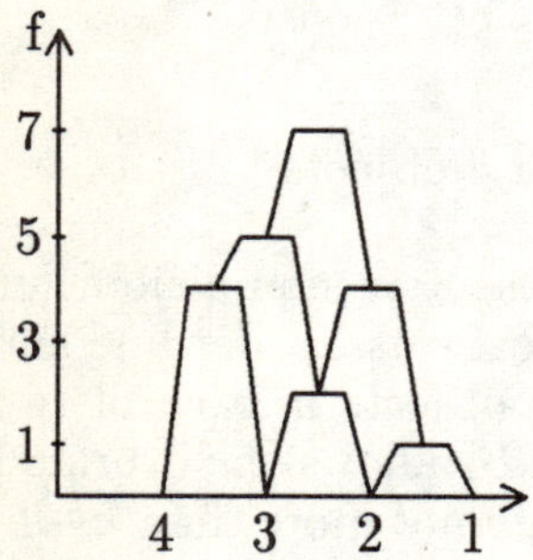

P = { {1},{2},{3},{4},

{1,2},{2,3},{3,4},

{1,2,3},{2,3,4},N }

Fig. 3: Indexed pyramid of 4 objects

2. Classification of Statistical Software Packages

After this short survey of different clumping techniques different overlapping classifications of 17 universal statistical software packages are discussed. The notion "universal statistical software" means a basic equipment which consists of univariate descriptive statistics, cross tabulation, correlation, multiple regression, tests, analysis of variance and graphics.

The data matrix consists of an ordinal variable **price level (PL)** defined from 0.25 (low) to 1.0 (high), 3 binary variables **host operating system (HS), coprocessor support (CS), keyboard macros (KM)** (with values 1 = yes, 0 = no) and 10 quantitative variables. The 10 quantitative variables **openness of system (OS), ease of use (EU), individual solutions (IS), descriptive statistics (DS), tests (T), multivariate**

	PL	HS	CS	KM	OS	EU	IS	DS	T	MS	SM	BG	SG
BMDP	1.00	1	1	0	0.81	0.50	0.83	0.93	0.56	0.59	0.49	0.00	0.20
CRUNCH	0.50	0	1	1	0.61	0.72	0.61	0.85	0.62	0.16	0.00	0.20	0.11
CSS	0.5	0	1	1	0.74	0.81	0.55	0.80	0.64	0.57	0.77	1.00	0.83
MICROSTAT II	0.25	0	1	0	0.62	0.43	0.41	0.62	0.79	0.14	0.03	0.00	0.09
MINITAB	0.50	1	1	0	0.53	0.40	0.55	0.62	0.64	0.15	0.31	0.00	0.28
NCSS	0.5	0	1	1	0.72	0.60	0.33	0.77	0.67	0.49	0.80	0.92	0.70
P-STAT	1.00	1	1	1	0.71	0.70	0.69	0.93	1.00	0.25	0.56	0.46	0.22
PC-STATISTIK	0.25	0	0	0	0.47	0.35	0.29	0.59	0.65	0.15	0.08	0.23	0.17
RS/1	1.00	1	1	0	0.80	0.68	0.83	0.54	0.73	0.10	0.28	0.49	0.51
S-PLUS	1.00	1	1	0	0.73	0.30	0.65	0.69	0.81	0.27	0.32	0.46	0.43
SAS	1.00	1	1	0	0.88	0.71	0.83	0.90	0.78	0.71	0.94	1.00	0.51
SPSS	1.00	1	1	0	0.89	0.71	0.41	0.89	0.86	0.61	0.52	0.86	0.21
STATA	0.50	1	1	1	0.72	0.57	0.83	0.66	0.68	0.27	0.43	0.89	0.36
STATGRAPHICS	0.75	0	1	1	0.74	0.86	0.32	0.87	0.78	0.34	0.56	0.97	0.42
STATISTIX	0.25	0	1	0	0.57	0.31	0.21	0.67	0.81	0.15	0.22	0.00	0.16
STATPAC GOLD	0.5	0	1	0	0.63	0.51	0.33	0.78	0.68	0.18	0.40	0.43	0.21
SYSTAT	0.75	1	1	0	0.88	0.58	0.67	0.87	0.64	0.61	0.45	0.89	0.85

Tab. 1: Datamatrix: 17 objects and 13 variables

statistics (MS), special methods (SM), business graphics (BG) and **statistical graphics (SG)** with scoring values between 0 and 1 are the result of a linear homogeneous aggregation of 228 variables. A scoring value of 1 means that the corresponding package performs all required features perfectly wheras a value of 0 represents very poorly implemented or missing features.

First different fuzzy s-classification with the exponent $r = 2$ in Tab. 2 and 3 are discussed, varying the number of classes s between 3 and 5. Considering the 3-class-solution, class 1 and class 3 contain typical representatives (marked by C) and most of the objects of class 2 (marked by B) belong to the intersection of two classes.

	3 Classes			4 Classes			
Object	K_1	K_2	K_3	K_1	K_2	K_3	K_4
BMDP	0.37B	0.45B	0.18	0.27	0.40B	0.14	0.19
CRUNCH	0.60B	0.31B	0.09	0.48B	0.33B	0.07	0.12
CSS	0.03	0.08	0.88C	0.02	0.03	0.87C	0.08
MICROSTAT II	0.90C	0.08	0.02	0.87C	0.08	0.02	0.03
MINITAB	0.82 C	0.14	0.04	0.73C	0.18	0.03	0.05
NCSS	0.05	0.11	0.84C	0.04	0.08	0.68B	0.20
P-STAT	0.16	0.68B	0.16	0.13	0.46B	0.12	0.29
PC-STATISTIK	0.86C	0.10	0.04	0.82C	0.11	0.03	0.04
RS/1	0.19	0.66B	0.15	0.14	0.60B	0.11	0.15
S-PLUS	0.22	0.68B	0.10	0.13	0.71C	0.06	0.10
SAS	0.07	0.19	0.74C	0.05	0.12	0.54B	0.29
SPSS	0.12	0.35B	0.52B	0.04	0.08	0.11	0.77C
STATA	0.11	0.58B	0.30B	0.09	0.35B	0.22	0.34B
STATGRAPHICS	0.10	0.26	0.64B	0.05	0.10	0.22	0.63B
STATISTIX	0.86C	0.10	0.04	0.83C	0.10	0.03	0.04
STATPAC GOLD	0.45B	0.44B	0.11	0.33B	0.42B	0.08	0.17
SYSTAT	0.07	0.21	0.72C	0.05	0.12	0.62B	0.21
$\frac{F(P) - 1/s}{1 - 1/s}$	0.36			0.33			

Tab. 2: Fuzzy s-classification $P = (p_{ij})$ with 17 objects,3 and 4 classes

Considering the 4- and 5-class-solution one can find typical representatives only in class 1 and most of the remaining objects could not be clearly classified. Also the values of the normalized partition coefficient F emphasize the assumption that not every object can be classified in exactly one class. Now the pyramidal classification based on an aggregated dissimilarity matrix

Object	5 Classes K_1	K_2	K_3	K_4	K_5
BMDP	0.19	0.36B	0.10	0.13	0.22
CRUNCH	0.31B	0.37B	0.05	0.08	0.19
CSS	0.01	0.02	0.88C	0.06	0.03
MICROSTAT II	0.83C	0.09	0.01	0.02	0.05
MINITAB	0.56B	0.26	0.03	0.04	0.11
NCSS	0.03	0.06	0.66B	0.18	0.07
P-STAT	0.10	0.33B	0.09	0.20	0.28
PC-STATISTIK	0.74C	0.14	0.02	0.03	0.07
RS/1	0.06	0.13	0.04	0.06	0.71C
S-PLUS	0.12	0.33B	0.06	0.08	0.41B
SAS	0.04	0.09	0.48B	0.27	0.12
SPSS	0.02	0.05	0.07	0.81C	0.05
STATA	0.06	0.15	0.15	0.21	0.43B
STATGRAPHICS	0.04	0.08	0.19	0.60B	0.09
STATISTIX	0.79C	0.11	0.02	0.03	0.05
STATPAC GOLD	0.18	0.55B	0.04	0.09	0.14
SYSTAT	0.04	0.09	0.54B	0.19	0.14
$\frac{F(P)-1/s}{1-1/s}$	0.31				

Tab. 3: Fuzzy 5-classification $P = (p_{ij})$ with 17 objects

is carried out. Using the complete linkage algorithm the resulting indexed pyramid can be seen in Fig. 4.

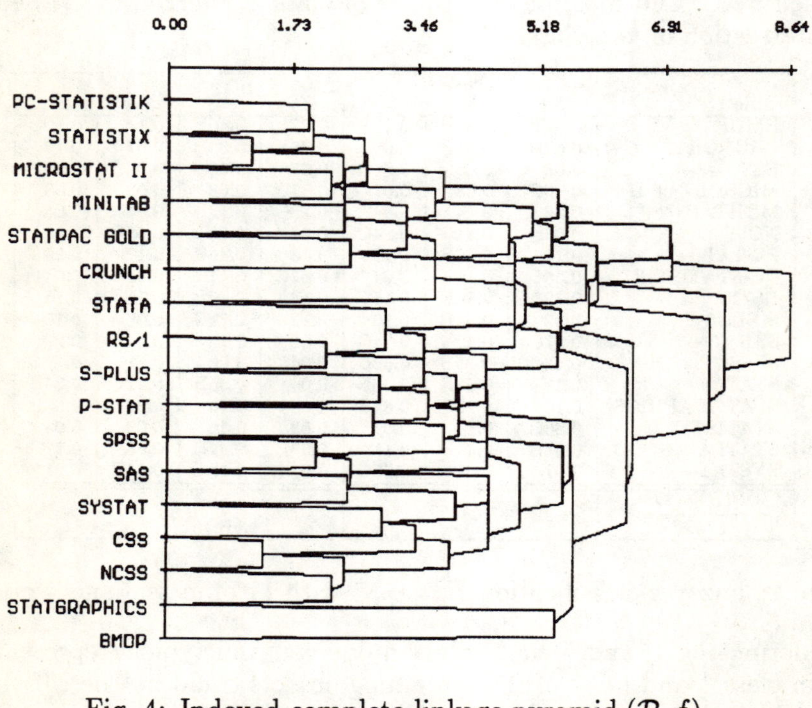

Fig. 4: Indexed complete-linkage-pyramid $(\mathcal{P}, f)$

From the indexed pyramid it is difficult to derive an appropriate number of classes. The modified elbow criterion of Fig. 5 makes the decision for a suitable number of classes easier, if the following two remarks are considered:

- It can be seen that there exist several overlapping classifications for a given number of classes s. Given $s = 3$ it is possible to construct 6 different solutions with index values from 5.63 to 6.97. To get homogeneous classes the minimal possible index 5.63 should be taken.
- The number of classes should be restricted to get interpretable solutions. The index 4.42 results in a 7- class-solution and with a slightly increase of the index to 4.45 one gets a 5-class-solution.

With this remarks only the dots on the dashed curve are relevant to the determination of an appropriate number of classes because this dots cause a maximal possible reduction of heterogenity if the number of classes increases. To meet class-homogenity and an interpretable number of classes a 5-class solution at the index 4.45 has been chosen.

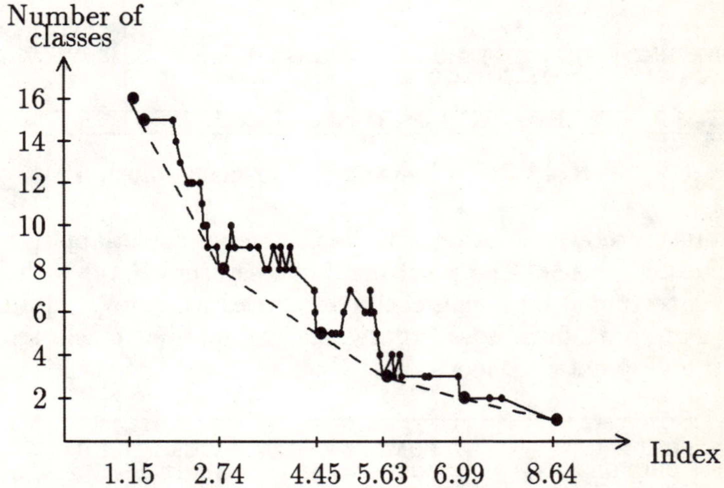

Fig. 5: Number of classes versus dissimilarity index

To show the advantages of a pyramidal classification with moderate overlappings the results of a quasi- hierarchical classification are presented. A quasi-hierarchy has no overlapping restrictions which entail more classes and overlappings at each threshold level c. A quasi-hierarchical classification of the software packages results in 117 fusion levels c with 244 different classes, a fact that makes a graphical representation and interpretation impossible. To get an impression of the number of classes and overlappings Tab. 4 contains all maximal c-cliques with $c = 4.45$. In contrast to the pyramidal 5-class solution there exist 12 different classes with numerous overlappings. For example the classes 9 and 10 consist of 6 objects and overlap in 5 objects.

Class 1	Class 2	Class 3	Class 4
CRUNCH MINITAB STATA STATPAC GOLD	CRUNCH MICROSTAT II MINITAB PC-STATISTIK STATISTIX STATPAC GOLD	MICROSTAT II MINITAB S-PLUS STATISTIX STATPAC GOLD	BMDP MINITAB RS/1 S-PLUS STATA STATPAC GOLD
Inner-class diss.: 3.70	Inner-class diss.: 3.81	Inner-class diss.: 4.05	Inner-class diss.: 4.33

Class 5	Class 6	Class 7	Class 8
BMDP P-STAT RS/1 S-PLUS SPSS STATA STATPAC GOLD	P-STAT S-PLUS SPSS STATA STATGRAPHICS STATPAC GOLD	BMDP P-STAT RS/1 S-PLUS SAS SPSS STATA SYSTAT	P-STAT S-PLUS SAS SPSS STATA STATGRAPHICS SYSTAT
Inner-class diss.: 4.42	Inner-class diss.: 4.42	Inner-class diss.: 4.43	Inner-class diss.: 4.43

Class 9	Class 10	Class 11	Class 12
CRUNCH CSS NCSS STATA STATGRAPHICS STATPAC GOLD	CSS NCSS SPSS STATA STATGRAPHICS STATPAC GOLD	CSS NCSS SAS SPSS STATA STATGRAPHICS SYSTAT	CRUNCH P-STAT STATA STATGRAPHICS STATPAC GOLD
Inner-class diss.: 4.24	Inner-class diss.: 4.45	Inner-class diss.: 4.45	Inner-class diss.:4.22

Tab. 4: Quasi-hierarchical 12-class-solution

To get an interpretable solution with less classes and overlappings the idea of Jardine/Sibson is used. The resulting 3-class-system $\mathcal{B}_k(c)$ with $c = 4.45$ in Tab. 5 is constructed by fusing each two classes which overlap in 5 or more objects. This procedure does not reduce the number of classes efficiently because the very heterogeneous class 1 consists of 14 of the 17 objects.

Class 1	Class 2	Class 3
BMDP CRUNCH CSS MINITAB NCSS P-STAT RS/1 S-PLUS SAS SPSS STATA STATGRAPHICS STATPAC GOLD SYSTAT	CRUNCH MICROSTAT II MINITAB PC-STATISTIK STATISTIX STATPAC GOLD	MICROSTAT II MINITAB S-PLUS STATISTIX STATPAC GOLD
Inner-class diss.: 6.45	Inner-class diss.: 3.81	Inner-class diss.: 4.05

Tab. 5: The system $\mathcal{B}_5(4.45)$

Finally the pyramidal 5-class-solution (Tab. 6) can be interpreted as follows:

Class 1	Class 2	Class 3
CRUNCH MICROSTAT II MINITAB PC-STATISTIK STATISTIX STATPAC GOLD	CRUNCH MINITAB STATA STATPAC GOLD	P-STAT RS/1 SAS S-PLUS SPSS STATA SYSTAT
Inner-class diss.: 3.81	Inner-class diss.: 3.70	Inner-class diss.:4.43

Class 4	Class 5
CSS NCSS SAS SPSS STATGRAPHICS SYSTAT	BMDP
Inner-class diss.: 4.45	Inner-class diss.: 0.00

Tab. 6: Pyramidal 5-class-solution

Class 1 **"Starting"-class:** Low price level; few possibilities in data management, user interface and import/export interfaces; poor features in graphics and statistical methods.

Class 2 **"Improved starting"-class:** Average pricelevel; in comparison to class 1 all remaining variables except tests are evaluated better.

Class 3 **"Individualist"-class:** Very good individual programming abilities; great number of data exchange formats; available at several hardware platforms; broad range of statistical tests; high price level.

Class 4 **"Graphic"-class:** Broad range of statistical and business graphics; good features in descriptive and multivariate statistics; manifold editing tools and data exchange formats; higher price level.

Class 5 **"Outlier"-class:** Highest price level; in comparison to class 3 and 4 only the variables individual solutions and descriptive statistics show slight better scoring values; the remaining scoring values are identical or smaller.

References:

BAUSCH, T., BANKHOFER, U., (1992): Statistical Software Packages for PCs - A Market Survey. *Statistical Papers, vol. 33, 283-306.*

BEZDEK, J. (1973): Fuzzy Mathematics in Pattern Classification. PhD Thesis, Cornell University, Ithaca, N.Y.

BEZDEK, J. (1974a): Numerical taxonomy with fuzzy sets. *Journal of Mathematical Biology 1, 57-74.*

BEZDEK, J. (1974b): Cluster validity with fuzzy sets. *Journal of Cybernetics 3, No 3, 58-71.*

JARDINE, N., SIBSON, R. (1971): Mathematical Taxonomy. Wiley, New York.

LASCH, R. (1993): Pyramidale Darstellung multivariater Daten. Josef Eul, Bergisch Gladbach, Köln.

Scenario Analysis with BASICS

—Testing the Representativity of the Results of the Dynamic Probability Adjustment of Scenario Components with the Help of Classification Methods—

M. Missler-Behr

Institut für Statistik und Mathematische Wirtschaftstheorie
Universität Augsburg, D-86135 Augsburg

Summary: In scenario analysis representative and meaningful scenarios can be obtained with the help of branch-and-bound algorithms, equation systems as well as optimization and simulation models. The computer program BASICS simulates scenarios by a stepwise probability adjustment of scenario components. BASICS uses a rational function for adjusting. Here the question arises why this special kind of adjustment function is chosen. We additionally use similar but easier adjustment functions, compare the results and use cluster analysis to test the representativity of the total scenario space.

1. Introduction to Scenario Analysis

Scenario analysis is used in strategic planning as a systematic method to develop possible pictures of the future called scenarios. Not only the pictures themselves but the logical progress of the scenarios from today to a future timepoint are the main aspects in scenario analysis. An advantage of scenario analysis to well-known forecasting methods is the possibility to use quantitative **and** qualitative variables to light up the future.

The scenario idea is depictured in Fig. 1 (see v. Reibnitz (1991)). We consider the period of time between t_0 and t_n. The total scenario space can be included in a funnel. The present situation is given at t_0 and can be represented by a point, the beginning of the funnel. The more t tends to t_n the more possibilities exist for future developments. The funnel is spread out. At t_n all possible scenarios are placed at the sectional plane of the funnel. The funnel is boarded by the extreme scenarios.

Consider scenario S_1 with its progress from t_0 to t_n. Now assume that at t_1 the development of S_1 is disturbed. The progress of S_1 has changed. Management has to recognize and to analyse the reasons for changing to install counter-measures to the disturbance. At t_2 the counter-measures begin to work and bring the changed progress of S_1 back as close to the original point in the sectional plane as possible. In Fig. 1 the changed scenario progress ends in S_2 quite close to S_1.

The aim of scenario analysis is not to explain all possible future pictures but to find some

few $(2-5)$, different, consistent and stable

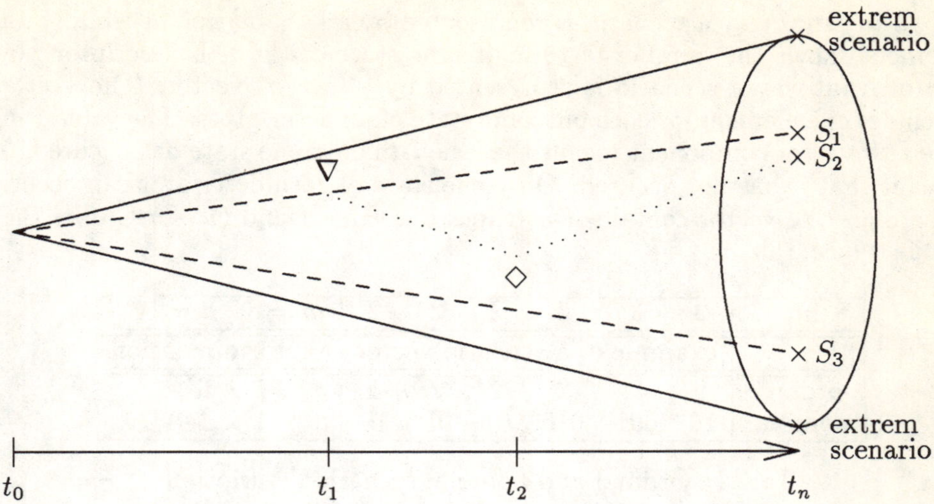

× scenario
– – – progress of a scenario
▽ disturbance
· · · the changed scenario progress after the disturbance
◇ installation of counter-measures

Fig. 1: The funnel model in scenario analysis

scenarios representing the total scenario space (Missler-Behr (1993)). Consistent means that there shall be no contradictions in scenario components. Stable means that a slight change in a scenario component shall not raise a goodness of fit criterion used to evaluate scenarios.

2. Formal Description

A scenario can be described by a vector representing the outcome states of the descriptors (see Tab. 1, Missler-Behr (1993)). The descriptors are the variables in scenario analysis. Descriptors can be trends, events or developments which describe the influencing factors of the examined problem (Honton, Stacey, Millett (1984)). Descriptors have different outcome states. The states describe the possible descriptor outcomes completely and are disjunctive to each other. Now let be:

m	the number of descriptors
n_i	the number of outcome states for descriptor i $(i = 1, \ldots, m)$
$n = \sum_{i=1}^{m} n_i$	the total number of outcome states over all descriptors
$\tilde{n}_i = \sum_{j=1}^{i} n_j$	the i-th partial sum of n

In an **ordinal** way a scenario is represented by a $(1 \times m)$ vector whose i-th element shows the number of the outcome state of the i-th descriptor. In a **nominal** way a scenario is represented by a $(1 \times n)$ vector. The vector includes one element for each outcome state of all descriptors. The value 1 in the i-th vector component means that the i-th outcome state has occured, 0 means that it has not occured. Only one state of each descriptor can occur. Therefore the vector contains (m)-times the value 1 and $(n-m)$-times the value 0 (see Tab. 1).

n_i	2	3	3	2	2	2	$m = 6 \quad n = 14$
	example of a scenario vector						vector dimension
ordinal	1	3	2	2	2	1	1×6
nominal	10	001	010	01	01	10	1×14

Tab. 1: Ordinal and nominal scenario description

3. The BASICS Computer Program

The BASICS (**Ba**ttelle **S**cenario **I**nputs to **C**orporate **S**trategy) computational program was developed and is applied by the Battelle-Institute (Honton, Stacey, Millett (1984)). It uses a simulation algorithm to generate scenarios which adjusts a priori probability estimations of the outcome states depending on the interactions of two states, one of them given.

The major inputs in BASICS are:

- conditional consistency estimations k_{ij}
- a priori probability estimations p_i

for all outcome states of the descriptors. The input data of the used example are given in Tab. 2.

Consistency estimations are subjective and ordinal compatibility valuations of the relationship of two outcome states. They express the intensity of the relationship and describe whether the outcome states have an increasing, decreasing or neutral effect to each other. The two outcome states have to belong to different descriptors because just one state of a descriptor can occur. The k_{ij}'s are conditional consistency estimations. We assume that state j occurs and evaluate its influence on the occurence of state i.

The p_i's are estimations of the a priori probability of each outcome state. Because a descriptor is completely described by its outcome states and the states are disjunctive the a priori probabilities of the states of each descriptor have to sum up to 1:

$$\sum_{i=\tilde{n}_{l-1}+1}^{\tilde{n}_l} p_i = 1 \qquad \text{for} \quad l = 1, \ldots, m \tag{1}$$

i	p_i	k_{ij}													
1	0.45	x	x	-1	0	1	0	0	2	-1	2	-1	1	0	-1
2	0.55	x	x	0	0	0	0	0	-1	1	-1	1	0	0	0
3	0.45	1	0	x	x	x	2	-2	1	-1	1	-1	-1	0	1
4	0.35	0	0	x	x	x	0	0	0	0	0	0	0	1	0
5	0.20	-1	1	x	x	x	-1	1	0	0	-2	2	1	0	-1
6	0.45	-2	2	0	0	0	x	x	1	0	2	-1	-1	0	1
7	0.55	0	0	0	0	0	x	x	0	1	-1	1	0	1	0
8	0.25	2	-2	2	0	-2	1	-1	x	x	1	0	3	-1	-3
9	0.75	-1	1	-1	0	1	0	0	x	x	-1	0	-3	1	2
10	0.45	0	0	1	0	-1	1	-1	-1	1	x	x	2	1	-2
11	0.55	0	0	-1	0	1	0	0	1	-1	x	x	-1	0	1
12	0.20	-2	1	2	-1	-2	1	-1	-2	2	-1	1	x	x	x
13	0.40	-1	1	0	0	0	0	0	1	-1	0	0	x	x	x
14	0.40	3	-2	-3	2	3	-1	1	2	-2	1	-1	x	x	x

i the infuenced descriptor state
j the conditioning descriptor state
p_i a priori probability estimation of descriptor state i
k_{ij} conditional consistency estimation of i assuming j occurs
x no estimation is given, because i and j belong to the same descriptor

Tab. 2: Example data

In BASICS k_{ij} is out of $\{-3,-2,-1,0,1,2,3\}$. $k_{ij}=-3$ means that the occurence of outcome state j effects extrem negatively the occurence of state i. They are more or less contradictory. Therefore the probability adjustment used in the simulation will reduce p_i drastically. $k_{ij}=3$ has the same meaning in the opposite direction. $k_{ij}=0$ denotes a neutral relationship between i and j. They do not effect each other. Therefore no adjustment takes place.

With the estimated k_{ij}'s BASICS also computes consistency numbers $k_{i\bar{j}}$. Here we assume that state j has not occured and evaluate that influence on the occurence of i.

These input data are used for the simulation which is shown in Fig. 2. Starting in succession with each outcome state, for its occurence and non-occurence $2n$ adjustment processes are driven for a complete simulation. Now we assume that state j is given and occures. In the adjustment process all a priori probabilities p_i belonging to a different descriptor than j are adjusted to $p_{i(j)}$ depending on k_{ij}. The $p_{i(j)}$'s are elements of [0, 1] but condition (1) is not satisfied in general. Therefore a standardization is carried out by dividing $p_{i(j)}$ by the sum of all adjusted probabilities belonging to the same descriptor as $p_{i(j)}$. Now the next given outcome state is chosen and decided whether it occurs or not. The adjusted probabilities become

- for all outcome states
  - by occurence and non-occurence
    - *ADJUSTMENT PROCESS*
      - adjust a priori probabilities
      - standardize the adjusted probabilities
      - choose next outcome state with occurence or non-occurence
      - the adjusted prob. become the new a priori prob.
    - repeat until a final scenario is reached

Fig. 2: The total simulation in BASICS

the new a priori probabilities and this series of adjustments is repeated until a final, consistent scenario is reached, which means until all probabilities are adjusted to 1 or 0. An exact mathematical decription of this procedure and an example are given by Missler-Behr (1993).

Maximal $2n$ different scenarios result from the total simulation in BASICS. The finally considered scenario space is so reduced from originally $\prod_{i=1}^{m} n_i$ to not more than $2n$ elements. At this point BASICS gives no concrete rule to select the few, different, representative scenarios out of the $2n$ elements.It uses the frequency of the scenarios as goodness of fit criterion for selection.

Tab. 3 shows the resulting scenarios of the total BASICS simulation using the data of Tab. 2. The scenarios are described in an ordinal way. 15 different scenarios are generated with the frequencies given in the last column of Tab. 3.

The shape and formula of the adjustment function used in BASICS is given in Fig. 3. It will be called F_1.

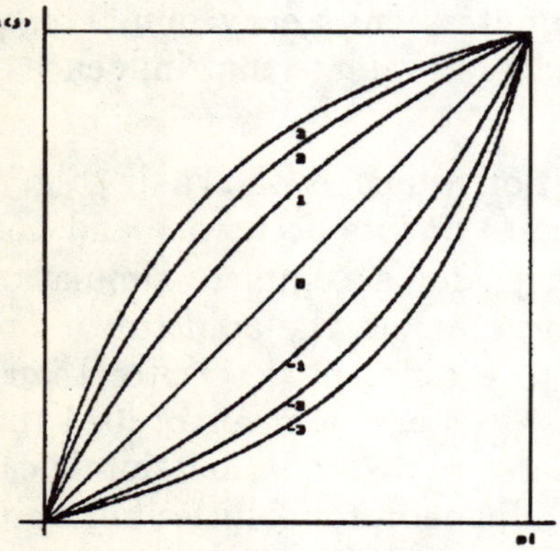

$$p_{i(j)} = \frac{p_i \cdot f_{ij}}{1 + p_i(f_{ij} - 1)}$$

with

$$f_{ij} = \begin{cases} k_{ij} + 1 & \text{for } k_{ij} \geq 0 \\ 1/(|k_{ij}| + 1) & \text{for } k_{ij} < 0 \end{cases} \tag{2}$$

Fig. 3: Adjustment function in BASICS: F_1

	descriptors 1 – 6						f
S 1	2	2	2	2	2	2	2
S 2	1	2	2	2	2	3	2
S 3	2	2	2	2	1	2	5
S 4	2	2	1	2	1	2	2
S 5	1	1	2	1	1	2	1
S 6	2	1	1	2	1	2	2
S 7	1	2	2	1	1	2	1
S 8	2	3	2	2	2	2	4
S 9	1	1	1	1	1	2	2
S 10	2	2	2	1	2	2	2
S 11	2	3	2	2	2	1	1
S 12	2	3	2	2	1	1	1
S 13	2	2	1	2	1	1	1
S 14	1	2	2	2	1	2	1
S 15	2	1	1	2	1	3	1

Tab. 3: Resulting scenarios of the total simulation

4. Alternative Adjustment Functions

In Fig. 3 we can see that the adjustment is particularly high for $|k_{ij}| = 1$ compared with the other possible values. No reason for this assumption is given. Also the adjustment is not in equal relation for different values of $|k_{ij}|$.

Just as rational is the reversed assumption that the adjustment for $|k_{ij}| = 1$ is pretty small because the consistency value is close to the neutral value 0, whereas the adjustment is particularly high for $|k_{ij}| = 3$. It is also reasonable to assume that the adjustment is equidistant for the different k_{ij}'s (see Fig. 4, F_2). Here we assume that each change in the consistency estimations by step one has the same influence on the adjustment process. We also can assume that the adjustment function is quite similar to F_1 but mathematically easier to describe. Such a function could be piecewise linear and equidistant at $p_i = 0.5$ (see Fig. 4, F_3). Further functions are reasonable.

5. Application of Classification Methods to Test the Representativity of BASICS Results

With the help of maximal different centers and maximal c-cliques around the most frequent scenario (Missler-Behr (1993); Opitz, Missler-Behr (1993)) a good representation of the scenario space can be found.

Now we use the example data of Tab. 2 and run BASICS using F_1, F_2, and F_3 in succession. The resulting scenarios of the three runs seem to be very different which means that the three adjustment functions bring out very different results. Five scenarios are generated by all adjustment functions in our example. They determine a common partial space. The common

adjustment function F_2:

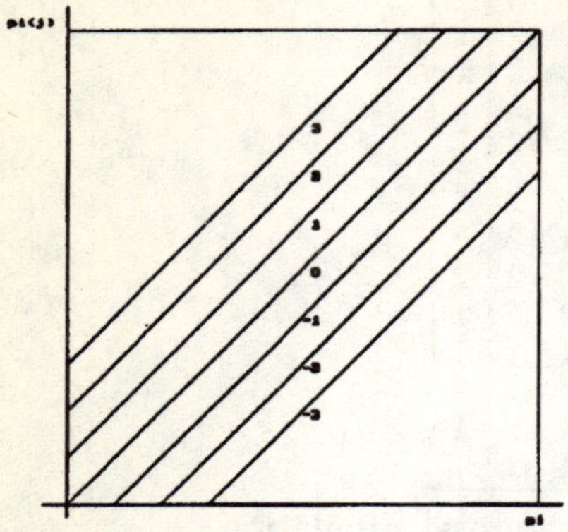

$$p_{i(j)} = \frac{k_{ij}}{10} + p_i \tag{3}$$

adjustment function F_3:

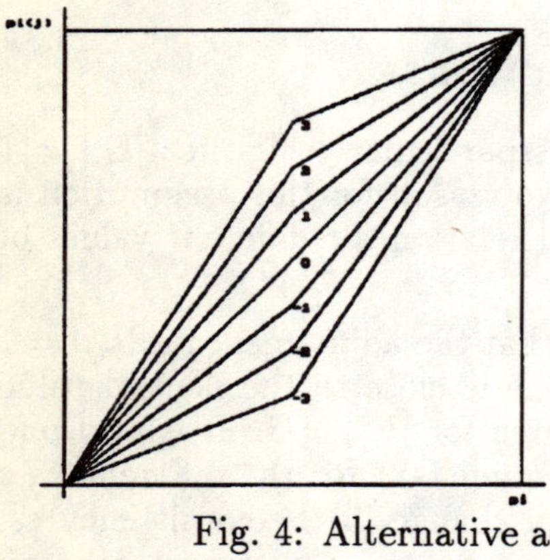

$$p_{i(j)} = \begin{cases} \left(1 + \frac{k_{ij}}{5}\right) p_i & \text{for } p_i \leq 0.5 \\ \frac{k_{ij}}{5} + \left(1 - \frac{k_{ij}}{5}\right) p_i & \text{for } p_i > 0.5 \end{cases} \tag{4}$$

Fig. 4: Alternative adjustment functions: F_2 and F_3

scenarios and their frequencies are shown in Tab. 4. They also contain the most frequent scenarios (boldfaced).

common scenarios	frequencies in F_1	F_2	F_3	total
2 2 2 2 2 2	2	2	3	7
2 2 2 2 1 2	**5**	2	3	10
2 1 1 2 1 2	2	**5**	**7**	14
2 3 2 2 2 2	4	2	1	7
1 2 2 2 1 2	1	2	2	5

Tab. 4: Common results of F_1, F_2, and F_3

In the next step we test the representation of the scenario space by F_1, F_2, F_3 using the frequency-criterion of BASICS and classification methods described at the beginning of chapter 5. We select three scenarios for each adjustment function with frequencies f. Then we determine the distances between the three scenarios. The distance measure d_{ij} counts 1 for each descriptor if the outcome states of a descriptor in scenario i and j are different and 0 if they are equal. Tab. 5 shows that the scenarios found by combing the results of maximal different centers and maximal c-cliques are more different and representative for the space as those found with BASICS. Considering the minimal distance, $\min d$, between a selected scenario and one of the common partial space (see Tab. 4) this result is confirmed. On the other hand we can see with $\min d$ that the cluster results are able to leave better the common partial space and emphasize the special shape of the adjustment function.

		Battelle				Classification			
F_i	no.	scenarios	f	d_{ij}	$\min d$	scenarios	f	d_{ij}	$\min d$
F_1	1	2 2 2 2 1 2	5	d|1 2	0	2 2 2 2 1 2	5	d|1 2	0
	2	2 3 2 2 2 2	4	2|2 –	0	1 1 1 1 1 2	2	2|4 –	2
	3	1 1 1 1 1 2	2	3|4 5	2	1 2 2 2 2 3	2	3|3 5	2
F_2	1	2 1 1 2 1 2	5	d|1 2	0	2 1 1 2 1 2	5	d|1 2	0
	2	2 1 1 2 1 1	3	2|1 –	1	1 1 2 1 2 3	1	2|5 –	4
	3	1 2 2 2 1 2	2	3|3 4	0	2 2 2 2 2 2	2	3|3 4	0
F_3	1	2 1 1 2 1 2	7	d|1 2	0	2 1 1 2 1 2	7	d|1 2	0
	2	2 2 1 2 1 2	5	2|1 –	1	1 2 2 2 2 3	2	2|5 –	2
	3	2 2 2 2 2 2	3	3|3 2	0	1 1 2 1 1 1	1	3|4 4	3

Tab. 5: Comparison of the results

Overall we can say the results of BASICS are depending on the special adjustment function. Up to now we can not say which function is the best one. Therefore the user should choose a individual function which reflects best her or his personal adjusting. For further improvements of the results of BASICS classification methods should be used instead of the frequency criterion to reach a better representation of the total scenario space after simulation. Here the use of maximal different centers and maximal c-cliques is recommended.

References:

HONTON, E.J., STACEY, G.S. and MILLETT S.M. (1984): Futures Scenarios: The BASICS Computational Method. *Economics and Policy Analysis Occasional Paper, No. 4, Battelle-Institute, Columbus/Ohio.*

MISSLER-BEHR, M. (1993): Methoden der Szenarioanalyse. DUV, Wiesbaden.

OPITZ, O., MISSLER-BEHR, M. (1993): Klassifikationsverfahren in der Szenarioanalyse. In: NIEDEREICHHOLZ, J. and SCHUHMANN, W. (eds.): *Wirtschaftsinformatik — Beiträge zur modernen Unternehmensführung.* Campus, Frankfurt/Main, New York, 240–256.

REIBNITZ, U. von (1991): Szenario-Technik: Instrumente für die unternehmerische und persönliche Erfolgsplanung. Gabler, Wiesbaden.

Analysis of Sales Data
A Neural Net Approach

F. Wartenberg, R. Decker

Institut für Entscheidungstheorie und Unternehmensforschung,
Universität Karlsruhe (TH), Postfach 6980, D-76128 Karlsruhe

Summary: As an increasing interest of the scientific community is devoted to the application of neural networks to pattern recognition, forecasting and optimization, we will demonstrate how so-called Backpropagation Networks can be used to specify market response functions from sales data. The proceeding will be demonstrated by market share analysis. Against this background it is also shown how market diagnostics (e.g., market shares and elasticities) which are well-known to marketing managers can be calculated from a calibrated neural network. Finally, results from an empirical application including a comparison of the new approach with a traditional market share model are sketched.

1. Introduction

Since more than two decades different types of sales data are regularly collected in market research, e.g. on the basis of point-of-sale scanning and diary panels, to support quantitative buying behavior analyses. At the same time an increasing interest of the marketing practice in flexible and easy-to-handle data analysis instruments is observable.

In this paper a new approach for sales data analysis which combines market share modeling with neural network techniques will be presented. Special attention is given to the adequate consideration of asymmetric competition and the calculability of meaningful market diagnostics. The model is embedded into the SOR-paradigm (SOR is the acronym for Stimulus-Organism-Reaction.) which is established in consumer behavior research for a long time. The resulting approach is called "SOR-network" in the following.

2. Selected Aspects of Market Share Modeling

Market share modeling (including model-based marketing mix analysis) is one of the most important areas in quantitative marketing research. Accordingly a large number of papers concerning this topic have been published in the past. A likewise literary and methodological framework for our own work is given in Fig. 1. Dependent on the number of alternatives (resp. brands in our context) to be integrated those models can be divided into univariate and multivariate approaches. As a rule econometric response models distinguish themselves by their compact structure and well-defined fields of application. But at the time the major part of these models is suffering from the so-called IIA-property. IIA stands for Independence from Irrelevant Alternatives. In market share analysis for example this can be very disadvantageous when asymmetric market competition has to be considered. The MNL (Multi-Nomial Logit) and the MCI (Multiplicative Competitive Interaction model as well are probably the best known representatives of this class of models. Therefore, we have selected the latter one as a kind of benchmark for our own approach. Modifications of the basic approaches (e.g. the N(ested)MNL model) with the objective to account for possible asymmetries often debit to the complexity of parameter estimation.

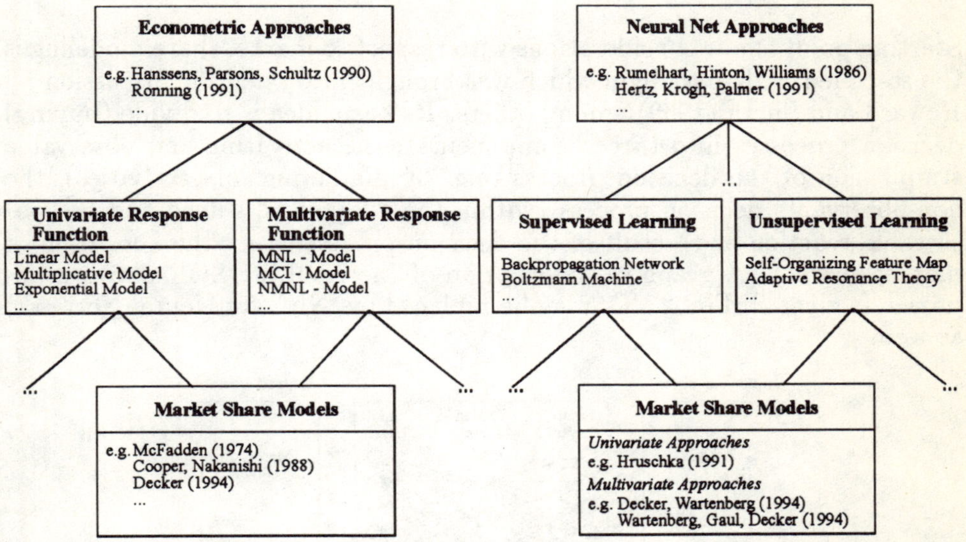

Fig. 1: Econometric and neural network-based market share modeling

All those approaches mentioned so far have in common that the basic structure of the respecting model has to be determined a priori. In contrast, neural networks are adaptable to very different interrelation structures. An introduction into the basic formalism of neural networks is given, e.g., in the textbook of Hertz et al. (1991). Classifications of neural networks (cf. Fig. 1) according to the underlying learning algorithm are surely still very rough, but they already stress a basic difference between existing representatives. Network calibrations using supervised learning are based on direct comparison of the neural network output with given external patterns. In contrast unsupervised learning gets along without such an "external support". Therefore, it has to be used when the relevant outputs are not known in advance. In the present application field explanatory variables (observed marketing mix activities) and explained variables (observed market shares) are given for a sufficient number of periods, so supervised learning can be used. Additionally, the well-known Multi-Layer-Perceptron which is based on supervised learning (cf. Rumelhart et al. (1986)), often also called Backpropagation-Network, has already been applied successfully in several application fields.

The large number of publications dealing with neural networks in general is faced by only a small number of papers concerning applications in marketing and marketing-data analysis. Hruschka (1991) for example emphasizes the similarity of neural networks and econometric (market) response models. In Decker and Wartenberg (1994) applications of neural networks in retail marketing are discussed whereas Wartenberg et al. (1994) give a behaviorist motivation of neural networks as a tool for marketing decision support.

3. Conceptual Framework

Starting point of our considerations with respect to market share modeling is the so-called SOR-paradigm which was brought into scientific discussion by Howard and Sheth (1969) among others. Its basic idea is to divide (buying) decision processes into three "components", the activating and observable stimulation of the decision process (e.g. by marketing mix activities), the non-observable decision process (within the "organism") itself and the observable reaction as a result of the decision process (e.g. the allocation of market shares). A graphical description of the resulting SOR-network approach is given in Fig. 2. 'O' may be replaced by 'NN' (for Neural Network) as well.

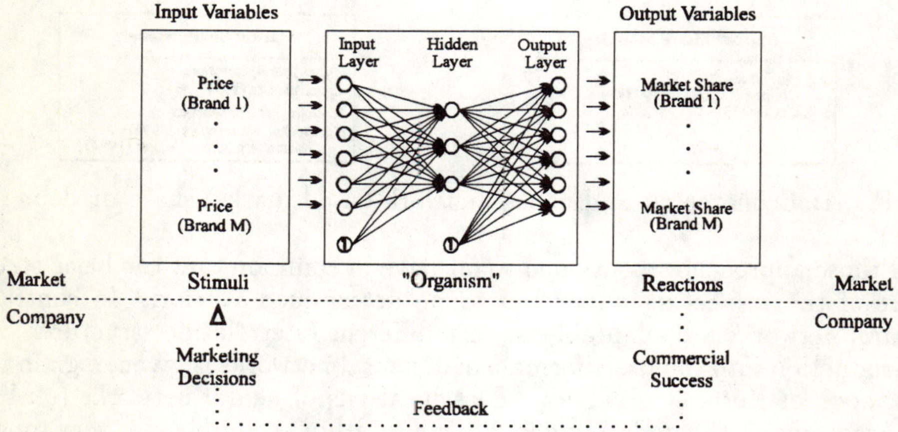

Fig. 2: Schematic representation of the SOR-network approach

With the SOR-network approach the explicit and – in consequence of the inherent complexity – only very incompletely possible modeling of the internal (decision making) process can be avoided. Instead the organism is regarded as a "black box" which does not allow looking inside and therefore – as one possibility – can be modeled with a neural network.

The number of hidden layers used in the SOR-network (cf. Fig. 2) can be adapted to different degrees of complexity concerning the purpose of modeling. Additionally the number of units in the hidden as well as in the input and output layer can be modified to attain an optimal adaptation of the neural network to the needs of special applications.

4. Mathematical Framework

The following model description starts from a neural network which has one input unit for each explanatory variable x_{mt} of brand m $(m = 1, \ldots, M)$ in period t $(t = 1, \ldots, T)$ – in the empirical study this will be the unit price – and three units in the hidden layer. Several preceding calibrations of similar neural networks have shown that the parsimonious specification of the hidden part of the network is sufficient for the present application. Finally, in the case of market share analyses the number of output units is fixed to the number of brands M.

If the weights between input and hidden units are denoted with $\mu_{k_H k_I}$ and those between hidden and output units with $\mu_{k_O k_H}$, and if we finally use $\tilde{y}_{k_O t}$ to identify the output of output unit k_O in period t, interactions between network output $\tilde{y}_{k_O t}$ and observed (standardized) explanatory variables $\tilde{x}_{mt}$ can be modeled as follows:

$$\tilde{y}_{k_O t} = \underbrace{s\left(\sum_{k_H=1}^{3} \lambda_{k_O k_H} s \left(\overbrace{\sum_{k_I=1}^{M} \mu_{k_H k_I} \tilde{x}_{k_I t} + \mu_{k_H M+1}}^{A} \right) + \lambda_{k_O 4} \right)}_{B} \tag{1}$$

with

$$\tilde{x}_{mt} = \frac{x_{mt} - 0.5(x_m^{max} + x_m^{min})}{x_m^{max} - x_m^{min}} \quad \text{and} \quad \begin{cases} x_m^{min} = \min\limits_{t=1,\ldots,T} \{x_{mt}\} \\ x_m^{max} = \max\limits_{t=1,\ldots,T} \{x_{mt}\}. \end{cases} \tag{2}$$

Following recommendations given in the relevant literature, the sigmoid function $s(a) = (1 + \exp(-a))^{-1}$ is used as transfer function. But output values calculated in this manner are only slowly approaching the lower (0) and upper (1) bound. Therefore, a further transformation of the observed market shares MA_{mt}^{obs} into the interval 0.2 to 0.8 is recommended (In the following we assume $k_O \equiv m$.):

$$\tilde{y}_{k_O t}^{target} = \frac{MA_{mt}^{obs} - MA_m^{min}}{MA_m^{max} - MA_m^{min}} 0.6 + 0.2 \quad \text{with} \quad \begin{cases} MA_m^{min} = \min\limits_{t=1,\ldots,T} \{MA_{mt}^{obs}\} \\ MA_m^{max} = \max\limits_{t=1,\ldots,T} \{MA_{mt}^{obs}\}. \end{cases} \tag{3}$$

Accordingly, the estimated market shares have to be re-transformed into the original interval [0; 1] by

$$y_{mt} = (\tilde{y}_{k_O t} - 0.2) \frac{MA_m^{max} - MA_m^{min}}{0.6} + MA_m^{min}. \tag{4}$$

The last transformation:

$$MA_{mt}^{NN} = \frac{y_{mt}}{\sum_{l=1}^{M} y_{lt}} \tag{5}$$

guarantees that the estimated market shares always sum to 1. With equation (1) to (5) we are able to derive a closed formula for market share elasticities:

$$ELA^{NN}_{MA_{mt},x_{jt}} = \frac{\partial MA^{NN}_{mt}}{\partial x_{jt}} \cdot \frac{x_{jt}}{MA^{NN}_{mt}} = \frac{\partial y_{mt}}{\partial x_{jt}} \cdot \frac{x_{jt}}{y_{mt}} - MA^{NN}_{mt} \sum_{l=1}^{M} \frac{\partial y_{lt}}{\partial x_{jt}} \cdot \frac{x_{jt}}{y_{mt}} \quad (6)$$

with

$$\frac{\partial y_{mt}}{\partial x_{jt}} = \frac{MA^{max}_{m} - MA^{min}_{m}}{0.6} s(B)(1-s(B)) \sum_{k_H=1}^{3} \lambda_{k_O k_H} s(A)(1-s(A)) \frac{\partial A}{\partial x_{jt}}. \quad (7)$$

Obviously this approach does not suffer from the IIA-property which is an important advantage when asymmetric competition on consumer goods markets has to be investigated. The second term of equation (6) accounts for the fact that changes in variables of a brand j may have different consequences for the market shares of brands m $(m \neq j)$.

The estimation of the unknown weights $\mu_{k_H k_I}$ and $\lambda_{k_O k_H}$ is carried out using a modified Backpropagation algorithm with momentum term $\alpha = 0.9$ (cf. Hertz et al. (1991, p. 123)). Furthermore – dependent on own experiences made in this area until now – the learning rate was set to 0.3 and the maximal number of learning cycles was fixed to 500.

5. Empirical Study

5.1 Calibration

The empirical comparison of the SOR-network approach and the MCI-model is based on three different data sets in order to increase reliability and validity of the final results. Because of the given limits with respect to the available space we have to restrict our considerations on one data set which contains real sales of five shampoo brands over 11 periods (corresponding unit prices included) collected via scanner at the point of sale of German drugstores in 1989.

Brand	MCI-Model		SOR-Net	
m	MRE_m	MT_m	MRE_m	MT_m
1	5.756	0.043	1.719	0.010
2	12.611	0.053	5.276	0.025
3	16.992	0.082	5.357	0.023
4	26.504	0.118	4.349	0.021
5	28.229	0.105	6.006	0.029

$$MRE_m = \frac{100}{T} \sum_{t=1}^{T} \frac{|MA^{obs}_{mt} - MA^{*}_{mt}|}{MA^{obs}_{mt}}$$

$$MT_m = \frac{\sqrt{\sum_{t=1}^{T}(MA^{obs}_{mt} - MA^{*}_{mt})^2}}{\sqrt{\sum_{t=1}^{T}(MA^{obs}_{mt})^2} + \sqrt{\sum_{t=1}^{T}(MA^{*}_{mt})^2}}$$

Tab. 1: Goodness-of-fit measures

Using this data, both, the SOR-network and the MCI-model can be calibrated. The goodness of parameter estimation is measured by two fit values, the Mean Relative Error MRE_m and the Modified Theil inequality coefficient

MT_m (see, e.g., Naert and Leeflang (1978)) referring to brand m. The possible range of MRE_m is [0, 100] whereas MT_m is restricted to [0, 1], both with optimum in 0. The actual results are depicted in Tab. 1.

The fit measures are confirming both approaches a good up to very good fit. At the same time the SOR-network is dominating the MCI-approach with respect to all brands. Concerning the other two data sets (not presented here) we got very similar results.

5.2 Selected Aspects of Asymmetric Competition

Before we turn to market share elasticities which – above all – are interesting from a competitive perspective, we want to present a special form of visualization of competition on consumer goods markets. Depictions like the one given in Fig. 3 facilitate comparisons of observed and estimated competition by plotting relative price changes $\Delta x_{mt} = \frac{x_{mt}-x_{m,t-1}}{x_{m,t-1}}$ of brand m ($m = 1,\ldots,5$) against the resulting relative changes in market share $\Delta MA^*_{mt} = \frac{MA^*_{mt}-MA^*_{m,t-1}}{MA^*_{m,t-1}}$ of these brands.

In Fig. 3 observed values were marked with a rhomb, MCI-results with a triangle and SOR-network results with a filled square. The x-coordinates of the points are representing changes in price whereas the y-coordinates are indicating changes in market share.

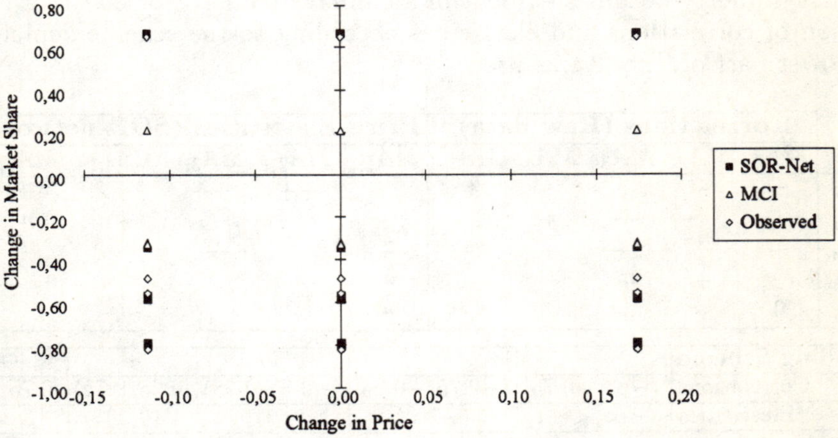

Fig. 3: Possible visualization of competitive structures for periods 1-2

It is obvious that the values estimated with the SOR-network are substantially closer to the observed values than those estimated with the MCI-model. Furthermore, the SOR-network is able to represent observable arrangements of some brands into groups as well. Information of this type gets lost when applying the MCI-model because of the IIA-property. In Fig. 3 only two MCI-points are visible for each change in price, one for direct and one for indirect effects. This can also be shown mathematically:

The attraction of brand m in period t is defined as $ATT_{mt} = \exp(\alpha_m)x_{mt}^{\beta_m}$

with parameters α_m and β_m and explanatory variable x_{mt}. The market share MA_{mt}^{MCI} of brand m in period t can then be calculated as a function of the weighted attraction of this brand:

$$MA_{mt}^{MCI} = \frac{ATT_{mt}}{\sum_{l=1}^{M} ATT_{lt}}. \tag{8}$$

A change of brand j's marketing variable from period $t-1$ to t – ceteris paribus – leads to the following relative changes of market share:

$$\Delta MA_{mt}^{MCI} = \frac{MA_{mt}^{MCI}}{MA_{m,t-1}^{MCI}} - 1 = \begin{cases} \frac{\sum_{l=1}^{M} ATT_{l,t-1}}{\sum_{l=1}^{M} ATT_{lt}} - 1 & m \neq j \\ \left(\frac{x_{mt}}{x_{m,t-1}}\right)^{\beta_m} \frac{\sum_{l=1}^{M} ATT_{l,t-1}}{\sum_{l=1}^{M} ATT_{lt}} - 1 & m = j. \end{cases} \tag{9}$$

Equation (9) shows that the relative change of market share is the same for all brands m ($m \neq j$) which leave own marketing activities unchanged.

A comparison of these results with correlations between price and market share calculated from periodical input data also yields a striking asymmetric structure. Turning from interval to point oriented contemplations we get point elasticities (with respect to market share) which are of special interest from a marketing practitioners perspective. Comparing the observed correlations with the elasticities calculated from the SOR-network (cf. Tab. 2) the cited asymmetry becomes evident in a similar manner. For clarity a classification of correlations and elasticities according to the scheme depicted in the lower part of Tab. 2 was used.

	Correlations (Raw data)					**Price Elasticities (SOR-network)**				
	$MA_{1\cdot}$	$MA_{2\cdot}$	$MA_{3\cdot}$	$MA_{4\cdot}$	$MA_{5\cdot}$	$MA_{1,1}$	$MA_{2,1}$	$MA_{3,1}$	$MA_{4,1}$	$MA_{5,1}$
$x_{1\cdot}$	−	+	+	+	+	−	++	++	+	+
$x_{2\cdot}$	+	−−	−−	+	+	+	−−	−−	+	+
$x_{3\cdot}$	+	−−	−−	+	+	+	−	−	−	−
$x_{4\cdot}$	0	+	+	−	−	+	+	+	−	−
$x_{5\cdot}$	0	+	+	−	−	+	+	+	−−	−

Coding Scheme	−−	−	0	+	++
Correlations	$(-\infty, -0.75]$	$(-0.75, -0.25]$	$(-0.25, 0.25)$	$[0.25, 0.75)$	$[0.75, \infty)$
Elasticities	$(-\infty, < -1]$	$(-1, 0)$	0	$(0, 1)$	$[1, \infty)$

Tab. 2: Structural comparison of correlations and elasticities in period 1

The asymmetry in the correlation matrix is essentially due to nearly identical price strategies within brand groups. Brand 2 and 3 as well as brand 4 and 5 are each constituting such a group because they belong to the same manufacturer and are following nearly the same pricing. As indicated by the very similar structure of both sign matrices, the asymmetric competition on the shampoo market was identified by the SOR-network approach in a convincing manner. The remaining discrepancies in a few cells of the matrices might be due to the linearity assumption of the correlation calculus.

6. Conclusions

In the present paper experiences with regard to a neural network-based model (called SOR-network) and a traditional approach for sales data analysis have been reported. The main results can be summarized as follows:

1. Both, the MCI approach as well as the neural network approach proved to be suitable instruments for market share modeling. Nevertheless, in our studies the MCI model was slightly dominated by the SOR-network.

2. With regard to the calculation of market diagnostics the SOR-network convinced by its special ability in reproducing asymmetric competition which, among other things, can be put down to the absence of the IIA-property.

3. Finally, a special advantage of the neural network approach is the non-necessity of an a priori specification of possible interactions (in variables). The degree of interaction is rather determined during calibration.

References:

COOPER, L.G., and NAKANISHI, M. (1988): Market-Share Analysis, Kluwer Academic Publishers, Boston.

DECKER, R. (1994): Analyse und Simulation des Kaufverhaltens auf Konsumgütermärkten, Lang, Frankfurt.

DECKER, R., and WARTENBERG, F. (1994): Neue Wege der Wettbewerbsanalyse im Einzelhandel. *Markenartikel*, 2, 46–51.

HANSSENS, D.M., PARSONS, L.J., and SCHULTZ, R.L. (1990): Market Response Models: Econometric and Time Series Analysis, Kluwer Academic Publishers, Boston.

HERTZ, J., KROGH, A., and PALMER, R.G. (1991): Introduction to the Theory of Neural Computation, Addison-Wesley, Redwood City.

HOWARD, J.A., and SHETH, J.N. (1969): The Theory of Buyer Behavior, Wiley, New York.

HRUSCHKA, H. (1991): Einsatz künstlicher neuraler Netzwerke zur Datenanalyse im Marketing. *Marketing ZFP*, 4, 217–225.

MCFADDEN, D. (1974): Conditional Logit Analysis of Qualitative Choice Behavior, in: P. Zarembka (ed.): *Frontiers in Econometrics*, Academic Press, New York, 105–142.

NAERT, P.A., and LEEFLANG, P.S.H. (1978): Building Implementable Marketing Models, Leiden, Stenfert Kroese.

RONNING, G. (1991): Mikroökonometrie, Springer, Berlin.

RUMELHART, D.E., HINTON, G.E., and WILLIAMS, R.J. (1986): Learning Internal Representation by Error Propagation. In: D.E. Rumelhart and J.L. McClelland (eds.): *Parallel Distributed Processing*, MIT Press, Cambridge, 318–362.

WARTENBERG, F., GAUL, W., and DECKER, R. (1994): Alternativ Modell: Neuronale Netze in der Kaufverhaltensforschung, Absatzwirtschaft, 7, 66–69.

Applications in Linguistics

On the Definition of Inflection

Pius ten Hacken

Institut für Informatik, Universität Basel, Petersgraben 51, CH-4051 Basel

Summary: A language-independent definition is proposed for the phenomenon of inflection. The definition is based on agreement between two elements in a certain configuration. It differentiates inflection on the one hand from derivation, and on the other hand from cliticization. This is especially urgent in Natural Language Processing systems where consistency has to be maintained even when many coders are involved. The definition can also be used in any theory of morphology, and in lexicography.

1. Introduction

In Natural Language Processing (NLP) systems, dictionaries are usually based on lexemes, i.e. a word with its inflectional paradigm. The input text consists of text words that have to be mapped to these lexemes to find the dictionary information necessary for further processing.

It is important to know whether something is inflection or not because of the difference in treatment. If a text word is an inflected wordform, the mapping to the lexeme is direct, otherwise the text word must be analyzed first.

In a large NLP-system, where several linguists and lexicographers collaborate, it is important that consistency in coding is supported by definitions and tests. In the case of inflection, people have intuitions on what should be included under this heading. However, these intuitions are not strong enough to decide in borderline cases, especially since intersubjective agreement tends to break down in these cases. Nevertheless it is preferable to search for the source of this intuition and make it explicit, rather than to stipulate an arbitrary list of inflectional features.

In ten Hacken (1993) I describe how definitions and tests can be related in a framework so that consistency and completeness are guaranteed. This framework imposes the following requirements on the definition of a phenomenon like inflection: it should be language-independent, and the criteria it consists of should be necessary conditions, formulated in positive rather than negative terms.

2. Grammatical Meaning vs. Conceptual Meaning

It is often difficult to draw an exact borderline between inflection and derivation. In both cases, there is a process affecting a base, e.g. by affixation. The intuition is that the result of the process has a different relation to the base, because derivation changes the conceptual meaning, whereas inflection only changes the grammatical meaning. This intuition can only be used as part of a definition if these notions are presented more formally. Two proposals from the literature will be discussed in sections 2.1 and 2.2, and in section 2.3 I will derive a partial solution from them.

2.1 Determinability

A first attempt to characterize change of grammatical meaning is the one by Matthews (1974). He proposes that when the choice between two forms is determined by a general grammatical rule, the forms are in an inflectional opposition. Thus, nominal Case in Latin is determined by the preposition in pairs like *ad montem* vs. *sub monte* ('towards/at the foot of [the] mountain'). Similarly, adjectival Case in Latin is determined by agreement with the modified noun, e.g. *monte alto* vs. *montem altum* ('[a] high mountain (abl./acc.)').

The choice between singular *ad montem* and the plural *ad montes*, however, is of a different kind. It is rather comparable to the choice of a preposition than to Case, because it is the result of an independent choice, not determined by a syntactic rule of any kind. Therefore, according to determinability, number is not an inflectional feature of nouns. For this reason, Matthews considers determinability a sufficient, non-necessary condition on inflection. Since this assumption violates a basic requirement of definitions in our framework, however, we cannot adopt this position.

2.2 Visibility

Anderson (1982) proposes that inflectional features are features that must be visible to syntactic rules for their correct application. This means that besides configurational properties such as nominal Case, and agreement properties such as adjectival Case in Latin, there are two further types of inflectional features.

One of them is the reverse of agreement, properties constituting the target for agreement. This includes number for nouns, because of contrasts like *monte alto* vs. *montibus altis* ('[a] high mountain' vs. 'high mountains' (abl.)). The other one includes features with direct consequences for phrases, realized at individual words. An example is verbal tense, determining whether a clause is finite or non-finite. Chomsky (1973) used this opposition in his Tensed S Condition to account for contrasts like *John is likely to go* vs. **John is likely that will go*.

A problem for this criterion is nominal gender in Latin. In the same way as number, it is a target for agreement, e.g. in *monte alto* vs. *arbore alta* ('[a] high mountain' vs. '[a] high tree' (abl.)). For Anderson this implies that gender is an inflectional feature, but it is hard to imagine what this means for the inflectional paradigm. It would be counterintuitive to call *monte* the ablative singular masculine of *mons*, but it is equally unsatisfactory to claim that there are inflectional features not represented in the inflectional paradigm.

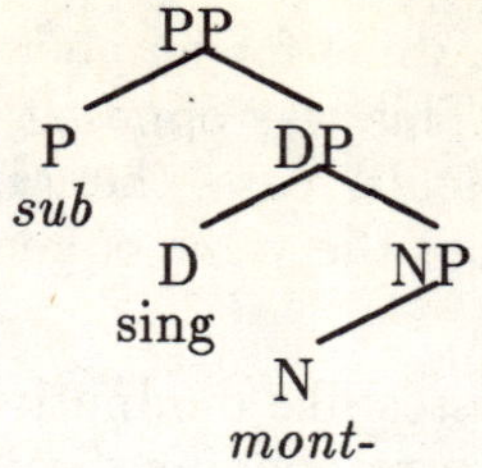

Fig. 1: Tree for *sub monte*

2.3 A Partial Solution

Recent developments in $\bar{X}$-theory make it possible to combine useful parts of determinability and visibility while leaving out their problems. According to $\bar{X}$-theory, a syntactic category X (or X^0) has two projections, X^1 (or $\bar{X}$) and X^2 (or XP), with complements attached to $\bar{X}$ and specifiers to XP. Chomsky (1970) originally proposed this structure only for the major categories N(oun), V(erb), and A(djective). Soon P(reposition)s and various minor categories were added to the scope of $\bar{X}$-theory. Stowell (1981) extended it to functional categories, such as I (or Infl), which in his analysis is the seat of tense and person/number agreement and the head of S=IP. VP is then a complement of I. Abney (1987) proposed a similar reanalysis for noun phrases, with D as a head of DP, and NP as a complement of D. Here D is the seat of person, number, gender, and Case, often lexically realized as a determiner or pronoun.

In this DP-analysis, the structure of *sub monte* is as in Fig. 1. This tree is slightly simplified because the unused level of $\bar{X}$ between X and XP has been left out. In the same way as *sub* is inserted as a P, and *mont-* as an N, the feature singular is inserted in D. P assigns ablative Case to DP, which is inherited by D. Masculine gender is a feature of *mont-* in the lexicon. By agreement between D and N, the features number, gender, and Case are shared. If we now define inflection as the realization of a feature whose value is determined by agreement, number and Case are inflectional features of the noun. Gender, however, is determined by the choice of the noun, rather than by agreement, so that it is not inflection.

3. Further Questions

There are some questions that the solution of 2.3 leaves unanswered. First, the reference to functional categories raises the question of which features originate in functional categories and which in the lexicon. This question is addressed in 3.1. Another question, addressed in 3.2, is how the distance between two elements in agreement can be constrained.

3.1 The Agreement Hypothesis

In 2.3 we assumed that the value for number appearing on nouns in Latin is originally inserted in D. This was opposed to gender, assigned to *mont-* in the lexicon. In this particular case, there is a strong motivation for the difference in analysis, because the value of gender depends on the choice of the noun.

More problematic are processes like diminutive formation in Dutch. Nearly all nouns in Dutch have a corresponding diminutive, e.g. *huis* ('house') vs. *huisje* ('small house'). Therefore, it is possible to describe the process as the realization of a feature [+/−dimin] originally inserted in D, parallel to number. It is more appropriate, however, to consider diminutive formation a process applied in the lexicon before insertion of the noun.

The difference between a feature in a functional category and a process in the lexicon is that only the former is available as information to influence syntax. Lexical processes may *affect* features visible to syntax, but they are not such features themselves. Thus, the agentive suffix *-er* in English changes the feature category (or part of speech) from V to N (e.g. *listen* vs. *listener*), but this is merely a side effect of the actual process it stands for.

It seems natural to assume that if it cannot be shown that a feature is in a functional category, it is not there. In other words, in order to be in a functional category, a feature must have direct syntactic influence that can be observed in at least some contexts. In this way it is not necessary to stipulate which features are in a functional category.

A prominent type of syntactic influence is agreement which is reflected overtly. There are two cases to be considered here. In the first case, the feature is the target of agreement realized at various words. Thus, number must be in D because it is not only realized at nouns, but also at adjectives agreeing with nouns. In the second case, the feature mediates between a lexical item and the word where it is realized. This is the case for Latin Case, assigned by a preposition.

For the alleged feature [+/−dimin] in Dutch, no such overt agreement can be shown. It is true that all diminutive nouns have neuter gender, and neuter gender is a target for agreement. Neuter gender is only a side effect of diminutive formation, however, and there is no agreement with a feature diminutive as such. In this context it is interesting to note that, according to Anderson (1982), the West-African language of Fula has an inflectional diminutive, involved in agreement. This illustrates the flexibility of the criterion presented here. It allows for the same process to be inflectional in one language and derivational in another one, depending on the differing properties it has.

Although we can now show that number for nouns in Latin and diminutive in Fula are inflectional, it remains very difficult to show that the diminutive in Dutch is not. Strictly speaking, we have to show that every conceivable syntactic rule functions correctly without reference to diminutive. This task is unrealistically big, if not entirely impossible. A way out is offered by the agreement hypothesis. The agreement hypothesis states that if a feature is syntactically active, it is at least involved in an overt agreement relation. By this hypothesis, the procedure to show which features are in functional categories is simplified quite drastically. Thus, for Dutch diminutives the above observation is sufficient evidence to show that diminutive is not an inflectional feature.

Another consequence of the agreement hypothesis is that we cannot use the opposition between finite and non-finite clauses to show that English tense is inflectional. This is not so serious, however, because we have contrasts like *I think John has left* vs. *I thought John had left*, where the value of tense in the subordinate clause agrees with tense in the matrix clause.

3.2 Clitics

The example of English tense at the end of the last section draws our attention to the problem of the distance between two elements in agreement. The English example is parallel to agreement between adjectives and nouns, in the sense that the relevant feature is realized at both items. A similar distance between the two elements, but with realization of the feature at one element only is found for French mood. In the same way as nominal Case in Latin may depend on the choice of the preposition, in French the choice between indicative and subjunctive mood of the subordinate verb may be determined by the lexical choice of the matrix verb, as in the following examples:

J'avoue que j'ai dormi. ('I admit that I have slept.')
Je nie que j'aie dormi. ('I deny that I have slept.')

If we do not restrict the distance of the two elements in agreement, a number of cases are classified as inflection that are more appropriately classified as involving clitics. An example is French weak object pronouns, as in the following example:

Il me/te reste fidèle. ('He to-me/to-you remains faithful.')

Me and *te* could in principle be analyzed as agreement of the verb *reste* with the person/number features of the DP argument of *fidèle*. It seems to me counterintuitive to do so, because *rester* would agree with features having no direct relation to this verb. Nevertheless, such an analysis is defended by Miller (1992). I think it is preferable to find a way of analyzing *me* and *te* as clitics, in line with the assumptions in traditional grammar and such classic works as Kayne (1975).

Another problematic example is the Czech complementizer with person/number marking, e.g. *abych* ('that-I'). The ending *-ch* could easily be analyzed as the realization of agreement with the subject DP, the more so because the verb of the clause introduced by *abych* lacks person/number features.

In order to formulate an appropriate constraint, I will refer to the generic tree in Fig. 2. As a general rule, X may agree with Y, but not with Z, for

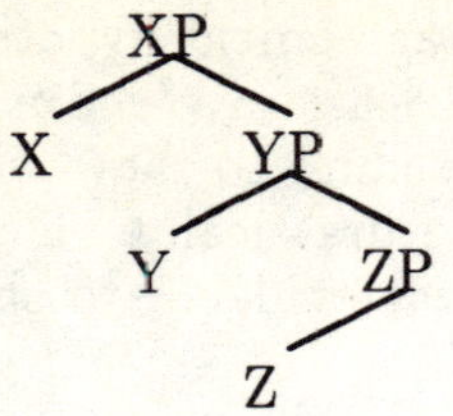

Fig. 2: Generic tree

the agreement to count as inflectional. This immediately solves the problem of French weak object pronouns. The tree for *me reste fidèle* matches Fig. 2 with X=V (*reste*), Y=A (*fidèle*), and Z=D (1sg.). Therefore, *me* cannot reflect inflectional agreement with the DP argument of *fidèle*.

A special case arises when Y is a functional category and ZP its complement. Adopting the terminology of Abney (1987), Y f-selects ZP and the maximal s-projection of Z becomes YP. If we formulate the relevant condition on inflection in terms of maximal s-projection instead of simply maximal projection, we can now solve the cases of English tense and French mood. In the latter case, the sentences used as examples above match Fig. 2 as follows: X=V (*avoue* or *nie*), Y=C (*que*) f-selects IP, whose head I in turn f-selects ZP=VP. The embedded verb Z=V is thus linked by a chain of f-selection to YP, the complement of X, so that the matrix verb may determine a feature of the embedded verb by inflectional agreement.

There is no explicit match for IP and I in Fig. 2, but since IP is f-selected by C and I f-selects VP, this does not matter. This is more than a trivial observation, because recent developments in $\bar{X}$-theory, initiated by Pollock (1989), tend to split up functional categories into more and more separate ones, e.g. T(ense) with TP. Since we only refer to maximal s-projections, the procedure is independent of any intermediate functional projection.

Our explanation of French weak object pronouns is not affected by the reference to maximal s-projections, because f-selection does not play a role there. The maximal s-projection of D is DP, and there is still the AP intervening between V and DP.

Finally, Czech *abych* would arise by agreement of C (*aby*) with the features '1sg' in the subject DP. Since the example involves a subject, i.e. a specifier, we cannot use the tree of Fig. 2. In Fig. 3, the $\bar{I}$ is represented explicitly. The maximal s-projection of D is DP, because DP is not f-selected. This means that what looks like agreement between C and D cannot be analyzed as agreement in the inflectional sense. Instead of being part of the inflectional paradigm of *aby*, *abych* is the result of another type of rule. I assume that, in the same way as French weak object pronouns, the ending *-ch* is the result of a clitic realization rule. Additional plausibility for such an analysis

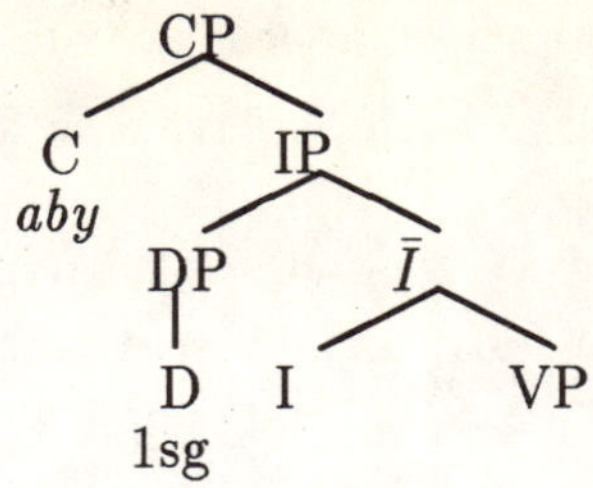

Fig. 3: Tree for *abych*

is supplied by Polish, because there the person-number ending can attach to almost any word in sentence-initial position instead of to the verb.

4. Conclusion

On the basis of the discussion in the preceding sections, we can formulate the following definition of inflection:

An inflection process is a process realizing a feature or combination of features F on a word W, such that:
• The value of F is determined by agreement with another word or with a functional category.
• If the two elements in agreement are in X and Y, either X and Y are in the same maximal s-projection, or the maximal s-projection of Y is the complement or the specifier of X.

Although a number of theory-internal notions are referred to in the definition, they do not affect its generality. As shown in the discussion, they are not primitive, but emerge as a result of observable properties of the language.

References:

ABNEY, Steven Paul (1987): *The English Noun Phrase in its Sentential Aspect*, Ph. D. Dissertation, Massachusetts Institute of Technology, Cambridge (Mass.).

ANDERSON, Stephen R. (1982): Where's Morphology?, *Linguistic Inquiry* 13, p. 571-612.

CHOMSKY, Noam (1970): Remarks on Nominalization, in Jacobs, Roderick A. & Rosenbaum, Peter S. (eds.): *Readings in English Transformational Grammar*, Ginn, Waltham (Mass.), p. 184-221.

CHOMSKY, Noam (1973): Conditions on Transformations, in Anderson, Stephen R. & Kiparsky, Paul (eds.), *A Festschrift for Morris Halle*, Holt, Rinehart & Winston, New York, p. 232-286.

TEN HACKEN, Pius (1993): *Defining Morphology: A Principled Approach to*

Determining the Boundaries of Compounding, Derivation, and Inflection, Ph. D. Dissertation, University of Basel, to appear at Olms Verlag, Hildesheim.

KAYNE, Richard S. (1975): *French Syntax: The Transformational Cycle*, MIT Press, Cambridge (Mass.).

MATTHEWS, Peter H. (1974): *Morphology: An Introduction to the Theory of Word Structure*, Cambridge University Press, Cambridge.

MILLER, Philip H. (1992): *Clitics and constituents in phrase structure grammar*, Garland, New York.

POLLOCK, Jean-Yves (1989): Verb Movement, Universal Grammar, and the Structure of IP, *Linguistic Inquiry* 20, p. 365-424.

STOWELL, Timothy (1981): *Origins of Phrase Structure*, Ph. D. Dissertation, Massachusetts Institute of Technology, Cambridge (Mass.).

Computer-Aided Analysis of Vocabulary Acquisition

Joachim Liedtke*

Fachbereich Sprach- und Literaturwissenschaften, Universität Bremen, Postfach 33 04 40, D-28334 Bremen (Telefax: 04 21/2 18-42 83)

Summary: After summing up the main parameters and problems of vocabulary acquisition in a foreign language, the paper discusses the possibilities of using a computer in order to optimize the learning process. As a concrete step towards finding a solution the concept and structure of the *VOCA-Research Project* is introduced, which attempts to make the conditions of second language vocabulary learning more explicit and effective, both from a theoretical and a practical point of view.

1. Introduction

Without any doubt vocabulary is – in addition to the grammar and the practice of communication in concrete situations – one of the most substantial components within the process of second language acquisition. Contrary to that importance there exists a great deal of uncertainty and ignorance concerning the factors and parameters which influence and determine vocabulary acquisition in a foreign language. For example, in the introduction of a modern learner's dictionary we find the following suggestions for working with the material:

"Every day learn a specific amount, taking breaks in between each session [...]; the actual amount of vocabulary depends on you. Don't forget to repeat and test yourself at regular intervals" (Langenscheidt (1991: X)).

At first glance these hints seem obvious and useful. Nevertheless they raise several questions as yet unanswered: For instance, how many items does the "specific amount" exactly contain, if students ask for the optimal volume? How long should the "regular intervals" last, and are there any limits to the breaks? Is it better to memorize the material more frequently in intervals of hours or days rather than in terms of weeks, months or even years? And finally: Is it really helpful and practicable to be given the advice: it "depends on you"?

The following discussion starts with a sketch of the main parameters and problems of vocabulary acquisition in a foreign language (chap. 2). After that the concept and structure of the *VOCA-Research Project* is introduced as a suggested solution (chap. 3). Illustrations of the evaluation of the parameters (chap. 4) and a conclusion (chap. 5) complete this paper.

*The author wishes to thank the *Dr. Meyer-Struckmann Stiftung* for supporting this work with a grant in association with the *Studienstiftung des deutschen Volkes.*

2. Vocabulary Acquisition: Parameters and Problems

2.1 Parameters of the Learning Process

If we concentrate on the parameters of the vocabulary within second language acquisition, we can distinguish at least four components of the learning process: (1) the learning subject, (2) the learning context, (3) the learning method, and (4) the learning person (cf. Fig. 1).

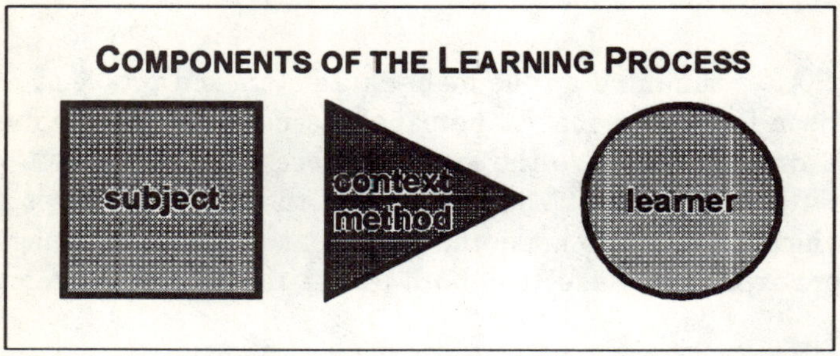

Fig. 1: Components of the learning process

The learning subject covers the lexical material, i.e. the vocabulary and its formal and semantic attributes. The learning context includes the temporal, the local and the personal conditions of the learning situation. The concrete course of the learning strategy is involved in the learning method, and the personal characteristics form part of the learner component.

Each of these groups can be separated into two subcategories: the quantitative and the qualitative parameters. For example, within the group of the quantitative factors of the *learning subject* the length, the number, and the frequency of the words are of prime importance. Qualitative characteristics are the formal and lexical complexity, the similarity, the iconicity and the usefulness of the expressions such as words, word groups and collocations.

Quantitative parameters of the *learning context* are the time of day, the premises with their ambience, and the persons. Qualitative influence results, for instance, from the distractions, the assistance, the encouragement, and the personal constellation and its hierarchy.

If we look at the quantitative parameters of the *learning method*, we find important factors like the volume and the duration of the learning sessions as well as the intervals and the frequency of the repeats. The didactic techniques, the technical instruments and tools and the psychological aids in vocabulary learning and memorising (e.g. mnemonics, imaginalization of verbal information, semantic evaluation of knowledge networks etc.) are elements of the qualitative subgroup.

Quantitative factors of the *learning persons* are their age and sex, among others. General and special prior knowledge, physiological and psychological conditions, motivation and, last but not least, the results and the success complete this final component of the learning process from a qualitative point of view (cf. Fig. 2).

LEARNING SUBJECT / MATERIAL (VOCABULARY)	LEARNING CONTEXT (SITUATION)	LEARNING METHOD (STRATEGY / TECHNIQUE)	LEARNING PERSON (INDIVIDUAL LEARNER)
quantitative parameters: • length • number and • frequency of the words	quantitative parameters: • time of day • premises • persons	quantitative parameters: • volume and • duration of the learning sessions • intervals and • frequency of the repeats	quantitative parameters: • age • sex (female / male)
qualitative parameters: • formal and lexical complexity • similarity • iconicity • usefulness	qualitative parameters: • distraction • assistance • encouragement • personal constellation and hierarchy	qualitative parameters: • (didactical) techniques and methods • (technical) instruments and tools • (psychological) learning aids	qualitative parameters: • general / special prior knowledge • physiological / psychological conditions • motivation • results and success

Fig. 2: Parameters of vocabulary acquisition

2.2 Problems of Evaluation and Optimization

Keeping in mind that – according to Chomsky (1981: 19) – the scientific analysis of empirical data should reach the levels of (i) observational, (ii) descriptive, and (iii) explanatory adequacy, it seems obvious that the extensive field of data as outlined above raises many unanswered questions.

First of all, we have a large number of parameters which must be taken into account if we try to analyse and optimize the vocabulary acquisition in an adequate manner. Second, it is necessary to find a way of quantifying such factors for which no exact measures exist, like the word complexity or the learner's motivation. Next, we have to consider the correlation between many variables. For instance, the success of the learning sessions probably depends on the time of day – e.g. morning vs. evening – (cf. Fig. 6), but undoubtedly an increase in motivation could make the learning effort less dependent on the external circumstances and conditions. And finally we have to heed the fact that there is no guarantee for the correctness of the general principles, predicted by foreign language didactics and learning psychology, are valid in view of the learner's individuality (cf. Fig. 3).

Considering that these problems have to be solved first if the research target is to analyse the conditions of vocabulary learning, we would like to introduce the VOCA-Program as a possible solution.

- **QUANTITY / COMPLEXITY:** the large number of factors of influence,
- **QUALITY / MEASURABILITY:** the question, how to quantify the qualitative parameters,
- **CORRELATION:** the dependency between many variables,
- **INDIVIDUALITY:** the fact, that the individual behaviour can considerably differ from general principles.

Fig. 3: Problems of parameter evaluation and optimization

3. Suggested Solution: The VOCA-Program

3.1 The Meanings of the VOCA-Program Concept

The concept of the VOCA-Program (abbreviation for v o c a b u l a r y a c q u i s i t i o n) includes at least three meanings: (i) It denotes our *research project*, which intends to accumulate theoretical knowledge about the process of vocabulary learning from the viewpoint of foreign language didactics and learning psychology. (ii) It is the name of the *software*, used for the extraction and analysis of the empirical data within this research project and (iii) it points to the practical value of the computer-assisted *learning method* in order to support the concrete learning process and to optimize the learning success (cf. Fig. 4).

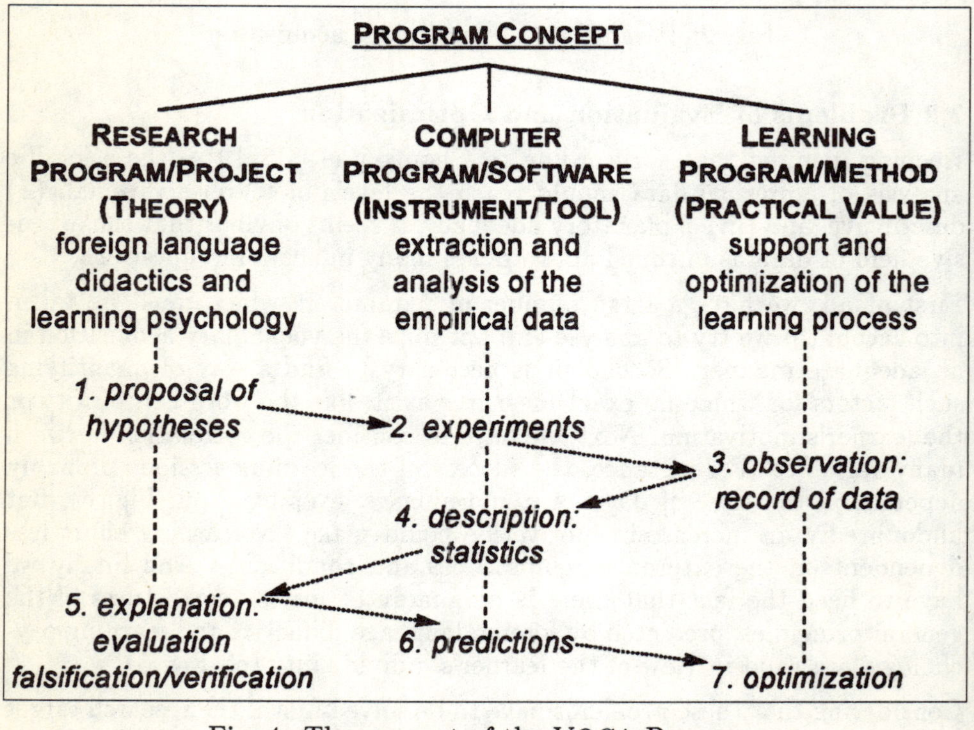

Fig. 4: The concept of the VOCA-Program

These focal points determine the sequence of the steps to be taken: (1) After the proposal of hypotheses on the basis of the traditional and current theory of foreign language didactics and learning psychology, (2) experiments are made using the specially constructed software to obtain empirical data. (3) By means of observing and recording the real learning behaviour, (4) a statistical description is prepared with the purpose of finding out the significant parameters. (5) On this background results are obtained by evaluating the empirical data in order to prove the hypotheses and (6) to deduce predictions about the leaning principles. (7) The work will be completed by the optimization of the learning behaviour (cf. Fig. 4). To make this process more explicit, we are going to outline the structure of the computer program in the next chapter.

3.2 The Structure of the VOCA-Program

According to the main parameter groups discussed above, the structure of the VOCA-Program has three units: (i) the vocabulary module, (ii) the learning process module, and (iii) the evaluation module (cf. Fig. 5).

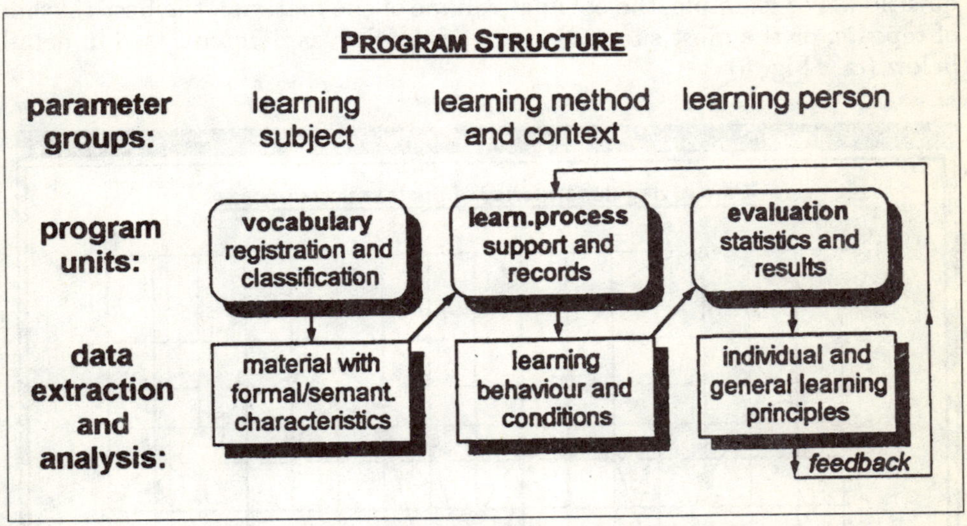

Fig. 5: The structure of the VOCA-Program

The first unit – the *vocabulary module* – serves the purpose of registering the lexical material and classifying its formal and semantic characteristics, like the length and number of words as well as their complexity, thematic assignment and synonyms (cf. the list of features in Fig. 2).

The second unit – the *learning process module* – has the task of supporting the (computer-assisted) learning process and recording the learning behaviour and its conditions, e.g. the time and duration of the learning sessions, the number of items, and the learning results.

And the third unit – the *evaluation module* – is conducive to working out the individual and general learning principles on the basis of collected facts.

To give an illustration two cases will be discussed in the next chapter: an example of quantitative and an example of qualitative parameter evaluation.

4. Illustrations and Discussion

4.1 The Time of Day-Dependency as an Example of Quantitative Parameter Evaluation

As was shown above, one of the main arguments for using the computer to analyse the conditions of the vocabulary acquisition is the enormous complexity of the parameters. Only the computer makes it possible to take the huge number of involved factors into account. For this reason the presented VOCA-Program records the quantitative parameters of the learning subject, the context, the method, and the learner during the real learning process. The resulting basis of data offers the possibility of explaining which characteristics are most influential. By means of this statistical material, the learners are encouraged to optimize their individual learning behaviour, which in turn lead to more general knowledge about the conditions of vocabulary acquisition. For example, the optimal volume of the material, the best rhythm of repeats, or the most successful time of the day, as demonstrated in detail below (cf. Fig. 6).

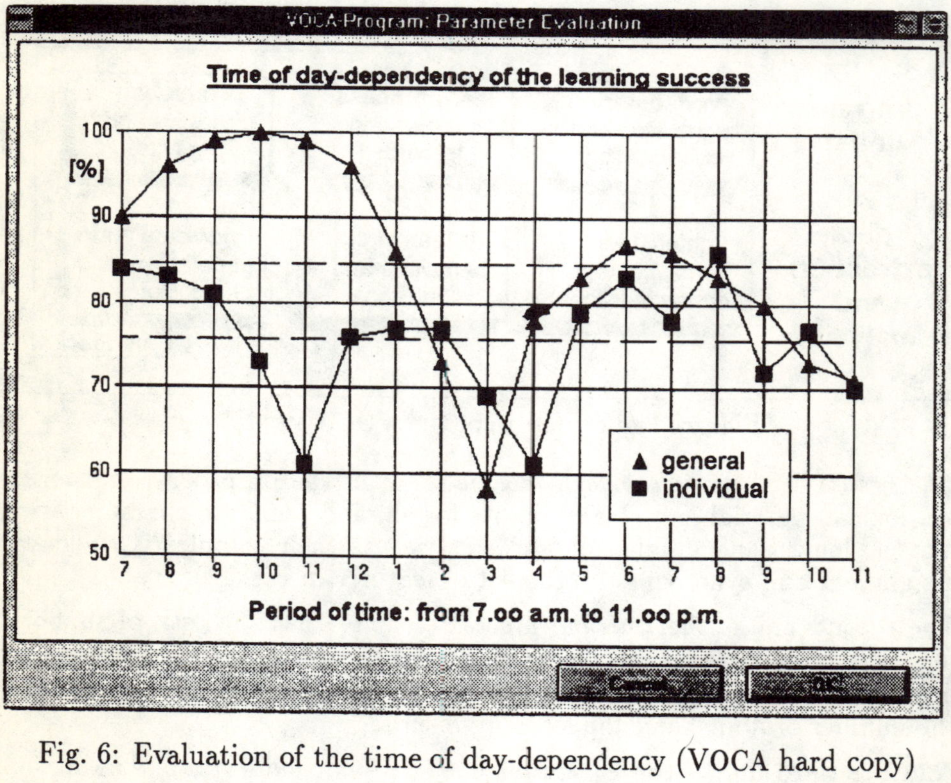

Fig. 6: Evaluation of the time of day-dependency (VOCA hard copy)

The graph – a hard copy of the VOCA-Program – illustrates the learning success of an individual learner depending on the hour of the day. As we can see, there is an approximate similarity in the space of time after 2 o'clock p.m. between individual learning capacity and the general values, predicted by research in the field of learning psychology (cf. Schräder-Naef (1992: 96)). However, there exists a substantial discrepancy in the hours of the morning, which underpins our assertion that only an individual evaluation of the parameters ensures the optimization of the learning behaviour. This argument is equally valid for the qualitative parameters like error analysis, exemplified in the next chapter.

4.2 The *Error Analysis* as an Example of Qualitative Parameter Evaluation

As discussed in Chapter 2, we need a qualitative analysis in addition to the quantitative evaluation of the parameters. To give an immediate impression of this program module, the following table shows a list of several errors, produced by an individual learner and recorded by the VOCA-Program during the concrete learning process (cf. Fig. 7).

(1) Item:	*Sicherheit*	(2) Item:	*etwas bereuen*
Translation:	*security*	Translation:	*repent*
Check 1:	*secur**e**ty*	Check 1:	*repen**d***
Check 2:	*secur**e**ty*	Check 2:	*repen**d***
Check 3:	*secur**e**ty*	Check 3:	*repen**d***
(3) Item:	*etwas beweisen*	(4) Item:	*gründlich*
Translation:	*prove*	Translation:	*thorough*
Check 1:	*pro**of***	Check 1:	***reliable***
Check 2:	*pro**of***	etc.	[= *zuverlässig*]
Check 3:	*pro**of***		
(5) Item:	*Angel*		
Translation:	*fishing-rod*		
Check 1:	***angel*** [= Engel]		
Check 2:	***angel*** [= Winkel; Aspekt, Gesichtspunkt]		
(6) Item:	*Er ist jetzt nur noch Haut und Knochen.*		
Translation:	*He's only skin and bone now.*		
Check 1:	*He's **still** skin and **beans only**.*		
Error analysis:	1st step: lexical choice (two times), word order		
Check 2:	*He's **still** skin and bone**s only**.*		
Error analysis:	2nd step: lexical choice (once), plural mistake		
Check 3:	*He's **still** skin and bone now.*		
Error analysis:	3rd step: lexical choice (once)		
Check 4:	*He **is** only skin and bone now.*		
Error analysis:	4th step: expression (within the tolerance limits)		

Fig. 7: Computer-assisted error analysis

The threshold of relevance in these cases is fixed at three. This means the condition for starting the computer-assisted qualitative error analysis is the appearance of more than two faulty repeats of an item.

One can easily recognize that there are typical categories of error, e.g. mistakes of similar (spoken/written) letters (cf. (1), (2)), of similar meanings of expressions (cf. (3), (4)), and of so-called false friends (cf. (5)).

The computer-assisted error analysis also allows for drawing conclusions as to the learning progress and the increase of grammatical knowledge. Example (6) indicates the development from a low level with several mistakes in the lexical choice and word order, to a higher level of advanced linguistic competence.

5. Conclusion

The wide range of problems associated with the question of how to optimize vocabulary acquisition requires the computer as a research tool and a learning aid. There is hardly any alternative if we try to do justice to the complexity, the quality, the correlation, and the individuality of the involved parameters. Only the computer-assisted recording of data during the learning process under realistic circumstances puts us in the position of describing the learning behaviour and establishing its conditions in an adequate manner.

The introduced and illustrated VOCA-Program as a proposed solution is meant to be a step forward, enabling us to gain knowledge about the principles of the vocabulary learning process: From a theoretical point of view, in order to promote second language didactics (e.g. the contrastive grammar), and learning psychology (e.g. the growth, the cognitive organization and mental representation of knowledge) as well as from a practical point of view, with the goal of making a learning tool available which supports the learning process and optimizes the learning success.

References:

CHOMSKY (1981): Regeln und Repräsentationen. Frankfurt am Main.

LANGENSCHEIDT (1991): Basic German Vocabulary. Berlin, Munich.

SCHRÄDER-NAEF, Regula D. (1992): Rationeller Lernen lernen. Weinheim, Basel.

Features and Tags

Sven Naumann

Linguistische Datenverarbeitung, Universität Trier, D–54286 Trier

Summary: In many natural language processing systems (NLP–systems) the analysis of linguistic data starts with the identification of word classes (*tagging*). This paper describes the knowledge–based tagger FTAG which was developed as part of a system for the acquisition of syntactic knowledge from corpora.

1. Overview

Since machine–readable corpora became available in the early sixties quite a number of different *tagging systems* have been proposed. FTAG is a knowledge–based *tagging system* (**tagger** from now on) which was designed as a flexible tool for the linguistic analysis of medium–sized corpora*. The following two sections describe the tag set used in FTAG and the approach taken to make sure that each word form gets a unique tag. Section 4 gives a detailed impression of the system architecture and the tagging algorithm. Finally, first results and necessary improvements are discussed.

2. Tags

Unlike most other taggers FTAG uses complex feature structures instead of a fixed set of simple tags. Using feature structures as tags offers several advantages:

1. As feature structures have turned out to be the most popular data structure for presenting linguistic information in NLP–systems, a tagger which uses feature structures can be easily combined with those systems.

2. Feature structures can be extended dynamically: Information produced by different modules of an NLP–system can be combined quite naturally in a single feature structure; i.e. the morpho–syntactic information generated by the tagger can be completed by information which come from the syntax or the semantic module.

3. Applications may differ with respect to the granularity of (lexical) analysis required. Feature structures make it possible to change the size of the tag set quite easily: If all features (and their values) are considered as relevant for determining the identity of two tags, maximum granularity is achieved. To reduce the size of the tag set, it is sufficient to mark a certain subset of features as irrelevant in this respect.

---

*At the moment, there is only a german version of FTAG. But the modular structure of the system makes is easy to adapt it to different languages.

4. Feature structures allow a flexible treatment of ambiguity, which is one of the main problems in tagging: If $\mathbf{C} = \{C_1, \ldots, C_n\}$ $(n > 1)$ is a set of candidate tags for a word form w, a unique tag for w can be formed either by simply disregarding all features whose value are not determined unambiguously (*underspecification*) or by using logical connectives (such as disjunction and negation) to combine the conflicting information within one feature structure[†].

FTAG uses typed feature structures which contain at least the following features:

$$\begin{bmatrix} \text{PHON} & \textit{word form} \\ \text{SYN} & \begin{bmatrix} \text{HEAD} & \begin{bmatrix} \text{MAJ} & \textit{category} \\ \text{TYP} & \textit{subclass} \\ \vdots & \vdots \end{bmatrix} \\ \text{LEX} & + \end{bmatrix} \end{bmatrix}$$

If w is a word form and f a feature structure generated by FTAG as tag for w, then the PHON–feature in f takes w itself as value. MAJ(or) determines the word class (verb, noun, adjectiv, ...) of w and TYP the subclass w belongs to. The feature specification <LEX +> marks w as a lexical item. Depending on the values of MAJ and TYP further feature specifications may be added. Features either take simple objects called *atoms* (e.g. PHON, LEX, etc.) or feature structures (e.g. SYN) as values. To minimize computational demands, the use of disjunction and negation is restricted to atomic values in FTAG.

Example (1)

Definite articles in German typically show a high degree of ambiguity concerning gender, number and case. There are, for example, eight readings for *die*:

$$\textbf{fem}\text{inine} \left\{ \begin{array}{l} \textbf{nom}\text{inative} \\ \textbf{acc}\text{usative} \end{array} \right\} \textbf{sing}\text{ular} \;\Bigg|\; \left\{ \begin{array}{l} \textbf{mas}\text{culine} \\ \textbf{fem}\text{inine} \\ \textbf{neu}\text{ter} \end{array} \right\} \left\{ \begin{array}{l} \textbf{nom}\text{inative} \\ \textbf{acc}\text{usative} \end{array} \right\} \textbf{plural}$$

Instead of generating eight candidate tags, FTAG proposes two tags:

$$\begin{bmatrix} \text{PHON} & \textit{die} \\ \text{SYN} & \begin{bmatrix} \text{HEAD} & \begin{bmatrix} \text{MAJ} & \text{article} \\ \text{TYP} & \text{definite} \\ \text{case} & \text{nom} \vee \text{acc} \\ \text{gen} & \text{fem} \\ \text{num} & \text{sg} \end{bmatrix} \\ \text{LEX} & + \end{bmatrix} \end{bmatrix}$$

[†]In some cases it may be still necessary to assign w more than one tag.

$$\begin{bmatrix} \text{PHON} & \textit{die} \\ \text{SYN} & \begin{bmatrix} \text{HEAD} & \begin{bmatrix} \text{MAJ} & \text{article} \\ \text{TYP} & \text{definite} \\ \text{case} & \text{nom} \vee \text{acc} \\ \text{gen} & \text{mas} \vee \text{fem} \vee \text{neu}^{\ddagger} \\ \text{num} & \text{pl} \end{bmatrix} \\ \text{LEX} & + \end{bmatrix} \end{bmatrix}$$

3. Disambiguation

When work on FTAG started the tagger was planned as a module of a system for the corpus–based acquisition of syntactic knowledge (cf. Naumann & Schrepp (1992)). Considered in this context, it seemed to be unnecessary to tune the tagger in a way which promised to minimize the number of ambiguously tagged word forms. Disambiguation was regarded as a function of the learning process, mainly driven by the evaluation of competing structural hypotheses. Only later, when it became obvious that a tagger could be used as an independent system in quite a number of different contexts, other ways of dealing with ambiguity became interesting. While nowadays most taggers use statistical information to resolve ambiguity (usually the transition probability between tags which is computed on the basis of large manually tagged corpora (cf. Garside, Leech & Sampson (1987))), a simpler solution is used in FTAG: A small number of *context rules* is used to resolve ambiguity by analysing the local context of word forms for which more than one candidate tag is proposed[§]. Context rules can be written as context–sensitive rewriting rules:

$$\textit{left context } \{C_1, \ldots, C_n\} \textit{ right context} \longrightarrow C_i$$

By applying a context rule, one of the candidate tags of a ambiguously tagged word form w is selected. In FTAG the total context considered by a context rule (*left context* + *right context*) is restricted to three word forms.

Example (2)
Consider the sentence *Sie liegt auf der Lauer* (She is on the look–out). There are a number of possible readings for *der*: It can be interpreted as a definite article, a relative pronoun, or a demonstrative pronoun. Disambiguation takes several steps:

1. By applying the following context rule, the article reading is selected:

$$\textit{preposition} \begin{Bmatrix} \text{RelPro} \\ \text{DefArt} \\ \text{DemPro} \end{Bmatrix} \textit{noun} \longrightarrow \text{DefArt}$$

   But there is still more than one candidate tag for *der*:

[‡] If underspecification is used (cf. 4.1), the gender specification is simply dropped.

[§] This work has been carried out by C.Schulz in her M.A. thesis (cf. Schulz (1993)).

$$\begin{bmatrix} \text{PHON} & \textit{der} \\ \text{SYN} & \begin{bmatrix} \text{HEAD} & \begin{bmatrix} \text{MAJ} & \text{article} \\ \text{TYP} & \text{definite} \\ \text{case} & \text{gen} \\ \text{gen} & \text{mas} \vee \text{fem} \vee \text{neu} \\ \text{num} & \text{pl} \end{bmatrix} \end{bmatrix} \end{bmatrix}$$

$$\begin{bmatrix} \text{PHON} & \textit{der} \\ \text{SYN} & \begin{bmatrix} \text{HEAD} & \begin{bmatrix} \text{MAJ} & \text{article} \\ \text{TYP} & \text{definite} \\ \text{case} & \text{gen} \vee \text{dat} \\ \text{gen} & \text{fem} \\ \text{num} & \text{sg} \end{bmatrix} \end{bmatrix} \end{bmatrix}$$

$$\begin{bmatrix} \text{PHON} & \textit{der} \\ \text{SYN} & \begin{bmatrix} \text{HEAD} & \begin{bmatrix} \text{MAJ} & \text{article} \\ \text{TYP} & \text{definite} \\ \text{case} & \text{nom} \\ \text{gen} & \text{mas} \\ \text{num} & \text{sg} \end{bmatrix} \end{bmatrix} \end{bmatrix}$$

2. The preposition *auf* is marked as either governing dative or accusative case:

$$\begin{bmatrix} \text{PHON} & \textit{auf} \\ \text{SYN} & \begin{bmatrix} \text{HEAD} & \begin{bmatrix} \text{MAJ} & \text{preposition} \\ \text{TYP} & \text{pre} \\ \text{case} & \text{dat} \vee \text{acc} \end{bmatrix} \end{bmatrix} \end{bmatrix}$$

   This information can be used to eliminate all but the second candiate tag for *der*. In return, the dative reading for *auf* is selected.

3. Finally, the agreement features of the determiner are instantiated in the noun tag which was drastically underspecified up to this moment:

$$\begin{bmatrix} \text{PHON} & \textit{lauer} \\ \text{SYN} & \begin{bmatrix} \text{HEAD} & \begin{bmatrix} \text{MAJ} & \text{noun} \\ \text{TYP} & \text{common} \\ \textit{case} & \textit{dat} \\ \textit{gen} & \textit{fem} \\ \textit{num} & \textit{sg} \end{bmatrix} \end{bmatrix} \end{bmatrix}$$

4. Tagging

4.1 The FTAG Modules

FTAG is written in Common Lisp and consists of three modules: an *input module*, a *tagging module* and an *output module.* These modules can be easily configured to adapt the system to different needs.

Input module. The input module controlls the interaction between FTAG and the corpus to be analysed. Codes for special characters used in the corpus (such as punctuation–marks, brackets, etc.) have to be declared in this module. It is possible to extend the internal lexicons which FTAG uses by providing additional lexical knowledge–bases and appropriate converters, which transform the information contained in these knowledge–bases into the feature notation used by FTAG.

Tagging module. The linguistic knowledge which guides the tagging process consists of a number of small lexicons, a deflection algorithm and a set of context rules which can be used for tag–disambiguation. If the disambiguation function is activated, one of the following three modes can be used:

- *Automatic Mode*
  Only the context rules are used to reduce the number of tags assigned to a word form.

- *Interactive Mode*
  In cases where there is more than one candidate tag for a word form, all candidate tags and the left and right context of the word form are displayed on the screen and the user is asked to select the right one(s).

- *Mixed Mode*
  The context rules are tried first and if necessary, remaining ambiguities are resolved like in the interactive mode.

Output module. By default, all information produced by the tagging module is directed to a file. There are a number of switches which controll the behavior of the system in cases where more than one tag is proposed for a word form w: It is possible to keep all candidate tags or to use underspecification and/or disjunction and negation to reduce the number of tags assigned to w. A protocol–function makes it possible to collect some usefull information about the progress of the analysis in a separate file. Among the data which can be recorded are: The number of sentences and word forms processed, the frequency of certain tags and the average sentence length.

4.2 Tag Selection

The selection of a tag for a word form w is determined by the lexical and the morphological knowledge available to the system.

Lexicons. In the German version of FTAG there are three lexicons which play respectively different roles for the tag selection:

- The main lexicon (size: 700 items) contains entries for the elements of the closed word classes (determiner, prepositions, pronouns, etc.).

- The verb–adjective lexicon (size: 350 items) contains the base form of morphological unanalizable adjectives. and a number of irregular verbs.

- The suffix lexicon (size: 85 items) contains the derivational and inflectional suffixes which can be used to identify nouns, verbs and adjectives.

The main lexicon and the verb–adjective lexicon are organized as hash tables, the items of the suffix lexicon are compiled into a letter tree (*trie*).

Morphological Analysis. A simple deflection algorithm is used to analyse a word form w not found in the main lexicon or in the verb–adjective lexicon: w is scanned form right to left to identify derivational and/or inflectional suffixes stored in the suffix–trie. All possible segmentations are evaluated and the one with the highest score is selected. By changing measures different search strategies can be realized (*longest match*, *shortest match*, etc.)

Lexical and morphological knowledge is combined in a straightforward way: Each time FTAG is confronted with a new word form w, analysis starts with a lexical lookup using the main lexicon. If this operation is successfull, the information found is used to generate a unique tag for w. Otherwise the capitalization test is executed: If w starts with a capital letter and is not the first word of a sentence, w is taken to be a noun by default and a highly underspecified noun–tag for w is generated. In a second step the deflection algorithm is used to confirm this hypothesis and to produce further information concerning gender, sex and case. If the capitalization test fails the verb–adjective lexicon is searched. In case the search is successfull, tag selection is solely determined by the available lexical knowledge. Otherwise the deflection algorithm is used to generate a plausible tag hypothesis.

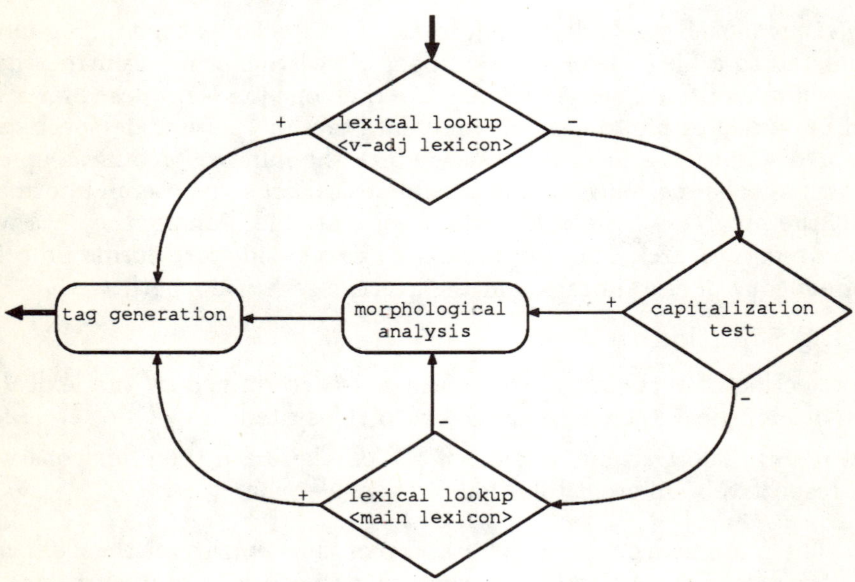

Fig. 1: Tag selection

5. Results and Further Work

To get a first impression of the performance of FTAG a number of small corpora (size: 1000–3000 word forms) containing short newspaper articles on politics and science have been tagged manually and using FTAG. In these tests FTAG has been applied twice: While in the first pass disambiguation information was disregarded, the context rules were used in the second pass to minimize the number of ambiguously tagged word forms. Tab. 1 shows the results obtained in both passes. The slight increase of the error rate when context rules were applied is due to incorrectly tagged word forms which fooled the disambiguation rules.

disambiguation	–	+
unambiguously tagged	30 %	45 %
unique MAJ value	28 %	37 %
multiple MAJ values	37 %	10 %
error rate	5 %	8 %

Tab. 1: Performance test

Although these first results are quite promising, there are a number of ways in which the efficiency and accuracy of the system could significantly be improved:

- The representation of linguistic information in the main and the verb–adjective lexicon suffers from unnecessary redundancy. First steps have been taken to design a suitable representation language using object–oriented methods.

- Texts usually contain abbreviations and idomatic expressions which cannot be processed correctly by FTAG at the moment. As the set of these expressions is strictly limited, it seems to be sensible to collect the most frequent idioms and abbreviations in a special lexicon.

- To analyse expressions containing numbers and formulas in an adequate way, typical patterns underlying this type of expressions have been coded by using regular expressions.

References:

GARSIDE, R., LEECH, G. & SAMPSON, G. (1987): The Computational Analysis of English. Longman, London and New York.

NAUMANN, S., SCHREPP J. (1992): An empirical approach to syntax learning. In: G.Görz (ed.), Konvens 92. Springer–Verlag: Berlin/Heidelberg, 209–17.

POLLARD, C., SAG, I. (1987): Information–Based Syntax and Semantics. Volume I: Fundamentals. Stanford University Press, CSLI Lecture Notes 13.

SCHOTT, G.,(1978): Automatische Deflexion deutscher Wörter unter Verwendung eines Minimalwörterbuchs. In: Sprache und Datenverarbeitung 2, 67–77.

SCHULZ, C. (1993): Wissensbasierte Identifikation von Wortarten. Unpublished M.A. thesis, University of Trier.

Semantically Based Universal Definitions of Grammatical Agreement and Agreement Domain Universals: A Critical Evaluation

Peter Schmidt

Linguistische Datenverarbeitung,
Universität Trier, D–54286 Trier

Summary: There exist two major proposals concerning the feasibility of a universal semantically based definition of grammatical agreement (GA), and the issue of a universal semantically based solution to the domain problem of GA, viz. Keenan's functional (dependency) principle and its elaborations, and the 'coreference/anaphoric' approach, first presented in Lapointe (1980). Both proposals have been put forward as absolute universal principles governing GA. This paper pursues several issues related to empirical testability of both approaches and shows, on the basis of Indo-European and Non-Indo-European evidence, that there exist significant counterexamples to both principles, which invalidate them as absolute universals and seriously undermine the corresponding universal definitions of GA.

1. Introduction

After some period of almost total neglect, morphosyntax has re-entered the focus of attention of theoretical linguistic research over the past two decades. Two of the major problems, with obvious relevance for both linguistic typology and the theory of grammar, are those of providing a cross-linguistically valid, preferably universal, definition of fundamental morphosyntactic notions, among them **grammatical agreement** (GA)/concord, and of finding a universal solution to the so-called **domain problem** of GA (cf. Barlow and Ferguson (1988)), viz. the question which constituents can/cannot stand in relations of GA in natural language (NL), and why.

There have been two major proposals, and corresponding lines of research, which aim at providing a semantically based solution to the aforementioned problems: (i) work inspired by the **functional (dependency) principle** (FDP), going back to Keenan (1974), which states that in NL, GA is possible only in two types of cases: (a) expressions (of arbitrary size, i.e., words, phrases, ...) denoting semantic **functions** may agree with expressions denoting their semantic **arguments**; (b) GA is likewise possible between an anaphoric expression and its antecedent. The FDP has been incorporated, in modified and elaborated form, into the G[eneralized]P[hrase]S[tructure]G[rammar] framework of Gazdar et al. (1985); cf. their C[ontrol]A[greement]P[rinciple]. (ii) the **'GA-as-morphologized-coreference'** approach (COR), going back to Lapointe (1980), who claims that GA is possible in NLs solely between expressions standing in a semantic relation of coreference/anaphor. This approach, which has been favoured and pursued by a number of authors, and which has been worked out in diverse ways (cf. e.g. Dowty and Jacobsen (1989), Barlow (1988)), has been incorporated, *inter alia*, into H[ead-Driven]P[hrase] S[tructure]G[rammar]; cf. Pollard and Sag (1987, 1992).

Both the FDP-based approach and the COR approach obviously provide - different - solutions to the domain problem, and have obvious consequences for a universal definition of GA, with the latter approach directly furnishing a semantically based GA definition, viz. as morphologically/grammatically manifested coreference.

Two questions naturally arise: (i) Which of these approaches is adequate, if any, and to what degree? (ii) To what extent can the empirical validity of these approaches be assessed in the first place?

In sections 2. and 3. we shall address various issues concerning the second problem, for the FDP and COR, respectively. In section 4. we shall present illustrative material from both Indo-European and 'exotic' languages which provides counterexamples of differing strength to both approaches, and which casts serious doubts on their universal validity. The concluding section 5. sums up our discussion and points out some consequences for further GA research.

For want of space, we can sketch only the barest outlines of this complex of problems, glossing over several important side-issues; a fuller presentation of this subject has to be deferred to another, more voluminous, publication; cf. also Schmidt and Lehfeldt (in preparation).

2. The FDP-Based Approach

Leaving aside here its uncontroversial and independently needed 'anaphoric' part, FDP restricts GA in NL to pairs of expressions standing in a semantic function-argument relation, presupposing a semantic description and classification of NL expressions as functions/arguments in the customary style of 'logical grammar', i.e., in terms of some type-theoretic language of semantic representation, or its associated model theory, respectively; cf., e.g., Montague (1974), Keenan and Faltz (1985). Note that this in no way restricts the choice of format for **syntactic** description; in particular, it does not necessitate a categorial syntax. Restricting ourselves here to phrase structure and categorial syntax, and making the usual assumptions about the syntax-semantics relation (esp. **compositionality**), FDP can be characterized, with obvious simplifications, as restricting GA to 'syntactic neighbours' (i.e., sister constituents, in phrase structure terms, or **syntactic** function-argument pairs, in terms of categorial syntax). Additionally, it imposes a directional constraint on GA, viz. that semantic arguments can only be GA **sources** (determining the grammatical features of other expressions), while semantic functions can only be GA **targets** (whose features are determined by other expressions), never *vice versa*.

Setting aside the considerable problem of determining the source/target status of agreeing expressions, the major obstacle to an empirical assessment of FDP is that, given the fact that function-argument relations can be arbitrarily reversed in semantic description without serious semantic consequences (by **type raising**; cf. e.g. Stechow (1991, 120ff.)), FDP is reduced to near-vacuity unless function/argument status can be motivated, and is being determined, independently of GA facts. Consequently, presupposing such independent motivation of semantic function/argument status, FDP would be refuted in its **strong form** already by cases of 'wrong' correspondence of function-argument and source-target ascriptions in agreeing pairs of expressions.

Certainly, FDP was motivated, in the first place, by a positive correlation in the prototypical cases under the most widely accepted function/argument ascriptions. Thus, attributive adjectives and determiners, which typically agree with the head nouns of their NPs, are usually taken to be semantic functions; verbs, which may agree with their subjects, direct and indirect objects, and other types of complements and adjuncts, are usually regarded as functions with the latter as their semantic arguments, etc; however, cf. e.g. Bartsch (1978) for the opposite view concerning function-argument status of verbs and their subjects and complements.

One could say that FDP would be refuted in its **weak form** if no conceivable function/argument ascription were reconcilable with the GA facts, i.e., in cases of mutually inverse GA between two phrases (in different feature categories).

3. The 'Coreference/Anaphoric' Approach

This approach, which includes anaphoric GA as a special case, obviously makes predictions that are very different from FDP's, and does not restrict GA with respect to syntactic adjacency or direction. As for its empirical content, obviously everything hinges here on the precise explication of "coreference". Note that this is a rather serious problem even in the most restrictive versions of the COR approach, such as Lapointe (1980), and that e.g. HPSG employs a rather loose notion of coreference here.

Furthermore, it must be pointed out that HPSG has introduced a dichotomy of those grammatical features that are relevant for the determination of referential properties, such as number and gender, and which are subject to the COR criterion, and syntactic GA features like case, which are exempt from it, thereby leaving room for the existence of non-semantic GA, and correctly so.

4. Empirical Counterevidence

Let us first concentrate on empirical counterevidence to COR, which is both easier to come by and less dependent on additional theoretical issues. Again leaving aside unclear or complicated cases which cannot be discussed here for want of space, at least two types of counterexamples to COR can be cited: (i) GA of **non-intersective** attributive adjectives, and of other prenominal modifiers of an adjectival type: The strictest version of COR, construing it as sharing the same **essential constant/variable** in semantic representation (cf. Lapointe (1980)), certainly excludes all cases of agreeing non-intersective As which cannot be re-analyzed as intersective; i.e., intensional As of the *alleged*, German *mutmaßlich* type: In German, where attributive As agree with their head Ns, a *mutmaßlicher Mörder* ("alleged murderer") is no more somebody who is at the same time alleged and a murderer (two predications of the same argument) than he is in English; rather, he is someone whom other people believe to be a murderer (one complex predication, in contrast to the intersective *red* in *red apple*, which can be analyzed, with some plausibility, as something which is at the same time red and an apple). Note that HPSG assimilates counterexmaples of this sort by weakening the coreference criterion and by the introduction of additional syntactico-semantic machinery for the treatment of adjuncts/attributes; cf. Pollard and Sag (1992, chapters 1, 2 and section 8.3).

Another even more recalcitrant case is provided by the Russian superlative marker *samyj*, which morpho-syntactically is an agreeing adjectival modifier of As:

samyj	*interesnyj*	*podxod*
"most-**m sg**	interesting-**m sg**	approach (**m sg**)"
samaja	*interesnaja*	*teorija*
"most-**f sg**	interesting-**f sg**	theory (**f sg**)",

while semantically it is most naturally construed as a function on the A-meaning (giving the degree of the A-property), with no reference to the individual the A and the N are taken to refer to. (ii) GA, with the NP in the absolutive case, of constituents of diverse syntactic and semantic status in a number of geographically connected, but not necessarily (clearly) genetically related, ergative languages in the North-Eastern Caucasus (Daghestan), e.g. **Archi**, **Lak**, and **Avar**. For literature on pertinent examples cf. the work of the Russian caucasiologist Kibrik (e.g. Kibrik (1992) and references therein), and the Caucasian volume of the handbook *Jazyki narodov SSSR*. In these languages, e.g. benefactive dative NPs, the NPs within local PPs, directional PPs, but also several types of adverbials, such as modal or temporal adverbials, indisputably agree with the absolutive NP. However, even presupposing a stretching of our semantic imagination and descriptive machinery to the limits of reason, these agreeing constituents can hardly be construed as standing in a coreference relation/'speaking about the same entity' as their GA source. Rather, semantically, they don't involve the referent of their GA source at all. For illustration, cf. the following examples:

Reɬ	*ɬa*-**r**	*dede*-**r**-*e*	**ʕiʧal**	*(-gi)*	**r**-*oʃun*	**r**-*oʔa*
she(II)	here-**IV**	father-**IV**-dat.	**apples (IV)**	ptc	**IV**-buy	**IV**-Aux

"She was buying apples for father here." (Avar)

dija	*horo:k-ej-***w***-u*	*ʧixir*	*(ø)c'arʃi*	*e***w***di*
father(I)	long ago-**Isg**	wine(IVsg)	IVsg-drinking	Aux-**Isg**

"Father has been drinking wine long ago." (Archi),

where Roman numerals designate noun classes (NCs), and GA sources are marked by bold type or underlining, with the relevant GA sources being here **apples**, and **father**, respectively, whose NC and number are replicated by GA in a number of semantically unmotivated syntactic positions. These latter cases also constitute potential counterevidence to the FDP line of research: If construed as GA with the - syntactically non-adjacent - absolutive NP source, to which they bear neither a function nor an argument relation, these cases clearly violate FDP. Granting even that, in most of these languages, it is GA with the finite verb, which in turn agrees with the absolutive NP, rather than direct GA with the latter, GA indiscriminately takes place with complements and adjuncts; now, while the latter are usually analyzed as semantic functions, the former are normally taken to be **arguments** of the V, which contradicts the directional constraint inherent in FDP, as they are GA **targets** in these cases. (Of course, this argument presupposes 'ordinary' syntactic structures to be fully valid; whether it might be weakened or obviated by 'unconventional' syntactic assumptions - provided these could be justified empirically - remains to be explored.).

May it suffice here to mention just one further problem case for FDP, viz. GA of possessive NPs: Extrapolating from Germanic, Romance,... evidence, the natural assumption seems to be to analyze 'possessive' NPs as **functions** of the 'possessed' Ns as arguments:

Haus **des Direktors**
("house of the director")

Now, data from Non-Indo-European languages, such as Turkish, suggest the opposite, as there one finds GA of the **possessed** N with the possessor NP:

müdür-ün	**ev-i**
("director-genitive	**house-his**"),

which leads Keenan to the opposite function-argument assignment in possessive constructions on a universal scale. This, however, is contradicted e.g. by Caucasian (e.g. Chamalal (Daghestan)) and Bantu evidence, where we find GA of the possessor NP with the possessed N; cf. Chamalal

ħek'vassu	*vac*
"man's-**Isg**	brother (**I sg**)"

ħek'vassvi	*jac*
"man's-**IIsg**	sister (**II sg**)"

where "I, II" mean 1st (male humans) and 2nd (female humans) NC, respectively, to which the possessed Ns *vac* and *jac* belong, and which are reflected by GA in the possessor's inflectional ending.

Now, in our view, a difference in syntactic-semantic analysis is to be rejected if it is motivated by nothing else than the very GA facts that prove to be problematic for FDP.

So, if there were no independent positive evidence for a GA-direction conforming difference in syntactic analysis of possessive constructions, the assumption of a universal uniform function-argument analysis of these constructions seems preferable, which would invariably contradict FDP in 'half' the cases under analysis.

5. Conclusions

For want of space, we must content ourselves with these illustrative examples of (potential) violations of both semantic GA principles. Summing up, it can be said that: (i) Both FDP (or its elaborations/offsprings, such as CAP) and the COR approach to GA can be shown to be invalidated as **absolute** universals by a considerable number of significant counterexamples. (ii) Nevertheless, both principles 'survive' as strong universal tendencies for GA. (iii) In view of (i), neither principle, at least as it stands, is a suitable candidate for the formulation of a semantically based universal definition of the concept of GA, and the classification of morphosyntactic phenomena as being of the GA or non-GA (e.g. morphosyntactic government) type. (iv) Finally, and more specifically, HPSG's move to distinguish two classes of GA features, and to exempt syntactic GA features from the 'coreference' condition, has been shown to be insufficient, as there does not exist a neat and stable universal partition of GA features into syntactic and semantic ones. Therefore, HPSG's coreference condition loses its explanatory force as a universal motivating principle governing GA in NLs: Since any grammatical feature involved in GA **could** do this in a **syntactic** capacity (clear contrary universal **tendencies** notwithstand ing), one is again left with the negative result that in principle 'anything goes' in GA – unless it could be shown that the cited counterexamples are correlated with other syntactic and/or semantic properties of NLs which would allow us to single out a well-defined restricted subtype of NLs with 'exceptional' GA characteristics.

References:

BARLOW, M.(1988): A situated theory of agreement. Ph.D. diss., Stanford University, Stanford.

BARLOW, M., and FERGUSON, C.A. (eds.) (1988): Agreement in Natural Language: Approaches, Theories, Descriptions. CSLI, Stanford.

BARTSCH, R. (1978): Infinitives and the control problem in categorial grammar. *Theoretical Linguistics, 5, 217-250.*

DOWTY, D., and JACOBSON, P. (1989): Agreement as a semantic phenomenon. In: J. Powers and K. de Jong (eds.): *Proceedings of the Fifth Eastern States Conference (1988).*

GAZDAR, G., KLEIN, E., PULLUM, G., and SAG, I. (1985): Generalized Phrase Structure Grammar. Blackwell, Oxford.

KEENAN, E. (1974): The functional principle: generalizing the notion 'subject of'. In: *Papers from the Tenth Regional Meeting of the Chicago Linguistic Society, 298-309.*

KEENAN, E. (1979): On surface form and logical form. In: B.B. Kachru (ed.): *Linguistics in the Seventies: Directions and Prospects.* University of Illinois, Dept. of Linguistics, Champaign-Urbana, 163-203.

KEENAN, E., and FALTZ, L. (1985): Boolean semantics for natural language. Reidel, Dordrecht.

KIBRIK, A.E. (1992): Očerki po obščim i prikladnym voprosam jazykoznanija. Izdatel'stvo Moskovskogo Universiteta, Moskva.

LAPOINTE, S. (1980): A theory of grammatical agreement. Ph.D. diss., University of Massachusetts.

MONTAGUE, R. (1974): The Proper Treatment of Quantification in Ordinary English. In: R.H. Thomason (ed.): *Formal Philosophy. Selected Papers of Richard Montague.* Yale University Press, New Haven, 247-270.

POLLARD, C., and SAG, I. (1987): Information-Based Syntax and Semantics. Vol. 1. Fundamentals. CSLI; Stanford.

POLLARD, C., and SAG, I. (1992): Head-Driven Phrase Structure Grammar. Penultimate Draft of April 28, 1992. To be published jointly by University of Chicago Press and CSLI Publications.

SCHMIDT, P., and LEHFELDT, W. (in preparation): Die zweigliedrigen Wortfügungen ("slovosočetanija") des Russischen. Begriffsanalyse und Explikation des Materials. O.Sagner, München.

STECHOW, A. von (1991): Syntax und Semantik. In: A. von Stechow and D. Wunderlich (eds.): Semantik. Semantics. de Gruyter, Berlin/New York, 90-148.

Towards a Hypermedia, Multilingual, On-Line Resource System for LSP Users/Learners

W. Wieden[1], K. Ronacher[2], A. Weiss[3], H. Goebl[4], K. Müller[2]

[1] Institut für Anglistik und Amerikanistik, Universität Salzburg, Akademiestrasse 24, A-5020 Salzburg
[2] Institut für Computerwissenschaften und Systemanalyse, Universität Salzburg, Jakob-Haringer-Strasse 5, A-5020 Salzburg
[3] Institut für Germanistik, Universität Salzburg, Akademiestrasse 20, A-5020 Salzburg
[4] Institut für Romanistik, Universität Salzburg, Akademiestrasse 24, A-5020 Salzburg

Summary: Multilingual access to expert knowledge is difficult to support because demands are specific. In this situation self-access resources, where the learner can autonomously decide on matters like object of study, pace of progression, type of retrieval, or style of learning, assume particular significance. The objective of this paper is to briefly survey current demands, to discuss selected problems involved in the use and acquisition of multilingual expert knowledge by potential users, to display some possibilities of machine representation of such knowledge, and to outline the software design of a first prototype which is at the stage of being implemented with multimedia data.

1. Introduction

In 1993 the above group of linguists and computer scientists in Salzburg (the FILCOM group) agreed to promote research in the field of computational linguistics by cooperating on a project in which multimedia technology was to be put at the service of people who want to communicate expert knowledge multilingually and learn the respective LSPs (languages for specific purposes). There were basically four questions the research group wanted to address:

- What kind of support is needed to satisfy the demands for multilingual access to expert knowledge?
- Which are the defining characteristics of the knowledge involved and which are possible principles/learner strategies for the use and learning of this knowledge?
- How should that knowledge be represented on machine to facilitate access according to the given needs?
- Which software design is likely to fulfill these demands?

Our expertise is still at an early stage of development: While research has so far had a focus in the fields of needs analysis, knowledge modelling, the study of multilingual use and learning of expert knowledge, and the development of a prototype system, it has not progressed beyond the programmatic stage in terms of the (large-scale) implementation of multimedia data and the testing of the prototype system for reliability, effectiveness, and ergonomy.

2. Specification of Needs

Sources of evidence derive from two different surveys, of which one was conducted by ourselves on 187 informants professionally involved in electronic data processing (EDP) in Austria and one by Brewster (1994), which was administered to 288 students in business-oriented education across Austria. The former survey shows, for example, that 80,5% of all informants use a PC when writing in an LSP; of these 64.5% use on-line resources occasionally, and 22.4% repeatedly when they have a problem with their software application. In the latter survey (Brewster (1994), p.34f.) presents data revealing, for example, that the students' awareness of what constitutes learning relates primarily to the variable "understand & apply" (81.2%), much less to "memorizing" (47.6%) or "familiarization" (19.8%), and that practice in LSP has a more prominent role in learning (76.0%) than conscious learning (38.9%). Further evidence (Brewster (1994), p.36) shows that self-access (77.4%) is considered as important for learning as practice in classroom (77.1%). This data highlights selectively user and use perspectives of self-access (multimedia, multilingual) on-line resource systems.

Concerning type of support the following major requests could be inferred:

- Quick access to information for which knowledge has more or less been acquired, but which is not readily or sufficiently available; example: finding the most appropriate native and foreign language terms for a particular expert concept.
- Repeatable and variable access to information for which knowledge has not yet been acquired; example: offer an individual help in assigning meaning to given (cross-linguistic) terms by offering him sufficient support for concept formation (model building).
- Give informants facilities to learn how to use that knowledge (bridging the competence-performance gap); example: give learners a chance to work on term-term, concept-concept, or term-concept relations.

The major problems involved appear to be first, the heterogeneity of the information requested/the knowledge to be acquired, and second, the difficulty of finding the means facilitating quick machine-user transfer of that information and catalyzing knowledge acquisition.

3. Knowledge Classification and Cognition

Knowledge has been studied (with different objectives) in a number of traditionally unrelated disciplines, such as philosophy, cognitive psychology, physiology, linguistics, or (in machine-related contexts) the computational sciences. Each field of study has proposed distinctions and procedures of analysis useful for its specific purposes, but often the various proposals are difficult to compromise: for example, the physiology-based distinction between declarative and procedural knowledge (Birbaumer and Schmidt (1991), p.539) does not correlate with any of the established categories of linguistic knowledge. The present study, oriented mainly at multilingual and expert knowledge, is therefore highly selective and tentative.

3.1 Classification

We assume that the organization of knowledge is modular. Under this view, a clear distinction between language knowledge and extra-linguistic knowledge is made. To this may be added skills of putting available knowledge into use (e.g. in the act of communication). A corollary of the modular view is that particular mechanisms, or relational knowledge (see Schmitz, forthcoming, for a related proposal) must be posited which relate the individual modules or structural elements within these modules.

Language knowledge is assumed to consist of various (in part mutually exchangeable) components such as rules and constraints operating on these rules, inventories of items, structures (formal or function relations between items).

Multilingual LSP knowledge is viewed in a two-dimensional context in which a first (native) language for specific purposes (LSP1) is related to a first (native) language for general purposes (LGP1), a second language for specific purposes (LSP2) and a second language for general purposes (LGP2), as displayed in Fig. 1. For example, in German (LSP1) *Rechner*, a metaphoric extension from an LGP1 word, competes with *Computer*, a borrowed term from (English) LSP2, which in turn originated as a term through metaphoric extension from an LGP2 word.

Extra-linguistic knowledge is viewed as a mental model of reality. It may be derived from sensory experiences as well as reasoning processes (Seel (1991), p.9f.). A relevant distinction in this context is that between common sense knowledge and expert (domain-specific) knowledge. The latter appears to be characteristically distinct from the former since it

- is available only to some members of a community;
- has an elaborate conceptual component;
- is adduced for the solution of problems arising only in specific fields of discourse;
- often derives (e.g. by analogical inferencing) from common sense knowledge.

There are reasons to assume that the organization of conceptual expert knowledge of multilingual individuals may differ from that of monolinguals: for the former group it is likely to be at least in part language-neutral.

Relational knowledge may be viewed in contrast to knowledge of the inventory type. As such it may not only constitute a substantial part of language and extra-linguistic knowledge, but also a link between language and extra-linguistic knowledge and thus an indispensable operative basis for communication. Such knowledge is assumed to include relations between

- linguistic units within a language, e.g. formal or functional term–term relations such as *Computer* and *Rechner* in German;
- linguistic units across languages, e.g. formal or functional term–term relations such as *Computer* in German and *computer* in English;
- extra-linguistic units of knowledge, e.g. concept–concept relations, such as INPUT DEVICE and OUTPUT DEVICE as conceptual elements of a computer;
- linguistic and extra-linguistic units; i.e. knowledge of the conventions that specify in which way terms and concepts are associated in one or more languages; multilingual access to expert knowledge faces specific problems with respect to cross-linguistic equivalences of term-concept relations; for example, the German term *Rechner* is used to denote computing devices including those performing only numerical calculations (*elektronische Rechenmaschine* according to DIN 9757E, October 1977), whereas the English term *computer* has a more limited denotation and contrasts with the term *(pocket) calculator*; on the other hand, the German terms *Datenbasis* and *Datenbank* refer to a possible conceptual distinction for which in English only the term *database* is available.
- linguistic and pragmatic units, i.e. knowledge of the conventions that specify which variety of language is appropriate for the communication of expert knowledge in a given situation (e.g. addressee, text type, or purpose).

3.2 Knowledge Representation

Although the issue of knowledge representation in the mind is controversial (see Seel (1991), p.18f.) there is reason to assume that people use multiple codes for representing their knowledge. In particular, evidence from linguistic studies testify to this. Monolinguals were found to differ, for example, in the way they assign word stress: some seem to prefer a supplementary entry for each lexical item, others seem to derive it preferably by derivational rules, while still others seem to rely on both (Kettemann (1988)). Multilingual individuals may be assumed to have either predominantly coordinate or predominantly compound organizations of language knowledge (Albert and Obler (1978)). However, as studies on second language learners show, cross-linguistic interdependencies may occur wherever certain (formal or functional) similarity criteria are met (Nemser (1993)).

3.3 Knowledge Use and Acquisition

Knowledge use may not only imply retrieval of existing units of knowledge, but also the generation of novel units, whether linguistic or extra-linguistic, e.g. by means of inferencing or derivation by rule. Complex knowledge-processing activity, such as multilingual communication of expert knowledge, is however likely to suffer from processing constraints, as studies in "mental workload" (Wieland-Eckelmann (1992)) show. Our own observations among students of the computer sciences with German as a native language revealed clear language deficits when they had to communicate expert knowledge in English, although the required second language (L2) competence had definitely been available, as subsequent inquiries showed.

On the basis of cognitive models of knowledge acquisition (Rumelhart and Norman (1978)) it is assumed that the acquisition of the respective knowledge is not only a process of knowledge growth (accretion), but also one of adding new types of knowledge representation (restructuring), and making existing knowledge less vague and more explicit (tuning). For example, studies in L2 acquisition suggest that learners with time of exposure shift from a predominantly form-based storage of the lexicon to a predominantly meaning- based storage (Meara (1984)). The setting up of firm associations between a language term and a concept may, on the other hand, be viewed as an instance tuning.

4. Knowledge Representation on Machine

The design of the planned software system rests on hypotheses which relate not only to diagnostic aspects of multilingual use and acquisition of expert knowledge (as specified above), but also on therapeutic aspects. It is therefore assumed that the system should offer

- support for multilingual access to domain-specific knowledge as well as (expert) knowledge-oriented access to the respective LSPs;
- multiple formats of knowledge representation which are highly adaptive to the learner's individual needs and learning style;
- (interactive) training facilities which help the individual learn to use his knowledge.

Therefore, the software system is largely hybrid with respect to knowledge representation. Different formats are (often alternatively) used. For example, units of expert knowledge can be displayed in terms of natural language description, by means of graphic models, and in terms of analogue sound samples (foreign language terms). Object-oriented formats are also included. For example, frames are used to display prototypical characteristics of terms and concepts. Relations are represented in a semantic network (concept–

concept relations) or specific link frames; for example, term–concept relations in a TC-frame, term–term relations within a language in a TTW frame, and term–term relations across languages in a TTA frame.

5. Software Design

Of the two modules (resource and learning) the former has already been implemented and can be described in detail, the latter is still in a state of development.

All data and data operations of the resource module are logically defined as object classes. The most important object classes are "term", "concept" and "links", henceforth called "term frame", "concept frame" and "link frame". Each frame is further specified by attributes and optionally subordinated object classes (typical subordinated object classes are graphic or sound objects). Connections among terms, among concepts, and between terms and concepts are established through link frames.

The term frame contains a defined number of slots, for example:

- term number, the object identifier (T-number);
- written natural language realization of term (T-name);
- acoustic natural language realization of term (T-sound);
- linguistic uses/constraints, such as collocations, word formations, idioms (T-use);
- pragmatic constraints, as originating from use across different fields of discourse, formal/informal registers, text types, addressees (T-constraints).

The structure of the term frame is independent of any particular language; however, the description of the terms is language-specific. Within a term frame the term receives a unique term number (object identifier).

The concept frame also contains a defined number of slots, for example:

- concept number, the object identifier (C-number);
- concept name (C-name):

  normally, but not necessarily, this name is a natural language term; eachconcept must have at least one name, but entries from a second or third language may be missing;
- explanation, by means of natural language and multimedia elements (C-explanation).

The concept frame represents knowledge that is independent of any natural language; however, the description of the concepts may be language-specific. The number assigned to a given concept within the concept frame is unique. Note that the concept number for a specific concept must be the same in each language.

The link frame "TC" specifies the relations between term frame and concept frame:

- concept number (C-number);
- term number (T-number);
- concept-based usage of the term.

Both concept number and term number constitute the object identifier of the link frame "TC".

Relations between different terms within a language ("TTW") to take care of synonyms, between different terms across languages ("TTA") to take care of cross-linguistic equivalence relations, and between concepts ("CC")to take care of semantic associations are specified by separate further link frames.

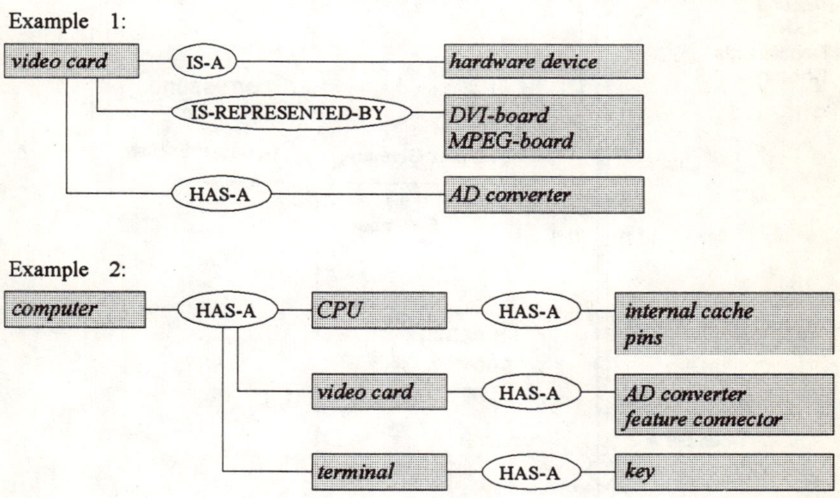

Fig. 1: Sample relations in the semantic network

Typical operations on these object classes are "find", "read", "insert", "delete" , "modify", "create a link", "delete a link", and "display multimedia representation of (concept)".

The semantic network represents semantic relations among concepts. These relations are possibly multidimensional and freely definable. Sample relations are IS_A, HAS_A, IS_REPRESENTED_BY (see Fig. 1). Thus each concept is semantically specified relative to the neighbouring concepts selected within a type of tree structure. Moreover, the diverse semantic relations of a concept can be listed.

All logical objects are stored in a relational DBMS according to SQL standard. The operating system is MS-Windows 3.1. MS FoxPro is used as the DBMS server. Toolbook 2.51 (Asymetrix) is used as the surface generator.

Both applications are connected through a DDE link.

The objects are decomposed into tables (according to first normal form assumption). Each table element (attribute) is represented either by a simple data type (such as a string of characters) or a number. The different tables are connected via record numbers.

Each language has ist own directory, in which the respective tables are stored. The tables for the generation of interlingual relations are stored in a separate directory. The relational structures are displayed in a (simplified) ER-model (see Fig. 2).

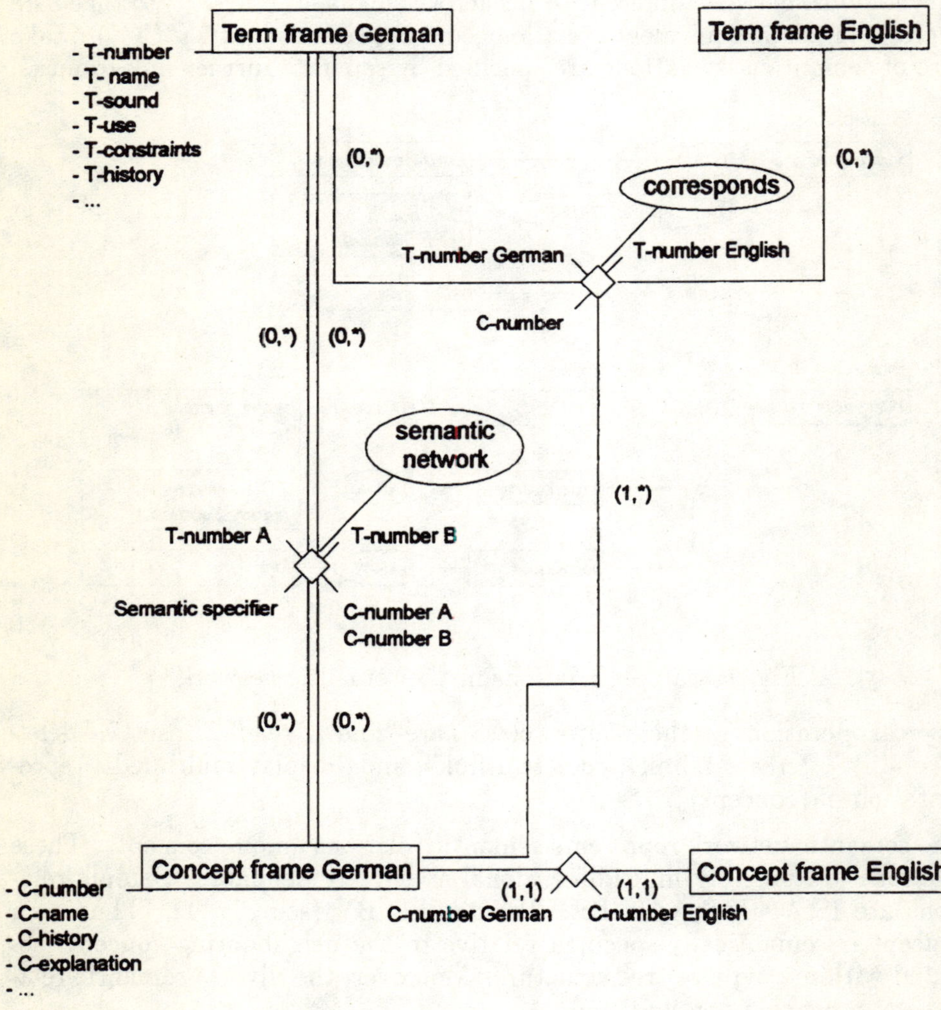

Fig. 2: Entity-Relationship (ER) model underlying the FILCOM prototype (as of spring 1994)

6. Conclusion and Outlook

This paper has attempted to show why knowledge-based approaches to software engineering increasingly require more sophisticated foundations of knowledge classification to provide more appropriate support for human learning endeavours. Although much remains to be done to make computer-based support of multilingual learning truly efficient, the direction of research and development seems to be correct, as preliminary tests in vocation-based learning suggest.

References:

ALBERT, M. and OBLER, L. (1978): The bilingual brain. Neuropsychological and neurolinguistic aspects of bilingualism. Academic Press, New York.

BIRBAUMER, N. and SCHMIDT, R. (1991): Biologische Psychologie. Springer, Berlin.

BREWSTER, E. (1994): Vocationally oriented language learning. Problems, possibilities, perspectives. PhD thesis, University of Salzburg.

KETTEMANN, B. (1988): Die Phonologie morphologischer Prozesse im amerikanischen Englisch. Tübinger Beiträge zur Linguistik 269. Narr, Tübingen.

MEARA, P. (1984): The study of lexis in interlanguage. In: A. Davies, C. Criper and A. Howatt (eds.): *Interlanguage proceedings of the seminar in honour of Pit Corder.* EUP, Edinburg, 225-235.

NEMSER, W. (1993): Experiencing a foreign lexicon: How Austrian learners of English cope with word poverty. In: W. Delanoy, J. Köberl, and H. Tschachler, (eds.): *Experiencing a foreign culture. Papers in English, American, and Australian studies.* Narr, Tübingen, 73-118.

RUMELHART, D. and NORMAN, D. (1978): Accretion, tuning, and restructuring: three modes of learning. In: J. Cotton and R. Klatzky (eds.): *Semantic factors in cognition.* Erlbaum, Hillsdale, N.J., 37-53.

SCHMITZ, K. (forthcoming): Verarbeitung fachsprachlichen Wissens in Terminologieverwaltungssystemen. In: GAL.Proceedings, Leipzig.

SEEL, N. (1991): Weltwissen und mentale Modelle. Hogrefe, Göttingen.

WIELAND-ECKELMANN, R. (1992): Kognition, Emotion und psychische Beanspruchung. Hogrefe, Göttingen.

Applications in Medicine and Biology

A New Methodologic Look at Describing the Performance of Diagnostic Classification Procedures in Medicine

Olaf Gefeller[1], Hermann Brenner[2]

[1] Abt. Medizinische Statistik, Universität Göttingen, Humboldtallee 32, D-37073 Göttingen, F.R.G.

[2] Institut für Med. Informationsverarbeitung, Biometrie und Epidemiologie, Ludwig-Maximilians-Universität, Marchioninistr. 15, D-81377 München, F.R.G.

Summary: A simple method is delineated for correcting the fundamental measures of validity of binary diagnostic tests, sensitivity and specificity, for the presence of chance agreement between classification result and true status. While recent proposals of such chance–correction techniques result in an undesirable dependence of the measures on the disease prevalence, our approach overcomes this deficit. Point and interval estimates of the chance–corrected measures are derived.

1. Introduction

The concepts of sensitivity and specificity of diagnostic classification procedures in medicine were introduced nearly half a century ago (Yerushalmy (1947)), and have become familiar to most physicians over the last decades. Sensitivity and specificity, representing the fundamental measures of validity for the evaluation of any binary diagnostic test in practice, are often regarded as the benchmarks of test performance and are used to compare the diagnostic value of alternative classification rules. Sensitivity describes quantitatively to what extent the disease characteristic of interest is actually detected by the diagnostic test; specificity focuses on the capability of the test to classify individuals without the disease correctly as test–negative. However, in such an assessment of test validity, the role of chance agreement between the diagnostic classification and the true disease status is completly ignored, although methodologic techniques to correct for chance agreement are well–established in other areas of diagnostic evaluation (e.g., in the context of reliability assessment (Cohen (1960)). In analogy to these approaches, we derive chance–corrected measures of sensitivity and specificity in this paper. We compare our measures with recent proposals by several authors (Bloch and Kraemer (1989), Coughlin and Pickle (1992), Jamart (1992)) who independently developed an algebraically equivalent type of chance correction, which suffers, however, from serious conceptual drawbacks. Furthermore, we present formulae for point and asymptotic interval estimates of the measures and analyse their small sample behaviour.

2. The Principle of Chance–Correction Procedures

Suppose a dichotomous disease characteristic (with outcomes denoted by D for the diseased and by $\overline{D}$ for undiseased) is to be detected by a binary diagnostic classification rule (T refers to test–positive, $\overline{T}$ to test–negative). Then, the (uncorrected) sensitivity (SE) is defined by the probability $P(T|D)$ and the specificity (SP) by $P(\overline{T}|\overline{D})$. The general form of chance–corrected measures, as being originally employed in the assessment of test reliability (Fleiss (1975)), has been introduced as $(P_O - P_E)/(1 - P_E)$, where P_O and P_E denote the probability parameters corresponding to the observed agreement between measured and true disease status (P_O) and to the agreement that would be expected by chance alone (P_E). Such measures of association quantify the excess agreement beyond chance relative to the maximum possible excess agreement. The crucial point in the application of this chance–correction principle to measures of validity relates to the specification of the expected agreement P_E. All previous proposals have taken $P_E = P(T)$ for chance–correction of the sensitivity and $P_E = P(\overline{T})$ for chance–correction of the specificity, respectively. In this approach, P_E is conceptualized by the corresponding quantity from the whole population under the assumption of independence between diagnostic classification and true disease status.

Simple calculus (e.g. Coughlin and Pickle (1992)) shows that the resulting chance–corrected measures depend on the prevalence of disease $P(D)$ in the population in which the test is employed. To overcome this major drawback, we suggest that, instead of deriving the expected agreement P_E from the whole population, different expressions for P_E are to be employed in the subpopulations D and $\overline{D}$. If a diagnostic test could not distinguish diseased individuals from undiseased individuals, the probabilities $P(T|D)$, $P(T|\overline{D})$ and $P(T)$ (as well as $P(\overline{T}|D)$, $P(\overline{T}|\overline{D})$ and $P(\overline{T})$) are identical, because in this situation the diagnostic classification is independent from the true disease status. The choice between these probabilities for defining P_E is thus arbitrary and the choice made in all previous proposals of chance–correction represents just one option. We can just as well quantify the probability of a positive test among truly diseased individuals that would be expected by chance alone by taking the probability of a false–positive diagnosis in truly undiseased individuals, i.e., $P_E^{pos} = P(T|\overline{D}) = 1 - SP$. Analogously, the probability of a negative test among truly undiseased individuals that would expected by chance alone can be expressed as the probability of a false–negative diagnosis in truly diseased subjects, i.e., $P_E^{neg} = P(\overline{T}|D) = 1 - SE$.

Chance–correction of the sensitivity involves only the positive classifications T in the diseased subpopulation D, whereas chance–correction of the specificity is based completely on the negative classifications $\overline{T}$ in the undiseased subpopulation $\overline{D}$. Hence, the parameters P_E^{pos} and P_E^{neg} may be employed separately for the purpose of adjusting sensitivity and specificity for chance agreement as follows:

$$SE_{cor} = \frac{P_O - P_E^{pos}}{1 - P_E^{pos}} = \frac{SE - (1 - SP)}{1 - (1 - SP)} = 1 + \frac{SE - 1}{SP},$$

$$SP_{cor} = \frac{P_O - P_E^{neg}}{1 - P_E^{neg}} = \frac{SP - (1 - SE)}{1 - (1 - SE)} = 1 + \frac{SP - 1}{SE}.$$

These measures do not depend on $P(D)$ since they are exclusively based on the uncorrected sensitivity and specificity of the diagnostic test. For diagnostic classification rules which cannot effectively distinguish diseased from undiseased subjects, i. e., in which $SE = 1 - SP$ and $SP = 1 - SE$, both measures equal zero (unless $SE = 0$ or $SP = 0$, in which case SE_{cor} or SP_{cor} is undefined, respectively), indicating that there is no agreement beyond chance alone. Any diagnostic test which is better than random classification leads to values of SE_{cor} and SP_{cor} which both lie between 0 and 1. The corrected measures SE_{cor} and SP_{cor} are identical to their uncorrected analogues only in the case of perfect specificity or perfect sensitivity. Otherwise, SE_{cor} is always lower than SE and SP_{cor} is always lower than SP. On the other hand, these chance–corrected measures yield always higher values than their prevalence–dependent counterparts described earlier, as the theoretical upper limits of the latter quantities equal SE_{cor} and SP_{cor}, respectively.

			previous proposals		new proposal	
SE	SP	$P(D)$	SE_{cor}	SP_{cor}	SE_{cor}	SP_{cor}
0.8	0.8	0	0.750	0	0.750	0.750
		0.333	0.667	0.500	0.750	0.750
		0.667	0.500	0.667	0.750	0.750
		1	0	0.750	0.750	0.750
	0.9	0	0.778	0	0.778	0.875
		0.333	0.700	0.700	0.778	0.875
		0.667	0.538	0.824	0.778	0.875
		1	0	0.875	0.778	0.875
0.9	0.7	0	0.857	0	0.857	0.667
		0.333	0.800	0.400	0.857	0.667
		0.667	0.667	0.571	0.857	0.667
		1	0	0.667	0.857	0.667
	0.9	0	0.889	0	0.889	0.889
		0.333	0.842	0.727	0.889	0.889
		0.667	0.727	0.842	0.889	0.889
		1	0	0.889	0.889	0.889

Tab. 1: Numerical illustration of chance–correction procedures for different values of uncorrected SE, SP and $P(D)$

These upper limits are attained at the boundaries of the interval [0,1] for the disease prevalence (at $P(D) = 0$ for the sensitivity and at $P(D) = 1$ for the specificity).

Table 1 provides a numerical illustration of the difference between the two approaches to chance–correction of measures of validity. The examples indicate that (i) the two types of chance–correction will often result in remarkably different values for chance–corrected sensitivity and specificity, (ii) our proposal leads to chance–corrected measures of validity which are independent of disease prevalence and which preserve their relative magnitude (compared to their uncorrected counterparts), whereas these properties do not hold for the previous proposals.

3. Estimation of the Chance–Corrected Measures

Formulae for point and asymptotic interval estimates of SE_{cor} and SP_{cor} are derived in this section, and their small sample behaviour is investigated in a simulation study. In Table 2, the notation is introduced for the observed counts within the 2×2 table representing the joint distribution of classification result and true status for a random sample of the population.

diagnostic test	true disease status positive (D)	 negative ($\overline{D}$)	Σ
positive (T)	n_{11}	n_{12}	$n_{1\cdot}$
negative ($\overline{T}$)	n_{21}	n_{22}	$n_{2\cdot}$
Σ	$n_{\cdot 1}$	$n_{\cdot 2}$	n

Tab. 2: Study sample representing the joint distribution of classification results from the diagnostic test and the true disease status

Principally, the statistical treatment of this issue has to distinguish between two sampling frameworks characterized by different distributional assumptions on the observed counts: (i) the column totals of true positive and negative classifications, respectively, are fixed prior to the study leading to the situation of two independent binomial distributions, (ii) only the total study size is fixed in advance resulting in a single multinomial distribution for the cell counts. Both sampling schemes can be found in practical applications of validation studies (White and Landis (1982)).

Under both sampling models, point estimates of the uncorrected measures of sensitivity and specificity are given as $\widehat{SE} = n_{11}/n_{\cdot 1}$ and $\widehat{SP} = n_{22}/n_{\cdot 2}$.

Point estimates of the chance–corrected measures of validity are then as follows:

$$\widehat{SE}_{cor} = 1 - \frac{n_{21} \cdot n_{\cdot 2}}{n_{22} \cdot n_{\cdot 1}}$$

and

$$\widehat{SP}_{cor} = 1 - \frac{n_{12} \cdot n_{\cdot 1}}{n_{11} \cdot n_{\cdot 2}}.$$

Expressions for the asymptotic variances of these estimates are derived by first order Taylor series expansions of the particular functions. Following a result given by Bennett (1983), in this special case the different distributional assumptions lead to the same formulae for the asymptotic variances which can be consistently estimated by:

$$\widehat{Var}\,\widehat{SE}_{cor} = \frac{n_{11} \cdot n_{21} \cdot n_{\cdot 2}^2}{n_{\cdot 1}^3 \cdot n_{22}^2} + \frac{n_{12} \cdot n_{21}^2 \cdot n_{\cdot 2}}{n_{22}^3 \cdot n_{\cdot 1}^2}$$

and

$$\widehat{Var}\,\widehat{SP}_{cor} = \frac{n_{12} \cdot n_{22} \cdot n_{\cdot 1}}{n_{11}^2 \cdot n_{\cdot 2}^2} + \frac{n_{12}^2 \cdot n_{21} \cdot n_{\cdot 1}^2}{n_{11}^3 \cdot n_{\cdot 2}^3}.$$

As the estimates $\widehat{SE}_{cor}$ and $\widehat{SP}_{cor}$ are asymptotically normal, symmetric $(1-\alpha)\%$ confidence intervals (CI) around them using the asymptotic variance expressions given above can be easily derived. In addition, confidence intervals may also be constructed based on a logarithmic transformation $(h(y) = \log(1-y))$ of the estimates to yield a better approximation to the normal distribution in finite samples. The resulting formulae for these interval estimates are as follows:

$$1 - \frac{n_{21} \cdot n_{\cdot 2}}{n_{22} \cdot n_{\cdot 1}} \cdot \exp\left(\pm z_{1-\alpha/2} \cdot \sqrt{\frac{n_{11}}{n_{\cdot 1} \cdot n_{21}} + \frac{n_{12}}{n_{\cdot 2} \cdot n_{22}}}\right)$$

for SE_{cor} and, analogously,

$$1 - \frac{n_{12} \cdot n_{\cdot 1}}{n_{11} \cdot n_{\cdot 2}} \cdot \exp\left(\pm z_{1-\alpha/2} \cdot \sqrt{\frac{n_{21}}{n_{\cdot 1} \cdot n_{11}} + \frac{n_{22}}{n_{\cdot 2} \cdot n_{12}}}\right)$$

for SP_{cor}, where $z_{1-\alpha/2}$ denotes the $(1-\alpha/2)$–quantile of the standard normal distribution. In an extensive simulation study covering 100 design situations (five different sample sizes, ranging from 100 to 10000, combined with ten structures of uncorrected sensitivity and specificity for each sampling framework) the performance of these point and interval estimates for SE_{cor} and SP_{cor} has been investigated. The accuracy of the point estimates for the chance–corrected measures was very high. No relevant differences between true value and point estimate (calculated as the mean of 100.000 simulation trials) could be observed for all design situations in the study.

Regarding the interval estimates, it turned out that the confidence intervals based on the untransformed estimates become anti–conservative for decreasing sample size. Table 3 shows an excerpt of the simulation results to point to this phenomenom. The general observation in the simulation study was that, given a fixed sample size, the discrepancy between nominal and actual confidence level increases as uncorrected SE and SP approach their maximal value of 1. In the extreme situation of $SE = 0.99$ and $SP = 0.99$, the actual coverage probability for a 95% confidence interval was only 86% for the untransformed interval estimates even for a sample size of 1000, indi-

			actual coverage probabilities			
sample	uncorrected		untransformed CI		log–transformed CI	
size	SE	SP	SE_{cor}	SP_{cor}	SE_{cor}	SP_{cor}
100	0.99	0.99	0.718	0.721	0.945	0.947
	0.90	0.90	0.899	0.887	0.951	0.950
	0.90	0.75	0.907	0.923	0.957	0.947
	0.75	0.90	0.927	0.901	0.946	0.957
	0.75	0.75	0.936	0.934	0.953	0.954
500	0.99	0.99	0.803	0.799	0.949	0.948
	0.90	0.90	0.939	0.938	0.950	0.952
	0.90	0.75	0.940	0.946	0.951	0.951
	0.75	0.90	0.947	0.940	0.950	0.951
	0.75	0.75	0.945	0.946	0.950	0.950
1000	0.99	0.99	0.862	0.859	0.950	0.949
	0.90	0.90	0.944	0.944	0.951	0.951
	0.90	0.75	0.947	0.949	0.951	0.949
	0.75	0.90	0.947	0.945	0.951	0.950
	0.75	0.75	0.947	0.948	0.951	0.951

Tab. 3: Actual coverage probabilities of interval estimates for SE_{cor} and SP_{cor} in a multinomial sampling framework derived from a simulation study with 100.000 repetitions (nominal confidence level: 95%)

cating a substantial degree of imprecision of the asymptotic formulae even for relatively large samples. The log–transformed interval estimate, however, does not show this deficiency. The results of the simulation study indicate that these interval estimates approximate the nominal level fairly well even for small samples. For all design situations, the actual coverage probabilities for a 95% confidence interval ranged between 94.5% and 96.1% when the sample size was only 100. Thus, from the experiences of the simulation study it can be concluded that in practical applications the logarithmic versions of the asymptotic interval estimates for SE_{cor} and SP_{cor}, respectively, should be given preference over their untransformed competitors.

The effect of the sampling framework ("product–binomial" versus "full multinomial") on the performance of point and interval estimates in small samples was negligible here; the results turned out to be nearly identical under both sampling models. Consequently, the methods for chance–correction in the estimation of sensitivity and specificity delineated in this paper can be equally applied to both designs of validation studies.

4. Discussion

The necessity of integrating chance–correction procedures in the assessment of the validity of diagnostic classification rules in medicine has been repeatedly emphasized in recent years. Whenever some diagnostic test is actually applied, some part of the correct test classifications is due to chance agreement between test result and true disease status. Thus, it seems reasonable to adjust for such chance agreement in quantification of the diagnostic capability of the classification rule. Chance–corrected estimates of sensitivity and specificity offer the opportunity to focus the evaluation of test validity separately on its two principal dimensions, i.e., the correct classification of truly diseased and truly subjects. These new measures of validity should, however, only supplement, not replace, traditional (uncorrected) measures of sensitivity and specificity, since the uncorrected and corrected quantities convey different information for the evaluation of a diagnostic test.

In this paper, we have outlined a new approach to chance–correction of sensitivity and specificity. Unlike other recent proposals for chance–correction in this context, our measures do not depend on the prevalence of the disease; they therefore characterize intrinsic features of the test itself rather than a combination of test and population features. The construction of point and asymptotic interval estimates for the chance–corrected parameters of interest has been delineated. On the basis of the results of the simulation study, it has to be emphasized that only the logarithmic modification of the interval estimate guarantees a reliable confidence level in small sample situations. The sampling framework of the validation study ("product–binomial" versus "full multinomial") showed, however, no influence on the properties of the estimates, they are equally well applicable under both sampling models.

For quantitative diagnostic tests, the chance–corrected measures of sensitivity and specificity, like their uncorrected counterparts, also vary considerably with the threshold employed for categorization of the test response. This phenomenon may be visualized by chance–corrected receiver operating characteristic (ROC) curves which can be constructed analogously to the case of the uncorrected measures of validity. The extension of the conventional ROC–curve approach to its chance–corrected analogue leads, however, to a substantial change in the graphical display which is illustrated in detail elsewhere (Gefeller and Brenner (1994)).

References:

BENNETT, B.M. (1983). Further results on indices of diagnostic screening II. *Biometrical Journal*, 25, 453-457.

BLOCH, D.A., and KRAEMER, H.C. (1989). 2×2 Kappa coefficients: measures of agreement or association. *Biometrics*, 45, 269-287.

COHEN, J. (1960). A coefficient of agreement for nominal scales. *Educational and Psychological Measurement*, 20, 37-46.

COUGHLIN, S.S., and PICKLE, L.W. (1992). Sensitivity and specificity–like measures of the validity of a diagnostic test that are corrected for chance agreement. *Epidemiology*, 3, 178-181.

FLEISS, J.L. (1975). Measuring agreement between two judges on the presence or absence of trait. *Biometrics*, 31, 651-659.

GEFELLER, O., and BRENNER, H. (1994). How to correct for chance agreement in the estimation of sensitivity and specificity of diagnostic tests. *Methods of Information in Medicine*, 33, 180-186.

JAMART, J. (1992). Chance–corrected sensitivity and specificity for three–zone diagnostic tests. *Journal of Clinical Epidemiology*, 45, 1035-1039.

WHITE, A., and LANDIS, J.R. (1982). A general categorical data methodology for evaluating medical diagnostic tests. *Communications in Statistics — Theory and Methods*, A11, 567-605.

YERUSHALMY, J. (1947). Statistical problems in assessing methods of medical diagnosis. *Public Health Reports*, 62, 1432-1449.

Xmed–DD
From Document Processing to Systematic Information Storage

W. Giere, A. Gregori, C. Luz

Zentrum der Medizinischen Informatik
Klinikum der J.W.Goethe-Universität, 60590 Frankfurt am Main

Summary: The processing of medical freetext, the structured storage of its meaning for later retrieval purposes and the instant generation of uniform internationally accepted codes as ICD, ICPM will be discussed. Xmed-DD is a new approach to this problem using the translation system TRANSOFT, a frame- and knowledge-based architecture for information processing and the document description language SGML for definition of structure and contents of text. The entire process is supported by lexicon-entries stored in a medical thesaurus.

1. Introduction

Freetext plays a significant role in medicine: a major part of observations made during the patient's treatment is recorded in narrative manner. The notes are kept either in paper form in the medical record or are stored as files in a general documentation- or information-system. Natural language processing (NLP) can be used to analyze the contents of those texts regarding to its relevancy to documentation purposes (Gell (1983), Sager et al. (1987)). One of those purposes is the classification of medical facts as diagnoses and procedures into uniform coding schemes like *ICD* (International Classification of Diseases) or *ICPM* (International Classification of Procedures in Medicine).

Our project's objective is to develop a system which supports this classification process. Discharge summaries are automatically segmented into semantical units and analyzed. The results of the analysis will be stored in a frame-oriented knowledge-base and classified according to ICD, in future also to ICPM.

We designed a so-called documentation workbench *Xmed*, which serves as a general guide for the processing of freetext. The translation system TRANSOFT builds the core of the workbench (Moore (1989)). According to this model a system for the translation of English medical texts into German (called *Xmed-ED*, "Englisch-Deutsch") has already been realized (Giere - (1994)).

Because of the way the Xmed-model's principles are used for our tasks, we call this approach *Xmed-DD* ("Deutsch-Deutsch"): the processing of discharge summaries can be regarded as the "translation" of an initially flexible notation into a standardized and structured representation. This representation is more suited for a later storage and classification of medical facts.

In this paper we will first outline the general principles of the model Xmed itself. The specific features of Xmed-DD and TRANSOFT are explained in section 3.

2. The Xmed-Model

Xmed is called a *model for an open documentation workbench.* We define an open workbench as a knowledge-based application, which

- accepts various kinds of information and document-types, and
- supports different analysis-purposes by alteration of the knowledge-base.

Documents in various structures are accepted and transfered into a standardized notation. A translation system (TRANSOFT) analyzes selected parts of the document and stores the results of the analysis in an output document. The entire analysis is performed by using specific knowledge about term- and grammar-structures stored in a thesaurus. The processing results are exported in the original format of the source document.
Fig. 1 visualizes the general documentation process:

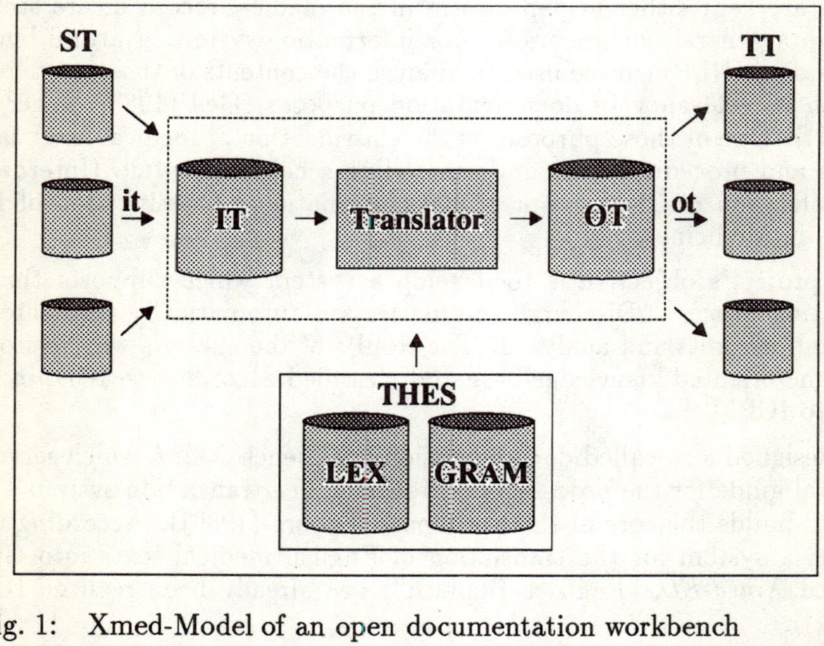

Fig. 1: Xmed-Model of an open documentation workbench

To summarize, four kinds of text are used:

ST source text, i.e. the text in its original format (freetext-field in an information-system, a wordprocessing text, a scanned document, etc.),
IT standardized input text,
OT output text as the result of the analysis, and
TT target text in the original format.

Two transformations *it* (input-transformation) and *ot* (output-transformation) convert between the different formats.

Depending on the complex of problems the model is applied to, the types of texts and transformations have to be defined and the processing mechanism must be specified. In the next paragraph we explain the specification of the Xmed-model regarding to the analysis and classification of medical documents.

3. Realisation of Xmed-DD

The system Xmed-DD has been developed to support the doctor in classifying medical statements according to ICD or ICPM. Since we don't want the doctor to change his documentation habits, we aim at gathering the facts relevant to the classification from existing clinical documents as discharge summaries. Here the doctor records diagnoses and procedures obtained during the patient's treatment. Our approach is to extract the diagnosis- and therapy-paragraphs out of the summaries and to classify the medical facts contained in those text-blocks.

The processing sequence contains five steps:

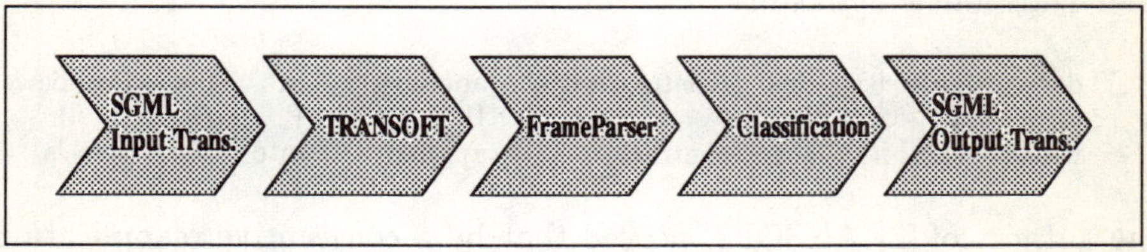

Fig. 2: Xmed-DD processing sequence for analyzing and classifying discharge summaries

3.1 Input-Transformation

Step 1 represents the input-transformation of the original text. This is done by utilizing the document description language SGML *(Standard Generalized Markup Language)* (Bryan (1988)) as standardized interface. Using this language permits to break medical documents down to logically related information-chunks (patient-name, -age, -address, diagnoses, therapies, histology, etc.) and to preformat them. This enables us to extract the diagnosis- and therapy-sections.

3.2 TRANSOFT

The text standardisation is performed in step 2 by using the translation system TRANSOFT. It was originally developed for the translation of English medical documents into German and is completely controlled by entries in a term- and grammar-lexicon. TRANSOFT translates a sentence by comparing its terms with entries in the term-lexicon. In a first phase each term in the source-language receives a default-translation in the target-language together with a *part-of-speech marker* (POS) designating its syntactic-semantical role in the sentence. Suggested POS were items as

- A adjective,
- B adverb,
- D determiner,
- N noun,
- P preposition, etc.

The appropriate POS assigned to a sentence form a *parsandum*, a consecutive sequence of class designators. For example, the following clause receives the parsandum "[DPDNUNV]":

,	der	das	in	die	Gefäße	applizierte	Kontrastmittel	aufweist	.
	[	D	P	D	N	U	N	V	]
,	which	the	into	the	vessels	applied	contrast medium	exhibits	.

The parsandum built in the first phase serves as a look-up in the grammar-lexikon. Here rules are defined, how a stereotypical word order in the source language is rearranged into an adequate word order in the target language. TRANSOFT uses parsing-rules notated as "DPDNUNV^2D5P6D7N4U3N1V" to define these rearrangement instructions. For the example above such a rule yields following result:

,	der	aufweist	das	Kontrastmittel	applizierte	in	die	Gefäße	.
	[	V	D	N	U	P	D	N	]
,	which	exhibits	the	contrast	applied	into	the	vessels	.

The authors of TRANSOFT proved that by a consecutive reapplication of these simple translation paradigms texts of any complexity can be processed. The efficiency of TRANSOFT was demonstrated by the translation of a German language medical textbook into a draft quality English language translation.

TRANSOFT used in Xmed-DD

In Xmed-DD, we have used TRANSOFT as a tool for transfering ("translating") German in everyday style written medical texts into a systematically ordered notation suited for further analysis; terms are standardized and the sentence structure is rearranged into a predefined format.

This was achieved by creating tables specific to the medical area we were concentrating on. The term-lexicon comprises an expandable list of general and technical terms (multiple-word terms also) used in clinical workaday routine. The construction of the lexicon followed the principles how a medical fact is generally described in diagnosis- and therapy-statements (Röttger (1969)): a medical fact can be represented by a *nucleus*

`S`	"Nierenresektion","Prostata-Karzinom"

Possible specifications as a *modificator* `M` or a *localisator* `L` establish *transformations* which expand the nucleus to a complex medical fact:

`M*(S)`	"am 7.12.93 durchgeführte Nierenresektion"
`L(M*(S))`	"am 7.12.93 durchgeführte Resektion der Niere"
`M*(L)(M*(S))`	"am 7.12.93 durchgeführte Resektion der linken Niere"

"`M*`" means that any number of modificators might be used; the same applies to the modification of localisators.
Complex facts can also be semantically linked by a *relation* `R`:

$$R(M^*(L)(M^*(S_1)), M^*(L)(M^*(S_2)))$$

We map the nucleus, its specifications and relations to following list of POS:

- X term relevant to a diagnosis,
- P term relevant to a procedure,
- L localisation,
- M modification,
- R relation,
- C conjunction, etc.

These POS are not supposed to structure the sentence according to syntactical features, but to conclude its contents from a semantical sentence structure (parsandum). In this case it is less relevant to consider the actual meaning of a single term than to analyze the sequence of POS built by all parts of the sentence.

TRANSOFT processes each sentence in two phases:

1. Standardisation of terms and construction of a parsandum
TRANSOFT uses the term-lexicon to replace the initial expressions in a sentence by preferred terms. The sequence of corresponding POS builds a syntactic-semantical representation of the sentence structure.
Example:

Pelvine	Lymphadenektomie	bds.	und	partielle	Resektion	der	Prostata
M	P	M	C	M	P	D	L
pelvin	lymphadenektomie	beidseitig	und	partiell	resektion	der	prostata

2. Standardisation of the parsandum
By using the grammar-lexicon TRANSOFT transfers the previously established parsandum into a standardized structure suitable for further text analysis. The parsandum describes a possible combination of medical terms and represents a sentence structure typically used in discharge summaries. Since

these sentence structures usually correspond to nominal phrases the number of actually used parsanda is limited. For example, the composition of a sentence whose medical contents can be characterized by the transformation (M*(L)((M*(S)) might be covered by grammar-entries as

- MXMDL^2M1X3M0D4L (results in XMML)
- XMDML^1X2M0D4M3L (results in XMLM)
- MMPDLM^2M3M1P0D4L5M (results in PMMLM)

The sentence will be restructured by grouping all specifications to a medical fact behind the nucleus. The rearrangement rules are established according to the assumption that specifications to a medical fact are mostly found in the neighborhood of the term itself (Baud (1992), Gabrielli (1991)).

The result of the standardisation can be formalized by following sequence of POS:

$$XM^*(LM^*)^*[R]\{X,P\}M^*(LM^*)^*[R]\ldots$$

"[R]" means the relation is optional; {X,P} means that either the POS X or P is selected. By this systematic structure the interrelation between several parts of a sentence is made obvious.
The rule "MPMCMPDL^2M1P3M0C5M4P0D6L" applies to the sentence used in phase 1 and yields following standardized parsandum:

lymphadenektomie	pelvin	beidseitig	resektion	partiell	prostata
P	M	M	P	M	L

3.3 Frame-Parser

Step 3 transfers the sentences processed in step 2 into a frame-oriented structure. The frame structure serves as a logical and systematic description of the statements revealed in the text (Sager et al. (1987), Scherrer et al. (1989)).

The composition of a frame itself follows the principles of TRANSOFT: a frame contains medical facts and consists of so-called diagnosis- and therapy-slots. These slots represent the nucleus of a statement and can be further specified by modification- and localisation-attributes. Complex medical facts are gathered from the parsandum and integrated into an adequate slot. Relations between facts are represented by relations between slot-entries. Here, a semantical differentiation of the marker R (causal, temporal, etc.) is performed. Beside the storage of the results acquired in the TRANSOFT phase, frames adequately visualize the processing sequence and render it transparent to the user.

3.4 Classification

In step 4 diagnoses are classified into ICD by analyzing the frame structure. In the thesaurus each entry in the diagnosis-slot has been assigned an appropriate ICD-code. In addition the codification of each entry can be put more precisely by considering modifications and localisations to a slot. Possible relations between slots are also taken into consideration. Examples:

- A slot-entry "Karzinom" specified by a localisation "Prostata" initially receives the ICD-code 239.9 ("Bösartige Neubildung unbekannten Charakters"). A rule states that the entire fact is represented more precisely by a code 185 ("Bösartige Neubildung der Prostata").

- The medical fact "Aortenklappenfehler" has the ICD-code 395. With the simultaneous existence of a diagnosis "Mitralklappenfehler" (ICD-code 394) both get assigned the code 396 ("Mitral- und Aortenklappenfehler"). This kind of codification rule can also be realized by using the frame structure.

3.5 Output-Transformation

Step 5 uses SGML again to export the results obtained in the previous steps (frame-entries, codes) from the internal representation into a standardized output format. The original document will be enriched by the results obtained from the analysis and receives by these means a formal description of its contents.

4. Results and Next Steps

The proceeding as described in section 3 has been examplarily realized for the urological area. Thereby a term-lexicon of about 5000 entries and a grammar-lexicon with about 6000 grammar structures have been established. These files have been embedded into a thesaurus with several semantical facets.

The system has been developed using the programming language M, which allows flexible string processing and data maintenance. As database we used FileMan of the Veterans Administration Decentralized Hospital Computer Program (*VA-DHCP*). A graphical system-interface assists the user during the documentation process.

The system has been tested on 40 medical letters of the urological department of the *Städtischen Kliniken Offenbach.* The diagnosis-blocks have been extracted, preformated and finally analyzed by *TRANSOFT* and *Frame-Parser.* Afterwards we compared the information contained in the frame structures with the statements made by the physician in the discharge summary. Then, for the diagnostic facts in the knowledge-base the ICD-codes have been automatically generated. Results are, that about 80% of the texts could be processed correctly. The remaining 20% derive either from unprecise statements or from sentence structures Xmed couldn't deal with.

These errors will be resolved in the main by a further specification of the lexicon.

The model's strengths lie in predefining and at the same time adapting itself to typical and common schemes for medical expressions and therefore covering most of the routine work. In case of rare and exceptional phrases the system asks for further clarification.

References:

BAUD, R.H., RASSINOUX, A.M., and SCHERRER, J.R. (1992): Natural Language Processing and Medical Records. In: K.C. Lun, P. Degoullet, T.E. Piemme, and O. Rienhoff (eds.): *MEDINFO-92.* North-Holland, Amsterdam.

BRYAN, M. (1988): SGML: an author's guide to the Standard Generalized Markup Language. Addison-Wesley, Wokingham.

GABRIELLI, E.R., and SPETH, D.J. (1991): Automated Analysis of Medical Text II - Cognitive Strategy. *J. Med. Sys., 15,* 65–78.

GELL, G. (1983): Free Text Processing in Medicine. In: J.H. van Bemmel, M.J. Ball, O. Wigertz (eds.): *MEDINFO-83.* North-Holland, Amsterdam.

GIERE, W., and MOORE, G.W. (1994): Xmed-ED. *Forschung Frankfurt, Wissenschaftsmagazin Universität Frankfurt am Main, 2,* 11–16.

MOORE, G.W., WAKAI, I., SATOMURA, Y., and GIERE, W. (1989): TRANSOFT: Medical translation expert system. *Artificial Intelligence in Medicine, 1,* 149–157.

RÖTTGER, P., REUL, H., SUNKEL, H., and KLEIN, I. (1969): Die vollautomatische Dokumentation und statistische Auswertung pathologisch-anatomischer Befundberichte. *Meth. Inform. Med., 8,* 19–26.

SAGER, N., FRIEDMAN, C., and LYMAN, M.S. (1987): Medical Language Processing: Computer Management of Narrative Data (eds.). Reading, Addison-Wesley, Mass..

SCHERRER, J.R., COTE, R.A., and MANDIL, S. (1989): Computerized Natural Medical Language Processing for Knowledge Representation (eds.). North-Holland, Amsterdam.

Ribosomal RNA Phylogeny Derived from a Correlation Model of Sequence Evolution

A. von Haeseler[1], M. Schöniger[2]

[1] Institute for Zoology, University of Munich, P.O. Box 202136, 80021 Munich, Germany

[2] Theoretical Chemistry, Technical University Munich, Lichtenbergstr. 4, 85747 Garching, Germany

Summary: Present-day stochastic models of DNA sequence evolution assume independent and identically distributed nucleotide sites. However, it is well known that there exists correlation among nucleotides in a DNA sequence. We present a stochastic model of sequence evolution that takes into account the correlation of non–overlapping pairs of nucleotides. This model enables us to estimate the number of substitutions per nucleotide. As a biological example we re–analyze the phylogenetic tree of Metazoa based on small ribosomal subunit RNA.

1. Introduction

DNA sequences are now widely used as a tool to elucidate the phylogenetic relationship among species. Especially ribosomal RNA has been proven useful to infer such relationships (Woese (1987), Field et al. (1988), Hedges et al. (1990), Van de Peer et al. (1993)). The outcome of such an analysis heavily depends on the validity of the assumptions that are implicit in the tree-building procedures. Counting nucleotide differences between an aligned pair of sequences and subsequently applying one of the frequently used correction methods to account for parallel and back substitutions (Jukes and Cantor (1969), Kimura (1980), Felsenstein (1981), Hasegawa et al. (1985)) assumes inter alia that each nucleotide position evolves independently of the others and that each site evolves at the same rate. This is the so–called iid (independent and identically distributed) assumption. In the statistical parlance the different classes (e.g. transitions, transversions, no differences) constitute a sample from a multinomial distribution that depends on the parameter substitution rate, which one wants to estimate.

Now data start to accumulate that show that the assumption of independence among sequence positions is generally not guaranteed (Elton (1975), Ohno (1988), Tavaré and Giddings (1989), Schöniger and von Haeseler (1994)). While in some instances (e.g. protein coding sequences of mammalian mitochondrial DNA) the violation of this assumption does not severely affect the estimation of the mutation rate, it does in the case of ribosomal RNA sequences, if opposite nucleotides in stem regions are considered as correlated non–overlapping doublets of nucleotides (Schöniger and von Haeseler (1994)). In this study we compare the effect of correction methods that take correlation into account with those that do not on tree building procedures.

More precisely, we analyze a collection of metazoan small ribosomal subunit RNA (SSU rRNA) (Neefs et al. (1993)) and compare the resulting trees.

2. Methods and Data

Due to limited space, we can give only a very brief description of the model of evolution of correlated sequences (see Schöniger and von Haeseler (1994) for details): The ingredients are a 16×16 rate matrix $\mathbf{R} = (r_{\nu,\mu})$ over the alphabet of dinucleotides (doublets) $\mathcal{A}_2 = \{AA, \ldots, TT\}$. $r_{\nu,\mu}\Delta t$ represents the probability that doublet ν will be substituted by doublet μ in an infinitesimally small time interval Δt. The entries of $\mathbf{R}$ are estimated from the data, by setting

$$\hat{r}_{\nu,\mu} = \begin{cases} \pi_\mu & \text{if } d(\mu,\nu) = 1 \\ -\sum_{\lambda \in \mathcal{A}_2, \lambda \neq \nu} r_{\nu,\lambda} & \text{if } \nu = \mu \\ 0 & \text{otherwise} \end{cases} \tag{1}$$

where π_μ is the frequency of doublet μ and $d(x,y)$ the number of nucleotide differences between sequences x and y, the so-called Hamming distance. Following Felsenstein's (1981) approach, we assume that the rate of substitution from all doublets is the same. It is only the rate of substitution to particular doublets, which is determined by π_μ.

The definition of $\mathbf{R}$ implies that only single substitutions at either position of a doublet are allowed to occur instantaneously. That is to say, each doublet has exactly six neighbors in the rate matrix. However, it should be clear from the definition of $\mathbf{R}$ that substitution between any pair of doublets is possible if $t > 0$.

Following standard approaches (cf. Karlin and Taylor (1975)) we can compute the corresponding probabilities to observe different doublets during a time interval as follows:

$$\mathbf{P}(t) = \exp(\mathbf{R}t) = I + \sum_{i=1}^{\infty} \frac{\mathbf{R}^i t^i}{i!} \tag{2}$$

provides the Markov transition probability matrix $\mathbf{P}(t) = (p_{\nu,\mu}(t))$, where $(p_{\nu,\mu}(t))$ is the probability that doublet ν is replaced by doublet μ during the time interval t. $\mathbf{P}(t)$ represents a continuous-time time-homogeneous Markov chain with a stationary distribution π_μ. Once $\mathbf{R}$ is given, the expected number of substitutions D is defined as

$$D(t) = tl \sum_{\nu \in \mathcal{A}_2} \sum_{\mu \in \mathcal{A}_2, \mu \neq \nu} \pi_\nu r_{\nu,\mu} \tag{3}$$

where l is the number of doublets in the sequences. Similarily, the number N of observed nucleotide differences between two sequences is readily computed from $\mathbf{P}(t)$.

	A	G	C	U
A	79	113	181	**3786**
G	107	47	**6582**	**1999**
C	71	**6422**	90	136
U	**4191**	**1954**	30	136

Tab. 1: Observed doublets in the stem regions of 60 metazoan SSU rRNA sequences. Watson–Crick and $G \cdot U$ base pair counts are boxed

N is given by

$$N = l\left(1 - \sum_{\nu \in \mathcal{A}_2} \pi_\nu p_{\nu,\nu}(t) + \sum_{\nu \in \mathcal{A}_2} \sum_{\mu \in \mathcal{A}_2, d(\nu,\mu)=2} \pi_\nu p_{\nu,\mu}(t)\right) \tag{4}$$

The more challenging problem to infer D from N could only be solved numerically for our model. This is different for Felsenstein's model of sequence evolution that conserves the base composition and is the analogue to our model, if independent sites are considered: here we compute

$$D = -2lh \ln\left(1 - \frac{N}{2lh}\right), \tag{5}$$

where $h = 1 - \sum_{X \in \{A,G,C,T\}} \pi_X^2$ and π_X is the frequency of the nucleotide X in the data.

Sequence data and structural alignment of the SSU rRNA sequences were retrieved from the database of small subunit RNA sequences (Neefs et al. (1993)). The rate matrix $\mathbf{R}$ was estimated from the analysis of the stem regions of 60 metazoan SSU rRNA sequences, by counting the number of doublets, which are elements of $\mathcal{A}_2$, that occur in the stem regions as determined by the secondary structure of the SSU rRNA (Neefs et al. (1993)). That is to say, opposite nucleotides in the stems are considered as doublets. Tab. 1 shows the counts of the corresponding 16 nucleotide pairs.

Since there is no natural order in the doublets of a stem region the counts in Tab. 1 are symmetrized, that is $c_{ij}^{sym} = \frac{c_{ij}+c_{ji}}{2}$, where c_{ij} are the corresponding entries from Tab. 1 and i and j are one of the four nucleotides. The marginal distribution of Tab. 1 gives the average nucleotide composition of the metazoan sequence, which is used for the Felsenstein model of sequence evolution. The neighbor joining method (Saitou and Nei (1987)) was used to infer trees for the various distance matrices obtained.

3. Results

3.1 Rate Estimate

The inspection of Tab. 1 shows that correlation within doublets is a good

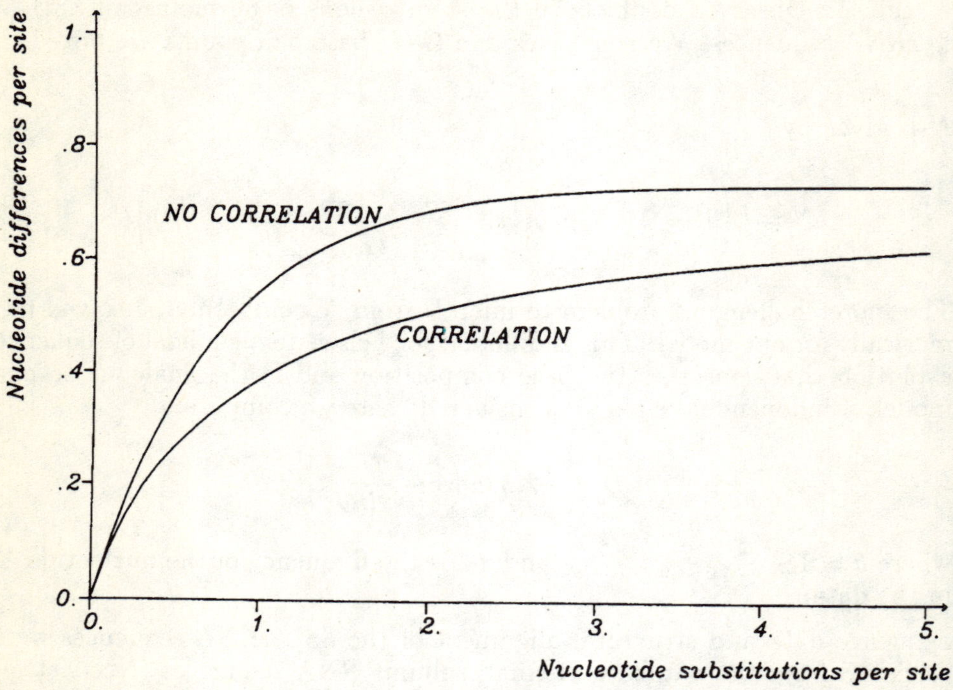

Fig. 1: Relations between the expected number of observed nucleotide differences per site and the number of substitutions that took place in the course of evolution for Felsenstein's model of sequence evolution (upper curve) and for the correlation model (lower curve). Estimates of the nucleotide distribution and the dinucleotide distribution are based on the observed numbers in the stem regions of 60 metazoan SSU rRNA sequences (see Tab. 1).

assumption. Based on the counts in Tab. 1 the assumption of independently evolving sites in the stems is soundly rejected. This is of course due to the fact that stem regions need to conserve the secondary structure of the molecule. This goal is accomplished by maintaining base–pairs in this region. According to the base pairing rules A pairs with U, G pairs with C, and G pairs with U in ribosomal sequences. However, evolution allows for a certain degree of sloppiness in the stems, that is the reason why not all doublets in a stem do constitute an admissible base pair. This sloppiness can be explained by substitutions that are selectively neutral and where the corresponding compensatory base change has not occurred yet.

Fig. 1 shows the effect of correlation on the estimation of expected substitutions per site from the observed number of differences per site if the Felsenstein model is used (top, see also equation (5)) or if the model of correlation within doublets in the stems is used (bottom, equations (3) and (4)). The difference in the shape of the curves is quite remarkable, especially if $N \geq 0.2$ the Felsenstein model underestimates the true evolutionary distance of two sequences substantially. Hence, using the wrong model of sequence evolution (i.e. Felsenstein's) will lead to a severe underestimation of true evolutionary distances. On the other hand one should be aware that both curves approach the same value h if D tends to infinity. This remark, however, is of purely theoretical interest and is not relevant for all practical purposes.

3.2 Application to Tree Reconstruction

This section is intended to elucidate the impact of an inappropriate model of distance correction on tree reconstruction methods. As the technique of choice to reconstruct a tree the neighbor joining method (Saitou and Nei (1987)) was applied to 30 metazoan sequences of SSU rRNA. The sequences stem from different metazoan phyla. Namely:

- $1, 4, 8, 9, 10, 11, 12, 13, 28, 29$ are from various vertebrates
- 41 is a tunicate sequence
- $44, 45, 46$ are insect sequences
- $49, 50, 51, 53, 54, 55, 57, 58$ are crustaceaen sequences
- $47, 48$ are sequences from spiders
- $75, 76, 77$ are molluscan sequences
- $65, 66$ are sequences from flatworms
- 80 is a cnidarian sequence.

The sequences were chosen from an article of Van de Peer et al. (1993) to facilitate comparison with their reconstruction. The sequences were analyzed as follows: For each pair of sequences the estimated number of substitutions assuming a Felsenstein model of sequence evolution was inferred using equation (5). All positions no matter if in a stem or loop region were treated equally. Based on the resulting distance matrix a neighbor joining tree was computed (Fig. 2).

For the correlation model sequences were split in stem and loop positions. For the former the number of substitutions based on the correlation model was inferred using numerical routines (Schöniger and von Haeseler (1994)), whereas for the latter the number of substitutions was inferred from equation (5). Then the weighted sum (according to the number of positions involved) was computed as the distance between a pair of sequences. The resulting neighbor joining tree is displayed in Fig. 3.

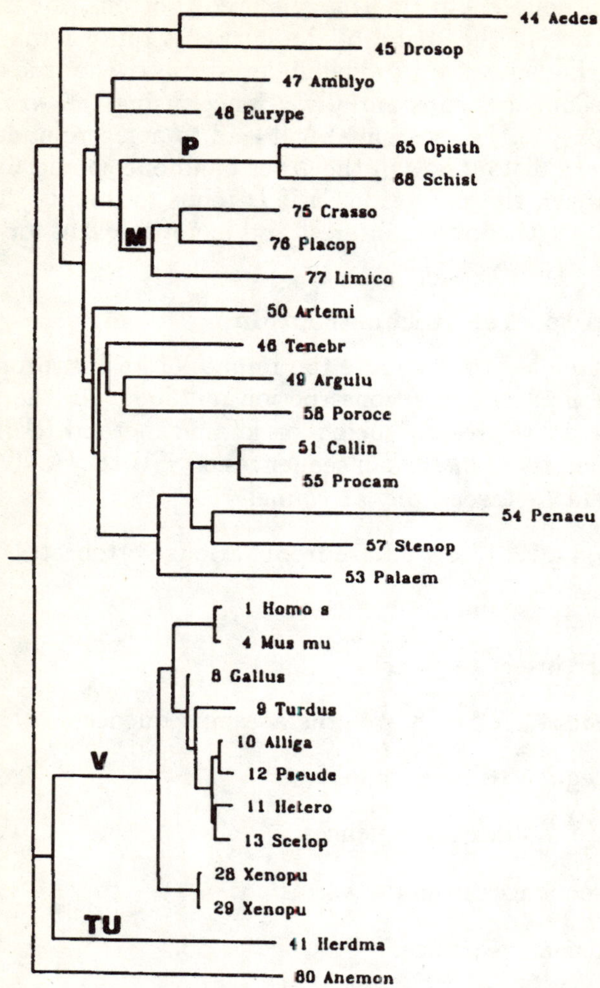

Fig. 2: The outcome of the metazoan phylogeny using neighbor joining on a simple estimate of substitution rate with Felsenstein's model. For details see text. Abbreviations are as follow: **A** = Arthropoda, **C** = spider, **M** = Mollusca, **P** = flatworms, **PM** = flatworms and Mollusca, **T** = Tracheata (insects and Crustacea), **TU** = Tunicata, **V** = vertebrates.

While on the first glance the two trees are not strikingly dissimilar, there are some subtle differences in the branching pattern that deserve attention. In the tree in Fig. 2 the arthropods do not constitute a monophyletic group, the insects (sequence numbers: 44, 45) and the spiders (sequence numbers: 47, 48) are separated from the remaining arthropods by the molluscs (**M**) and the flatworms (**P**). The biological more reasonable assumption of monophyly of arthropods is realized in the tree displayed in Fig. 3. This tree also shows the correct branching pattern within the arthropods, were spiders (**C**) and Tracheata (**T**) constitute sister groups. The placement of the flatworm Mollusca group (**PM**) also agrees with trees obtained by different methods (Van de Peer et al. (1993)).

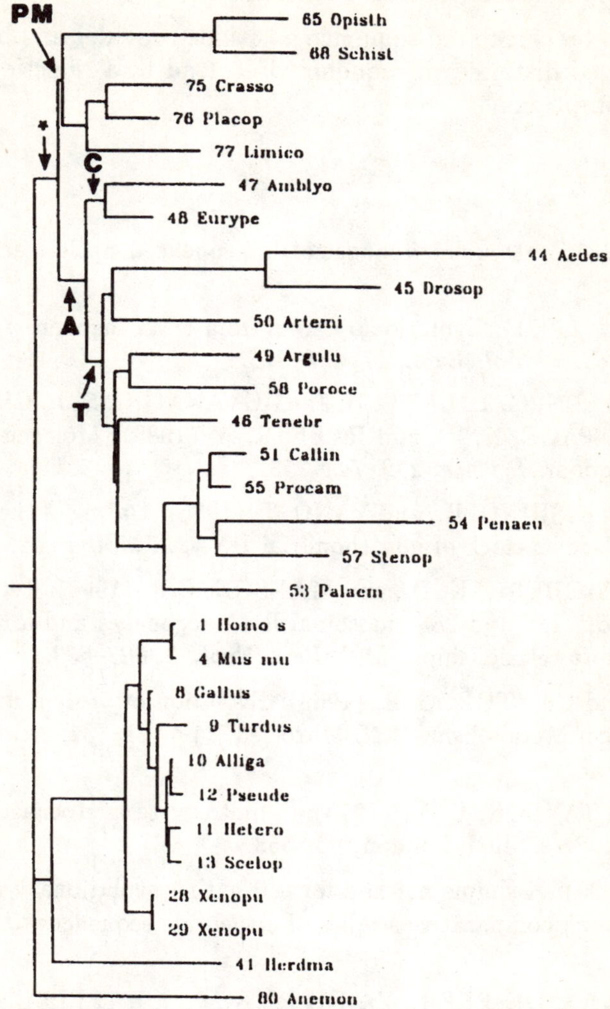

Fig. 3: Neighbor joining tree for distances of 30 metazoan SSU rRNA calculated with the model that assumes correlation between potentially base paired nucleotides in the stem regions of the molecules. For details see text. Abbreviations are as in Fig. 2.

In other regions both trees exhibit the same branching pattern, for example the subtree of the vertebrates is identical in both instances. Our distance correction method based on correlation is not able to force the three insects into one cluster. It is possible that the correlation method cannot compensate for the rapid accumulation of substitutions in the sequences of the flies (sequence numbers: 44, 45). Although there is a tendency towards the biologically more reasonable tree, we do not claim that this will be always the case. The result presented here is promising. But certainly more work (especially computer simulations) is necessary to elucidate the effect of correlation of sequence positions on the outcome of tree reconstruction methods that assume independence between positions.

Our model based on correlated sequence positions provides an objective tool to correct observed distances if sequence data lead to a rejection of the independence assumption.

References:

ELTON, R. A. (1975): Doublet frequencies in sequenced nucleic acids. *J. Mol. Evol., 4, 323–346.*

FELSENSTEIN, J. (1981): Evolutionary trees from DNA sequences: a maximum likelihood approach. *J. Mol. Evol., 17, 368–376.*

FIELD, K. G., OLSEN, G. J., LANE, D. J., GIOVANNONI, S. J., CHISELIN, M. T., RAFF, E. C., PACE, N. R., and RAFF, R. A. (1988): Molecular phylogeny of the animal kingdom. *Science, 239, 748–753.*

HASEGAWA, M., KISHINO, H., and YANO, T. (1985): Dating of the human–ape splitting by a molecular clock of mitochondrial DNA. *J. Mol. Evol., 22, 160–174.*

HEDGES, S. B., MOBERG, K. D., and MAXSON, C.R. (1990): Tetrapode phylogeny inferred from 18S and 28S ribosomal RNA sequences and a review of the evidence for amniote relationships. *Mol. Biol. Evol., 7, 607–633.*

JUKES, T. H., and CANTOR, C. R. (1969): Evolution of protein molecules. In: Mammalian Protein Metabolism. H.N. Muro (ed.), pp. 21–132. Academic Press, New York.

KARLIN, S., and TAYLOR, H. M. (1975): A first course in stochastic processes. 2nd ed. Academic Press, Inc., London, pp. 558.

KIMURA, M. (1980): A simple method for estimating evolutionary rates of base substitutions through comparative studies of nucleotide sequences. *J. Mol. Evol., 16, 111–120.*

NEEFS, J. M., VAN DE PEER, Y., DE RIJK, P., CHAPELLE, S., and DE WACHTER, R. (1993): Compilation of small ribosomal subunit RNA structures. *Nucl. Acids Res., 21, 3025–3049.*

OHNO, S. (1988): Universal rule for coding sequence construction: TA/CG deficiency — TG/CT excess. *Proc. Natl. Acad. Sci. U.S.A., 85, 9630–9634.*

SAITOU, N., and NEI, M. (1987): The neighbor–joining method: a new method for reconstructing phylogenetic trees. *Mol. Biol. Evol., 4, 406–425.*

SCHÖNIGER, M., and VON HAESELER, A. (1994): A stochastic model for the evolution of autocorrelated DNA sequences. *Molecular Phylogenetics and Evolution, 3(3), in press .*

TAVARÉ, S., and GIDDINGS, B. W. (1989): Some statistical aspects of the primary structure of nucleotide sequences. In: Mathematical methods for DNA sequences. M. S. Waterman (ed.), pp. 117–132, CRC Press, Boca Raton.

VAN DE PEER, Y., NEEFS, J.–M., DE RIJK, P., and DE WACHTER, R. (1993): Reconstructing evolution from eukaryotic small–ribosomal–subunit RNA sequences; calibration of the molecular clock. *J. Mol. Evol., 37, 221–232.*

WOESE, C. R. (1987): Bacterial evolution. *Microbiol. Rev., 51, 221–271.*

SALBIDH2 – Modifications of the LBI-Method for Automated Lexicon-Based Indexing of Diagnoses

K. Hofmann, B. Brigl, E. Glück, R. Haux

University of Heidelberg, Institute for Medical Biometry and Informatics,
Department of Medical Informatics,
Im Neuenheimer Feld 400, D-69120 Heidelberg

Summary: We present the modified LBI-method, called MLBI-method, for automated indexing of diagnoses based on the Systematized Nomenclature of Medicine (SNOMED II). The MLBI-method consists of three steps - preprocessing, morphological analysis and semantic analysis - and is defined as composition of functions. We expect to achieve both an indexing and retrieval quality, good enough to use the 'System for Automated Lexicon-Based Indexing of Diagnoses Heidelberg 2' (SALBIDH 2) based on the MLBI-method in routine.

1. Introduction

The German legislation requests the documentation not only of discharge diagnoses and admission diagnoses of all inpatients but also diagnoses of all outpatients. Furthermore, all these diagnoses must be classified into the International Classification of Diseases (ICD-9) to derive diagnoses statistics. For the retrieval of patients with certain diagnoses or combinations of diagnoses we however need nomenclatures like SNOMED (Wingert (1984)). With it the content of a diagnosis can be described as a tuple of indices. Heyen (1990) already showed that SNOMED itself causes errors when indexing automatically. Because manual indexing is to complicated we constructed an application system (SALBIDH) based on a simple formal method (LBI-method) for automated indexing of diagnoses (Brigl et al. (1994a)). An evaluation study showed that the retrieval based on SNOMED-indices is even at least as good as the retrieval based on ICD classes despite a lot of indexing errors (Brigl et al. (1994b), Mieth et al. (1993)). We expect considerable improvements for the retrieval if we reach a higher indexing quality. This can be achieved by a few modifications on the LBI-method, which are described below, and the revision of the lexicons used. A detailed description of the MLBI-method can be found in Hofmann (1994).

2. The MLBI-Method

The MLBI-method is based on the German edition of SNOMED II and is constructed for the German medical language. For that all examples to illustrate the MLBI-method are in German language. We assume the diagnoses to be noun phrases, which in general do not contain more than ten words.

2.1 The Modified Model for the Indexing of Diagnoses

Let $\underline{\mathtt{D}}$ denote the set of input strings – usually diagnoses –, $\underline{\iota}$ the set of indices of any nomenclature. We define indexing as mapping of any input string $\mathtt{d} \in \underline{\mathtt{D}}$ into a tuple of indices $\iota^\star :\subseteq \underline{\iota}^\star$, i. e. as mapping

$$\underline{\mathtt{D}} \rightarrow \underline{\iota}^\star$$

where $\underline{\iota}^\star$ denotes the set of tuples of indices. This model differs from the model used in the LBI-method by mapping a diagnosis into a tuple of indices instead of a set of indices. The model had been changed this way because the sequence of indices can be of great importance to express possible relations between two or more indices.

2.2 Definition of the MLBI-Method

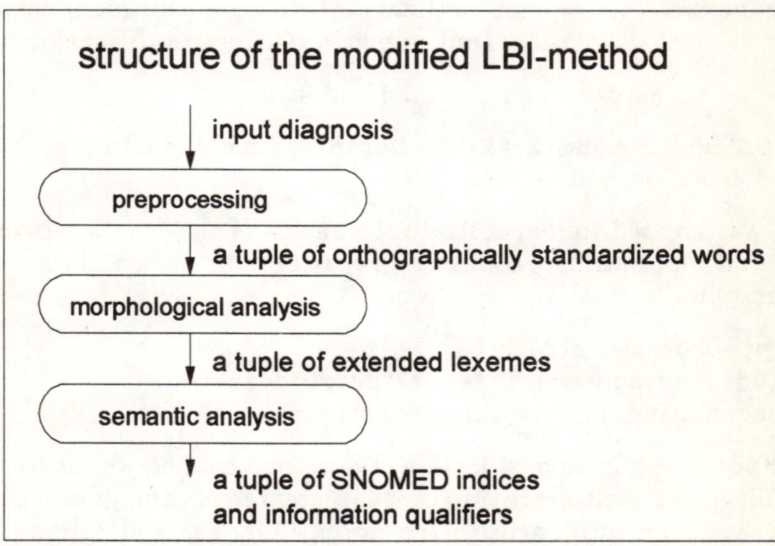

Fig. 1: Structure of the modified LBI-method

The MLBI-method which is based on the model explained below is defined as follows: let $\underline{\mathtt{I}}^\star$ denote the set of tuples of indices containing both SNOMED indices and the so-called information qualifiers like 'Verdacht auf' or 'Zustand nach'. The MLBI-method is defined as a function **indexing** and consists of three subfunctions (see Fig. 1). As mentioned above, the function **indexing** is defined as a mapping

indexing: $\underline{\mathtt{D}} \rightarrow \underline{\mathtt{I}}^\star$

and can be described as composition of three subfunctions

indexing($\mathtt{d}$) := **sem_analysis**(**morph_analysis**(**preprocessing**($\mathtt{d}$))).

2.3 Step 1: Preprocessing

The function **preprocessing** adapts an input string $\mathtt{d} \in \underline{\mathtt{D}}$ to the requirements for the morphological and semantic analysis. Therefore we need six substeps to substitute information qualifiers by special terms (**modification**), to remove date expressions (**filter**), to mark separate diagnoses being part of an enumeration (**enumeration**), to separate the input string into words (**separation**), to substitute abbreviations by their meanings (**abbreviations**) and to standardize the word spelling (**standardization**). The subfunctions filter and enumeration were created to eliminate indexing errors uncovered by the evaluation study. The subfunctions **modification** and **separation** has been modified due to the new subfunctions. Let $\underline{\mathtt{D}}$ denote the set of input strings, $\underline{\mathtt{W}}^*$ the set of tuples of orthographically standardized words. The function preprocessing is defined as a mapping

preprocessing: $\underline{\mathtt{D}} \rightarrow \underline{\mathtt{W}}^*$

and can be described as composition of six subfunctions

$$\textbf{preprocessing}(\mathtt{d}) := \textbf{standardization}(\textbf{abbreviation}(\textbf{separation}(\textbf{enumeration}(\textbf{filter}(\textbf{modification}(\mathtt{d})))))).$$

Let us have a look on the following input string d

```
d = '1.V.a.  Appendizitis, 2.Dermoidzyste Augenbraue, LA,
     3.Dyspepsie 3/91',
```

which was created to demonstrate the effects of the function **preprocessing** on an input string. By using **preprocessing** we obtain a tuple $\mathtt{w}^*$ of orthographically standardized words

```
w* := preprocessing(d) =
    ($d1,$va,appendicitis,$d2,dermoidcieste,
    augenbraue,linces,auge,$d3,diespepsie).
```

Thus `'$d1'`, `'$d2'` and `'$d3'` are information qualifier terms to mark separate diagnoses (**enumeration**), `'$va '` represents the information qualifier `'V.a. '` (**modification**), the words `'linces'` and `'auge'` represent the decomposed meaning of `'LA'` (**abbreviation**) and the date expression `'3/91'` was removed (**filter**).

2.4 Step 2: Morphological Analysis

The morphological analysis serves for the decomposition of compounds, determination of lexemes by removing suffixes and the removal of words which are not relevant for the indexing. To improve this second step we now use two different word models (see Fig. 2) for analyzing each orthographically standardized word as well as the principle of longest match and the principle of backtracking. Thus we obtain two possible results for each orthographically standardized word. Only one or possibly none of them will be selected using two selection criterions. Let $\underline{\mathtt{W}}^*$ denote the set of tuples of orthographically standardized words, $\underline{\mathtt{L}}'^{*}$ the set of tuples of extended lexemes. The set of extended lexemes contains both lexemes and information qualifier terms. The function **morph_analysis** is defined as a mapping

morph_analysis: $\underline{\mathtt{W}}^{*} \rightarrow \underline{\mathtt{L}}'^{*}$

and can be described as composition of three subfunctions

$$\textbf{morph\_analysis}(\mathtt{w}^{*}) := \textbf{selection} \; (\textbf{morph\_forward}(\mathtt{w}^{*}), \textbf{morph\_backward}(\mathtt{w}^{*})).$$

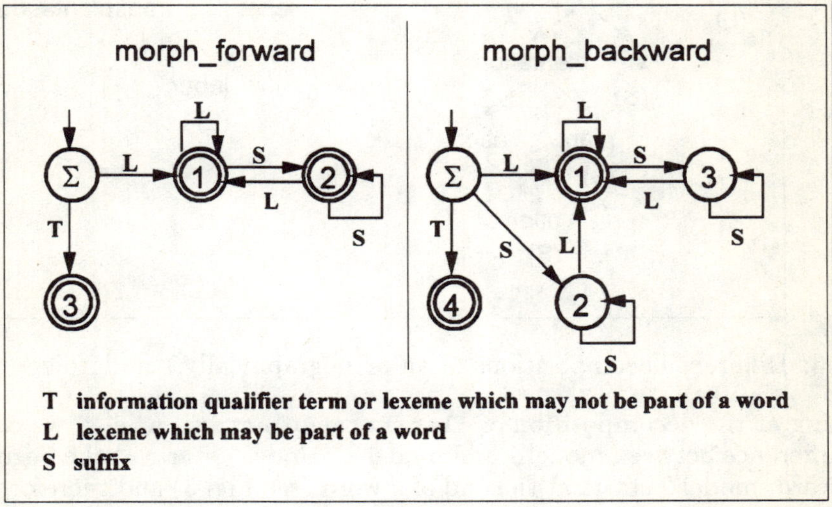

Fig. 2: Word model 1 represented as automaton used by **morph_forward**, word model 2 represented as automaton used by **morph_backward**

The word model 1 used by the subfunction **morph_forward** assumes that each word is generated from the left to the right. The subfunction **morph_backward** uses word model 2 assuming that each word is generated from the right to the left. Both subfunctions use the principle of longest match and the principle of backtracking. We obtain two possibly different decompositions and the number of removed suffixes for each alternative decomposition. We then select the result of **morph_forward** if there is no difference between the two alternatives or if less lexemes were found by using model 1. If the number of lexemes found is equal we select that alternative with less suffixes removed. We choose the result of **morph_forward** if none of the two criterions lead to a decision. In case of an unsuccessful analysis, i. e. neither **morph_forward** nor **morph_backward** can decompose the word, it is omitted. Both model 1 and model 2 can be described by a grammar corresponding to a determined finite automaton (see Fig. 2).

Let us now have a look on a tuple $\mathtt{w}^{*}$ of orthographically standardized words to illustrate this function:

$\mathtt{w}^{*}$ = `($zn,lebertransplantation,quadranten,enterothorax)`.

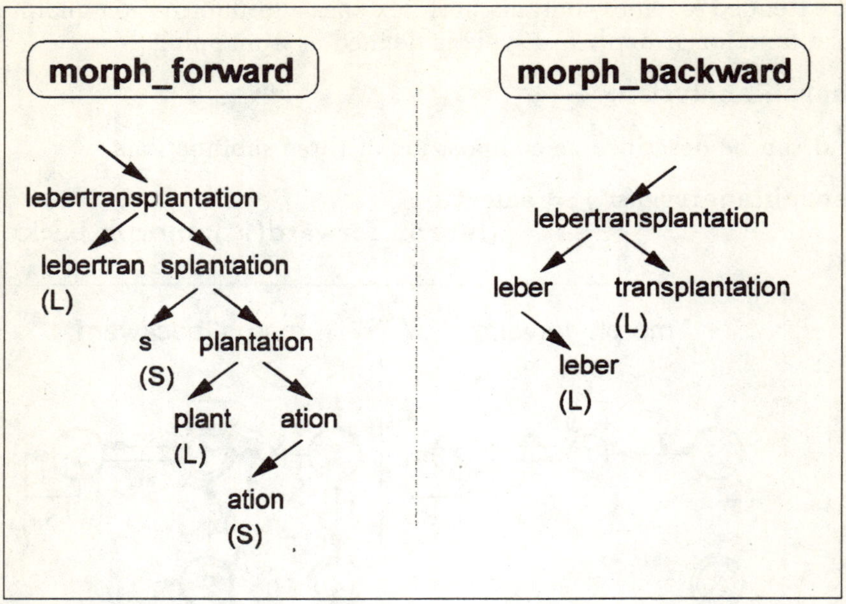

Fig. 3: Different decompositions of an orthographically standardized word

Looking at the decomposition of `'lebertransplantation'` e. g. we can see the difference between model 1 and model 2. Model 1 starts at the beginning of a word, model 2 starts at the end of a word (see Fig. 3) and therefore they generate two different results for `'lebertransplantation'`. Each alternative contains two lexemes, thus the first selection criterion cannot be used. However the second criterion makes the decision. When using model 1 two suffixes were removed but none when using model 2. Therefore the result of subfunction **morph_backward** is selected.

By using the subfunction **morph_forward** we obtain a tuple $\texttt{lf}'^{**}$ of tuples of extended lexemes. By using the subfunction **morph_backward** we obtain a tuple $\texttt{lb}'^{**}$ of tuples of extended lexemes.

$\texttt{lf}'^{**}$ = `(($zn), (lebertran,plant), (quadrant), (enter,thorax))`,

$\texttt{lb}'^{**}$ = `(($zn), (leber,transplantation), (quadr,ten), (enter,thorax))`.

By using the subfunction **selection** we obtain the tuple $\texttt{l}'^{*}$ of extended lexemes:

$\texttt{l}'^{*}$ = `($zn,leber,transplantation,quadrant,enter,thorax)`.

2.5 Step 3: Semantic Analysis

The semantic analysis assigns a tuple of extended indices to a tuple of extended lexemes. This tuple of extended lexemes represents the input diagnosis. Let $\underline{L}'^{*}$ denote the set of tuples of extended lexemes, $\underline{I}^{*}$ the set of tuples of extended indices. Then the function **sem_analysis** is defined as mapping

sem_analysis: $\underline{L}'^{*} \rightarrow \underline{I}^{*}$

and can be described as composition of six subfunctions

sem_analysis(l′*) :=
assignment(**modifiernumbers**(**lexemenumbers**(l′*)),
indices(**summarize**(**snomednumbers**(**lexemenumbers**(l′*))))).

We created the subfunctions **modifiernumbers** and **assignment** and modified the other subfunctions. Therefore we can use the modifiers of SNOMED and we are able to represent information qualifiers in the final result. Now we describe in an informal manner how the third step **sem_analysis** maps a tuple of extended lexemes into a tuple of extended indices.
First the subfunction **lexemenumbers** maps extended lexemes to extended lexeme numbers. The resulting tuple of extended lexeme numbers is used by the subfunctions **modifiernumbers** and **snomednumbers** as input. The subfunction **modifiernumbers** searches for all lexeme numbers all modifiers – these are represented by modifier numbers –, which are attached to the lexemes. The resulting sets of modifier numbers are processed recursively: we successively build intersections of these sets until we get the empty set. This corresponds to the principle of longest match. Now we take the last non-empty set and search for those modifiers which comprise exactly the concerned lexemes, i. e. no additional lexemes. If such a modifier cannot be found we use a backtracking algorithm: We take the last but one non-empty intersection and try to find again an appropriate modifier etc. If backtracking does not lead to any success we omit the first lexeme concerned and apply this algorithm once again. The subfunction **snomednumbers** searchs for all lexeme numbers all SNOMED entries – these are represented by SNOMED numbers –, which are attached to the lexemes. The SNOMED entries are determined in the same way as the modifiers. The present tuple of SNOMED entries contains redundancies because there may exist more than one SNOMED entry for the same SNOMED index. Therefore we remove redundant SNOMED entries. Frequently multi-word entries, which are more specific than several one-word entries, cannot be found because these entries mostly consist of preferred terms of a medical term. Generally a diagnosis is not made up of preferred terms. Therefore the subfunction **summarize** tries to summarize several SNOMED entries of the dimension topography to one SNOMED entry if they directly succeed one another within the tuple of SNOMED numbers. The subfunction indices determines the corresponding SNOMED indices. The subfunction **assignment** tries to modify a SNOMED entry by using one of the modifiers. If a modifier is attached to such a SNOMED entry, this entry will be changed. Then we examine whether an index resulted out of the same lexemes as the modifier. In this case the index concerned is omitted to avoid redundancy. This last substep needs information about both the attachment between the lexemes

and modifiers and the attachment between the lexemes and indices. We obtain this information first by using the subfunctions **modifiernumbers** and **snomednumbers**. When using **summarize** and **indices** we make this information topical.

Let us now have a look on a tuple `l'*` of extended lexemes:

`l'*` = `(ophphen,dislociert,phractur)`.

By using the subfunction **lexemenumbers** we obtain the tuple `(81,96, 8479)` containing extended lexeme numbers attached to the extended lexemes of tuple `l'*`. Out of these lexeme numbers the subfunction **modifiernumbers** results in three modifiers attached to the lexeme `'ophphen'` and one modifier attached to the lexeme `'dislociert'`. By using **snomednumbers**, **summarize** and **indices** we obtain a tuple `(MX0004,M12000)` of indices, thereat `MX0004` represents 'offen' and `M12000` represents 'Fraktur'. Up to now we got four modifiers and two indices. By using the subfunction **assignment** the final step is done. Both `'ophphen'` and `'dislociert'` are correct modifiers attached to the SNOMED entry `M12000`. As final result we then obtain the tuple `(M12202)` with `'ophphen'` modifying the third position by changing it into `'2'` and with `'dislociert'` modifying the fifth position by changing it into `'2'`. Because the SNOMED entry `MX0004` resulted out of the first lexeme as well as one of the modifiers representing `'offen'` we omit this redundant SNOMED entry now.
The final result is a tuple `i*` of extended indices:

`i*` := **sem_analysis** (`l'*`) = `(M12202)`.

3. Realization

The MLBI-method is realized as '**S**ystem for **A**utomated **L**exicon-**B**ased **I**ndexing of **D**iagnoses **H**eidelberg **2**' **(SALBIDH 2)** using the 4th generation language NATURAL2/ADABAS. It is integrated in the application system for the 'fundamental documentation' of the Heidelberg hospital information system.
We carry out an evaluation study to determine both indexing quality and retrieval quality. With all modifications on the LBI-method and the modifications of the lexicons we expect to reach a quality of SALBIDH2 good enough to use it in routine.

References:

BRIGL, B., MIETH, M., HAUX, R., GLÜCK, E. (1994a): The LBI-method for the automated indexing of diagnoses using SNOMED Part 1: Design and Realisation. To appear in: Int. J. Bio-Med. Comp.

BRIGL, B., MIETH, M., HAUX, R., GLÜCK, E. (1994b): The LBI-method for the automated indexing of diagnoses using SNOMED Part 2: Evaluation. To appear in: Int. J. Bio-Med. Comp.

HEYEN, P. (1990): Erfahrungen mit der automatischen Indexierung von Volltexten. In: G. Giani and R. Repges (eds.): *Biometrie und Informatik*, 34. Jahrestagung der GMDS. Springer, Berlin, 172 - 177.

HOFMANN, K. (1994): Automatische Indexierung von Diagnosen auf der Basis von SNOMED - Modifikation der LBI-Methode, Realisierung und Evaluation. Thesis, Medical Informatics Program, University of Heidelberg/School of Technology Heilbronn. In preparation.

MIETH, M., BRIGL, B., GLÜCK, E., HAUX, R. (1993): Zur automatischen Indexierung von Diagnosen nach SNOMED und ihrer Anwendbarkeit in der Neurochirurgie. *Informatik, Biometrie und Epidemiologie in Medizin und Biologie*, 24, 101 - 109.

WINGERT, F. (1984): Systematisierte Nomenklatur der Medizin: SNOMED, Band I und II. Springer, Berlin.

Record Linkage of Anonymous Data by Control Numbers

W. Thoben, H.-J. Appelrath, S. Sauer

OFFIS, Westerstr. 10-12, D-26121 Oldenburg

Summary: The processing of personal data in information systems is of decisive importance in regard to data protection. Several applications exist where the personal identifying data make no difference but are needed for the record linkage with other data sets. This paper describes a concept which enables to compare anonymous data and to fulfil the complex requirements of data protection in the processing.

1. Introduction

The right of access to personal files means that everyone decides, in principle, for himself on the revelation and use of his personal data. This right is one of the basic rights protected by constitution and should protect a person against the abuse of h is personal data while the information technology is expanding more and more.

However there are applications in which this right competes with the need of a population-based data aggregation with the possibility to identify individuals and where no personal identifying data is needed for the processing. In this case a data record c an be coded, i.e. to encrypte the personal identifying data or to separate it from the other data and to use only the anonymous data.

A decisive problem which follows from such a proceeding is to join different anonymous data records. The primary criteria for searching in personal identifying data are generally identifying features. But as well as they are of no account for the processi ng their use is not allowed in regard to data protection, so it is unalterable to develop a concept which does not use these features but nevertheless permits a record linkage.

2. Cancer Registry Lower-Saxony

An example for an application in which the personal identifying data makes no difference is an epidemiological cancer registry. The different cancer registration models are described in (Schrage (1991), Gruner et al. (1989)). In cooperation with the insti tute OFFIS (Oldenburger Forschungs- und Entwicklungsinstitut für Informatik-Werkzeuge und -Systeme) the concept of Prof. Michaelis (Michaelis and Krtschil (1992), Schmidtmann et al. (1993)) is proved in the years 1993/94 in Lower-Saxony (Appelrath et al. (1993), Brand et al. (1993)). This concept, which is taken into account in the actual bill of an Act for cancer registries (KRG (1994)), contains a registration of cancer divided in a position of trust and a register. Personal reports to the cancer re gistry are collected in the position of trust, where the personal identifying data are separated from the epidemiological data. Applying a public-key cryptosystem (using two keys - public key for encryption and secret key for decryption (Rivest et al. (19 78)) the personal identifying data is encrypted. These anonymous reports are transfered to the register and condensed to a population-based epidemiological cancer registry.

Because each cancer case is repeatedly reported to the cancer registry and to increase the quality and quantity of the registry by integration of external data sets - e.g. information on the death certificates, reports from pathologists - a record linkage of the anonymous data records has to be realized in the register. At this point the problem is to recognize several anonymous data records, which have to be attached to the same respectively different persons as the same respectively different reports.

3. Record Linkage System

3.1 Robustness of the Record Linkage System

A decisive aspect for the record linkage of data records is to guarantee the robustness of the record linkage system, if there should be errors in the plaintext. In this case several data records, which describe the same person, but include some different identifying data, are matched to the same person. This so-called fault tolerance is opposed to the aim of a high discriminating power, i.e. to attach data records of different persons even with negligible differences in their identifying data to differen t persons. Typical causes for differences of natural language character strings, especially in personal identifying data, which result from errors during the registration or transfer of the personal identifying data or incomplete or incorrect information about the person, are (Ultsch (1987), Gruner et al. (1989)):

- **Typing error:** Writing or typing errors are mainly independent of an application and can be subdivided in four elementary error categories (missing, additional, incorrect character or transposition of characters).
- **Phonetic misunderstandings and ambiguouses:** Phonetic misunderstandings can result during verbal transfer of information, especially during the registration of personal data records from other than written sources. A similar problem is represented by different styles of equally sounding names (e.g. Meier or Meyer), where confusions conduct to ambiguouses.
- **Differing information:** Another error cause is differing information in reports, which describe the same person. E.g. in one data record a person with double name is registrated completely while in another only one name is available. A similar probl em are different addresses when a person is migrated or has several places of residence.

- **Missing information:** In this case the question is to recognize the similarity of data records and furthermore, which and how much missing information in a data record is tolerable.

3.2 Control Numbers

In principle there is the possibility of a record linkage through the ciphertexts of single reports whereby already little errors in the plaintext cannot be detected in the ciphertexts. Using a nondeterministic cryptosystem (adding a random number to the encryption to prevent chosen plaintext attacks), which is realized in the Cancer Registry Lower-Saxony, several identical data records of one person are enciphered to different ciphertexts and, thus, they cannot be matched to one data record.

One approach for a record linkage of data records is the use of control numbers (Michaelis and Krtschil (1992), Schmidtmann et al. (1993), Appelrath et al. (1993)), a deterministic encryption of characters of personal data records, e.g. name, birthname, a ddress. The control numbers are generated in parallel to the encryption of the personal identifying data by using a one-way function (deterministic function for which the inverse function cannot be practically developed). The identity of a person cannot b e inferred from these control numbers. Decisive for the work with control numbers is the fact that the one-way function is deterministic, because several reports of the same person must be represented by the same control numbers. Two types of errors occur in record linkage (Gruner et al. (1989)):

- **Homonyms:** Reports of different cancer cases are matched to the same person, i.e. the control number (variable) of different persons is assigned with the same value.

- **Synonyms:** Different reports of one cancer case are represented to different persons, i.e. one control number (variable) is assigned with different values, in spite of the equality of the person.

Causes for these errors are the typing errors, phonetic misunderstandings and ambiguouses represented in section 3.1, but also the conception of the control numbers - extracts of personal data records - can generate homonyms itself.

The aim of the record linkage system is to find attributes or parts of them for the specification of control numbers, which produce minimal synonyms and homonyms. Therefore the similarity of data records must be defined in form of metrics respectively the development of suitable heuristics has to be forced (Newcombe (1985)).

4. Examinations

4.1 Examinated Control Numbers

During the first examinations with the prototype of the record linkage system in the course of the Cancer Registry Lower-Saxony 16 control numbers have been generated (Appelrath et al. (1993)), which can be - if required - adapted respectively reduced or completed. The following 6 control numbers are presented exemplary:

(1) surname, first name, date of birth, sex
(2) surname[1], first name[1], date of birth, sex
(3) soundex(surname), soundex(first name), date of birth, sex
(4) ASCII-$\sum$(surname) + LEN(surname), ASCII-$\sum$ (first name) + LEN(first name), date of birth, sex
(5) surname[3], first name[3], month/year of birth, sex
(6) surname[3], first name[3], year of birth, sex

The digit in the parentheses behind the attributes shows the number of viewed places from the beginning of a word and the functions ASCII-$\sum$ and LEN are definded as follows:

- ASCII-$\sum$(string): sum of the ASCII-encoding of the characters,
- LEN(string): number of characters of the string.

On the one hand the control numbers are generated directly from the attributes of the data records (see (1), (2)), on the other hand by using functions (see (3), (4)), which try to recognize errors like a transposition of characters (in this case the res ults of the functions SUM and ASCII-$\sum$ are equal) or phonetic misunderstandings with phonetic codes like soundex (Mresse (1984)).

4.2 Rates

Suitable measures to judge the quality of the control numbers have to be determined. In this case that are the rate of homonyms and the rate of synonyms, which are defined as follows:

Given are n reports without synonyms, i.e. all n reports are from different cases. A control number is generated for each of these cases and then the number of differently generated control numbers k is determined. The rate of homonyms H results f rom $H = \frac{n-k}{n}$, the ratio of equally generated to the total number of all control numbers.

To determine the rate of synonyms of a control number there are given n different cases with 2 reports for each case. A control number is generated for each of these $2n$ reports and the number of correct matches l is determined, i.e. in l of n ca ses the control numbers of both reports of a case process the same value. The rate of synonyms S of a control number results from $S = \frac{n-l}{n}$, the ratio of all falsely matched to the total number of different cases.

4.3 Examinated Data Sets and Examination Process

In the first step of the project of Cancer Registry Lower-Saxony 3 data sets are examinated (Appelrath et al. (1993)):

- Nachsorgeleitstelle Oldenburg (NLS): 13.269 data records,
- death certificates from 1990-92 of the city Oldenburg (DC): 4.587 data records,
- Ma.Ca.-death certificates from 1990-92 of the city Oldenburg and the district Ammerland (Ma.Ca.): 88 and 47 data records.

For all data records of the 3 data sets 16 control numbers are generated and are saved with the data records in a database. The control numbers of every data record are compared with those of all other data records and in the case of identity they are sav ed in a special database. Homonyms are calculated by the comparison of each data record of a data set with every other data record of the same data set. Therefore, it is important that there are no synonyms in these data sets. To determine the rates of sy nonyms by using the double registrated reports the 3 data sets are compared with each other.

For the evaluation of the examinated control numbers all equal entities of a control number are visually controlled by using the plaintext of the data sets. Thereby it is checked if both data records really describe the same person. In the following it ca n be determined for each control number,

- how many similarities have been correctly recognized,
- how many data records have been detected as similar but do not belong to the same person,
- how many data records have not been detected as similar although they should describe the same person.

5. Results

5.1 General Survey

We have to distinguish between two kinds of record matching process: internal record linkage (see Tab. 1) to determine the rates of homonyms and external record linkage (see Tab. 2) to determine the rates of synonyms. Under the section matches there is noted how many reports are related to at least another one and with how many control numbers this correspondence has been ascertained. In the table of internal matches every match is of course a fault. The total number of examined data records produces the denominator for calculating the rate of homonyms for each control number (e.g. NLS: 13.269).

The table of external matches shows also the absolute numbers of matches.

	NLS	DC	Ma.Ca.
matches data records control numbers	 387 596	 74 123	 0 0

Tab. 1: internal matches

	NLS-DC	NLS-Ma.Ca.	DC-Ma.Ca.
matches data records control numbers	 519 4.512	 73 854	 90 1.120
errors data records control numbers	 181 291	 3 5	 2 4

Tab. 2: external matches

These cases have been visually controlled and the discovered wrong matches (synonyms) have been determined. The difference between matches and faults (e.g. NLS-DC at Tab. 2: 519-181 = 338) is the denominator for calculating the rate of synonyms for each control number because these are the data records for which exactly two reports exist.

5.2 Rates of Homonyms and Synonyms

On the base of the denominator from Tab. 1 and Tab. 2 and with further detailed examinations of each control number the rates of homonyms and synonyms have been determinated by using the definitions of chapter 4.2.

Tab. 3 and Tab. 4 show the percentages for the control numbers given

control number	NLS	DC	Ma.Ca.
1	0	0	0
2	0,12	0,13	0
3	0	0	0
4	0	0	0
5	0,17	0,07	0
6	1,31	0,7	0

Tab. 3: rates of homonyms

in chapter 4.1. The numbers in parentheses within the table of synonyms mark the part of faults that are caused by a non-uniform registration of first names in the different data sets if more than one is entered in the original report.

In principle the rates of synonyms are much worse than the rates of homonyms. One reason is that these rates are a result of a comparison of two different data sets and therefore different guidelines for registration. This fact shows especially the faults that are exclusively caused by the non-uniform regis-

control number	NLS-DC	NLS-Ma.Ca.	DC-Ma.Ca.
1	30,77 (20,41)	28,57	28,41 (11,36)
2	7,69	4,29	0
3	18,34 (11,24)	8,57	19,32 (5,68)
4	30,18 (19,82)	28,57	28,41 (11,36)
5	10,36	7,14	5,68
6	8,88	5,71	5,68

Tab. 4: rates of synonyms

tration of first names.

Besides, you have to recognize that the rates influence each other. Control numbers with a low rate of homonyms (e.g. control number (1), (3), (4)) have a considerably higher number of synonyms or vice versa. This is affected by the fact that a control nu mber with a slight rate of homonyms is characterized by a high discriminating power while it generates more synonyms at the same time. In contrast a control number with a low rate of synonyms has a high fault tolerance but it neglects the discriminating p ower. Therefore, the aim has to be to find a suitable mixture of control numbers that combine both qualities.

6. Prospect

The present examinations of anonymous data records have been exclusively made on the base of isolated control numbers. In fact the approach by means of control numbers seems to be promising but the results in chapter 5.2 show that isolated control numbers cannot ensure reliable record linkage with minimal rates of homonyms and synonyms. So in a second step combinations (conjunctions of single control numbers) will be examined which guarantee both the aspect of discriminating power and a sufficiently high fault tolerance. The third step will then be the construction of rules (disjunctions) out of single control numbers and combinations of them. As a result there will be a system of rules that ensures minimal rates of homonyms and synonyms.

Furthermore, we have to consider other data sources to improve the validity of previous results. Especially the examination of synonyms should be forced, i.e. to find data sources with double registration of individuals and apply the methods of record lin kage on them.

Another aspect of further efforts is improving the quality of registration. The results show remarkably the problems that arise by recording personal data and so they form the basis for an improvement of the quality of data.

References:

APPELRATH, H.-J., THOBEN, W., RETTIG, J., and SAUER, S. (1993): CARLOS (Cancer Registry Lower-Saxony): Tätigkeitsbericht für den Zeitraum 1.4.-

1.11.1993. Oldenburg.

BRAND, H., REICHLING, I., APPELRATH, H.-J., ILLIGER, H.-J., UNGER, I., and WINDUS, G. (1993): CARLOS (Cancer Registry Lower-Saxony) - Pilotstudie für ein bevölkerungsbezogenes Krebsregister in Niedersachsen. In: S.J. Pöppl, H.-G. Lipinski and T.Mansky (eds.): *Medizinische Informatik - Ein integrierender Teil arztunterstützender Technologien.* MMV Medizien Verlag, München, 404-406.

GRUNER, G., HARTMANN, S., MEISNER, C., PIETSCH-BREITFELD, B., and SELBMANN, H.K. (1989): Forschungsvorhaben "Epidemiologisches Krebsregister Baden-Württemberg". Bericht Nr. 6/1989, Institut für Medizinische Informationsverarbeitung, Universitä t Tübingen.

KRG (1994): Entwurf eines Gesetzes über Krebsregister (Krebsregistergesetz KRG). Stand: 26.5.1994, Drucksache 12/6478, Bonn.

MICHAELIS, J., and KRTSCHIL, A (1992): Aufbau des bevölkerungsbezogenen Krebsregisters für Rheinland-Pfalz. *Ärzteblatt Rheinland-Pfalz*, 45, 434-438.

MRESSE, M. (1984): Information Retrieval - eine Einführung. Leitfäden der angewandten Informatik, Teubner, Stuttgart.

NEWCOMBE, H.B. (1985): Handbook of Record Linkage - Methods for health and statistical studies, administration, and business. Oxford University Press, Oxford.

RIVEST, R.L., SHAMIR, A., and ADLEMAN, A. (1978): A method for obtaining digital signatures and public-key cryptosystems. *Communications of the ACM*, Vol. 21, No. 2, 120-126.

SCHMIDTMANN, I., POMMERENING, K., and MICHAELIS, J. (1993): Pilotstudie zum Aufbau eines bevölkerungsbezogenen Krebsregisters in Rheinland-Pfalz, In: S.J. Pöppl, H.-G. Lipinski and T.Mansky (eds.): *Medizinische Informatik - Ein integrierender Teil arztunterstützender Technologien.* MMV Medizin Verlag, München, 399-403.

SCHRAGE, R. (1991): Zur Krebsregisterfrage - modifiziertes Melderechtsmodell zur Verbesserung des Datenschutzes. *Öffentliches Gesundheits-Wesen*, 53, Georg Thieme Verlag, Stuttgart, 746-752.

ULTSCH, A.G.H. (1987): Control for knowledge-based information retrival. Dissertation, no. 3, institute of computer science, ETH Zürich.

Information Systems and Knowledge Organization

Processing Partial Information in Decision Support Systems

F. Dellmann

FB Wirtschaftswissenschaften, Wirtschaftsinformatik und BWL, Universität GH Essen, Altendorfer Str. 97-101, D-45143 Essen

Summary: Often in economic reality only partial information (PI) is available. Decision problems under PI can be dealt with Decision Support Systems (DSS). A theoretical foundation of processing linear and nonlinear PI in DSS is attained by combining comparative/interval probability theory, probabilistic logic, and decision theory. The consistency of the decision maker's subjective statements is examined by a system of probability axioms. A representation of the PI as a system of equations/inequalities is shown. Bounds of unknown probabilities are determined by solving a linear or signomial programming problem. Applying the $MaxE_{min}$ principle even to nonlinear PI is proposed to resolve decision problems.

1. Introduction

Economic decisions in reality often depend on uncertainty caused by incomplete information. In these situations the information is given to the decision maker rather partially than exactly. DSSs can be used to process PI. This thesis follows from the definition of DSSs, which provide "computer-based support for management decisionmakers who are dealing with semistructured problems" (Keen and Scott Morton (1978), 97).

A problem is called semistructured, if at least one phase of the decision process is unstructured, and if at least one is structured (Gorry and Scott Morton (1971)). Problems are structured, if they are repetitive and routine. In this case an automated procedure can be worked out to solve the problem. However, unstructured problems are too elusive or not well-defined enough to be solved by an algorithm. Stabell (1979) advances three characteristic features of semistructured problems: ambiguous or conflicting objectives, difficulty to determine the causes of changes in consequences, and uncertainty about the consequences of the actions taken.

One may conclude that uncertainty and especially PI yield semistructured decision problems. They can be solved by the interaction between the decision maker and the DSS. The algorithms regarding the structured part of the decision problem are implemented in the DSS to support the decision maker dealing with the unstructured part.

Decision theory is used here to define uncertainty and PI. The underlying decision model is generally described by a set of actions $A = \{a_1, \ldots, a_m\}$, a set of states $S = \{s_1, \ldots, s_n\}$, a function $c : A \times S \to C$ yielding the consequences $c_{ij} = c(a_i, s_j)$, $c_{ij} \in C$, and a cardinal utility function $u : C \to \mathbf{R}$ where $\mathbf{R}$ denotes the set of the real numbers.

A decision under certainty is characterized by the exact knowledge of the true state of world and the consequence resulting deterministically from the selected action. Uncertainty is defined by the lack of knowledge about both the states and the consequences yielded by actions. Extreme cases of uncertainty are risk and ignorance. Risk is characterized by the knowledge of the exact numerical probability distribution of the states and the consequences. In case of ignorance, the decision maker only knows or presumes the existence of some states and consequences, but he is not provided with any probability information.

Decisions under PI are placed between risk and ignorance (Kofler and Menges (1976)). In this case, the decision maker knows a part, but he does not know the whole of the underlying probability distribution. For example, PI can be given in form of comparative probabilities (e.g. the $ exchange rate is more probable to rise than to fall), interval probabilities (e.g. the probability of a loan interest rate of 9 % is between 0.2 and 0.4), or linear combinations of probability information (e.g. the probability of a cash flow less than $1m and the probability of a cash flow more than $10m sum up 0.1). The decision maker is often provided with this type of information in economic reality.

To build a DSS, which is closer to reality by processing PI, a theoretical foundation needs to be established, what has not been done yet this way. The resulting system should make use of the whole extent of the available information. The objective is to support the decision maker by determining bounds of unknown probabilities or even by proposing a decision.

The framework presented here combines comparative and interval probability theory, probabilistic logic, and decision theory to realize such a DSS. Axiom systems of probability theory are used to check the consistency of the subjective probability informat ion the decision maker states (Ch. 2). On this consistent base, linear and nonlinear PI are represented by means of probabilistic logic, and unknown probabilities can be determined (Ch. 3). A solution of the decision problem can be proposed on the base of the $MaxE_{min}$ principle taken from decision theory (Ch. 4). Since the information is represented in accordance with an axiom system, this framework is founded on a mathematically consistent base. The flow of processing PI in DSSs is depicted in Fig. 1.

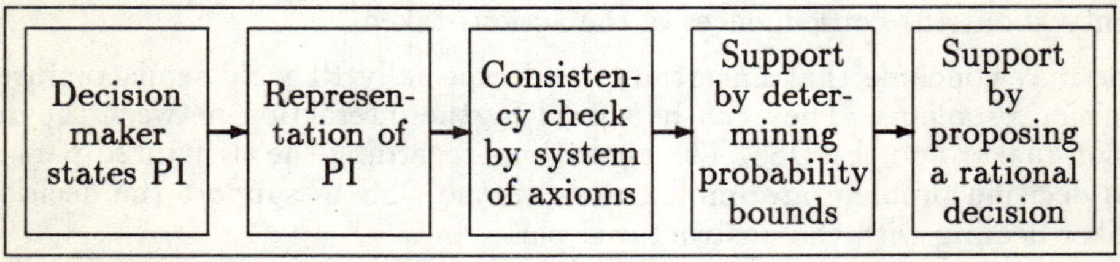

Fig. 1: Flow of processing PI in DSS.

Major theories related to this framework are fuzzy theories (Zadeh (1965)), Dempster-Shafer theory of evidence (Shafer (1976)), and certainty factors (Shortliffe and Buchanan (1975)). Compared to these theories, the framework presented here has the crucial advantage that the decision maker does not need to specify his prior probability information completely.

A system processing linear PI and determining unknown probabilities is realized by SIMUNC (Rimscha (1991)), which can be rated as a DSS generator. Probabilistic logic is used in this system to represent only linear PI. It lacks of integrating proposals based on principles taken from decision theory.

2. Consistency Check of Subjective Probabilities

The PI the decision maker is provided with often comes from personal assessments. For this reason it is affected with subjectivity, what may lead to contradictions or inconsistencies. In order to avoid them, it is necessary to check the consistency of thesubjective probability information.

The Dutch Book Theorem by Ramsey (1926) and de Finetti (1937) expresses that the decision maker cannot have a book made against him, if and only if his subjective probabilities are compatible with a system of axioms of probability theory. It has been show n by these authors that, if the condition above is met, the decision maker cannot suffer a sure loss independently of the occurring state through a constructed bet. Following this theorem, the compatibility of the subjective probabilities with a system ofaxioms must be checked.

In case of exact numerical subjective probabilities, the well known Kolmogorov axioms can be applied. But, if the available information of the decision maker is partial, appropiate systems of axioms are needed. Fishburn (1986) provides an overview and a discussion of axiom systems for comparative or interval probabilities and also for linear combinations of probability information.

In order to check the consistency of the PI in a DSS, it has to be represented. This task will be solved by probabilistic logic.

3. Probabilistic Logic

Whereas truth values of a logical formula are binary in propositional logic, they are real numbers in the interval $[0;1]$ in probabilistic logic (Nilsson (1986)). For this reason they can be interpreted as probabilities of the formula. If an atomic proposition is regarded as a state, the corresponding truth value can be interpreted as state probability.

To represent the states, binary truth values are assigned to them. Such an assignment of truth values to the states is called a possible world. It describes a possible combination of the occurrence of the states. The representation of n states $s_1, \ldots, s_n$ leads to $N = 2^n$ possible worlds. Exact knowledge and column generation methods as proposed by Nilsson (1986) reduce this exponential growth.

Note that the states need not be mutually exclusive in this setting. This enhances the possibilities of decision theory, which deals only with mutually exclusive states. The integration of decision theory in probabilistic logic is treated in Ch. 4. The next sections show that any PI can be represented adequately by means of probabilistic logic.

The assignment of truth values $t = (t_1, \ldots, t_N)$ to the states yields an $n \times N$ matrix T with elements t_{jk} indicating the truth value of state s_j in the possible world k. A probability vector $\pi = (\pi_1, \ldots, \pi_N)$ can be assigned to the possible worlds with

$$\sum_{k=1}^{N} \pi_k = 1 \tag{1}$$

$$\pi \geq 0. \tag{2}$$

The truth values of the states in the interval $[0; 1]$ are now interpreted as state probabilities p. They can be determined by

$$p = T \cdot \pi'. \tag{3}$$

3.1 Representation of Linear Partial Information

Any PI can now be represented by means of equation (3). Supposing the existence of only two states, the matrix T is given by

$$T = \begin{pmatrix} 0 & 0 & 1 & 1 \\ 0 & 1 & 0 & 1 \end{pmatrix}. \tag{4}$$

A comparative probability $s_1 \succsim s_2$ meaning "state s_1 is at least as probable as state s_2" can be expressed in form of state probabilities by $p_1 - p_2 \geq 0$. The application of (3) yields $\pi_3 - \pi_2 \geq 0$. Similarly, an interval probability $P_*(s_1) = a, P^*(s_1) = b$ is represented by $\pi_3 + \pi_4 \geq a$ and $\pi_3 + \pi_4 \leq b$ where $a, b \in \mathbb{R}$.

In general, the decision maker can express his available PI by M equations and inequalities of linear combinations of state probabilities and real numbers. They can be represented by the following systems:

$$G \cdot \pi' \begin{Bmatrix} = \\ \geq \end{Bmatrix} g \tag{5}$$

where G: $M \times N$ and g: $M \times 1$.

(1), (2), (5) can be represented graphically as a convex polyhedron in the N-dimensional space. Every consistent additional probability information cuts off a part of the polyhedron. PI, which is not compatible with the underlying system of axioms, leads to an empty polyhedron indicating inconsistency.

Given the case of a nonempty polyhedron Hailperin (1965) shows that it is possible to determine the upper and lower bounds of unknown probabilities by solving a linear programming problem. This determination means a remarkable support to the decision maker, since he is provided with additional information about uncertain states.

The upper probability of a state s_j can be determined by maximizing the sum of the products of the truth values t_{jk} and the probabilities π_k over all possible worlds subject to the convex polyhedron. The lower probability is given by minimization analogously. Formally, one has to solve the linear programming problem

$$p_j = \sum_{k=1}^{N} t_{jk} \cdot \pi_k \rightarrow \begin{cases} \max \\ \min \end{cases} \tag{6}$$

$$\text{s.t. (1), (2), (5).}$$

3.2 Representation of Nonlinear Partial Information

It is shown in section 3.1 how to represent linear PI. But the decision maker often disposes of other important types of information like conditional comparative probabilities or (conditional) independence of states.

A conditional comparative probability $s_1|s_3 \succsim s_2|s_4$ has the meaning of "state s_1 given state s_3 is at least as probable as state s_2 given state s_4". Applying the definition of conditional probabilities results in

$$\frac{P(s_1 \cap s_3)}{P(s_3)} \geq \frac{P(s_2 \cap s_4)}{P(s_4)}. \tag{7}$$

Independence of states is defined by

$$P(s_1 \cap s_2) = P(s_1) \cdot P(s_2). \tag{8}$$

To model conditional independence, Andersen and Hooker (1994) link probabilistic logic and Bayesian networks calling it Bayesian logic. Two states s_1 and s_2 are defined as conditional independent, given $s_3, \ldots, s_n$, if

$$P(s_1 \mid s_2 \cap s_3 \cap \ldots \cap s_n) = P(s_1 \mid s_3 \cap \ldots \cap s_n). \tag{9}$$

Applying (3) to (7)–(9) conditional comparative probabilities, independence, and conditional independence of states are represented as nonlinear equations and inequalities after some transformations (Rimscha (1990); Andersen and Hooker (1994)):

$$\sum_{h=1}^{q} \alpha_h \cdot \pi_{k_h} \cdot \pi_{l_h} \begin{Bmatrix} = \\ \geq \end{Bmatrix} 0 \tag{10}$$

where $k_h, l_h \in \{1, \ldots, N\}$.

More generally, it is possible to represent nonlinear PI by nonlinear equations and inequalities in the following way:

$$\sum_{h=1}^{q} \alpha_h \cdot \prod_{k=1}^{N} \pi_k^{\beta_{kh}} \begin{Bmatrix} = \\ \geq \end{Bmatrix} \gamma. \tag{11}$$

In addition to the mentioned possibilities (7)–(9) the decision maker may also express his information directly in form of (11). Huschens (1985), for example, provides a possibility to get nonlinear, nonconvex PI (see Ch. 5).

Bounds of unknown probabilities result again from an optimization problem:

$$p_j = \sum_{k=1}^{N} t_{jk} \cdot \pi_k \to \begin{cases} \max \\ \min \end{cases} \tag{12}$$

$$\text{s.t. } (1), (2), (5), (11).$$

This optimization problem is nonlinear, because the probabilities π_k in (11) are linked multiplicatively and the exponents β_{kh} can take values not equal to 0 or 1. Furthermore, it is nonconvex, because the parameters α_h can be positive and negative. This type of nonconvex, nonlinear programming problems should be solved by a signomial programming algorithm (Andersen and Hooker (1994), 208).

Dembo (1978) provides an overview of several algorithms for signomial programs with inequality constraints. But if one deals with equality constraints, it is necessary to split each of them into two inequality constraints leading to a greater problem. Gonen and Avriel (1986) propose an algorithm to solve signomial programs with equality constraints. But numerical difficulties result, if the problem contains inequality constraints, which are transformed to equality constraints by introducing artificial variables.

4. Decisions under Partial Information

In addition to support the decision maker by determining the bounds of unknown probabilities it is possible to give him wider support by proposing which action to choose. This suggestion bases on a rational decision principle coming from decision theory under PI (Kofler and Menges (1976)).

Each possible probability distribution $p = (p_1, \ldots, p_n)'$ regarding the states $s_1, \ldots, s_n$ is represented as a probability point $\psi(p) \in \mathbf{R}^n$. These probability points span an $(n-1)$-dimensional distribution simplex $S^{(n)} = \{(p_1, \ldots, p_n) \mid p_j \geq 0, \sum_{j=1}^{n} p_j = 1\}$. Decisions under PI are characterized by the knowledge of a subset $T(p) \subset S^{(n)}$.

PI is defined as linear in this case, if the corresponding subset $T(p)$ of $S^{(n)}$ can be represented as a system of linear equations and inequalities:

$$G \cdot p' \begin{Bmatrix} = \\ \geq \end{Bmatrix} g \tag{13}$$

$$\sum_{j=1}^{n} p_j = 1 \tag{14}$$

$$p \geq 0. \tag{15}$$

Obviously, this representation is the same as (1), (2), (5), which were derived from probabilistic logic, except that the resulting system is formulated on the base of p and not of π. But decision theory requires that all states s_j are mutually exclusive and their union is the universal set S. Applying this exact knowledge to the modeling of the possible worlds yields $p = \pi$ giving exact correspondence of the representations.

Just as in probabilistic logic, linear PI is represented as a convex polyhedron with certain corner points $\psi^{(f)} = (p_1^{(f)}, \ldots, p_n^{(f)})', f = 1, \ldots, r$. If the column vectors of these corner points are joined together, one gets an $n \times r$ matrix $M[LPI]$ of these points, another representation form of linear PI. The determination of the extreme points can be relatively time-consuming, since n of the $M + n + 1$ restrictions of (13)–(15) must be fulfilled as equations to yield a solution, that requires solving $\binom{M+n}{n-1}$ systems. Kofler and Menges (1976, 180ff.) present special cases regarding the weak ordering and interval probabilities.

To resolve the decision problem under PI, the $MaxE_{min}$ principle is used. It is an extension of the Bernoulli principle following the maximin principle. Due to the variability of the underlying probability distribution a minimal expected utility valueis calculated for the subspace of PI. In the linear case, the $MaxE_{min}$ principle states:

$$a^* \text{ optimal} \Leftrightarrow \max_i \min_f \{\sum_{j=1}^{n} u_{ij} \cdot p_j^{(f)}\}. \tag{16}$$

Since the extremum of a linear function over a convex polyhedron contains at least one corner point, the expected utility value can be determined by the product of the $m \times n$ utility matrix U of elements u_{ij} with the matrix of the extreme points $M[LPI]$ of elements m_{jf}. This yields an $n \times r$ matrix of expected utility values for each corner point $\psi^{(f)} = (p_1^{(f)}, \ldots, p_n^{(f)})'$ (Kofler and Menges (1976, 142)):

$$a^* \text{ optimal} \Leftrightarrow \max_i \min_f \{\sum_{j=1}^{n} u_{ij} \cdot m_{jf}\}. \tag{17}$$

Nonlinear PI as described in (7)–(9) cannot occur in decision theory, because the states are mutually exclusive resulting in empty intersections. In these cases a support is only possible by computing the bounds of unknown probabilities by solving (12). But it is still possible to express nonlinear PI in form of (11) (with $\pi = p$) and to support the decision maker by applying the $MaxE_{min}$ principle, which is proposed in the following.

If the PI is nonlinear, the minimum expected utility value has to be computed for each action a_i with respect to the nonlinear and nonconvex subspace represented by (11) and (13)–(15). Formally, one has to solve the optimization problems (for $i = 1, \ldots, m$)

$$\begin{aligned} E_i(u) &= \min \textstyle\sum_{j=1}^{n} u_{ij} \cdot p_j \\ &\text{s.t. } (11), (13), (14), (15). \end{aligned} \tag{18}$$

(Note that in (11) $\pi = p$ holds.) This problem should be solved by applying a signomial programming algorithm as in section 3.2.

In this case, the optimal action according to the $MaxE_{min}$ principle is given by

$$a^* \text{ optimal} \Leftrightarrow \max_i E_i(u). \tag{19}$$

The $MaxE_{min}$ principle is chosen here, because it is justified axiomatically extending the axioms establishing the Bernoulli principle. It has been shown by Kofler and Menges (1976, 138ff.) that a rational decision under PI has to be made with respectto the $MaxE_{min}$ principle.

5. Case Study

A small case study is presented to demonstrate the applicability of this framework in a DSS. The different stages of the PI during its processing in the DSS are specified (see Fig. 1 in Ch. 1).

A decision maker has to decide between two exclusive investments. The decision depends on whether the interest rates are high ($X_1 = 1$) or low ($X_1 = 0$) and whether the investment returns are high ($X_2 = 1$) or low ($X_2 = 0$). The decision maker assumes to know that these two binary variables are independent and identically distributed with

$$P(X_i = 1) = w, \qquad P(X_i = 0) = 1 - w \tag{20}$$

where $i = 1, 2$, and $0 \leq w \leq 1$ (case of ignorance regarding X_1 and X_2, respectively). The common probability distribution of $X = (X_1, X_2)'$ is

$$\begin{aligned} P(X_1 = 1, X_2 = 1) = w^2,\ P(X_1 = 0, X_2 = 0) = (1-w)^2, \\ P(X_1 = 1, X_2 = 0) = P(X_1 = 0, X_2 = 1) = w \cdot (1-w), \end{aligned} \tag{21}$$

which is a nonlinear and nonconvex PI (Huschens (1985), 43ff.) regarding the states $s_1 = (X_1 = 1, X_2 = 1)'$, $s_2 = (X_1 = 1, X_2 = 0)', s_3 = (X_1 = 0, X_2 = 1)'$, $s_4 = (X_1 = 0, X_2 = 0)'$. This PI is the input into the DSS and has to be represented therein by means of probabilistic logic (see Ch. 3). Because the states are mutually exclusive, $\pi = p$ holds. Now this PI can be represented subject to (14) and (15) by equations of state probabilities

$$p_1 \cdot p_4 = p_2 \cdot p_3 \tag{22}$$

$$p_2 = p_3. \tag{23}$$

This PI has to be checked by the DSS for consistency, because it stems from subjective assessments of the decision maker (see Ch. 2). The corresponding subspace is not empty, because, for example, $p = (1, 0, 0, 0)'$ is a feasible solution of (14), (15), (22), (23). It follows that the available PI is consistent.

The decision maker can now be supported by computing bounds of unknown state probabilities (see Ch. 3). For example, the upper bound for the probability of state s_2 (high interest rate and low investment returns) can be computed by the DSS solving the optimization problem

$$\begin{aligned} & p_2 \rightarrow \max \\ \text{s.t. } & (14), (15), (22), (23). \end{aligned} \tag{24}$$

The solution of this optimization problem is $p_2 = 0.25 = p_1 = p_3 = p_4$. Suppose that the decision maker states a corresponding utility matrix

$$U = \begin{bmatrix} 10 & 5 & 12 & 6 \\ 16 & 2 & 18 & 3 \end{bmatrix} \tag{25}$$

consisting of net present values. This matrix is used as input into the DSS, which applies the $MaxE_{min}$ principle to it. Doing so, the DSS supports the decision maker by proposing which investment to realize according to this rational decision principle (see Ch. 4). Because of (18) it is possible to compute the minimum expected utility values for the two investments by

$$\begin{aligned} & E_i(u) = \min \textstyle\sum_{j=1}^{4} u_{ij} \cdot p_j \\ \text{s.t. } & (14), (15), (22), (23), \end{aligned} \tag{26}$$

where $i = 1, 2$. The solutions of these optimization problems are $E_1(u) = 6$ and $E_2(u) = 3$ for the actions a_1 and a_2, respectively. Note that expected utility values of 5 and 2 corresponding to the maximin principle cannot be realized, because $p = (0, 1, 0, 0)'$ is incompatible with the available PI. Because of $\max\{6, 3\} = 6$ the DSS returns action a_1 as a rational decision to the decision maker.

6. Conclusion and Suggestions for Future Research

This paper shows the representation of PI on a consistent base. The whole extent of the available information is used, since no part of it has been lost while doing so. It is also shown how to support the decision maker by determining bounds of unknown probabilities and suggesting which action to choose. The data structures to represent PI in probabilistic logic, the linear and signomial programming algorithms and the algorithms to determine the optimal action according to the $MaxE_{min}$ principle can be implemented in a DSS. Such a DSS processing PI gives wide support to the decision maker dealing with semistructured problems caused by uncertainty.

To improve the presented framework, an efficient signomial programming algorithm has to be found, which works with equality and inequality restrictions without numerical difficulties. Furthermore, a more efficient procedure to determine the extreme pointsof the convex polyhedron in the linear case is needed.

References:

ANDERSEN, K.A., and HOOKER, J.N. (1994): Bayesian logic. *Decision Support Systems, 11, 191–210.*

DE FINETTI, B. (1937): Foresight: Its Logical Laws, Its Subjective Sources. In: H.E. Kyburg jr. and H.E. Smokler (eds.): *Studies in subjective probability.* Krieger, Huntington, 2nd ed., 1980, 53–118.

DEMBO, R.S. (1978): Current State of the Art of Algorithms and Computer Software for Geometric Programming. *Journal of Optimization Theory and Applications, 26, 149–183.*

FISHBURN, P.C. (1986): The Axioms of Subjective Probability. *Statistical Science, 1, 335–358.*

GONEN, A., and AVRIEL, M. (1986): A Primal-Dual Newton-Type Algorithm for Geometric Programs with Equality Constraints. *Journal of Optimization Theory and Applications, 49, 239–269.*

GORRY, G.A., and SCOTT MORTON, M.S. (1971): A Framework for Management Information Systems. *Sloan Management Review, 13, 55–70.*

HAILPERIN, T. (1965): Best possible inequalities for the probability of a logical function of events. *The American Mathematical Monthly, 72, 343–359.*

HUSCHENS, S. (1985): Entscheidungen bei Unsicherheit: Die Berücksichtigung inferentieller und strukturierter statistischer Information. Fischer, Frankfurt.

KEEN, P.G.W., and SCOTT MORTON, M.S. (1978): Decision Support Systems. Addison–Wesley, Reading.

KOFLER, E., and MENGES, G. (1976): Entscheidungen bei unvollständiger Information. Springer, Berlin.

NILSSON, N.J. (1986): Probabilistic Logic. *Artificial Intelligence, 28, 71–87.*

RAMSEY, F.P. (1926): Truth and Probability. In: D.H. Mellor (ed.): *Foundations.* Routledge & Kegan Paul, London, 1978, 58–100.

RIMSCHA, M.v. (1990): The Determination of Comparative and Lower Probability. In: Proceedings of the Workshop "Uncertainty in Knowledge-Based Systems", FAW-report 90025, Vol. 2, 344–376.

RIMSCHA, M.v. (1991): Introduction to SIMUNC. Working Paper, FH Flensburg.

SHAFER, G. (1976): A Mathematical Theory of Evidence. Princeton.

SHORTLIFFE, E.H., and BUCHANAN, B.G. (1975): A Model of Inexact Reasoning in Medicine. *Mathematical Biosciences, 23, 351–379.*

STABELL, C.B. (1979): Decision Research: Description and Diagnosis of Decision Making in Organizations. Working Paper No. 79.006, Institute for Information Systems Research, Norwegian School of Economics and Business Administration, Bergen.

ZADEH, L.A. (1965): Fuzzy Sets. *Information and Control, 8, 338–353.*

Consistency Conditions for the Classification in LIS/Cl

Wolfgang Lenski, Michael M. Richter, Elisabeth Wette-Roch

Fachbereich Informatik, Universität Kaiserslautern,
Postfach 3049, D-67653 Kaiserslautern, Germany

Summary: The classification component LIS/Cl of the Logic Information System LIS is established to supervise the classification process of the literature on mathematical logic. We define a formal model for LIS/Cl which is based on the distinction of a top-down and a bottom-up approach to representing the field of mathematical logic. This model gives rise to classification rules which capture conditions for a classification that is consistent with the mathematical contents of documents. The approach is extended by a model for (semi-)automatic document classification specifying formal conditions to determine descriptors that are suitable for classification.

1. Introduction

The task of classifying certain objects is a scientific objective in many situations. The classification, however, is in all relevant cases not something which is of interest for itself, but it is supposed to be useful for some additional purposes. These vary from situation to situation.

In this article we focus on some features concerning classification of the literature on mathematical logic. The background is an enterprise of the Heidelberg Academy for the Humanities and the Sciences which grew out of an attempt to cover the field of mathematical logic through a series of research monographies. Its basic objective is to achieve a complete survey of the field of mathematical logic; this is considered as a cultural task. As mathematical logic is a comparatively young and small mathematical discipline (its origins are in the late 19th century), it is to be hoped that such a complete survey still may be achievable.

Before the current enterprise started in 1993, 'complete survey' meant a complete, classified bibliography of mathematical logic. In this context, 'structurizing' meant that the field was covered by a classification scheme with additional comments illustrating the understanding of the respective items. The classification codes were assigned by a group of editors of the Ω-bibliography (Müller and Lenski (1987)).

We have widened the scope of the bibliography project from the classification of literature to a more comprehensive view of including aspects like concepts, terminology, etc. As a consequence we add a classification component LIS/Cl to the Logic Information System LIS to supervise the classification task and to support editors in assigning adequate classification codes.

2. Classification Hierarchies and Descriptors

From an abstract point of view one is given a classification hierarchy with certain codes attached to its nodes and has to assign one or more codes to each of the objects to be classified. The set of objects belonging to a node in this sense can be regarded as its semantics. As long as there are no rules and laws for such a semantics, it is completely arbitrary, in particular, it cannot be correct or false. In many cases, including our situation, there are certain rules which are, however, not formally specified and only in the background knowledge of certain experts (in our case: the editors).

The classification process can now be divided into two parts:

1.) With each object certain pieces of knowledge are associated; these pieces will be called *descriptors*. Descriptors have a name and contain additional (more or less structured) information.
2.) On the basis of suitable combinations of descriptors the corresponding classification codes are obtained.

This yields a second semantics for the classification hierarchy: with each classification node sets of (suitable combinations of) descriptors are associated. We regard this view of the classification process as a *bottom-up* approach, whereas we call the direct assignment of codes to objects *top-down*.

Usually, the methods to establish a classification hierarchy yield at the same time a guideline for the classification process. A possible formalization of a top-down approach was obtained in the CECoS-System described in Reinartz and Schmalhofer (1994). Here the background knowledge was used via explanation based learning methods. The bottom-up approach usually runs under the name *conceptual clustering*; a typical representative is the COBWEB-System described in Fisher (1987). It presupposes a set of descriptors and includes methods for checking consistency and redundancy.

In our approach, the aspects of establishing a classification hierarchy and of classifying documents are considered as being independent: A classification hierarchy and a (normed) set of descriptors reflect different points of view. We assume that a fixed classification system is given. The classification system then constitutes the top-down-structure. It is established by an intellectual insight into areas, topics, proof strategies and techniques, etc. which on the one hand constitute a scientific discipline and on the other hand mark the frontiers to neighbor disciplines. It connects descriptors to a formulation of the scientific research interests. The descriptors themselves, however, rather lead to a bottom-up-structure: They list the specific topics the field covered by the classification scheme is dealing with and are mainly used to describe the contents of publications. Moreover, they may be embedded into a network of various relationships. In the sequel we will show how these different approaches may be combined into one single system.

In the two step process of classifying mathematical literature described above, the first phase of obtaining descriptors is split again into two parts:

a) assigning the descriptor name (e.g. 'modus ponens')
b) finding the full descriptor (e.g. the properties of 'modus ponens').

Step a) seems to be easier because these names are usually mentioned by the authors themselves. The other steps, however, require the real expert knowledge which includes insight into the structure of the descriptors, e.g. their definitions and their relations to each other, as well as a deep understanding of the principles underlying the classification system. These latter aspects are what descriptors distinguish from simple keywords.

3. The Model for the Classification Component LIS/Cl

We define classification as the assignment of labels of a classification scheme to bibliographic items. The labels are designators for entities which comprise mathematical properties, define subfields of mathematical logic, specify views on topics, etc.. It turns out that the assigment of descriptors is by no means trivial, if there is also a classification system involved. To capture all the aspects, the classification part LIS/Cl of the Logic Information System mainly consists of three components.

3.1 The Representation Component of LIS/Cl

The representation component of LIS/Cl is considered as the bottom-up part of the general task of overlooking and representing the field of mathematical logic. That is, we first identify the smallest units bearing mathematical significance. These are 'conceptual atoms' denoting the specific subjects of a scientific field, the methods under application, the proof strategies, etc.. Such concepts are usually found in mathematical textbooks. They have a clear mathematical definition. According to that definition they may be related to each other through a broader-narrower relation.

This implies that the concepts are embedded into networks which basically are directed graphs and thus represent conceptual neighborhood. In general, these graphs are acyclic; hence each linking relation r yields a hierarchical structure on subsets $D_r \subseteq D$ (D being the set of all descriptors) which may be interpreted as a partial ordering $\leq_r$ on D. If these partial orderings do not mutually contradict (as it is the case in LIS/Cl), we may amalgamate them into one single structure (see Hodges (1985); note that consistency conditions are needed to guarantee that the union can be (partially) ordered, too). The resulting structure is again a partial ordering $\leq$ on D which may be considered as the (set-theoretic) union of the partial orderings $\leq_r$ for all relations r occurring in the representation component.

Definition 1: *The representation component is a (first-order) structure of the form $\mathcal{D} = \langle D, \leq^{\mathcal{D}} \rangle$ where $D = \{d_1, \ldots, d_m\}$ is the set of descriptors and $\leq^{\mathcal{D}}$ is the canonical order on $\mathcal{D}$ determined by the polyhierarchy of the individual relations on D (which can be considered as the set-theoretic union of the single hierarchy types).*

Being a constituent of LIS/Cl, the representation component may thus serve as a structured collection of descriptors to determine the contents of articles in the field of mathematical logic. The representation will be formalized within a terminological logic (cf. e.g. KL-ONE (Brachmann and Schmolze (1985)) or TAXON (Hanschke, Abecker et al. (1991))). This should allow the construction of the canonical order on $\mathcal{D}$ by automatic subsumption algorithms and may also be used for consistency tests.

3.2 The Classification Scheme

The classification of the references is based on the *Omega Classification Scheme* which is essentially the 03/04-part of the *Mathematical Subject Index* used by the *Mathematical Reviews* and the *Zentralblatt für Mathematik und ihre Grenzgebiete.* Fig. 1 lists an extract of the classification scheme.

B25 Decidability of theories and sets of sentences

C60 Model-theoretic algebra
+ Applications of model theory to specific algebraic theories

C65 Models of other mathematical theories
+ Theories of orderings

D25 Recursively enumerable sets and degrees

D35 Undecidability and degress of sets of sentences

Fig. 1: Extract from the Omega Classification Scheme

From an analytical point of view the scheme consists of three parts. The first part is a label, e.g. 'C60', the second part is a name for the label, e.g. 'Model-theoretic algebra', and the third part (together with the name of the classification point) defines and explains the content of the classification item, e.g. 'Applications of model theory to specific algebraic theories'. These latter parts of a classification point are called the *classification units.*

While the representation component links the conceptual items according to their inherent (or definitorial) mathematical interrelations, the classification units may be understood as setting 'land-marks' in the scientific landscape in that they identify scopes, topics, and lines of research, comprehend uniform approaches, denote applications of specific proof strategies or techniques, etc.. The classification scheme altogether is designed to cover the whole subject area (mathematical logic), but the perspective is on research. The classification items are establish from the viewpoint of detecting the scientific problem fields in which investigations have been made. Hence this approach may be considered as a top-down approach to represent the field.

From a systematic point of view the classification units may be considered as combinations of descriptors and thus constitute entities of a different kind. In general, they are not interrelated by conceptual dependencies and do not reflect conceptual neighborhood except in case the specific area in question bears scientific significance in its own right. Instead, these classification units may be viewed as a *skeleton* inside the representation component which reflects research categories rather than conceptual dependencies.

The single descriptors $d \in D$ are composed to connectives which are the units determining the classification. From a linguistic point of view these classificational units are noun phrases which compose conceptual units (descriptors) of the representation component $\mathcal{D}$ to *classificational units* k of the classification component. In figure 2 the frames mark those descriptors which are composed to classificational units (represented by their labels) via the dotted lines.

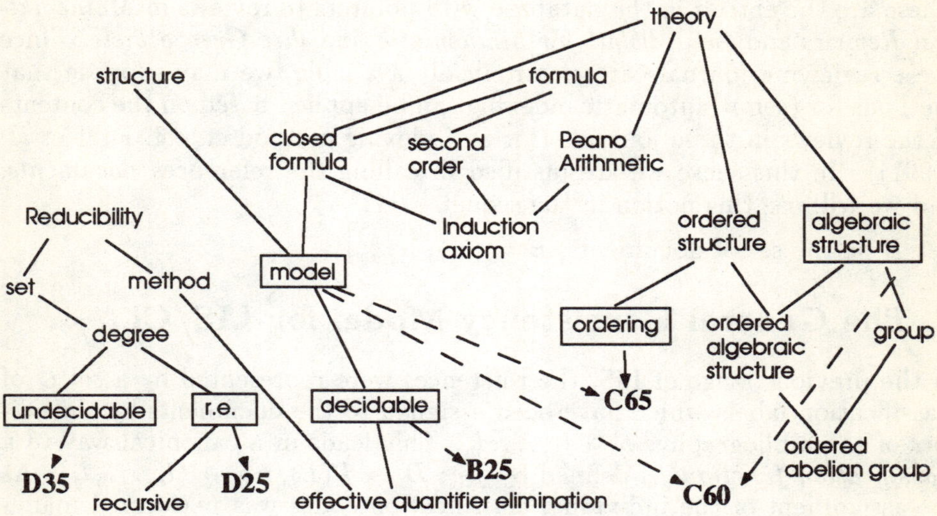

Fig. 2: Extract from the representation component of LIS/Cl with embedded classification structure

Let K be the set of these classificational units, $K = \{k_1, \ldots, k_m\}$.

Example: *model * ordered abelian group * theory* $= k \in K$.

In this context it suffices to view these units as set-theoretic comprehensions of descriptors, i.e. we need not consider their linguistic structure.

Definition 2:

i) A classification item E is a set of classification units, i.e. $E \subset K$.

ii) A classification structure $\mathcal{C}$ is a collection of mutually disjoint classification items, i.e. $\mathcal{C} \subset \mathcal{P}(K)$ *or* $\mathcal{C} = \{E_1, \ldots, E_n\}$ *for suitable* $E_1, \ldots, E_n$.

In order to establish a suitable organization of the classification structure and to ease its use, every classification item has uniquely attached

- a label l from a (hierarchically) ordered system L, e.g. 'C60'
- a designator m out of a set M, e.g. 'model-theoretic algebra'

Thus, we have a surjective monomorphism $\gamma : \mathcal{C} \to L \times M, \quad E \mapsto \langle l_E, m_E \rangle$ where $\pi_i \circ \gamma$, $i = 1, 2$, are (surjective) monomorphisms, and π_1, π_2 are the corresponding projections.

Now a *classification system* or a *system of classification codes* CLC is a set of labels and designators, $CLC = \{\langle l, m\rangle_E \mid E \in \mathcal{C}\}$.

By definition 2 there exists a function $\kappa : K \to L \times M$, $k \mapsto \langle l_E, m_E\rangle$ where E is the (uniquely determined) classification item satisfying $k \in E$.

3.3 The Bibliographic Items

These are the entries in the database with pointers to reviews in *Mathematical Reviews* and *Zentralblatt für Mathematik und ihre Grenzgebiete.* Since these reviewing journals are electronically available, we may suppose that methods for (semi-)automatic indexing can be applied based on the contents of the reviews in these journals (for an indexing method cf. e.g. Fuhr et al. (1991)). In this sense we are justified in calling the references *documents*, and we will use this notion in the sequel.

Let B be the set of documents, $B = \{b_1, \ldots, b_n\}$.

4. The General Consistency Model for LIS/Cl

In the previous stage of LIS, the references were represented by a set C of classification labels which have been assigned to the documents by the editors of the bibliography: $b \rightsquigarrow l_1, \ldots, l_n$. This leads in a canonical way to a *classification function* f_C defined by $f_C : B \to \mathcal{P}(L)$, $b \mapsto \{l_1, \ldots, l_j\}$. As the assignment of the individual classification codes was performed manually, the classification function depends on the (cumulative) opinions of the editors. 'Correctness' of an assignment of a classification code cannot be checked, and no criteria are deducible from the data to judge adequacy.

In reaction to this situation the system LIS/Cl was designed to establish (more objective) criteria for the classification function depending on the mathematical contents of the references.

4.1 Consistency Conditions for the Classification

To elucidate the need for more objective criteria we introduce the *extensional viewpoint* to the classification system: Accordingly, a classification item may alternatively be defined by the totality of all the references sharing the same classification code. Hence let E be a classification item. Then $C_E = \{b \mid l_E \in f_C(b)\}$ where $l_E = \pi_1 \circ \gamma(E)$ is the label corresponding to E.

Using this definition we now have at our disposal a formal setting to check consistency: Determine the relation between the classification structure $\mathcal{C}$ and $\{C_E \mid E \in \mathcal{C}\}$.

In section 3.2 we have seen that the classification units may be viewed as a skeleton in the representation part. The single references now deal with mathematical objects: they discuss or give proofs of relationships between mathematical entities or, in the terminology of section 3.1, they deal with descriptors. Now as a first step towards a correlation between C_E and E

we have to relate classification units and references. This is done via the predicate $t_e(b,k)$ expressing that reference b *treats essentially* the classification unit k.

Now the predicate t_e gives rise to *classification rules*:

Rule 1: *If a document b treats essentially the classification unit k and k is element of the classification item E, then we require that b be contained in the set of all E-classified documents:* $t_e(b,k) \wedge k \in E \rightarrow b \in C_E$ (1)

Note that $b \in C_E$ is equivalent to $l_E \in f_C(b)$.

Rule 2: *If b is E-classified, then there should be some reason for this assignment based on the definition of E:* $b \in C_E \rightarrow \exists u(u \in E \wedge t_e(b,u))$ (2)

Rule 2 yields a criterion when a document b is 'falsely' classified.

On the basis of these implications we may define a classification function cl as follows: $cl : B \rightarrow \mathcal{P}(L), \quad b \mapsto \{\pi_1 \circ \gamma(E) \mid \text{there exists } k, t_e(b,k), k \in E\}$.

For the following we assume that the mathematical content of a document is defined by a list of descriptors.

Definition 3: *A classification is called consistent with the mathematical contents of the documents (or simply consistent for short), if the canonical classification function f_C is cl.*

Theorem 1: *(1) and (2) are satisfied, if the classification is consistent.*

Proof: Suppose the classification is consistent. Then the canonical classification f_C with $b \leadsto l_1, \ldots, l_m$ is identical to cl.
Let $b \in B$, k a classification unit, E a classification item, and $k \in E$ such that $t_e(b,k) \wedge k \in E$. From the definition of cl we deduce $l_E = \pi_1 \circ \gamma(E) \in cl(b) = f_C(b)$. Hence by definition of C_E we get $b \in C_E$, and (1) is satisfied. Now let $b \in C_E$. Hence $\pi_1 \circ \gamma(E) = l_E \in f_C(b) = cl(b)$. By the definition of cl there exists $k \in E$ such that $t_e(b,k)$, i.e. condition (2) is satisfied, too. □

4.2 A Model for (Semi-)Automatic Classification

In this section we extend the model established in section 3 and describe how the representation component together with the embedded classification structure can be used for (semi-)automatic classification of documents. Especially, we give conditions to evaluate which descriptors are suitable and which ones should be rejected. For this purpose we generalize the relation t_e to descriptors and add the following rule: $d \le d_1 \ \wedge \ t_e(b,d) \rightarrow t_e(b,d_1)$.

The next rule links descriptors and classification units and shows how they are related: $t_e(b,d_1) \wedge t_e(b,d_2) \rightarrow t_e(b, d_1 * d_2)$.

Now we are able to define conditions.

Definition 4: *A descriptor d is called conflict-free, if for all $d_1, d_2 < d$, for all classification units k_i including d_i (in symbols: $k_i \succeq d_i$), $i = 1,2$, and all classification items E holds: $k_1 \in E \Leftrightarrow k_2 \in E$.*

A descriptor is conflict-free, if it does not already give rise to different classification items below itself. It follows immediately from the definition that all least elements of the representation component $\mathcal{D}$ are conflict-free. In figure 2 'degree' is not conflict-free, because 'undecidable degree' and 'r.e. degree' are more specific descriptors leading to different classification items.

Definition 5: *A descriptor d is called determined, if there exists $d_1 \geq d$ and a classification item E such that $d_1 \in E$.*

Definition 6: *A descriptor d is called predetermined, if there exists $d_1 \geq d$, a classification unit $k \succeq d_1$, and a classification item E such that $k \in E$.*

Clearly, *predetermined* is a more general notion than *determined.* In figure 2 'effective quantifier elimination' and 'decidable' are determined; 'model' is predetermined, but not determined; 'degree' is not predetermined. A descriptors d satisfying $d \not\leq d^*$ for all $d^* \in D_E$ is not predetermined.

Theorem 2: *For every descriptor d which is not conflict-free there exist predetermined descriptors $d^* < d$.*

Proof: Let d be not conflict-free. Then there exist $d_1, d_2 < d$, classification units $k_i \succeq d_i$, and classification items $E_i \ni k_i$, $i = 1, 2$. By definition, d_1 and d_2 are already predetermined. □

5. Conclusions

To gain criteria to judge the classification of literature one has to classify the documents according to the classification function *cl* (cf. theorem 1). From the viewpoint of a bottom-up approach the direct assignment of classification codes must be replaced by the assignment of predetermined and conflict-free descriptors. LIS/Cl is able to detect not conflict-free and not predetermined descriptors which will be rejected. However, in the case of not conflict-free descriptors d more specific descriptors exist which are predetermined and hence may be used instead. These can be found below d according to theorem 2. In this sense LIS/Cl is able to supervise the classification process.

A principle limitation is given by the fact that the classification component cannot guarantee the existence of a classification unit, even if a document has assigned only conflict-free and predetermined descriptors. This is, because classification units in general do not reflect conceptual relationships, i.e. in general they are not 'ideals' in the representation component $\mathcal{D}$ but rather constitute a skeleton. Thus the classification is decided on the level of documents and not on the level of descriptors.

Example: 'Algebraic structure' is predetermined and conflict-free (see fig. 2), but does only lead to a classification code, if 'model' is assigned to the document, too. This is inevitable: In the absense of aspects of model theory the respective document will not belong to mathematical logic!

Hence the best LIS/Cl can actually perform is to reject descriptors which are not conflict-free and predetermined. Moreover, LIS/Cl is able to detect whether a classification unit k is achievable by the assignment or not, and the system will interactively demand further descriptors in case no classification unit can be composed.

References:

BIEBRICHER, P., FUHR, N., LUSTIG, G., SCHWANTNER, M., and KNORZ, G. (1988): The automatic indexing system AIR/PHYS – From research to application. In: Y. Chiaramella (ed.): Proceedings of the 11th International Conference on Research and Development in Information Retrieval. Presses Universitaires de Grenoble, 333–342.

BOOKSTEIN, A. (1988): Set oriented retrieval. In: Y. Chiaramella (ed.): Proceedings of the 11th International Conference on Research and Development in Information Retrieval. Presses Universitaires de Grenoble, 583–596.

BRACHMANN, R.J., and SCHMOLZE, J.G. (1985): An overview of the KL-ONE knowledge representation system. *Cognitive Science, 9(2), 171–216.*

FISHER, D.H. (1987): Knowledge acquisition via incremental conceptual clustering. *Machine Learning, 2, 139–172.*

FUHR, N., HARTMANN, S., KNORZ, G. et al. (1991): AIR/X – a rule-based multistage indexing system for large subject fields. In: Proceedings of the RIAO '91, Barcelona, 606–623.

HANSCHKE, P., ABECKER, A., and DROLLINGER, D. (1991): TAXON: a concept language with concrete domains. In: H. Boley and M.M. Richter (eds.): Processing declarative knowledge '91. Springer-Verlag, Heidelberg, 411–413.

HODGES, W. (1985): Building models by games. University Press, Cambridge.

MÜLLER, G.H., LENSKI, W., et al. (1987): Ω-Bibliography of Mathematical Logic I-VI. Springer-Verlag, Heidelberg.

REINARTZ, Th., and SCHMALHOFER, F. (1994): An integration of knowledge acquisition techniques and EBL for real world planning. To appear in: *Knowledge Aquisition Journal.*

RICHTER, M.M. (1993): Some recent developments in the representation and processing of knowledge. *Fundamenta Informaticae, 18, 233–248.*

SCHÄUBLE, P. (1988): An information structure dealing with term dependence and polysemy. In: Y. Chiaramella (ed.): Proceedings of the 11th International Conference on Research and Development in Information Retrieval. Presses Universitaires de Grenoble, 519–533.

WATTERS, C.R. (1989): Logic Framework for Information Retrieval. *Journal of the American Society for Information Science, 40(5), 311–324.*

WETTE-ROCH, E. (1994): Conceptual structures in mathematical logic and their formal representation. In: H.H. Bock, W. Lenski, and M.M. Richter (eds.): Information Systems and Data Analysis. Proceedings of the 17th Annual Conference of the Gesellschaft für Klassifikation. Springer-Verlag, Heidelberg, 206–216.

Using Hypertext for Information Retrieval in STEP/EXPRESS Schemata

Horst Lührsen, Hartmut Wedekind

Lehrstuhl für Datenbanksysteme, Friedrich-Alexander-Universität
Erlangen-Nürnberg, Martensstrasse 3, D-91058 Erlangen

Summary: At present, the new ISO standard 10303 STEP (Standard for the Exchange of Product Model Data) is under development. The language EXPRESS, which is part of STEP, allows the formal description of product model data to exchange data between CAD/CAM applications. For the management of STEP/EXPRESS schemata, currently only simple text editors and some specialized EXPRESS-G graphical editors are available. They all lack ease of use. In this paper, a mapping from STEP/EXPRESS schemata to HTML hypertext is described. HTML hypertext is the input of advanced hypertext browsers like Mosaic that allow a convenient information retrieval on EXPRESS schemata.

1. STEP

In the past, a lot of different CAD/CAM-systems with distinct data formats have been introduced. These proprietary formats were not compatible, and data exchange between applications became a severe problem. To solve this problem, standards for the exchange of product model data between applications have been developed, e.g. IGES, SET, VDAFS, and others. Each standard defines a common product data model, and data described using such a model can be easily exchanged between applications from different vendors.

The IGES geometry standard was the first widely accepted standard and is today one of the most popular data exchange formats in the CAD/CAM area. Unfortunately, IGES has some disadvantages restricting its usage. The physical file format of IGES has some restrictions and modern applications like CSG modellers are badly supported. Therefore, other standards have been developed by national standard bodies.

Examples are SET and VDAFS. The French Standard d'Echange et Transfert (SET) has an improved physical file format, better support for CSG and surface representation. The German VDAFS also supports CSG and surface representation but has a restricted physical file format. Because of the diverging development of these national standards, the ISO decided to develop an international standard. The new standard should bring together different developments and should combine the positive features of the different standards.

The new standard was called STEP (Standard for the Exchange of Product Model Data, ISO TC184/SC4 (1993)). In addition to the exchange of data between different CAD applications, STEP provides support for the whole product data life cycle, including CIM applications like computer aided planning (CAP), computer aided quality assurance (CAQ), and others. To do so, STEP supports a product data model which goes far beyond geometric and topologic data, and allows the description of product structure configuration, materials, tolerances, form features etc.

To make possible the exchange of product model data, STEP must first provide means to describe product model data. For this purpose, a formal language called EXPRESS has been developed. EXPRESS is a schema description language for product model data and the core part of the whole ISO standard STEP (Part 11, ISO TC184/SC4 (1993)).

The other parts of STEP are divided in 3 series, called the partial models. Most of these models are descriptions of product model data written in EXPRESS. The 40s series (41-49) is called the *integrated generic resources* and provides the basic means for product data description, while the 100s series (101-105, the *integrated application resources*) allows the description of more application-specific data. Finally, the 200s series (201-219, the *application protocols*) describes data used in end-user applications.

For more information about the ISO standard STEP/EXPRESS, the reader should refer to section 2, where some details about EXPRESS are outlined.

2. EXPRESS

2.1 EXPRESS and EXPRESS-G

EXPRESS is a formal object-oriented language for the description of product model data schemata and the core part of the new coming ISO standard 10303 STEP. A number of basic data types like INTEGER, NUMBER, REAL, STRING, BINARY, BOOLEAN, and LOGICAL and the aggregations like ARRAY, LIST, SET, and BAG are available.

All data types, aggregations, and self-defined data types (TYPE) are used to build objects (ENTITY) that consist of several attributes of these simple types, aggregations, and other entities. For entities, inheritance can be formulated using the keywords SUPERTYPE and SUBTYPE. Because EXPRESS is a schema description language and not a programming language, there is no executable main program like in programming languages. However, EXPRESS allows the description of algorithms used in local and global RULES and also for the computation of DERIVEd attributes.

To summarize, EXPRESS is a schema description language rich of constructs which make the description of product model data feasible. In EXPRESS, related data is put in one SCHEMA. This is similar to the *conceptual schema* known from the database area, which describes data and their interdependencies.

EXPRESS-G provides a graphical representation of EXPRESS. EXPRESS-G diagrams are understood intuitively and therefore are suitable in particular for inexperienced users. However, EXPRESS-G has some serious disadvantages, which should be mentioned here:

- There is no way to represent the algorithm used for computation of derived attributes in an EXPRESS-G diagram.

- Local RULEs restricting the domain of an attribute are only represented by marking the restricted attribute. The RULE itself is not expressed within EXPRESS-G.

- Similar to local RULEs, global RULEs which impose restrictions on entities also are only represented by marking the entities as being restricted, but the global rule itself is not represented. The user only knows, that there are restricted entities but has no information on what the interdependencies between these entities are.

2.2 Tools for EXPRESS and EXPRESS-G

A lot of processing tools for EXPRESS have been developed in the past few years, including syntax checkers, parsers, text editors for EXPRESS schemata, and graphical editors for EXPRESS-G. Wilson (1993) gives a detailed overview about the actual development. Despite of the EXPRESS-G disadvantages mentioned above, a number of EXPRESS-G graphical editors have been built and are currently used by developers of EXPRESS schemata. The usage of these editors shows up other disadvantages of EXPRESS-G:

- EXPRESS schemata tend to be complex and consist of lots of data types, entities, and rules. Therefore, the graphical representation easily becomes very big and is difficult to handle and to overlook by the user.

- Some graphical editors like EXPREME (CADLAB (1994)) allow to enter algorithms for derived attributes and local and global rules in special text editor windows. This can be used to generate a complete EXPRESS description from the graphical representation, but offers an inconsistent user interface which separates graphical and textual parts of EXPRESS.

These problems and the inherent problems of EXPRESS-G outlined in the previous subsection result in a difficult usage of graphical editors. EXPRESS-G is well suited for small examples, but unsuitable for real schemata. A fact emphasizing this point is that nearly all STEP partial models are written in EXPRESS and EXPRESS-G is rarely used. In practice, usually simple text editors or specialized text editors for EXPRESS are used to write and modify schemata. They also have some drawbacks:

- If a user wants to know where defined data types, functions, procedures etc. are referenced, he or she can only do a simple textual search.

- Complex interdependencies, which are visible using EXPRESS-G, are hidden in the text and therefore hardly noticeable to the user.

Despite of this inconveniences, the usage of text editors for EXPRESS schemata offers some advantages which make them first choice for developers of EXPRESS schemata at present:

- The EXPRESS schema is complete.
- An EXPRESS schema can be easily modified.
- Since the presentation is the source code, it is compact and easier to overlook than the graphical representation.

As shown before, both approaches have pros and cons. EXPRESS-G editors offer an intuitively understandable layout which allows to follow complex interdependencies by arrows. However, rules and algorithms can not be represented and complex schemata can be difficult to overlook. EXPRESS text editors provide a compact textual representation, but hide the interdependencies from the user.

2.3 Interdependencies in EXPRESS Schemata

As mentioned before, plain EXPRESS schemata contain complex structural interdependencies. A rough classification of these interdependencies is given in the following:

- Data types are defined only once in a schema and can be multiply used elsewhere, e.g. as domains of attributes.
- Entities form an inheritance hierarchy with sub- and supertypes.
- Entities can be component entities, i.e. they can be used as an attribute of another entity.
- Functions and procedures are defined in one place of a schema and used in different other places like entities and rules.
- Simple attributes of entities and whole entities can be referenced in entities, algorithms, and rules.
- Entities defined in other schemata can be USEd or REFERENCEd in a schema.

These complex interdependencies given above form the semantics of a schema. Unfortunately, most of the interdependencies can not be represented either in EXPRESS-G editors nor in text editors for EXPRESS. EXPRESS-G does not provide means to represent derived attributes and the restrictions laid on attributes and entities by local and global rules.

Simple text editors for EXPRESS provide much less functionality. Definitions of used data types, functions, and procedures have to be found by doing simple textual search. The same applies to the inheritance hierarchy, component entities, and rules. Support for the recognition of complex interdependencies in EXPRESS schemata is very poor, which means the task of finding the interdependencies is left to the user.

3. Using Hypertext for EXPRESS Schemata

We propose hypertext to support EXPRESS usage. Hypertext integrates the advantages of a compact EXPRESS representation with the management and representation of interdependencies available in EXPRESS-G. In particular, the following assets apply:

- Since hypertext is presented as text, a compact presentation of the schemata can be given.
- Hypertext links can be added to represent the complex structural interdependencies. This allows the user to follow these links, e.g. from the usage of data types to their definition, from global rules to referenced entities and so on.
- If modifications to EXPRESS schemata presented as hypertext are made, a new EXPRESS description can easily be generated by stripping of the hypertext portion of the source.
- After the transformation from an EXPRESS schema to a hypertext EXPRESS schema, hypertext tools like browsers and editors can be used to browse through and edit these schemata.

Because of the reasons mentioned above, a project has been started at the University of Erlangen, the aim of which is to convert EXPRESS schema descriptions into hypertext by transforming the EXPRESS source code into hypertext. This is done by adding the appropriate hypertext links, formatting commands etc. to the source code. For the project, the HTML hypertext markup language (Berners-Lee and Conolly (1993)) was chosen because this language is freely available, well-suited for the available hard- and software and offers some further opportunities:

- A collection of appropriate browsers like Mosaic (Mosaic (1994)) is freely available, and the rapid development leads to an increasing number of available tools.
- The HTML language allows the use of the hypertext transport protocol HTTP (Berners-Lee (1993)). HTTP enables remote browsing of HTML documents in computer networks.

Translation from EXPRESS to HTML can be done straightforward. In simple cases, where no references outside a schema are present, the translation process is carried out in two phases. First, the ending points for the hypertext links have to be found and marked with an unique anchor. Then, the hypertext links referencing the anchors are entered.

Given the case that a schema has external references to one or more other schemata, the translation is more sophisticated. In a first step, the external anchors have to be identified. Then, processing is similar to the simple case. In the case of cyclic references between multiple schemata, processing must be done in two parts. First, anchors have to be found and marked in all schemata, then the hypertext links can be added.

The process described above is illustrated in the following simple EXPRESS schema example, consisting of one self-defined data type and an inheritance hierarchy with three entities.

```
SCHEMA example;

TYPE date = ARRAY[1:3] OF INTEGER; END_TYPE;

ENTITY person SUPERTYPE OF ( ONEOF ( male, female ) );
  name:          STRING;
  birthday:      date;
END_ENTITY;

ENTITY male SUBTYPE OF ( person );
  wife:          OPTIONAL female;
END_ENTITY;

ENTITY female SUBTYPE OF ( person );
  husband:       OPTIONAL male;
END_ENTITY;

END_SCHEMA;
```

In this example a lot of interdependencies are hidden. Four anchors can be identified, the self-defined data type *date*, and the entities *person*, *male*, and *female*. Seven hypertext links describe the structural relationships. The resulting hypertext document appears as follows:

```
<PRE>
SCHEMA example;

TYPE <A NAME="date">date</A> = ARRAY[1:3] OF INTEGER; END_TYPE;

ENTITY <A NAME="person">person</A> SUPERTYPE OF (
        ONEOF ( <A HREF="#male">male</A>,
                <A HREF="#female">female</A> ) );
  name:          STRING;
  birthday:      <A HREF="#date">date</A>;
END_ENTITY;
```

```
ENTITY <A NAME="male">male</A>
        SUBTYPE OF ( <A HREF="#person">person</A> );
  wife:         OPTIONAL <A HREF="#female">female</A>;
END_ENTITY;

ENTITY <A NAME="female">female</A>
        SUBTYPE OF ( <A HREF="#person">person</A> );
  husband:      OPTIONAL <A HREF="#male">male</A>;
END_ENTITY;

END_SCHEMA;
</PRE>
```

The four anchors are in the form <A NAME="*anchor's name*">*target*</A>. They are referenced by the seven hypertext links in the form <A HREF= "*#anchor's name*">*source*</A>. Browsers allow the user to follow this hypertext links to the anchor, which can be at another place in the document or in another hypertext document. The hypertext document begins with the tag <PRE>, which marks the document as preformatted, so that browsers like Mosaic represent the schema in a fixed width font.

4. Implementation

The parser Fed-X (Libes (1993)), developed at the National Institute of Standards and Technology (NIST), was selected as a base for mapping EXPRESS to HTML. Fed-X allows the integration of self-written backends into the parser. As a result of this, the work of writing a parser is reduced to writing an appropriate backend. Syntax checking comes for free.

The usage of HTML hypertext also has some disadvantages. HTML currently only offers one type of hypertext link, so that all classes of interdependencies listed in subsection 2.3 have to be mapped on this type of link. If there were different classes of hypertext links in HTML, the complex interdependencies inherent in EXPRESS schemata could be represented much better.

Another drawback of HTML is the lack of an appropriate software engineering environment for writing HTML hypertext. A wide variety of browsers is currently available, but tools for writing HTML are rare. However, considering the rapidly ongoing development in the research community, we are convinced that suitable tools will be available within the next few months

The implementation of *EXP2HTML* was finished in May 1994. The source code has a size of about 1600 lines of code and can be found on our WWW-Server using the URL *http://www6.informatik.uni-erlangen.de:1200 /Express/exp2html.html.* Further information and instructions for obtaining and installing EXP2HTML can also be found using the URL given above.

5. Conclusion

In this paper, we proposed hypertext to support browsing and manipulation of EXPRESS schemata. Advantages of this approach include a compact representation of schemata, the representation of structural interdependencies, and others.

Our future research directions include the application of hypertext also on EXPRESS instance level and integration of multi media techniques. Since there are multi media product data types like photos or movies (e.g. crash tests in automotive production) which have not yet been integrated into current product data models, there are still opportunities for further research.

The authors would like to thank Olaf Will for his programming work on the EXPRESS to HTML mapping and Richard Lenz and Thomas Kirsche for making this paper more readable.

References:

BERNERS-LEE, T., CONOLLY, D. (1993): Hypertext Markup Language: A Representation of Textual Information and Metainformation for Retrieval and Interchange. Internet Draft, 13. July 1993

BERNERS-LEE, T. (1993): Hypertext Transfer Protocol: A Stateless Search, Retrieve and Manipulation Protocol. Internet Draft, 5. November 1993

LIBES, D. (1993): The NIST EXPRESS Toolkit – Introduction and Overview. NIST, Gaithersburg, 1993.

ISO TC184/SC4 (1993): ISO IS 10303 Part 1, Overview and Fundamental Principles

ISO TC184/SC4 (1993): ISO IS 10303 Part 11, The EXPRESS Language Reference Manual

CADLAB (1994): EXPREME: An EXPRESS-G Editor Supporting ISO 10303-11 EXPRESS. Product Description

WILSON, P. R. (1993): EXPRESS Tools and Services. Rensselaer Polytechnical Institute, Revised Version September 1993

MOSAIC (1994): Mosaic Version 2.4 available from ftp.ncsa.uiuc.edu as /Web /Mosaic-source/Mosaic-2.4.tar.Z

Two Software Tools Supporting Enduser Oriented Information Retrieval in Physics

L. Weisel[1] and B. Diekmann[2]

[1]Fachinformationszentrum Karlsruhe, Postfach 2465, D-76012 Karlsruhe, E-mail: elw0c@fizvax.kfk.de

[2]FB Physik, C.-v.-O. Universität Oldenburg, Postfach 2503, D-26111 Oldenburg, E-mail: diekmann@merlin.physik.uni-oldenburg.de

Summary: With a pilot project of the Federal Minister for Research and Technology (BMFT) 5000 scientists and students of 40 faculties of physics at universities in Germany have been trained during the last three years in using bibliographic databases. A significant contribution to the continuous acceptance of electronical specialized information came from two developments: the search- and administration-software "UNIrech" and a problem oriented "Userguide". In this paper both tools will be presented and the perspectives for an interdisciplinary use at universities or industry will be assessed.

1. Introduction

In physics as in other disciplines an enormous increase of scientific literature can be observed currently. The information offered doubles within ten years. In 1992 160.000 journal articles were published worldwide, more than 400 per day, as De Waard (1993) reports. In contrast to chemistry the use of specialized information from external databases as an answer to this information flood remained almost unknown in physics till the beginning of the nineties. Though the young physicists are agile users of PCs, workstations or mainframes and communicators via worldwide academic networks, the administrative and financial structures of a decentralized information retrieval were missing. Concerning capacity and competence the classical information providers, the central libraries, did not advertise a database search service. Therefore an independent use by faculties failed due to the ignorance of the landscape of information retrieval and the necessary techniques.

Meanwhile electronical information retrieval is regularly used mostly by doctorate and diploma candidates, stimulated by the pilot project "Electronical Specialized Information on Physics at Universities". The extensive and persisting use of this service with 180 searches in bibliographic databases per year and per typical physics faculty was enabled by the concept of accessing from the working place of the scientist.

We will now show the offer and the special conditions for the access to external physics-related specialized information. The needs for a problem oriented search- and administration-software and specific userguides are deduced.

2. Information Retrieval in Bibliographic Databases

The Fachinformationszentrum Karlsruhe (FIZ) and the 40 physics departments supported by the project signed contracts for the use of the bibliographic database PHYS, based on a flat rate of DM 4500.- . In PHYS references to the literature published worldwide since 1979 on all topics of physics, astronomy and adjacent disciplines are made with 1.8 Mio. citations, mostly with abstracts. They are updated yearly by 130.000 citations from journals and series, books, research reports and conference proceedings.

The above-mentioned cost model does not allow FIZ Karlsruhe to pay the expenses of the database. On the other hand, the faculties are not able due to cost cuts at the basic fundings to bear higher flat rates. What is needed are regulations, which prohibit the misuse of commercial databases, but allow the widespread access from the working areas. This conflict of interest was solved from FIZ Karlsruhe by granting only two passwords to a faculty and the obligation to do 50% of the activities after 5 pm (local time), when fullprice paying customers do significantly less searching. So specialized information for faculties is not free of costs but the endusers can access to information without costs. Therefore the education of the endusers has focused on high efficiency by the information retrieval: the aim is to get the most relevant information, it is not to get much information.

The results of the project show, that the information retrieval is mainly used to observe the progress of research, but less to infer new research topics. Endusers perform only few and simply structured search types (basic index, controlled terms or authors search). Fig. 1 shows the number of users under a password-splitting system (for example dept. of physics, TH-Darmstadt, Germany).

3. The Tools

3.1 UNIrech

"UNIrech" was designed with the aim to facilitate the access to databases in the university and to record and protect database accesses. This software will be made available to all universities. So far there has been no commercial software serving these functions.

History:
In a model project "Information transfer" of the BMFT *MVS Recherch* was developed in 1986-1989 which was run with the operation system MVS. Database accesses of the Physics Department at Oldenburg University were supervised with this programme.

In the initial phase of the BMFT project (1990-1993) this programme was rewritten and extended so that it could be managed centrally but was available to all physics departments. It met with a great demand of the departments.

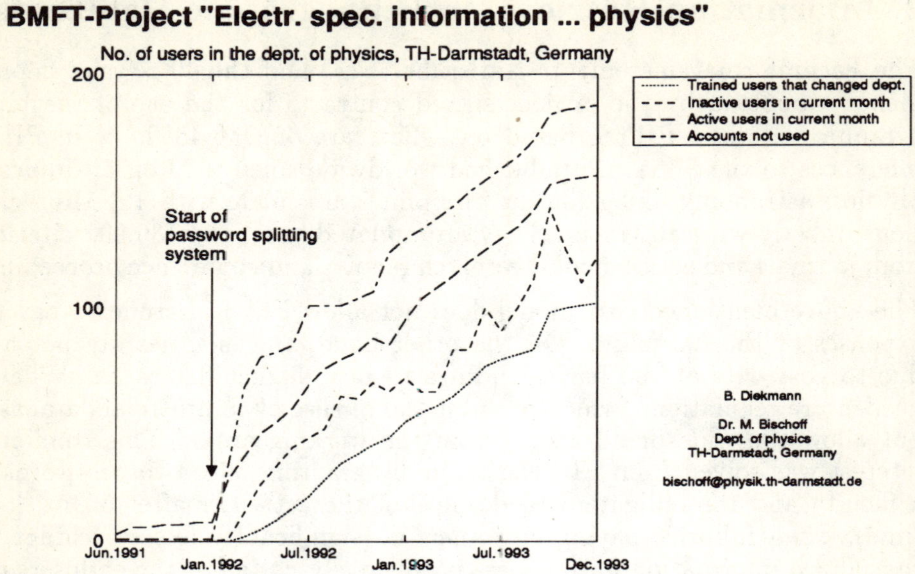

Fig. 1: No. of users in dept. of physics, TH-Darmstadt, Germany
This figure shows the increasing number of users after connecting their computers to a password-splitting system. These users have access-permissions to the bibliographic-database PHYS. As you can see, the number of active users is similar constant. The total number of users increases because many trained users changed their working group, department etc. .

When the MVS computer was turned off, 44 departments or institutes actively used the programme. Within the time period between 2 January 1993 and 22 June 1993 a total of 2950 database inquiries were performed. There were another 34 interested institutes permitted to use the programme but not able to use it for mainly technical reasons. The departments using the programme belong to the fields of physics and mathematics.

Simultaneously the programme was moved to the operation system UNIX, since there was an increasing number of UNIX computers available at the universities and the operation system MVS or the hosts on which it was run, were replaced. The UNIX version was to be installed directly in the departments. In the middle of 1993 the "UNIrech" version 1.0 was available to all departments.

Service:
"UNIrech" allows a simple and controlled access to external databases. Each performed database inquiry is nominally recorded with sub-passwords, which can be issued by the administrators in the departments. Inquiries are on-line and possible by command files. The connection to the database suppliers is established automatically. The user gets directly to the retrieval system. Filters may be defined by programme adjustments forestalling certain retrieval

commands. Following an inquiry the inquiry protocol may be dispatched via e-mail or stored as a file. The user inputs are supported by a menu. The programme unit is freely available to universities and may be obtained as a compiled programme unit for the current UNIX workstations from an Anonymus-ftp-Server via the internet. The current version is "UNIrech 2.1". The programme unit is presently used by about 75 departments (35 physics departments, 35 mathematics departments and 5 informatics departments. The local administrators in the departments were supported by training courses. Furthermore, the departments were given support concerning internet (e-mail) and direct installation by the programmers.

This programme can be used wherever database inquiries with costs are part of the scientific work and where this service should be available directly from the normal place of work. All necessary controls are guaranteed. All process parameters are in situ and are adjustable to most different database suppliers. In case of available UNIX computers every place of work equipped with a computer becomes an easily operable place of inquiry.

3.2 The Userguide

User aids, handbooks or electronical media for teaching the knowledge of information retrieval in external databases is offered in a luxurious and comprehensive form by commercial information suppliers and for a wide spectrum of users. The "Userguide" discussed below contrasts with this standard manuals by

- only teaching the basic knowledge with realistic physics-based examples,
- using the experiences of physics users for the benefit of users in physics,
- supplying printed and electronical versions,
- doing modules and therefore allowing an adaptation to changes in the landscape of information retrieval and to local needs,
- directing critique and suggestions to advancements in an active dialog.

Four modules are offered, which may be chosen according to the personal retrieval know-how. We recommend to use these instruments for navigation to sporadic users of external databases or newcomers of physics-related specialized information.

The "Navigator to Use External Databases for Beginners" consists of the basic information to start with information retrieval: it gives consulting for the prior conditions, the preparation, the conducting of searches and the post-processing of results. It is restricted to a simple representation of the external database and an overview over the retrieval language. It shows the benefit of electronical information with an authors search. First focal points for problems are the locally based faculty information specialist or the Help Desk staff of the host. This guide is produced only in a printed version and will be distributed to all German physics departments at universities.

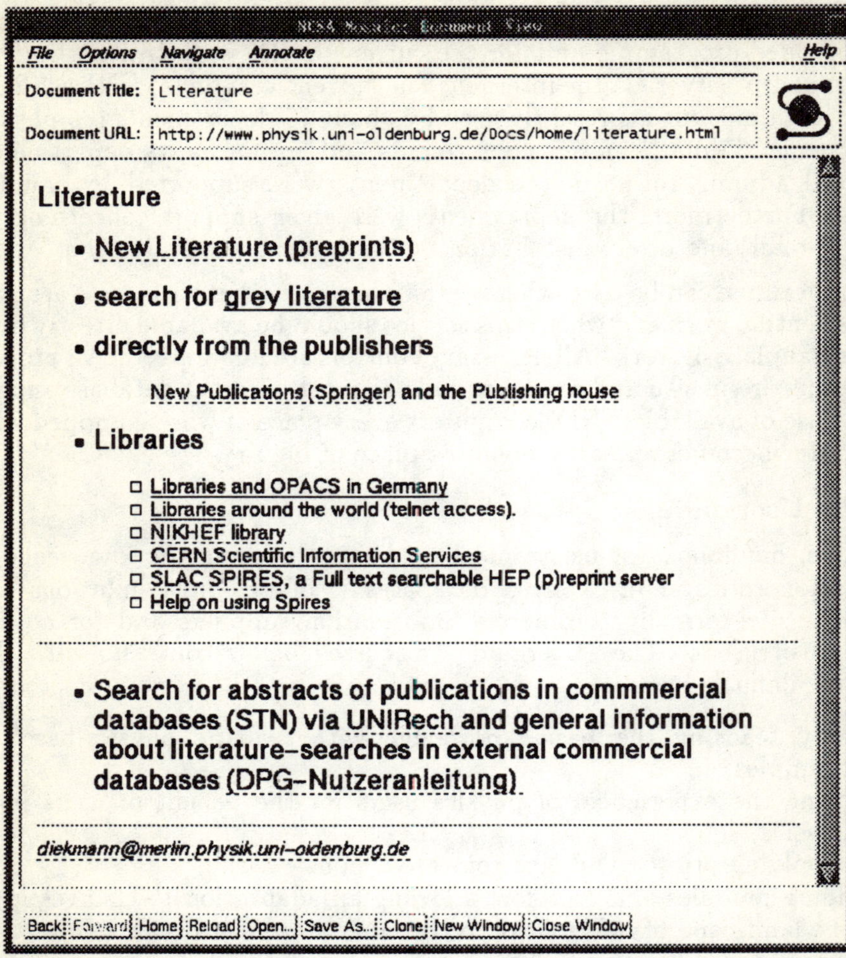

Fig. 2: *WWW*-Page with literature-relevant information
This figure shows a page of the *WWW-Server* of department of physics of Oldenburg University. The contents of this page are literature-relevant information and services. As you can see, the tools "UNIrech" with online-help and the "Userguide" (Nutzeranleitung) are integrated. The scientists use this server to access databases and get help for using these services.

The Manual "Electronical Specialized Information – An Introduction" describes basic searches: it deals with the structure of a publication, the most important journals, the physics-relevant information sources and their suppliers and the specific search-steps are explained.

The "Guide for Advanced Searchers in Databases" shows more sophisticated searches: on 70 pages the chances and limits of searches in external databases and the development of retrieval strategies are discussed. The set of commands and of search- and display-fields is explained with examples.

The retrieval for lecture books or review articles, for energy- or temperature-ranges or in specific disciplines like high energy physics or solid state physics is performed. Electronical teaching media and additional database user-guides are mentioned.

The brochure "Electronical Specialized Information and Communication" gives an overview over the existing and an outlook to the future landscape of information retrieval in physics: the basic knowledge of information retrieval shall be secured for the faculties. The experience of the just finished pilot project will be documented, the future needs, the sources, channels and the information habits will be investigated. A recommendation for concepts of the future "Information and Communication" and the necessary measures will be worked out. The brochure is intented to be a platform of discussion for interested physicists and also for other disciplines.

3.3 Access and Use of the Tools

The "Userguide" is electronically obtainable on a *Gopher-Server* and will be available on a *WWW-Server* from 1 May 1994. The address of the Gopher-Server is:

physinfo.uni-augsburg

The "Userguide" is found under

DPG-Projekt "Elektronische Fachinformation" / Nutzeranleitung

The address (URL) of the WWW-Server is

http://elfikom.physik.uni-oldenburg.de

We will encourage readers to get familiar with these tools in an electronical dialog. Institutions at universities can obtain these products free of costs. Only the expense for installation and support will be placed to account.

4. Perspectives of an Interdisciplinary Use

Research on overlapping areas of classical disciplines is currently a hot topic. Examples are biophysics, physical chemistry, material sciences or biochemistry. The classical publication system proves completely insufficient in hyperactive research times for the following science- and technology-transfer process. By the initiative of the BMFT to increase the utilization of specialized information in physics, mathematics, informatics, chemistry and by further institutions in and outside of the universities, an experience exchange process started, which supports the competition and the research cooperation. "UNIrech" operates in 35 faculties of mathematics and 5 faculties of informatics and plenty of locally adapted versions of "Userguides" have been triggered. "UNIrech" also stimulated the development or optimization of locally based solutions of search- and administration-software. These non-commercial solutions are always preferred, if the direct benefit by the near contact of the "concerned" scientist is obvious. Therefore "UNIrech" in connection with effective "Userguides" will have its chance also in commercial information retrieval areas, as the trend goes to "campus-licensing" or other flat-rate price schemes for using of electronical information sources.

5. Outlook

Currently the process of communication in natural science is in a rapid phase transition. The classical role distribution within the information transfer process is loosening. Electronic products are concurring to print products at information providers. Publishers become hosts by electronical transferring fulltexts directly to endusers. Scientists and students use new "grey" information services and distribute their electronical manuscripts via networks. The classical information mediators and providers search their new profile. Typical of this situation is a widespread uncertainty about risks and chances of the new possibilities and the pressure to cut the costs. Answers to the topics like copyright, quality of "grey information services" or costs of information supply are missing. Meanwhile the German Physical Society (DPG) has named a working group to deal with questions like the referee process, rules and standards of the future specialized communication in physics. The definition of an integrating electronical interface at the working place of the scientist plays an important role in getting access to worldwide distributed information by simple unified means and to distribute results globally. Only the information relevant for physicists shall be retrieved. The conflict of interest between the new worldwide based information sources, which have, however, not been quality-checked up to now, and the classical highly qualified external commercial databases has to be resolved. We need tools to make the external databases or catalogues of the libraries more complete and more up-to-date, to ease the access to the original literature and to simultaneously integrate the advantages of the information retrieval features of commercial information providers. Fig. 2 shows a first attempt for an integrated management information system at the working place of a physicist at Oldenburg University, using "World Wide Web", developed by the CERN in Geneva. "UNIrech" and the "Userguide" are here connected as modules for an extended information retrieval and accessible via a fully graphical user interface. This will enable physicists to get better knowledge from data in future.

6. Conclusion

The basis has been prepared for a routinely operating supply of scientific literature at the working places in physics faculties, currently in form of bibliographic abstracts. A significant contribution was made by the development and performance of the database search- and administration software "UNIrech" and the production and offer of a printed and electronical "Userguide". "UNIrech" allows a distributed and menue driven access to external databases from the working places of the scientists, a central, instantaneous cost-control and -splitting and a statistical manipulation of the search results. The "Userguide" produced by and for endusers describes the entrance for persons interested in information retrieval, who did not search up to now or who did not benefit from know-how transfer. "UNIrech" is already in use in faculties of mathematics and informatics. "UNIrech" and the concept of the "Userguide" can be important moduls of the future landscape of information retrieval in research, product development and teaching.

Acknowledgements

The pilot project of the BMFT "Electronical Specialized Information on Physics at Universities" has been performed by the German Physical Society, using its bonds to research and technical sciences. "UNIrech" was developed and is managed at the University of Oldenburg, Faculty of Physics, by a team of Prof. Dr. E. Hilf. The "Userguide" is based on the experiences of Prof. Dr. D. Hasselkamp, Fac. of Physics, Giessen University. We have to thank Dipl.-Phys. R. Utermann, Fac. of Physics, Augsburg University; Dr. F. Wuensch, Fac. of Physics, Regensburg University; Dr. W. Ziegler, Fac. of Physics, Jena University; Dipl.-Phys. K. Barghorn, cand-phys. J. Stamerjohanns and cand-soc. A. Wallaschek, Fac. of Physics, Oldenburg University, who examined the set of locally adapted search assistance guides, imported useful hints and enabled the access to printed and electronical versions. Special thanks to Dr. M. Bischoff, Fac. of physics, Darmstadt University, who provided us Fig. 1. With the production of the "Userguide", the installation, management, and advancement of "UNIrech", the worldwide research network internet was used for communication of the participating working groups. Thanks to all unknown supporters of the nets.

References:

DE WAARD, A. (1993): Physicists write too much. Europhys.News 24, 86.

WEISEL, L. (1994): Elektronische Datenbanken in der Physik. Cogito 2-94 (submitted).

WEISEL, L. and URBAN, K. (1993): Fachinformation und Fortschritte in der Physik. Nachrichten f. Dokumentation 44, 362-380.

From Verbal Data to Practical Knowledge

Josef Zelger

Institut für Philosophie, Universität Innsbruck,
Innrain 52, A-6020 Innsbruck

Summary: It is one of the central tasks of the social sciences and humanities to show ways and means of creatively processing complex problems in our society. Such problems are generally unstructured and difficult to define. Frequently they are tied to dynamic situations affecting a large number of individuals or groups of individuals. On the other hand the experiences gained by those affected, e.g. the employees of an institution, contain a creative problem solving potential. The GABEK method (Ganzheitliche Bewältigung sprachlich erfaßter Komplexität) suggests itself when trying to make systematic use of this potential. GABEK provides a new method which can be classified as a qualitative research method (c.f. Lamnek 1988). GABEK can both be used for a detailed description and explanation of a problem situation as well as the practical development of aims and measures and their correlation. It is a problem solving procedure that includes the experiences and aims of all participants and individuals affected. The presentation of the GABEK method uses an example taken from the evaluation of an open-ended question presented to South Tyrolean primary school teachers concerning their experiences with the Italian primary school reform in the school year 1991/92.

1. Knowledge Creation Founded on Open-Ended Questions

As a great amount of problem solving potential is contained in the knowledge and the experiences of the teachers we are interested in joining up the many partial aspects. In the GABEK procedure the variety of opinions can be used to draw a holistic picture of the reform. It is to be shown that unsorted statements in everyday language can be analysed holistically, filtered in various ways, and evaluated accordingly. The results of analysis can be checked against the authentic texts, whilst every step of procedure can be reconstructed in an intersubjective way.

Using PC-support we develop a conceptual network, which is used as an *indexing system* for the assertions of the teachers. We proceed as follows: first we decompose the text into small units - ordinarily sentences.Then we mark the most relevant key concepts in each sentence. Thus the possible nodes in the network are determined. In our example we take each teacher's answer as one 'sentence'. In the computer the sentence *A01 (The shared responsibility provides less strain on the teacher, more classes and greater variety.)* is also saved as a relation between the expressions *shared, responsibility, less strain on teachers, more classes, variety.* In Fig. 1 we can see that the answers *A92 (Pleasant experiences: less load on certain subjects, better methods of differentiation, shared responsibility.)* and *B41 (Less strain on teachers, looking after weaker pupils is easier; a greater variety of school life for the child; no such intensive relationship between pupils and teacher.)* are related to the answer *A01* via shared key concepts.

When a larger data base is presented in this way, we obtain a very complex

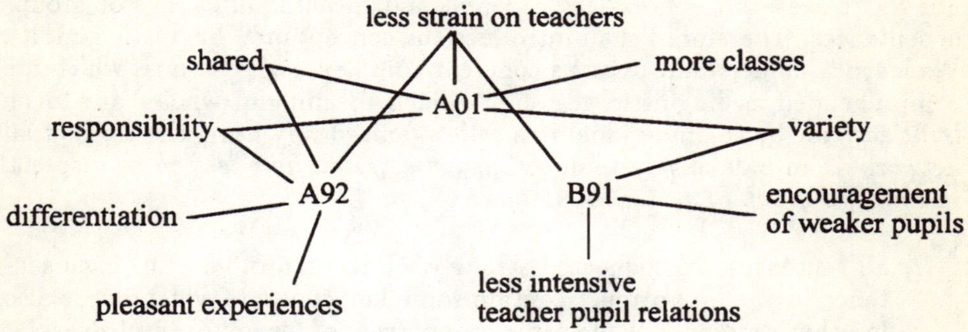

Fig. 1: Three sentences related via key concepts

conceptual network. How can we analyse such a network? GABEK represents a pragmatically applicable method encouraging holistic, networked thinking. This is supported by the PC-program WINRELAN (Windows-Relation-Analysis) which provides for opportunities to gain an overview of all the unsorted answers of the teachers. In contrast to other software applied in content analysis we don't preselect few words, topics or symbols but some hundred concepts which seem to come into question. We don't scan large volumes of material, don't cross-tabulate large lists of the frequencies of words and don't evaluate them statistically (c.f. Wheeler 1988). WINRELAN (1993) is used for deep analysis of less verbal material in ordinary language. It is used to support the reasoning and learning process where the PC provides for various guises of overviews, data organization and retrieval opportunities. The results of analysis are meaningful chunks of sentences, ordered hierarchically according to priority and importance. Results are structured in order of relevance (Zelger 1994). Cluster analysis and other automatic procedures of analysis are used by WINRELAN only to support personal search- and navigation processes (Zelger 1993). It is our aim to form *meaningful* groups of sentences for a better understanding of complex situations. And, we can order the text basis more efficiently if we start from PC-processed clusters. But we don't accept clusters as final results. In the GABEK method we judge every cluster according to consistency and adequacy criteria, delete some sentences or add other ones step by step. In particular we judge the applicability of groups of sentences as orientation-, explication- or action patterns.

2. Understanding Through Construction of Linguistic Gestalten

If we try to explain the conditions of understandig a complex situation it can be useful to investigate the process of perception. We do not perceive our surroundings as isolated sense data but in the form of organized gestalten (c.f. Stumpf 1939). Similarly we understand texts neither at one glance nor as isolated concepts but as coherent and meaningful chunks or groups of sentences. Therefore I shall introduce the concept of a 'linguistic gestalt'. We learn to understand texts as coherent complexes of sentences which are comprehended as linguistic gestalten. Such meaningful wholes are to be built actively by the individual in a self-organized way using the conceptual network as an indexing system. *A linguistic gestalt here is seen as a special meaningful group of sentences so that:*

(i) all sentences are connected strongly whith each other. I.e. each sentence in the group must contain some key concepts which occur also in other sentences of the same group *(rule of formal connectedness).*

(ii) The sentences must be distinguishable clearly from each other within the group. I.e. the content of any two sentences may not be to similar. Each sentence must contain something new so that it is a kind of complement to all other sentences in the group. Therefore the following rules are to be fulfilled

(iii) (rules of formal differentiation):

  (a) Each sentence in the group must be distinguished from all other sentences in the same group by more than one key concept. As a consequence any pair of sentences containing the same key concepts is forbidden.

  (b) The key concepts of one sentence in the group may not be included in the set of the key concepts of another sentence in the group.

(iv) The group of sentences may not contain too many sentences so that all relations between the sentences can be reconstructed and they can be conceived as a unit of meaning. The necessary steps to reach from each sentence to each other in the group may not exceed the maximal value of 2 steps *(rule of the maximal formal distance).*

(v) A group of sentences is meaningful only if it is possible to explain or demonstrate all the relations between the sentences intersubjectively. Therefore there must exist an ideal paradigmatic example which fulfils all the conditions assumed by the sentences. Especially in communicative learning situations it is necessary to point to models, examples, applications in a given modelling facility. As suggested by Pask (1976) 'modelling facility' refers to the material immediately available to the individual, with which he can produce something observable, e.g. models which represent all the relations assumed by the sentences in the group *(rule of semantic demonstrability)*.

(vi) From a pragmatic point of view a group of sentences is relevant for an individual x in the situation s and at the point in time t only if the individual x believes in s and t that all sentences together with their relations between each other are applicable as perception, orientation or action patterns *(rule of pragmatic applicability)*.

In the procedure of grouping sentences we observe that it is not easy to comply with the above conditions. Especially the rules of formal variety and formal connectivity oppose each other and lead to a kind of equilibrium when they are fulfilled. The more the sentences are internally differentiated, the more likely it is that they are internally not sufficiently networked. And the closer the network, the more likely it is that internal differentiation does not suffice. The PC-program WINRELAN (1993) provides for interactive methods to choose meaningful groups of sentences i.e. linguistic gestalten from the given verbal data base. As we give priority to the relevance of the semantic and pragmatic rules the formal ones are sometimes infringed in practical text analysis. We use the formal rules more as rules of thumb: we concentrate our attention field first on one concept, one sentence or a small cluster of sentences. Then we navigate through the data base, searching for further sentences which can be connected to the former one. The next operation is a process of evaluation: the user controls whether the selected sentences are internally consistent and are coherent with the user's system of beliefs. Also, intersubjective criteria of evaluation and selection should be applied. Finally the fulfilment of the above conditions is controlled. In this way we achieve the integration of opinions into more complex wholes (Seiler 1994), i.e. meaningful linguistic gestalten. These allow a better understanding of the problem situation.

3. Hierarchical Organization of Knowledge

Starting from 360 answers of teachers in our example we formed 32 linguistic gestalten - in the way described. As groups of sentences they are too complex to be conceived in few instances. We must focus on one sentence first, switch to the next to apprehend one after the other. Therefore we summarize each linguistic gestalt by one short statement or few very short ones. Besides, the new statement must follow logically from the group of sentences. If these are true, then the new statement - I call it a *selective representation of the linguistic gestalt* - must also be true.

Finally we ask whether the new formed statement - the selective representation - is well founded by the whole context of the data base. The selective representation was built on the basis of only some sentences selected from the data base. So we look now, wether there are contradictions with other sentences (external consistency) or whether there are more sentences supporting the selective representation. (For details of the procedure refer to Zelger 1994)

Until now we have formed, as a result of analysis, 32 statements (selective representations of linguistic gestalten) describing briefly the situation of the school reform as seen by teachers. These statements are themselves very strongly connected with each other, and form a new linguistic network, which is less complex than the original data base of all answers. But this new net may still seem to be too complicated for presentation to a bigger audience. So we can apply the same procedure of gestalt building again to the selective representations of the 32 linguistic gestalten in order to construct a still less complex net of *linguistic hypergestalten.* Therefore we apply the procedure recursively until we obtain a single linguistic unit - in our case a *linguistic hypergestalt of second order.* In our example the final summary of 360 answers of the teachers runs as follows:

Hg 2nd Ord.
Swapping teachers leads to strain on teachers and to restlessness especially in lower classes and in small classes. In lower classes the children need a fixed person of confidence not a team. Careful planning of team teaching is necessary. Some teachers shift the responsibility to the other team partners.

Thus we have organized the answers to an open-ended question in pyrami-

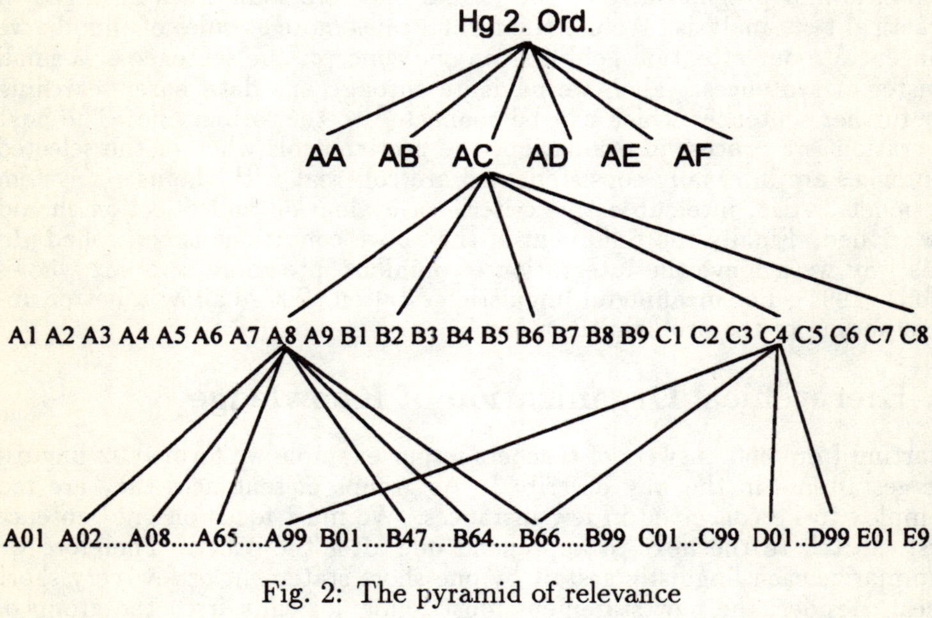

Fig. 2: The pyramid of relevance

dal form. We also call it a *pyramid of relevance* because the statements on the top have more applications in a modelling facility than lower ones. Each statement on a higher level is applicable as orientation, explanation or action pattern in all cases where statements subordinated to the higher ones are applicable. For that reason statements on the top of the pyramid of relevance are suited for short information to executive individuals.

The pyramid of relevance can be discussed rationally and can by reconstructed intersubjectively. If someone does not accept the selective representation on a higher level it can be *explicated* by the sentences which together form the corresponding linguistic gestalt. The opposite process supports holistic *un-*

derstanding: we understand a sentence if we can place it in a bigger system, i.e. if we show the connections with other sentences or if we can formulate a linguistic gestalt of higher order containing the sentence in question. Further, we can *test* a selective representation by showing all sentences in the data base which contradict or which support the selective representation. This can be done by checking all texts in the context which contain at least one key concept ocurring also in the selective representation. Furthermore, *we can choose the level of complexity* we want and can return to the more complex or less complex level if we are interested in greater or less differentiation of the results.

The pyramid of relevance is also a means for goalseeking and decision processes. We can choose the level of complexity and select the sentences which express causal assumptions. These contain the most useful knowledge for planning. Using the PC-Program WINRELAN we can master the complexity and can take advantage of the big potential of ordinary knowledge and experiences of all those affected by the problem.

4. A Qualitative Causal Network for the Evaluation of Side-Effects, Long Term and Backlash Effects

Let us select the opinions concerning causal relations. To do this we have to read all the texts again. We don't apply quantitative methods of data analysis like correlation analysis etc., but we judge for each sentence qualitatively whether it expresses an assumed causal relation and select only these ones. This means that we only take a partial network into account. If a statement from any level of complexity claims a causal connection between two variables (key concepts), we record this in the appropriate diagram. By joining together the individual graphs we receive the causal network which contains all the assumed causal relations which were discovered in the course of the analysis. Vester (1980) and many followers have shown the usefulness of such causal networks. By GABEK we build such a net - diferentiated much more in detail - on the base of experiences and opinions of employees or those affected by the problem. Fig. 3 shows a small part of the whole causal network. It may at first sight create a confusing impression like a paper cutting. Consequently I shall give a few hints on how it can be applied as an aid to orientation in practice.

First a simple comment: if we select a random variable (a key concept) in the causal network, then we can see in which other variables changes can be expected according to the opinions of the teachers. The arrows show the direction of an assumed effect. If a minus '-' is found at the arrow-head then the corresponding cause contributes to the opposite of the variable at the arrow-head.

Time pressure $\longrightarrow$ - *quiet* signifies, for example, that restlessness results from time pressure. Of course the assumed causal relationships have not been scientifically proven and investigated. But we must not ignore the surprising fact that the network of opinions concerning the assumed causal relations is largely consistent. There are only very few incompatibilities to be noticed. This makes it appear plausible to accept the results as a rough

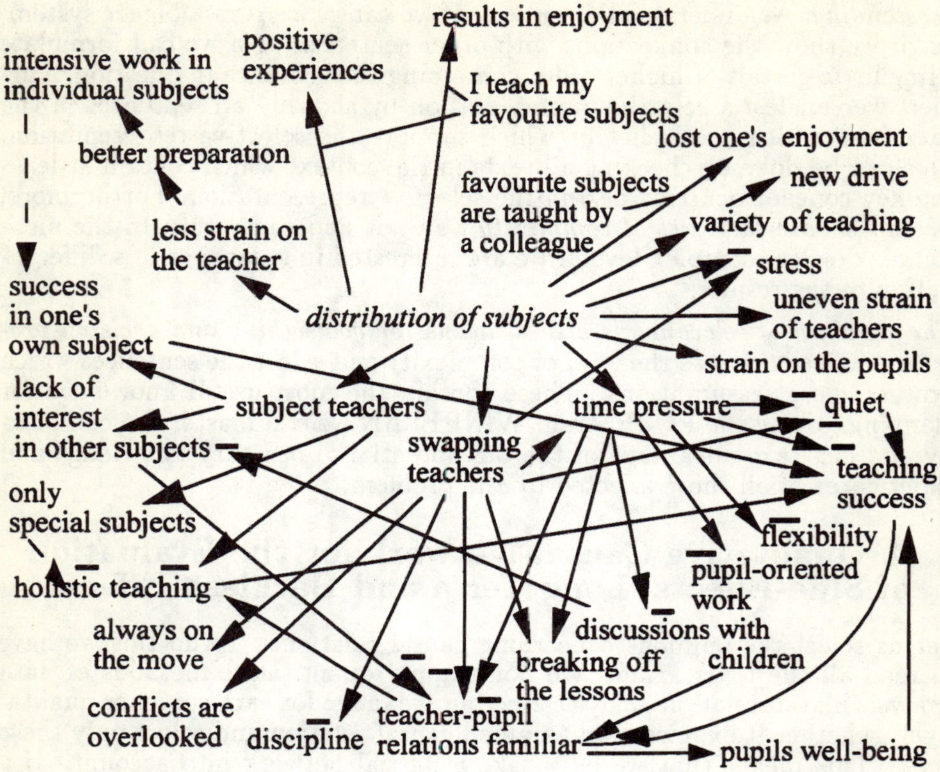

Fig. 3: Part of the causal network. The whole net includes about a 100 variables.

guide for decisions. Foremost the assumed causal network should be tested as to its completeness. Experts can muster the network and add supplementary hypotheses, i.e. assumed causes and effects. Dotted arrows relate to additions to the causal network for which no sentence is to be found in the data base.

Some variables in the causal network shall now be declared goals. For instance *less strain on pupils, independence of pupils, encouragement of pupils, teacher- pupil relation more familiar, less strain on teachers etc.* When determining the goals, the evaluation of the teachers are to be taken into account. They judge on the basis of their experience. Thus one should only ignore the wishes of the majority of the teachers with good reason. (On the development of an evaluation profile providing arguments for the weighting of the aims on the basis of the analysis: cf. Zelger 1993, p 154ff) Executives now have a clear overall picture of the situation and are able to attribute their own weightings to the variables.

The causal network shows on a kind of map which information is going to play a part in planning. Some of the aims are not immediately observable. I.e. we cannot simply observe whether a development for better or for worse is taking place. Therefore one will determine observable variables in the causal network which allow us indirectly to observe favourable or unfavourable changes in the overall system. Passive variables are particularly suited as indicators, i.e. variables to which a large number of arrows point but are the origin of few or none. Yet it is still to be determined how these variables are to be operationalised. We are therefore interested in determining variables whose development can constantly be monitored so that one can anticipate favourable or unfavourable changes in the overall system. The indicators represent a kind of *early warning system.* They should allow to determine in time whether measures are required or not.

The causal network at first sight shows assumed immediate results of the measures. Let us, for example, select the independent variable *distribution of subjects.* It is independent because the distribution of subjects among the teachers is reintroduced by decisions, that is through the school reform. In the causal network the *distribution of subjects* appears as an active variable, from which many arrows emanate but to which none lead. Every possible set of measures such as the distribution of subjects above, has to be evaluated in regard to its side effects.

The qualitative causal network is an effective means of dealing with current problems. In case we wish to change some selected variables we can identify those variables which we assume to have an effect on the intended variables so that we can also consider appropriate measures. The control of a complex system requires complex measures. We will therefore have to determine which variables can be manipulated and aim at manipulating these variables by parallel measures. We are concerned with the question of which nodal variable can act as a pivot for the causal network. Active variables are most suited to measures, i.e. those from which a large number of arrows originate. Another option is presented by critical variables from wich many arrows originate and to which many point (cf. Vester 1980). In our example we find the following variables as appropriate for measures: *less swapping of teachers especially in small and lower forms, further education, supervision, discussions on educational aims, less distribution of subjects especially in lower forms, joint planning, time table, team teaching, holistic teaching, pupil-oriented work, group work, differentiation of teaching.*

Summarising, we recommend in complex problem situations not to proceed by deciding on certain suggestions but to use all the employees and some of those affected as informants and to ask them about the situation. Only after this step, should the problem be developed by the experts. Planning thus becomes easier. Due to the simple method of obtaining information (some open-ended questions) and the relatively few hours of expert advice the expenses incurred with GABEK are less than those of conventional methods. The great advantage lies not only in the holistic and more comprehensive view of the situation and the more realistic evaluation of the consequences, but also that those affected are included in the planning process, whether these are municipalities, offices, patients or people in business. The possibility of introducing authentic opinions is taken seriously. A plan is justified by authentic quotations of colleagues and others affected. Many individuals are able to identify with the plans and are therefore ready to support their realization.

References:

LAMNEK, S. (1988): *Qualitative Sozialforschung. Band 1. Methodologie.* Psychologie Verlags Union, München, Weinheim.

PASK, G. (1976): *Conversation Theory*, Elsevier, Amsterdam.

SEILER, Th. B. (1994): Ist Jean Piagets strukturgenetische Erklärung des Denkens eine konstruktivistische Theorie? Delfin, in print. STUMPF, C. (1939): *Erkenntnislehre.*, 1, Johann Ambrosius Barth, Leipzig.

VESTER, F. (1980): *Neuland des Denkens.* Stuttgart.

WHEELER, D.R. (1988): Content Analysis: An Analytical Technique for International Marketing Research. *International Marketing Review*, Winter, 34-40.

WINRELAN (1993) - developed for GABEK by J. Schoenegger and J. Zelger, Institut für Philosophie der Universität Innsbruck, Innrain 52, A-6020 Innsbruck.

ZELGER, J. (1993): GABEK. A New Method for Qualitative Evaluation of Interviews and Model Construction with PC-Support. In: E. Stuhler and M.O. Silleabhin (eds): *Enhancing Human Capacity to Solve Ecological and Socio- economic Problems.* Hampp, Mnchen, Mering, 128-172.

ZELGER, J. (1994): Qualitative Auswertung sprachlicher Äußerungen. Wissensvernetzung, Wissensverarbeitung und Wissensumsetzung durch GABEK. In: R. Wille and M. Zickwolff (ed.): *Begrifflichen Wissensverarbeitung.* Grundfragen und Aufgaben, B.I. Mannheim, Wien, Zürich, 239 - 265.

I wish to thank to Dr. Philip Herdina (University of Innsbruck) for suggestions and criticism.

Index

Springer-Verlag and the Environment

We at Springer-Verlag firmly believe that an international science publisher has a special obligation to the environment, and our corporate policies consistently reflect this conviction.

We also expect our business partners – paper mills, printers, packaging manufacturers, etc. – to commit themselves to using environmentally friendly materials and production processes.

The paper in this book is made from low- or no-chlorine pulp and is acid free, in conformance with international standards for paper permanency.